D1574969

Technician mathematics 3

Second edition

Series Editor:

D. R. Browning, B.Sc., F.R.S.C., C.Chem., A.R.T.C.S.
Formerly Principal Lecturer and Head of Chemistry, Bristol
Polytechnic

Other titles in this series:

Technician mathematics 1 Third edition   *J O Bird and A J C May*
Technician mathematics 2 Third edition   *J O Bird and A J C May*
Technician mathematics 4/5 Second edition   *J O Bird and A J C May*
Mathematics for electrical technicians 4/5 Second edition
  J O Bird and A J C May

Other mathematics titles:

Mathematical formulae Second edition   *J O Bird and A J C May*
Mathematics for scientific and technical students   *H G Davies and
  G A Hicks*

3 Technician mathematics

Second edition

J. O. Bird
B.Sc.(Hons), C.Math., F.I.M.A., C.Eng., M.I.E.E., F.Coll.P., F.I.E.I.E.

A. J. C. May
B.A., C.Eng., M.I.Mech.E., F.I.E.I.E., A.M.B.I.M.

 LONGMAN

Addison Wesley Longman Limited
Edinburgh Gate, Harlow,
Essex, CM20 2JE, England
and Associated Companies throughout the world

First published 1978
Second edition 1994
Second impression 1995
Third impression 1997

British Library Cataloguing in Publication Data
A catalogue entry for this title is available from the British Library.

ISBN 0-582-23424-7

Set by 16SS in 10/12 pt Times

Produced through Longman Malaysia, VVP

Contents

Preface

Technician mathematics 3 is the third in a series which deals simply and carefully with the fundamental mathematics essential in the development of technicians and engineers.

The material for the second edition of this successful textbook has been updated and expanded to cover (together with *Technician mathematics 2*) the main areas of the Business and Technology Education Council's 'Mathematics for Engineering' module for National Certificate and Diploma courses likely to be studied in a 'Mathematics' lesson. Four-figure tables are assumed to be redundant; the electronic calculator is now an accepted tool in technician and engineering studies. The modifications in this second edition reflect this change.

This book also fully covers the BTEC unit (4684B) which was prepared in collaboration with the Engineering Council, the Council for National Academic Awards and the Standing Conference on University Entrance, after taking account of wide-ranging advice from education and industry. The unit corresponds with the mathematics module envisaged in the Engineering Council's Standards and Routes to Registration (SARTOR) document.

It is widely recognised that facility in the use of mathematics on entry to an engineering degree course is a key element in determining subsequent success. First year undergraduates who need some remedial mathematics will also find that this book meets their needs.

The definition and solution of engineering problems relies on the ability to represent systems and their behaviour in mathematical terms, which in turn depends on the use of various mathematical tools. This textbook deals with some of those mathematical skills and concepts relevant to effective performance in engineering employment at technician level.

The aim of this book is to model simple engineering systems, to generate numerical values for system parameters, to manipulate data to determine system response in defined conditions, to evaluate the effects on systems of changes in variables and to communicate ideas mathematically.

Each topic considered in the text is presented in a way that assumes in the reader only the knowledge attained in *Technician Mathematics 2*. This practical mathematics book contains nearly 400 detailed worked problems, followed by some 1400 further problems with answers. Although specifically written for the latter stages of BTEC National Certificate and Diploma Mathematics and for Engineering Access courses at Universities, the book will also be suitable for NVQ and gNVQ studies and for students studying A level Mathematics.

I would like to express my appreciation for the friendly co-operation and helpful advice given by the publishers, and to Mrs Elaine Woolley for the excellent typing of the manuscript.

Finally, I would like to pay tribute to my colleague, friend and co-author Tony May who sadly died in 1991 after a short illness.

John O. Bird
Highbury College, Portsmouth

Chapter 1

The binomial expansion

1.1 The expansion of $(a + b)^n$, where n is a small, positive integer, using Pascal's triangle

The word binomial indicates a 'two-number' expression. When n is a small positive integer, up to, say, 10, the expansion of $(a + b)^n$ can be achieved by multiplying $(a + b)$ by itself n times. The result of doing this is shown in Table 1.1 for values of n from 0 to 6.

Table 1.1

Term	Expansion
$(a + b)^0 =$	1
$(a + b)^1 =$	$1a + 1b$
$(a + b)^2 =$	$1a^2 + 2ab + 1b^2$
$(a + b)^3 =$	$1a^3 + 3a^2b + 3ab^2 + 1b^3$
$(a + b)^4 =$	$1a^4 + 4a^3b + 6a^2b^2 + 4ab^3 + 1b^4$
$(a + b)^5 =$	$1a^5 + 5a^4b + 10a^3b^2 + 10a^2b^3 + 5ab^4 + 1b^5$
$(a + b)^6 =$	$1a^6 + 6a^5b + 15a^4b^2 + 20a^3b^3 + 15a^2b^4 + 6ab^5 + 1b^6$

An examination of Table 1.1 shows that patterns are forming and the following observations can be made.

(i)   The power of a, when looking at each term in any of the expansions and moving from left to right, follows the pattern:

$n, n - 1, n - 2, \ldots, 2, 1, 0$ (since $a^0 = 1$)

Thus for $n = 5$, the 'a' part of each term is:

a^5, a^4, a^3, a^2, a and 1

(ii) The power of b, when looking at each term in any of the expansions and moving from left to right, follows the pattern:

0, (since $b^0 = 1$), 1, 2, 3, $\ldots$, $n - 2$, $n - 1$, n

For $n = 5$, the 'b' part of each term is:

$1, b, b^2, b^3, b^4$ and b^5

(iii)   The values of the coefficients of each of the terms in any of the expansions are symmetrical about the middle coefficient when n is even and symmetrical about the middle two coefficients when n is odd. This can be seen in Table 1.1, where for n being an even number, say 4, the coefficients of $(a + b)^4$ are 1, 4, 6, 4, 1, i.e. symmetrical about 6, the middle coefficient. When n is odd, say 5, the coefficients of $(a + b)^5$ are 1, 5, 10, 10, 5, 1, i.e. symmetrical about the two 10's, the middle two coefficients.

Table 1.2 shows the coefficients of the expansions of $(a + b)^n$, where n is a positive integer and varies from 0 to 6.

Table 1.2

Term	Values of coefficients in the expansion
$(a + b)^0$	1
$(a + b)^1$	1 1
$(a + b)^2$	1 2 1
$(a + b)^3$	1 3 3 1
$(a + b)^4$	1 4 6 4 1
$(a + b)^5$	1 5 10 10 5 1
$(a + b)^6$	1 6 15 20 15 6 1

The coefficient of, say, the fourth term in the expansion of $(a + b)^6$ is obtained by adding together the 10 and 10 immediately above it, giving a result of 20, this being shown by the triangle in the table. An examination of Table 1.2 shows that the first and last coefficients of any expansion of the form $(a + b)^n$ are 1's. To obtain the coefficients of, say, $(a + b)^4$, the first and last coefficients are 1's. The second coefficient, 4, is obtained from adding the 1 and 3 from the line above it. Similarly the third coefficient, 6, is obtained by adding the 3 and 3 from the line above it, and so on. The configuration shown in Table 1.2 is called **Pascal's triangle**. It is used to determine the coefficients of the expansion of $(a + b)^n$ when n is a relatively small positive integer.

Worked problems on the expansion of $(a + b)^n$, where n is a small positive integer, using Pascal's triangle

Problem 1. Find the expansion of $(a + b)^7$

The coefficients of the terms are determined by producing Pascal's triangle as far as the seventh power and selecting the last line.
From Table 1.2, the coefficients of $(a + b)^6$ are:

1, 6, 15, 20, 15, 6, 1

The coefficients of $(a + b)^7$ are therefore:

1, since the first and last coefficients are always 1's;
7, obtained by adding the first and second coefficients of $(a + b)^6$, i.e. $1 + 6$;
21, obtained by adding the second and third coefficients of $(a + b)^6$, i.e. $6 + 15$;
and so on, giving the coefficients of $(a + b)^7$ as:

1, 7, 21, 35, 35, 21, 7, 1

The 'a' terms are a^7, a^6, a^5, ..., a, 1
The 'b' terms are 1, b, b^2, ..., b^6, b^7
Combining these results gives:

$$(a + b)^7 = a^7 + 7a^6b + 21a^5b^2 + 35a^4b^3 + 35a^3b^4 + 21a^2b^5 + 7ab^6 + b^7$$

A check for blunders can be made by adding the powers of a and b for each term. These should always be equal to n. In this problem adding the powers of a and b together for each term gives:

$7 + 0 = 7$, $6 + 1 = 7$, $5 + 2 = 7$, $4 + 3 = 7$, $3 + 4 = 7$, $2 + 5 = 7$, $1 + 6 = 7$ and $0 + 7 = 7$

Thus no blunder has been made in determining the powers of a and b

---

Problem 2. Find the expansion of $(1 - 3x)^{12}$ as far as the term in x^4

Comparing $(1 - 3x)^{12}$ with $(a + b)^n$ shows that $a = 1$, $b = (-3x)$ (note the minus sign) and $n = 12$. Hence the 'a' terms are
1^{12}, 1^{11}, 1^{10}, ..., 1^1, 1^0
The 'b' terms are $(-3x)^0$, $(-3x)^1$, $(-3x)^2$, $(-3x)^3$, $(-3x)^4$ and only five terms of the expansion are required since the expansion is as far as the term in x^4. Also, only the first five coefficients of each line of Pascal's triangle are necessary. Taking the first five coefficients of the seventh-power expansion from Problem 1 and calculating the first five coefficients as far as the twelfth power gives:

			1		7		21		35		35
		1		8		28		56		70	
	1		9		36		84		126		
1		10		45		120		210			
1	11		55		165		330				
1	12	66		220		495					

Thus

$$(1 - 3x)^{12} = 1(1)^{12} + 12(1)^{11}(-3x) + 66(1)^{10}(-3x)^2$$
$$+ 220(1)^9(-3x)^3 + 495(1)^8(-3x)^4$$

as far as the term in x^4, i.e.

$(1 - 3x)^{12} = 1 - 36x + 594x^2 - 5940x^3 + 40095x^4$ as far as the term in x^4

Problem 3. Expand $\left(-1 - \dfrac{3}{y}\right)^7$ to five terms.

Comparing $\left(-1 - \dfrac{3}{7}\right)^7$ with $(a + b)^n$ shows that $a = (-1)$,
$b = \left(-\dfrac{3}{y}\right)$ and $n = 7$. Note that the minus signs must be included for both the 'a' and 'b' terms. The first five coefficients of the seventh power are obtained using Pascal's triangle as shown in Problem 1 and are:

$$1, \quad 7, \quad 21, \quad 35 \quad \text{and} \quad 35$$

The 'a' terms are $(-1)^7$, $(-1)^6$, $(-1)^5$, $(-1)^4$ and $(-1)^3$

The 'b' terms are $\left(\dfrac{-3}{y}\right)^0$, $\left(\dfrac{-3}{y}\right)^1$, $\left(\dfrac{-3}{y}\right)^2$, $\left(\dfrac{-3}{y}\right)^3$ and $\left(\dfrac{-3}{y}\right)^4$

Hence the first five terms of the expansion of $\left(-1 - \dfrac{3}{y}\right)^7$ are:

$$(1)(-1)^7\left(\dfrac{-3}{y}\right)^0 + (7)(-1)^6\left(\dfrac{-3}{y}\right)^1 + (21)(-1)^5\left(\dfrac{-3}{y}\right)^2$$
$$+ (35)(-1)^4\left(\dfrac{-3}{y}\right)^3 + (35)(-1)^3\left(\dfrac{-3}{y}\right)^4$$

that is,

$$(1)(-1)(1) + (7)(1)\left(\dfrac{-3}{y}\right) + (21)(-1)\left(\dfrac{9}{y^2}\right) + (35)(1)\left(\dfrac{-27}{y^3}\right)$$
$$+ (35)(-1)\left(\dfrac{81}{y^4}\right)$$

i.e. $-1 - \dfrac{21}{y} - \dfrac{189}{y^2} - \dfrac{945}{y^3} - \dfrac{2835}{y^4}$

Problem 4. Determine the expansion of $\left(2x + \dfrac{y}{2}\right)^9$ as far as the term in y^5

Taking the first six coefficients of the seventh-power expansion from Problem 1 and calculating the first six coefficients as far as the ninth power gives:

```
     1    7    21    35      35      21
  1    8    28    56      70      56
1    9    36    84    126      126
```

The 'a' terms in the $(a + b)^n$ expansion are replaced by $2x$ and are:

$(2x)^9$, $(2x)^8$, $(2x)^7$, $(2x)^6$, $(2x)^5$, $(2x)^4$

The 'b' terms in the $(a + b)^n$ expansion are replaced by $\dfrac{y}{2}$ and are:

$\left(\dfrac{y}{2}\right)^0, \left(\dfrac{y}{2}\right)^1, \left(\dfrac{y}{2}\right)^2, \left(\dfrac{y}{2}\right)^3, \left(\dfrac{y}{2}\right)^4$ and $\left(\dfrac{y}{2}\right)^5$

Combining these three results gives:

$$\left(2x + \frac{y}{2}\right)^9 = 1(2x)^9\left(\frac{y}{2}\right)^0 + 9(2x)^8\left(\frac{y}{2}\right)^1 + 36(2x)^7\left(\frac{y}{2}\right)^2$$

$$+ 84(2x)^6\left(\frac{y}{2}\right)^3 + 126(2x)^5\left(\frac{y}{2}\right)^4 + 126(2x)^4\left(\frac{y}{2}\right)^5$$

as far as the term in y^5

Thus, $\left(2x + \dfrac{y}{2}\right)^9 = 2^9x^9 + 9(2)^7x^8y + 36(2)^5x^7y^2 + 84(2)^3x^6y^3$

$$+ 126(2)x^5y^4 + \frac{126}{2}x^4y^5$$

as far as the term in y^5. That is,

$$\left(2x + \frac{y}{2}\right)^9 = 512x^9 + 1\,152x^8y + 1\,152x^7y^2 + 672x^6y^3 + 252x^5y^4$$

$$+ 63x^4y^5, \text{ as far as the term in } y^5$$

Further problems on the expansion of $(a + b)^n$, where n is a small positive integer, using Pascal's triangle, may be found in Section 1.4 (Problems 1 to 10), page 20.

1.2   The general expansion of $(a + b)^n$

The value of the coefficients of the expansion of $(a + b)^n$ for integer values of n from 0 to 6 are shown in Table 1.2. This table shows that the coefficients for $(a + b)^6$ are:

1, 6, 15, 20, 15, 6 and 1

Instead of using Pascal's triangle to derive these coefficients, they could have been obtained using a factor method from the relationships:

$$1, \quad \frac{6}{1} = 6, \quad \frac{(6)(5)}{(1)(2)} = 15, \quad \frac{(6)(5)(4)}{(1)(2)(3)} = 20, \quad \frac{(6)(5)(4)(3)}{(1)(2)(3)(4)} = 15,$$

$$\frac{(6)(5)(4)(3)(2)}{(1)(2)(3)(4)(5)} = 6 \text{ and } \frac{(6)(5)(4)(3)(2)(1)}{(1)(2)(3)(4)(5)(6)} = 1$$

Replacing $(a + b)^6$ by $(a + b)^n$ and building up the coefficients by a factor method using those for $(a + b)^6$ as a pattern gives:

$$1, \quad n, \quad \frac{n(n - 1)}{(1)(2)}, \quad \frac{n(n - 1)(n - 2)}{(1)(2)(3)}, \quad \frac{n(n - 1)(n - 2)(n - 3)}{(1)(2)(3)(4)} \text{ and so on.}$$

For example, the value of the third coefficient of $(a + b)^5$ is obtained from $\frac{n(n - 1)}{(1)(2)}$ where n is 5, and is $\frac{(5)(4)}{(1)(2)}$, i.e. 10. Similarly, the value of the fourth coefficient of $(a + b)^4$ is determined using $\frac{n(n - 1)(n - 2)}{(1)(2)(3)}$ where n is equal to 4, and is $\frac{(4)(3)(2)}{(1)(2)(3)}$, i.e. 4.

Combining this factorial method of writing coefficients with the observations previously made for $(a + b)^n$ shows that the terms in 'a' are

a^n, a^{n-1}, $a^{n-2}, \ldots$, and the terms in 'b' are
b^0, b, b^2, $b^3, \ldots$

Thus the general expansion of $(a + b)^n$ is:

$$(a + b)^n = a^n + na^{n-1}b + \frac{n(n - 1)}{(1)(2)} a^{n-2}b^2 + \frac{n(n - 1)(n - 2)}{(1)(2)(3)} a^{n-3}b^3$$
and so on.

The product $(1)(2)(3)$ is usually denoted by 3!, called 'factorial 3'. In general, $(1)(2)(3)(4) \cdots (n)$ is denoted by $n!$ (factorial n). Hence,

$$(a + b)^n = a^n + na^{n-1}b + \frac{n(n - 1)}{2!} a^{n-2}b^2$$

$$+ \frac{n(n - 1)(n - 2)}{3!} a^{n-3}b^3 + \cdots$$

This expansion is the **general binomial expansion** of $(a + b)^n$.

With the binomial theorem, n may be a positive or negative integer, a fraction or a decimal fraction.

If $a = 1$ and $b = x$ in the general expansion $(a + b)^n$ then

$$(1 + x)^n = 1 + nx + \frac{n(n-1)}{2!}x^2 + \frac{n(n-1)(n-2)}{3!}x^3 + \cdots$$

which is valid for $-1 < x < 1$.

Practical problems can arise, for example, in the binomial distribution in statistics, where it is required to find the value of just one or two terms of a binomial expansion. The fifth term of the expansion of $(a + b)^n$ is

$$\frac{n(n-1)(n-2)(n-3)}{4!}a^{n-4}b^4$$

It can be seen that in the fifth term of any expansion the number 4 is very evident. There are four products of the type $n(n-1)(n-2)(n-3)$; 'a' is raised to the power $(n-4)$; 'b' is raised to the power of 4, and the denominator of the coefficient is 4!. For any term in a binomial expansion, say the rth term, $r - 1$ is very evident. The value of the coefficient of the rth term is given by:

$$\frac{n(n-1)(n-2)}{(r-1)!} \cdots \text{ to } (r-1) \text{ terms}$$

The power of 'a' for the rth term is $n - (r - 1)$ and the power of 'b' is $(r - 1)$. Thus the rth term of the expansion of $(a + b)^n$ is:

$$\frac{n(n-1)(n-2) \cdots \text{ to } (r-1) \text{ terms}}{(r-1)!}a^{n-(r-1)}b^{(r-1)}$$

For example, to find the fifth term in the expansion of $(a + b)^{15}$, n is 15 and r is 5 and $(r - 1)$ is 4. Hence the fifth term is

$$\frac{(15)(14)(13)(12)}{4!}a^{15-4}b^4, \text{ i.e. } 1\,365a^{11}b^4$$

Worked problems on the general expansion of $(a + b)^n$

Problem 1. Use the binomial theorem to determine the expansion of $(2 + x)^6$

The binomial expansion of

$$(a + x)^n = a^n + na^{n-1}x + \frac{n(n-1)}{2!} a^{n-2}x^2 + \cdots$$

When $a = 2$ and $n = 6$,

$$(2 + x)^6 = 2^6 + 6(2)^5x + \frac{(6)(5)}{2!} (2)^4x^2 + \frac{(6)(5)(4)}{3!} (2)^3x^3$$

$$+ \frac{(6)(5)(4)(3)}{4!} (2)^2x^4 + \frac{(6)(5)(4)(3)(2)}{5!} (2)^1x^5$$

$$+ \frac{(6)(5)(4)(3)(2)(1)}{6!} x^6$$

i.e. $(2 + x)^6 = 64 + 192x + 240x^2 + 160x^3 + 60x^4 + 12x^5 + x^6$

Problem 2. Expand $(x + y)^{20}$ as far as the fifth term.

The general binomial expansion for $(a + b)^n$ is

$$a^n + na^{n-1}b + \frac{n(n-1)}{2!} a^{n-2}b^2 + \frac{n(n-1)(n-2)}{3!} a^{n-3}b^3 + \cdots$$

Substituting in this general formula, $a = x$, $b = y$ and $n = 20$ gives:

$$(x + y)^{20} = x^{20} + 20x^{(20-1)}y + \frac{20(20-1)}{(2)(1)} x^{(20-2)}y^2$$

$$+ \frac{20(20-1)(20-2)}{(3)(2)(1)} x^{(20-3)}y^3$$

$$+ \frac{20(20-1)(20-2)(20-3)}{(4)(3)(2)(1)} x^{(20-4)}y^4 + \cdots$$

That is:

$$(x + y)^{20} = x^{20} + 20x^{19}y + \frac{20(19)}{2} x^{18}y^2 + \frac{20(19)(18)}{6} x^{17}y^3$$

$$+ \frac{20(19)(18)(17)}{24} x^{16}y^4 + \cdots$$

i.e.
$(x + y)^{20} = x^{20} + 20x^{19}y + 190x^{18}y^2 + 1\,140x^{17}y^3 + 4\,845x^{16}y^4$
when expanded as far as the fifth term

Problem 3. Determine the expansion of $\left(p - \dfrac{4}{p^2}\right)^{15}$ as far as the term containing p^3

Substituting $a = p$, $b = \left(\dfrac{-4}{p^2}\right)$ and $n = 15$ in the general expansion of $(a + b)^n$ gives:

$$\left(p - \frac{4}{p^2}\right)^{15} = (p)^{15} + 15(p)^{14}\left(\frac{-4}{p^2}\right) + \frac{15(14)}{(2)(1)}(p)^{13}\left(\frac{-4}{p^2}\right)^2$$

$$+ \frac{15(14)(13)}{(3)(2)(1)}(p)^{12}\left(\frac{-4}{p^2}\right)^3$$

$$+ \frac{15(14)(13)(12)}{(4)(3)(2)(1)}(p)^{11}\left(\frac{-4}{p^2}\right)^4 + \cdots$$

i.e. $\left(p - \dfrac{4}{p^2}\right)^{15} = p^{15} + 15p^{14}\left(\dfrac{-4}{p^2}\right) + 105p^{13}\left(\dfrac{16}{p^4}\right)$

$$+ 455p^{12}\left(\frac{-64}{p^6}\right) + 1\,365p^{11}\left(\frac{256}{p^8}\right) + \cdots$$

i.e. $\left(p - \dfrac{4}{p^2}\right)^{15} = p^{15} - 60p^{12} + 1\,680p^9 - 29\,120p^6 + 349\,440p^3$

when expanded as far as the term in p^3

Problem 4. Determine the sixth term of the expansion of $\left(\dfrac{1}{m} + \dfrac{m^2}{2}\right)^{14}$

The rth term of the expansion of $(a + b)^n$ is given by

$$\frac{n(n - 1)(n - 2) \cdots \text{ to } (r - 1) \text{ terms}}{(r - 1)!}\, a^{n-(r-1)}b^{(r-1)}$$

Substituting $a = \dfrac{1}{m}$, $b = \dfrac{m^2}{2}$, $n = 14$ and $(r - 1) = 5$ (since $r = 6$) in this expression gives:

$$\frac{(14)(13)(12)(11)(10)}{(5)(4)(3)(2)(1)}\left(\frac{1}{m}\right)^{14-5}\left(\frac{m^2}{2}\right)^5 = 2\,002\left(\frac{1}{m}\right)^9\left(\frac{m^2}{2}\right)^5$$

$$= \frac{1\,001}{16}m$$

Thus the sixth term of the expansion of $\left(\dfrac{1}{m} + \dfrac{m^2}{2}\right)^{14}$ is $\dfrac{1\,001}{16}m$

Problem 5. Find the middle term of the expansion of $\left(3u - \dfrac{1}{3v}\right)^{18}$

In any expansion of the form $(a + b)^n$ there are $(n + 1)$ terms. Hence, in the expansion of $\left(3u - \dfrac{1}{3v}\right)^{18}$ there are 19 terms. The middle term is the 10th term. Using the general expression for the rth term, where $a = 3u$, $b = \left(-\dfrac{1}{3v}\right)$, $n = 18$ and $(r - 1) = 9$, gives:

$$\frac{18(17)(16)(15)(14)(13)(12)(11)(10)}{9(8)(7)(6)(5)(4)(3)(2)(1)}(3u)^9\left(-\frac{1}{3v}\right)^9$$

$$= 48\,620(3)^9(u^9)\frac{(-1)^9}{3^9v^9} = -48\,620\left(\frac{u}{v}\right)^9$$

Thus the middle term of the expansion of $\left(3u - \dfrac{1}{3v}\right)^{18}$ is

$$-48\,620\left(\frac{u}{v}\right)^9$$

Problem 6. Derive the term containing y^{12} in the expansion of $\left(y^2 - \dfrac{x}{4}\right)^{10}$

The y terms are $(y^2)^{10}$, $(y^2)^9$, $(y^2)^8$, $(y^2)^7$, $(y^2)^6$ and so on. Hence the term involving y^{12} is the fifth term. Using the expression for the rth term, where

$a = y^2$, $b = \left(-\dfrac{x}{4}\right)$, $n = 10$ and $(r - 1) = 4$, gives

$$\frac{10(9)(8)(7)}{4(3)(2)(1)}(y^2)^{10-4}\left(-\frac{x}{4}\right)^4 \text{ i.e. } \frac{105}{128}y^{12}x^4$$

Thus the term containing y^{12} in the expansion of $\left(y^2 - \dfrac{x}{4}\right)^{10}$ is

$$\frac{105}{128}y^{12}x^4$$

Problem 7. (a) Expand $\dfrac{1}{(1 + 2x)^4}$ in ascending powers of x as far as the term in x^4, using the binomial theorem. (b) State the limits of x for which the expansion in part (a) is valid.

(a) Using the binomial theorem for $(1 + x)^n$ when $n = -4$ and x is replaced by $2x$ gives:

$$\frac{1}{(1 + 2x)^4} = (1 + 2x)^{-4} = 1 + (-4)(2x) + \frac{(-4)(-5)}{2!}(2x)^2$$

$$+ \frac{(-4)(-5)(-6)}{3!}(2x)^3$$

$$+ \frac{(-4)(-5)(-6)(-7)}{4!}(2x)^4$$

$$= 1 - 8x + 40x^2 - 160x^3 + 560x^4 - \cdots$$

(b) The expansion is valid provided $|2x| < 1$,

i.e. $|x| < \dfrac{1}{2}$ or $-\dfrac{1}{2} < x < \dfrac{1}{2}$

Problem 8. (a) Expand $\dfrac{1}{(3 - x)^2}$ in ascending powers of x as far as the term in x^3 using the binomial theorem. (b) What are the limits of x for which the expansion in part (a) is true?

(a) $\dfrac{1}{(3 - x)^2} = \dfrac{1}{\left[3\left(1 - \dfrac{x}{3}\right)\right]^2} = \dfrac{1}{3^2\left(1 - \dfrac{x}{3}\right)^2} = \dfrac{1}{9}\left(1 - \dfrac{x}{3}\right)^{-2}$

Using the expansion for $(1 + x)^n$,

$$\frac{1}{(3 - x)^2} = \frac{1}{9}\left(1 - \frac{x}{3}\right)^{-2}$$

$$= \frac{1}{9}\left[1 + (-2)\left(-\frac{x}{3}\right) + \frac{(-2)(-3)}{2!}\left(-\frac{x}{3}\right)^2\right.$$

$$\left. + \frac{(-2)(-3)(-4)}{3!}\left(-\frac{x}{3}\right)^3 + \cdots\right]$$

$$= \frac{1}{9}\left[1 + \frac{2}{3}x + \frac{1}{3}x^2 + \frac{4}{27}x^3 + \cdots\right]$$

(b) The expansion in part (a) is true provided $\left|\dfrac{x}{3}\right| < 1$,

i.e. $|x| < 3$, or $-3 < x < 3$

Problem 9. Use the binomial theorem to expand $\sqrt{(4 + y)}$ in ascending powers of y to four terms. Give the limits of y for which the expansion is valid.

$$\sqrt{(4 + y)} = \sqrt{\left[4\left(1 + \frac{y}{4}\right)\right]} = \sqrt{4}\sqrt{\left(1 + \frac{y}{4}\right)} = 2\left(1 + \frac{y}{4}\right)^{\frac{1}{2}}$$

Using the expansion of $(1 + x)^n$,

$$2\left(1 + \frac{y}{4}\right)^{\frac{1}{2}} = 2\left[1 + \left(\frac{1}{2}\right)\left(\frac{y}{4}\right) + \frac{\left(\frac{1}{2}\right)\left(-\frac{1}{2}\right)}{2!}\left(\frac{y}{4}\right)^2 \right.$$

$$\left. + \frac{\left(\frac{1}{2}\right)\left(-\frac{1}{2}\right)\left(-\frac{3}{2}\right)}{3!}\left(\frac{y}{4}\right)^3 + \cdots\right]$$

$$= 2\left(1 + \frac{y}{8} - \frac{y^2}{128} + \frac{y^3}{1\,024} + \cdots\right)$$

$$= 2 + \frac{y}{4} - \frac{y^2}{64} + \frac{y^3}{512} - \cdots$$

This expression is valid when $\left|\frac{y}{4}\right| < 1$,

i.e. $|y| < 4$ or $-4 < y < 4$

Problem 10. Expand $\dfrac{1}{\sqrt{(1 - 3x)}}$ in ascending powers of x as far as the term in x^3. State the limits of x for which the expansion is valid.

$$\frac{1}{\sqrt{(1 - 3x)}} = (1 - 3x)^{-\frac{1}{2}} = 1 + \left(-\frac{1}{2}\right)(-3x) + \frac{\left(-\frac{1}{2}\right)\left(-\frac{3}{2}\right)}{2!}(-3x)^2$$

$$+ \frac{\left(-\frac{1}{2}\right)\left(-\frac{3}{2}\right)\left(-\frac{5}{2}\right)}{3!}(-3x)^3 + \cdots$$

$$= 1 + \frac{3}{2}x + \frac{27}{8}x^2 + \frac{135}{16}x^3 + \cdots$$

The expansion is valid when $|3x| < 1$,

i.e. $|x| < \frac{1}{3}$ or $-\frac{1}{3} < x < \frac{1}{3}$

Problem 11. Simplify $\dfrac{\sqrt{(1 - 4x)}\ \sqrt[3]{(1 + x)}}{\left(1 + \dfrac{x}{2}\right)^4}$ given that the powers

of x above the first may be neglected.

$$\frac{\sqrt{(1 - 4x)}\ \sqrt[3]{(1 + x)}}{\left(1 + \dfrac{x}{2}\right)^4} = (1 - 4x)^{\frac{1}{2}}(1 + x)^{\frac{1}{3}}\left(1 + \frac{x}{2}\right)^{-4}$$

$$\approx \left[1 + \left(\frac{1}{2}\right)(-4x)\right]\left[1 + \left(\frac{1}{3}\right)(x)\right]\left[1 + (-4)\left(\frac{x}{2}\right)\right]$$

when expanded by the binomial theorem as far as the x term only,

$$= (1 - 2x)\left(1 + \frac{1}{3}x\right)(1 - 2x)$$

$$= 1 - 2x + \frac{1}{3}x - 2x$$

when powers of x higher than unity are neglected

$$= 1 - \frac{11}{3}x$$

Problem 12. Express $\dfrac{\sqrt[3]{(1 - 3t)}}{\sqrt{(1 + 2t)}}$ as a power series as far as the

term in t^2. State the range of values of t for which the series is convergent.

$$\frac{\sqrt[3]{(1 - 3t)}}{\sqrt{(1 + 2t)}} = (1 - 3t)^{-\frac{1}{3}}(1 + 2t)^{-\frac{1}{2}}$$

$$(1 - 3t)^{\frac{1}{3}} = 1 + \left(\frac{1}{3}\right)(-3t) + \frac{\left(\frac{1}{3}\right)\left(-\frac{2}{3}\right)}{2!}(-3t)^2 + \cdots$$

$$= 1 - t - t^2 - \cdots$$

which is valid for $|3t| < 1$, i.e. $|t| < \dfrac{1}{3}$

$$(1 + 2t)^{-\frac{1}{2}} = 1 + \left(-\frac{1}{2}\right)(2t) + \frac{\left(-\frac{1}{2}\right)\left(-\frac{3}{2}\right)}{2!}(2t)^2 + \cdots$$

$$= 1 - t + \frac{3}{2}t^2 - \cdots$$

which is valid for $|2t| < 1$, i.e. $|t| < \dfrac{1}{2}$

Hence $\dfrac{\sqrt[3]{(1 - 3t)}}{\sqrt{(1 + 2t)}} = (1 - 3t)^{\frac{1}{3}}(1 + 2t)^{-\frac{1}{2}}$

$$= (1 - t - t^2 - \cdots)\left(1 - t + \frac{3}{2}t^2 - \cdots\right)$$

$$= 1 - t + \frac{3}{2}t^2 - t + t^2 - t^2$$

neglecting terms of higher power than 2

$$= 1 - 2t + \frac{3}{2}t^2$$

The series is convergent if $-\dfrac{1}{3} < t < \dfrac{1}{3}$

Further problems on the expansion of $(a + b)^n$ may be found in Section 1.4 (Problems 11 to 30), page 21.

1.3 The application of the binomial expansion to determining approximate values of expressions

The general binomial expansion of $(a + b)^n$ is:

$$(a + b)^n = a^n + na^{n-1}b + \frac{n(n - 1)}{2!}a^{n-2}b^2$$

$$+ \frac{n(n - 1)(n - 2)}{3!}a^{n-3}b^3 + \cdots$$

When $a = 1$ and $b = x$, then

$$(1 + x)^n = (1)^n + n(1)^{n-1}x + \frac{n(n - 1)}{2!}(1)^{n-2}x^2$$

$$+ \frac{n(n - 1)(n - 2)}{3!}(1)^{n-3}x^3 + \cdots$$

i.e. $(1 + x)^n = 1 + nx + \dfrac{n(n-1)}{2!} x^2 + \dfrac{n(n-1)(n-2)}{3!} x^3 + \cdots$

as obtained in Section 1.2.

If the first term of the expression on the right of this equation is u_1, the second term u_2, the third term u_3, the nth term u_n, and so on and if the sum S_N is given by $S_N = u_1 + u_2 + u_3 + \cdots + u_n$, then provided that S_N approaches a definite finite value when N is large, the series is said to be **convergent**. An example of a convergent series is $1 + \dfrac{1}{1!} + \dfrac{1}{2!} + \dfrac{1}{3!} + \dfrac{1}{4!} + \cdots$ which approaches a value of $2.718\,28\ldots$ when N is large.

The series for $(1 + x)^n$ given above is convergent provided x lies between -1 and 1. For example, if x is 0.6 and n is 0.5,

$(1 + x)^n = 1.6^{0.5} = \sqrt{1.6} \approx 1.265$. The series for $(1 + x)^n$ is

$$1 + (0.5)(0.6) + \frac{(0.5)(-0.5)(0.6)^2}{2} + \frac{(0.5)(-0.5)(-1.5)(0.6)^3}{3 \times 2} + \cdots$$

that is, $1 + 0.3 - 0.045 + 0.013\,5 - \cdots \approx 1.269$. Thus, when N is only 4 terms, the series is already approaching its definite finite value of $\sqrt{1.6}$.

If an expression is written in the form $(1 + x)^n$ where x is small compared with 1, then terms such as $x^2, x^3, x^4, \ldots$ become very small and can be ignored if only an approximate result is required. Approximate values of expressions which could be written in this form used to be found in this way before electronic calculators came into widespread use. However solving problems of this sort using the binomial expansion, assists the understanding and provides practice in the expansion of two numbers into a series. A series of the form $1 + ax + bx^2 + cx^3 + \cdots$ where $a, b, c, \ldots$ are constants is called a **power series** since it is expressed in terms of powers of x. Thus the binomial expansion is used to produce a power series for a two-number expansion. Using this method to find the value of, say, $(1.002)^7$ correct to 4 decimal places, the expression is written as $(1 + 0.002)^7$ and since 0.002 is small compared with 1, only a few terms of the binomial expansion are required. Thus

$$(1 + 0.002)^7 \approx 1 + 7(0.002) + \frac{7(6)}{2}(0.002)^2 + \cdots$$

$$\approx 1 + 0.014 + 21(0.000\,004) + \cdots$$

$$\approx 1 + 0.014 + 0.000\,084 + \cdots$$

$$= 1.0141 \text{ correct to 4 decimal places}$$

The fourth term of the expansion is $\dfrac{(7)(6)(5)}{(1)(2)(3)}(0.002)^3$ and does not affect the result, to the accuracy required.

In experimental work, measurements are taken in the workshop or laboratory under the conditions prevailing at the time and corrections are subsequently made to enable results to be obtained more accurately. For example, the radius and height of a cylinder are measured and the volume is calculated. Later on, corrections are made due to temperature fluctuations or inherent inaccuracies within the measuring devices. The measured value of the radius has an error of $2\frac{1}{2}\%$ too large and the measured value of the height has an error of $1\frac{1}{2}\%$ too small. The binomial expansion can be used to find an approximate value of the error made in calculating the volume, when the other errors are known.

Let the correct values be volume V, radius r and height h. Then the correct value of volume is given by $V = \pi r^2 h$ for a cylinder. The uncorrected value of the radius is $\dfrac{102.5}{100} r$ or $(1 + 0.025)r$, since the radius is $2\frac{1}{2}\%$ too large. The uncorrected value of the height is $\dfrac{98.5}{100} h$ or $(1 - 0.015)h$ since the measured value of the height is $1\frac{1}{2}\%$ too small. Thus the uncorrected value of the volume, V_1, based on these measurements is given by:

$$V_1 = \pi[(1 + 0.025)r]^2(1 - 0.015)h$$

$$= (1 + 0.025)^2(1 - 0.015)\pi r^2 h$$

Using the binomial expansion to evaluate $(1 + 0.025)^2$ and ignoring the term containing $(0.025)^2$, since $(0.025)^2 = 0.000\,625$, which is small compared with 1, gives:

$$V_1 \approx [1 + 2(0.025)](1 - 0.015)\pi r^2 h$$

$$\approx (1 + 0.05)(1 - 0.015)\pi r^2 h$$

$$\approx [1 + 0.05 - 0.015 + (0.05)(-0.015)]\pi r^2 h$$

When approximate values are required, it is also usual to ignore the products of small terms. In general, in any binomial expansion, both products of small terms and powers of small terms can be ignored. This is because numbers less than unity get progressively smaller both when multiplied together and when they are raised to larger powers.

Hence $V_1 \approx (1 + 0.05 - 0.015)\pi r^2 h$

$$\approx 1.035\pi r^2 h \text{ or } 1.035V \text{ or } \frac{103.5}{100} V$$

That is, the uncorrected value V_1 is approximately 3.5% larger than the correct value.

Worked problems on determining approximate values using the binomial expansion

---

Problem 1. Find the value of $(1.003)^{10}$ correct to: (a) 3 decimal places, and (b) 6 decimal places, using the binomial expansion.

Writing $(1.003)^{10}$ as $(1 + 0.003)^{10}$ and substituting $x = 0.003$ and $n = 10$ in the general expansion of $(1 + x)^n$ gives:

$$1 + 10(0.003) + \frac{10(9)}{(2)(1)}(0.003)^2 + \frac{10(9)(8)}{(3)(2)(1)}(0.003)^3 + \cdots$$

or $(1.003)^{10} = 1 + 0.03 + 0.000\,405 + 0.000\,003\,24 + \cdots$

Hence $(1.003)^{10} = 1.030$ correct to 3 decimal places and $1.030\,408$ correct to 6 decimal places (which may be checked using a calculator).

Problem 2. Find the value of $(0.98)^7$ correct to 5 significant figures by using the binomial expansion.

$(0.98)^7$ is written as $(1 - 0.02)^7$. Using the $(1 + x)^n$ expansion gives:

$$(1 - 0.02)^7 = 1 + 7(-0.02) + \frac{7(6)}{(2)(1)}(-0.02)^2 + \frac{(7)(6)(5)}{(3)(2)(1)}(-0.02)^3$$

$$+ \frac{(7)(6)(5)(4)}{(4)(3)(2)(1)}(-0.02)^4 + \cdots$$

$$= 1 + 7(-0.02) + 21(0.000\,4) + 35(-0.000\,008)$$

$$+ 35(0.000\,000\,16) + \cdots$$

$$= 1 - 0.14 + 0.008\,4 - 0.000\,28 + 0.000\,005\,6 - \cdots$$

$$= 0.868\,13$$

correct to 5 significant figures (which may be checked using a calculator).

Problem 3. The radius of a cylinder is reduced by 3% and its height is increased by 4%. Determine the approximate percentage change in (a) its volume, and (b) its curved surface area (neglecting the products of small quantities).

If r and h are the original values of radius and height respectively, then volume of cylinder $= \pi r^2 h$

The new values are $0.97r$ or $(1 - 0.03)r$ and $1.04h$ or $(1 + 0.04)h$

(a)   New volume $= \pi[(1 - 0.03)r]^2[(1 + 0.04)h]$
$$= \pi r^2 h(1 - 0.03)^2(1 + 0.04)$$

Now $(1 - 0.03)^2 = 1 - 2(0.03) + (0.03)^2 = 1 - 0.06,$

neglecting products of small terms

Hence new volume $\approx \pi r^2 h(1 - 0.06)(1 + 0.04)$
$$\approx \pi r^2 h(1 - 0.06 + 0.04)$$

neglecting products of small terms,
$$\approx \pi r^2 h(1 - 0.02) \text{ or } 0.98(\pi r^2 h)$$

i.e. 98% of the original volume

Hence the volume is reduced by 2%

(b)   Curved surface area of cylinder $= 2\pi r h$

New surface area $= 2\pi[(1 - 0.03)r][(1 + 0.04)h]$
$$= 2\pi r h(1 - 0.03)(1 + 0.04)$$
$$\approx 2\pi r h(1 - 0.03 + 0.04)$$

neglecting products of small terms,
$$\approx 2\pi r h(1 + 0.01) \text{ or } 1.01(2\pi r h)$$

i.e. 101% of the original surface area

Hence the curved surface area is increased by 1%

Problem 4. Pressure p and volume v are related by the expression $pv^3 = C$, where C is a constant.
Find the approximate percentage change in C when p is increased by 2% and v decreased by 0.8%

Let p and v be the original values of pressure and volume.

The new values are $\dfrac{102}{100}p$ or $(1 + 0.02)p$ and $\dfrac{99.2}{100}v$ or $(1 - 0.008)v$

Let the new value of C be C_1, then

$C_1 = (1 + 0.02)p[(1 - 0.008)v]^3$
$= (1 + 0.02)(1 - 0.008)^3 pv^3$
$(1 - 0.008)^3 \approx 1 - (3)(0.008) + \cdots$
$\approx 1 - 0.024$

Hence $C_1 \approx (1 + 0.02)(1 - 0.024)C$

and neglecting the products of small terms, this becomes

$C_1 \approx (1 + 0.02 - 0.024)C$
$\approx (1 - 0.004)C$

Hence **the value of C is reduced by approximately 0.4%** when p is increased by 2% and v decreased by 0.8%

Problem 5. The resonant frequency of a vibrating shaft is given by $f = \dfrac{1}{2\pi}\sqrt{\left(\dfrac{k}{I}\right)}$, where k is the stiffness and I is the inertia of the shaft. Determine the approximate percentage error in determining the frequency using the measured values of k and I, when the measured value of k is 3% too large and the measured value of I is 1.5% too small.

Let f, k and I be the true values of frequency, stiffness and inertia respectively. Since the measured value of stiffness, k_1 is 3% too large, $k_1 = \dfrac{103}{100} k = (1 + 0.03)k$. The measured value of inertia, I_1 is 1.5% too small, hence $I_1 = \dfrac{98.5}{100} I = (1 - 0.015)I$

The measured value of frequency,

$$f_1 = \frac{1}{2\pi}\sqrt{\left(\frac{k_1}{I_1}\right)} = \frac{1}{2\pi}\frac{k_1^{\frac{1}{2}}}{I_1^{\frac{1}{2}}} = \frac{1}{2\pi}k_1^{\frac{1}{2}}I_1^{-\frac{1}{2}}$$

$$= \frac{1}{2\pi}[(1+0.03)k]^{\frac{1}{2}}[(1-0.015)I]^{-\frac{1}{2}}$$

$$= \frac{1}{2}(1+0.03)^{\frac{1}{2}}k^{\frac{1}{2}}(1-0.015)^{-\frac{1}{2}}I^{-\frac{1}{2}}$$

$$= \frac{1}{2\pi}k^{\frac{1}{2}}I^{-\frac{1}{2}}(1+0.03)^{\frac{1}{2}}(1-0.015)^{-\frac{1}{2}}$$

i.e. $f_1 = (1+0.03)^{\frac{1}{2}}(1-0.015)^{-\frac{1}{2}}f$

$$\approx \left(1 + \left(\frac{1}{2}\right)(0.03)\right)\left(1 - \left(-\frac{1}{2}\right)(0.015)\right)f$$

$$\approx (1+0.015)(1+0.007\,5)f$$

Neglecting the products of small terms,

$f_1 \approx (1 + 0.015 + 0.007\,5)f \approx 1.022\,5f$

Thus, the percentage error based on the measured values of k and I is approximately $(1.022\,5)(100) - 100$, i.e. **2.3% too large.**

Further problems on determining approximate values, using the binomial expansion, may be found in the following Section 1.4 (Problems 31 to 44), page 22.

1.4   Further problems

Expansions of $(a + b)^n$ where n is a small positive integer using Pascal's triangle

1. Determine the expansion of $(a + b)^8$
    $[a^8 + 8a^7b + 28a^6b^2 + 56a^5b^3 + 70a^4b^4 + 56a^3b^5$
    $\qquad\qquad\qquad\qquad\qquad + 28a^2b^6 + 8ab^7 + b^8]$

2. Find the expansion of $(x - y)^5$
    $[x^5 - 5x^4y + 10x^3y^2 - 10x^2y^3 + 5xy^4 - y^5]$

3. Find the expansion of $(2p - 3q)^6$
    $[64p^6 - 576p^5q + 2\,160p^4q^2 - 4\,320p^3q^3 + 4\,860p^2q^4$
    $\qquad\qquad\qquad\qquad\qquad\qquad\quad - 2916pq^5 + 729q^6]$

4. Expand $(p + 3q)^{11}$ as far as the term containing q^4
    $[p^{11} + 33p^{10}q + 495p^9q^2 + 4\,455p^8q^3 + 26\,730p^7q^4]$

5. Find the expansion of $(x - 2y)^{10}$ as far as the term containing y^5
    $[x^{10} - 20x^9y + 180x^8y^2 - 960x^7y^3 + 3\,360x^6y^4 - 8\,064x^5y^5]$

6. Determine the expansion of $\left(-m - \dfrac{n}{2}\right)^7$ as far as the term containing n^4
    $$\left[-m^7 - \frac{7}{2}m^6n - \frac{21}{4}m^5n^2 - \frac{35}{8}m^4n^3 - \frac{35}{16}m^3n^4\right]$$

7. Determine the expansion of $\left(\dfrac{w}{2} - \dfrac{x}{3}\right)^4$
    $$\left[\frac{1}{16}w^4 - \frac{1}{6}w^3x + \frac{1}{6}w^2x^2 - \frac{2}{27}wx^3 + \frac{1}{81}x^4\right]$$

8. Find the expansion of $(3 + 4y)^6$ and express the result in the form $a + by + cy^2 + \cdots$, where $a, b, c, \ldots$ are constants.
    $[729 + 5\,832y + 19\,440y^2 + 34\,560y^3 + 34\,560y^4 + 18\,432y^5$
    $\qquad\qquad\qquad\qquad\qquad\qquad\qquad\qquad\qquad + 4\,096y^6]$

9. Determine the expansion of $\left(\dfrac{x}{4} - 7\right)^5$ and express the result in the form $ax^5 + bx^4 + cx^3 + \cdots$, where $a, b, c, \ldots$ are constants.
    $$\left[\frac{x^5}{1\,024} - \frac{35}{256}x^4 + \frac{245}{32}x^3 - \frac{1\,715}{8}x^2 + \frac{12\,005}{4}x - 16\,807\right]$$

10. Expand $\left(-5 - \dfrac{p}{3}\right)^8$ as far as the term containing p^4
    $$\left[390\,625 + \frac{625\,000}{3}p + \frac{437\,500}{9}p^2 + \frac{175\,000}{27}p^3 + \frac{43\,750}{81}p^4\right]$$

Expansions of the type $(a + b)^n$

11.   Find the expansion of $(a + b)^{12}$ as far as the term containing b^5
      $[a^{12} + 12a^{11}b + 66a^{10}b^2 + 220a^9b^3 + 495a^8b^4 + 792a^7b^5]$

12.   Determine the expansion of $\left(x - \dfrac{y}{2}\right)^{16}$ as far as the term
      containing y^4
      $$\left[x^{16} - 8x^{15}y + 30x^{14}y^2 - 70x^{13}y^3 + \frac{455}{4}x^{12}y^4\right]$$

In Problems 13–15, find the first four terms of the expansions of the expressions given.

13. $\left(m - \dfrac{n^2}{2}\right)^{13}$ $\left[m^{13} - \dfrac{13}{2}m^{12}n^2 + \dfrac{39}{2}m^{11}n^4 - \dfrac{143}{4}m^{10}n^6\right]$

14. $(-p^2 - 2q)^{17}$ $[-p^{34} - 34p^{32}q - 544p^{30}q^2 - 5440p^{28}q^3]$

15. $\left(-\dfrac{1}{x} + \dfrac{3}{y}\right)^{19}$ $\left[-\dfrac{1}{x^{19}} + \dfrac{57}{x^{18}y} - \dfrac{1539}{x^{17}y^2} + \dfrac{26163}{x^{16}y^3}\right]$

16.   Determine the middle term of the expansion of $(x^2 - y^2)^{14}$
      $[-3432x^{14}y^{14}]$

17.   Find the eleventh term of the expansion of $\left(2p - \dfrac{q}{2}\right)^{21}$
      $[705432p^{11}q^{10}]$

18.   Find the value of the middle term of the expansion of
      $\left(2c^2 - \dfrac{1}{2c^2}\right)^{12}$ $[924]$

19.   Determine the two middle terms of the expansion of $\left(2p - \dfrac{1}{3p}\right)^9$
      $\left[\dfrac{1344p}{27}, \ -\dfrac{224}{27p}\right]$

20.   Find the term involving a^{12} in the expansion of $\left(a^3 - \dfrac{b}{2}\right)^{14}$
      $\left[\dfrac{1001}{1024}a^{12}b^{10}\right]$

In Problems 21–26, expand in ascending powers of x as far as the term in x^3 using the binomial theorem. State in each case the limits of x for which the series is valid.

21. $\dfrac{1}{(1 + x)}$ $[1 - x + x^2 - x^3 + \cdots, \ |x| < 1]$

22. $\dfrac{2}{(1 + x)^2}$ $[1 - 2x + 3x^2 - 4x^3 + \cdots, \ |x| < 1]$

23. $\dfrac{1}{(1 + 2x)^3}$ $\left[1 - 6x + 24x^2 - 80x^3 + \cdots \quad |x| < \dfrac{1}{2}\right]$

24. $\dfrac{1}{(4 - x)^2}$ $\left[\dfrac{1}{16}\left(1 - \dfrac{x}{2} + \dfrac{3}{16}x^2 + \dfrac{1}{16}x^3 + \cdots\right) \quad |x| < 4\right]$

25. $\sqrt{(3 + x)}$ $\left[\sqrt{3}\left(1 + \dfrac{x}{6} - \dfrac{x^2}{72} + \dfrac{x^2}{216}\cdots\right) \quad |x| < 3\right]$

26. $\dfrac{1}{\sqrt{(1 - 2x)}}$ $\left[1 + x + \dfrac{3}{2}x^2 + \dfrac{5}{2}x^3 + \cdots \quad |x| < \dfrac{1}{2}\right]$

27. Express $\dfrac{1}{(2 - 3t)^5}$ to three terms. For what value of t is the expansion valid?

$\left[\dfrac{1}{32}\left(1 + \dfrac{15}{2}t + \dfrac{45}{2}t^2 + \dfrac{135}{4}t^3 + \cdots\right) \quad |t| < \dfrac{2}{3}\right]$

28. When x is very small show that:

(a) $\dfrac{1}{(1 + x)^2\sqrt{(1 + x)}} \approx 1 - \dfrac{5}{2}x$

(b) $\dfrac{(1 - 3x)}{(1 + 2x)^4} \approx 1 - 11x$

29. If x is very small such that x^2 and higher powers may be neglected, determine the power series for $\dfrac{\sqrt{(x + 3)}\sqrt[3]{(8 - x)}}{\sqrt[4]{(1 + x)^3}}$

$\left[\dfrac{2}{\sqrt{3}}\left(1 + \dfrac{7x}{8}\right)\right]$

30. Express the following as a power series in ascending powers of x as far as the term in x^2. State in each case the range of x for which the series is valid.

(a) $\sqrt{\left(\dfrac{1 + x}{1 - x}\right)}$ (b) $\dfrac{(1 - x)\sqrt[3]{(1 - 3x)^4}}{\sqrt{(1 + x^2)}}$

(a) $\left[1 + x + \dfrac{1}{2}x^2 \quad |x| < 1\right]$ (b) $\left[1 - 5x + \dfrac{11}{2}x^2 \quad |x| < \dfrac{1}{3}\right]$

Determining the approximate values of expressions

31. Use the binomial expansion to calculate the value of $(0.995)^{12}$ correct to (a) 4 decimal places; and (b) 6 decimal places.   [0.941 6, 0.941 623]

32. Find the value of $(1.05)^3(0.98)^4$ correct to 5 significant figures by using binomial expansions.   [1.067 8]

33. Determine the value of $(3.036)^3$ correct to 3 decimal places by using the binomial expansion. [27.984]

34. Use the binomial expansion to find the value of $(2.018)^5$ correct to 6 significant figures. [33.466 2]

35. An error of 3.5% too large was made when measuring the radius of a sphere. Ignoring the products of small quantities, determine the approximate error in calculating: (a) the volume; and (b) the surface area when they are calculated using the incorrect radius measurement.
[(a) 10.5% too large   (b) 7% too large]

36. The area of a triangle is given by $A = \frac{1}{2}ab \sin C$, where C is the angle between the sides a and b of a triangle. Calculate the approximate change in area (ignoring the products of small quantities), when: (a) both a and b are increased by 2%; and (b) a is increased by 2% and b is reduced by 2%
[(a) 4% increase   (b) no change]

37. The moment of inertia of a body about an axis is given by $I = kbd^3$ where k is a constant and b and d are the dimensions of the body. Determine the approximate percentage change in the value of I when b is increased by 5% and d reduced by 1%, if products of small quantities are ignored. [2% increase]

38. The radius of a cone is reduced by 4.5% and its height increased by 1.5%. Determine the approximate percentage change in its volume, neglecting the products of small quantities. [7.5% reduction]

39. The power developed by an engine is given by $I = kPLAN$ where k is a constant. Find the approximate percentage increase in power when P, L, A and N are each increased by 3.5% [14%]

40. The modulus of rigidity G is given by $G = \dfrac{R^4\theta}{L}$ where R is the radius, θ the angle of twist and L the length. Find the approximate percentage error in G when R is measured 1.5% too large and θ is measured 5% too small.
[1% too large]

41. The degree of hydrolysis of methyl ethanoate is given by the equation $\alpha = \sqrt{[K_w/(K_aC)]}$, where K_w is a constant. If there is an error of $+1.2\%$ in K_a and -2.5% in C, determine the resultant error in α [0.65%]

42. The viscosity (η) of a liquid is given by $\eta = \dfrac{ar^4}{vl}$, where a is a constant. If there is an error in r of $+3\%$, in v of $+4\%$ and l of -2%, what is the resultant error in η? [10%]

43. The energy W stored in a flywheel is given by $W = kr^5N^2$, where k is a constant, r is the radius and N the number of revolutions. Determine the approximate percentage change in W when r is increased by 1.5% and N is decreased by 3% [1.5% increase]

44. In a series electrical circuit containing inductance L and capacitance C, the resonant frequency f_r is given by $f_r = \dfrac{1}{2\pi\sqrt{(LC)}}$. If the values of L and C used in the calculation are 2.5% too large and 1% too small respectively, determine the approximate percentage error in the frequency. [0.75% too small]

Chapter 2

Exponential functions and Napierian logarithms

2.1 The exponential function

In calculus and more advanced mathematics, a mathematical constant e is frequently used. This constant is called the exponent and has a value of approximately 2.718 3

A function containing e^x is called an **exponential function** and can also be written as exp x.

e^x is a function which increases at a rate proportional to its own magnitude (see Chapter 3).

The natural laws of growth and decay are of the form $y = A(1 - e^{kx})$ and $y = A e^{kx}$ (see Chapter 3) and therefore the exponent is of considerable importance in engineering and science.

2.2 Evaluating exponential functions

The value of e^x may be determined by using

(a) the power series for e^x
or (b) a calculator
or (c) tables of exponential functions.

The most common method of evaluating an exponential function is by a scientific notation calculator, this now having replaced the use of tables. However, let us first look at the power series for e^x.

The power series for e^x

The value of e^x can be calculated to any required degree of accuracy since it is defined in terms of the following power series:

$$e^x = 1 + x + \frac{x^2}{2!} + \frac{x^3}{3!} + \cdots \tag{1}$$

which is valid for all values of x (where $3! = 3 \times 2 \times 1$ and is called 'factorial 3' – see Chapter 1). This series is said to **converge**; that is, if all the terms are added an actual value for e^x is obtained, where x is a real number. The more terms that are taken the closer will be the value of e^x to its actual value.

The value of the exponent e – correct to, say, 4 decimal places – may be determined by substituting $x = 1$ in the power series of equation (1). Thus

$$e^1 = 1 + 1 + \frac{(1)^2}{2!} + \frac{(1)^3}{3!} + \frac{(1)^4}{4!} + \frac{(1)^5}{5!} + \frac{(1)^6}{6!} + \frac{(1)^7}{7!} + \frac{(1)^8}{8!}$$

$$= 1 + 1 + 0.5 + 0.166\,67 + 0.041\,67 + 0.008\,33$$
$$+ 0.001\,39 + 0.000\,20 + 0.000\,02$$
$$= 2.718\,28$$

i.e. $e = 2.718\,3$, correct to 4 decimal places

The value of $e^{0.01}$ – correct to, say, 9 significant figures – is found by substituting $x = 0.01$ in the power series for e^x. Thus

$$e^{0.01} = 1 + 0.01 + \frac{(0.01)^2}{2!} + \frac{(0.01)^3}{3!} + \frac{(0.01)^4}{4!} + \cdots$$

$$= 1 + 0.01 + 0.000\,05 + 0.000\,000\,167 + 0.000\,000\,000\,4$$

and by adding, $e^{0.01} = 1.010\,050\,17$, correct to 9 significant figures. In this example successive terms in the series grow smaller very rapidly and it is relatively easy to determine the value of $e^{0.01}$ to a high degree of accuracy. However, when x is near to unity or larger than unity, a very large number of terms are required for an accurate result.

If, in the series of equation (1), x is replaced by $-x$, then

$$e^{-x} = 1 + (-x) + \frac{(-x)^2}{2!} + \frac{(-x)^3}{3!} + \cdots$$

i.e. $e^{-x} = 1 - x + \dfrac{x^2}{2!} - \dfrac{x^3}{3!} + \cdots$

In a similar manner the power series for e^x may be used to evaluate any exponential function of the form $a\,e^{kx}$, where a and k are constants. In the series of equation (1) let x be replaced by kx, then

$$a\,e^{kx} = a\left\{ 1 + (kx) + \frac{(kx)^2}{2!} + \frac{(kx)^3}{3!} + \cdots \right\}$$

$$\text{Then } 3\,e^{2x} = 3\left\{ 1 + (2x) + \frac{(2x)^2}{2!} + \frac{(2x)^3}{3!} + \cdots \right\}$$

$$= 3\left\{ 1 + (2x) + \frac{4x^2}{2 \times 1} + \frac{8x^3}{3 \times 2 \times 1} + \cdots \right\}$$

$$= 3\left\{ 1 + 2x + 2x^2 + \frac{4}{3}x^3 + \cdots \right\}$$

Use of a calculator

Most **scientific notation calculators** contain an 'e^x' function which enables all practical values of e^x and e^{-x} to be determined, correct to 8 or 9 significant figures. For example,

$$e^1 = 2.718\ 281\ 83$$

$$e^{2.5} = 12.182\ 494\ 0$$

and $e^{-1.732} = 0.176\ 930\ 195$, correct to 9 significant figures

In practical situations the degree of accuracy given by a calculator is often far greater than is appropriate. The accepted convention is that the final result is stated to one significant figure greater than the least significant measured value. Use your calculator to check the following:

$e^{0.16} = 1.173\ 5$, correct to 5 significant figures

$e^{-1.52} = 0.218\ 7$, correct to 4 decimal places

$e^{-0.632} = 0.531\ 53$, correct to 5 decimal places

$e^{7.5} = 1\ 808.04$, correct to 6 significant figures

$e^{-4.673} = 0.009\ 344\ 2$, correct to 7 decimal places

Worked problems on evaluating exponential functions

Problem 1. Determine the value of $2\,e^{0.3}$ correct to 5 significant figures by using the power series for e^x

Substituting $x = 0.3$ in the power series

$$e^x = 1 + x + \frac{x^2}{2!} + \frac{x^3}{3!} + \cdots \text{ gives}$$

$$e^{0.3} = 1 + 0.3 + \frac{(0.3)^2}{(2)(1)} + \frac{(0.3)^3}{(3)(2)(1)} + \frac{(0.3)^4}{(4)(3)(2)(1)} + \frac{(0.3)^5}{(5)(4)(3)(2)(1)}$$

$$= 1 + 0.3 + 0.045 + 0.004\ 5 + 0.000\ 338 + 0.000\ 020$$

$$= 1.349\ 86, \text{ correct to 6 significant figures}$$

Hence $2\,e^{0.3} = 2.699\ 7$, correct to 5 significant figures

---

Problem 2. Determine the value of $-4\,e^{-1}$ correct to 4 decimal places using the power series for e^x

Substituting $x = -1$ in the power series

$$e^x = 1 + x + \frac{x^2}{2!} + \frac{x^3}{3!} + \cdots \text{ gives}$$

$$e^{-1} = 1 + (-1) + \frac{(-1)^2}{(2)(1)} + \frac{(-1)^3}{(3)(2)(1)} + \frac{(-1)^4}{(4)(3)(2)(1)} + \cdots$$

$$= 1 - 1 + 0.5 - 0.166\,667 + 0.041\,667$$

$$- 0.008\,333 + 0.001\,389 - 0.000\,198 + \cdots$$

$$= 0.367\,858, \text{ correct to 6 decimal places}$$

Hence $-4\,e^{-1} = (-4)(0.367\,858)$
$$= -\mathbf{1.471\,4}, \text{ correct to 4 decimal places}$$

Problem 3. Expand $e^x(x^2 + 1)$, as far as the term in x^5

The power series for e^x is $1 + x + \frac{x^2}{2!} + \frac{x^3}{3!} + \frac{x^4}{4!} + \cdots$

Hence $e^x(x^2 + 1) = \left(1 + x + \frac{x^2}{2!} + \frac{x^3}{3!} + \frac{x^4}{4!} + \cdots\right)(x^2 + 1)$

i.e. $e^x(x^2 + 1) = \left(x^2 + x^3 + \frac{x^4}{2!} + \frac{x^5}{3!}\right)$

$$+ \left(1 + x + \frac{x^2}{2!} + \frac{x^3}{3!} + \frac{x^4}{4!} + \frac{x^5}{5!} + \cdots\right)$$

Grouping like terms gives:

$$e^x(x^2 + 1) = 1 + x + \left(1 + \frac{1}{2!}\right)x^2 + \left(1 + \frac{1}{3!}\right)x^3$$

$$+ \left(\frac{1}{2!} + \frac{1}{4!}\right)x^4 + \left(\frac{1}{3!} + \frac{1}{5!}\right)x^5 + \cdots$$

i.e. $e^x(x^2 + 1) = 1 + x + \frac{3}{2}x^2 + \frac{7}{6}x^3 + \frac{13}{24}x^4 + \frac{7}{40}x^5$
when expanded as far as the term in x^5

Problem 4. Evaluate the following correct to 4 decimal places using a calculator:

(a) $\frac{1}{7}e^{3.7681}$ (b) $0.013\,e^{-0.173}$ (c) $\frac{3(e^{0.32} + e^{-0.32})}{(e^{0.32} - e^{-0.32})}$

(a) $\frac{1}{7}e^{3.7681} = \frac{1}{7}(43.297\,72\ldots)$
$$= \mathbf{6.185\,4}, \text{ correct to 4 decimal places}$$

(b) $0.013\,e^{-0.173} = 0.013(0.841\,137\,6\ldots)$
$$= \mathbf{0.010\,9}, \text{ correct to 4 decimal places}$$

(c) $\dfrac{3(e^{0.32} + e^{-0.32})}{(e^{0.32} - e^{-0.32})} = \dfrac{3(1.377\,127\ldots + 0.726\,149\ldots)}{(1.377\,127\ldots - 0.726\,149\ldots)}$

$$= \dfrac{3(2.103\,276\ldots)}{(0.650\,978\ldots)}$$

$$= \mathbf{9.692\,8}, \text{ correct to 4 decimal places}$$

Problem 5. Determine the value of voltage v, given $v = Ve^{-t/CR}$, correct to 4 significant figures, when $V = 150$, $t = 20 \times 10^{-3}$, $C = 5 \times 10^{-6}$ and $R = 40 \times 10^3$

$v = Ve^{-t/CR} = 150\,e^{[(-20 \times 10^{-3})/(5 \times 10^{-6} \times 40 \times 10^3)]}$

$$= 150\,e^{-0.10} = 150(0.904\,837\ldots)$$

$$= \mathbf{135.7\ volts}$$

Further problems on exponential functions may be found in Section 2.5 (Problems 1 to 10), page 32.

2.3   Napierian logarithms

A **logarithm** of a number is defined as the power to which a base has to be raised to be equal to the number. Thus, if $y = a^x$ then $x = \log_a y$.

Logarithms having a base of 10 are called **common logarithms** and the common logarithm of x is written as lg x. Logarithms having a base of e are called **hyperbolic**, **Napierian** or **natural logarithms** and the Napierian logarithm of x is written as $\log_e x$ or, more commonly, ln x.

The most logical base when using calculus and when dealing with problems involving the natural growth or decay laws is the exponent e.

2.4   Evaluating Napierian logarithms

The value of a Napierian logarithm may be determined by using

(a)   a calculator
or (b)   a relationship between common and Napierian logarithms
or (c)   Napierian logarithm tables.

The most common method of evaluating the logarithmic function is by a scientific notation calculator, this now having replaced the use of 4-figure tables and also the relationship between common and Napierian logarithms, $\log_e y = 2.302\,6 \log_{10} y$.

Most scientific notation calculators contain a 'ln x' function which gives the value of a Napierian logarithm of a number displayed when the appropriate key is pressed. Using a calculator,

ln $5.321 = 1.671\,661\ldots = 1.671\,7$, correct to 4 decimal places
and ln $21.43 = 3.064\,791\ldots = 3.064\,8$, correct to 4 decimal places

Use a calculator to check the following values:

$\ln 1.739 = 0.553\,31$, correct to 5 significant figures

$\ln 1 = 0$

$\ln 0.4 = -0.916\,3$, correct to 4 decimal places

$\ln 673 = 6.511\,7$, correct to 4 decimal places

$\ln 1\,600 = 7.377\,8$, correct to 5 significant figures

$\ln 0.12 = -2.120\,264$, correct to 6 decimal places

$\ln 0.001\,5 = -6.502\,3$, correct to 5 significant figures

$\ln e^2 = 2$

$\ln e^5 = 5$

We can conclude from the latter two examples that

$$\log_e e^x = x$$

This is useful when solving equations involving exponential functions. For example, to solve $e^{2x} = 5$ take Napierian logarithms of both sides, which gives

$$\ln e^{2x} = \ln 5$$

i.e. $\qquad 2x = \ln 5$

from which $x = \frac{1}{2}\ln 5 = \mathbf{0.804\,7}$, correct to 4 decimal places.

Worked problems on evaluating Napierian logarithms

Problem 1. Use a calculator to evaluate the following, each correct to 5 significant figures:

(a) $\frac{1}{6}\ln 3.682\,9$ (b) $\dfrac{\ln 4.921}{4.921}$ (c) $\dfrac{3.48 \ln 21.47}{e^{-0.738}}$

(a) $\frac{1}{6}\ln 3.682\,9 = \frac{1}{6}(1.303\,700\,4\ldots)$

$\qquad\qquad = \mathbf{0.217\,28}$, correct to 5 significant figures

(b) $\dfrac{\ln 4.921}{4.921} = \dfrac{1.593\,511\,7\ldots}{4.921}$

$\qquad\qquad = \mathbf{0.323\,82}$, correct to 5 significant figures

(c) $\dfrac{3.48 \ln 21.47}{e^{-0.738}} = \dfrac{3.48(3.066\,656\,6\ldots)}{(0.478\,069\,0\cdots)}$

$\qquad\qquad = \mathbf{22.323}$, correct to 5 significant figures

Problem 2. Evaluate the following:

(a) $\dfrac{\ln e^{1.6}}{\lg 10^{3.2}}$ (b) $\dfrac{5\, e^{1.72}\, \lg 1.72}{\ln 1.72}$ (correct to 4 decimal places)

(a) $\dfrac{\ln e^{1.6}}{\lg 10^{3.2}} = \dfrac{1.6}{3.2} = \mathbf{0.5}$

(b) $\dfrac{5\, e^{1.72}\, \lg 1.72}{\ln 1.72} = \dfrac{5(5.584\,528\ldots)(0.235\,528\ldots)}{(0.542\,324\ldots)}$

$= \mathbf{12.126\ 6}$, correct to 4 decimal places

Problem 3. Solve the equation $5 = 3\, e^{-2x}$ to find x correct to 4 significant figures.

Rearranging $5 = 3\, e^{-2x}$ gives: $\frac{5}{3} = e^{-2x}$

Taking the reciprocal of both sides gives: $\frac{3}{5} = \dfrac{1}{e^{-2x}} = e^{2x}$

i.e. $0.60 = e^{2x}$

Taking Napierian logarithms of both sides gives:
$\ln 0.60 = \ln e^{2x} = 2x$
Hence $x = \frac{1}{2} \ln 0.60 = \frac{1}{2}(-0.510\,83)$
i.e. $x = \mathbf{-0.255\ 4}$

Problem 4. Given $36 = 72(1 - e^{-t/3})$ determine t, correct to 3 significant figures.

Rearranging $36 = 72(1 - e^{-t/3})$ gives: $\frac{36}{72} = 1 - e^{-t/3}$

and $e^{-t/3} = 1 - \frac{36}{72} = 0.50$

Taking the reciprocal of both sides gives: $e^{t/3} = \dfrac{1}{0.50} = 2.0$

Taking Napierian logarithms of both sides gives:

$\dfrac{t}{3} = \ln 2.0$

from which $t = 3 \ln 2.0 = 3(0.693\ 1)$
i.e. $t = 2.079\ 3$ or $\mathbf{2.08}$, correct to 3 significant figures

---

Problem 5. Solve the equation $2.58 = \ln\left(\dfrac{4.92}{x}\right)$ to find x

From the definition of a logarithm, since $2.58 = \ln\left(\dfrac{4.92}{x}\right)$, then

$$e^{2.58} = \dfrac{4.92}{x}$$

Rearranging gives: $x = \dfrac{4.92}{e^{2.58}} = 4.92\,e^{-2.58} = (4.92)(0.075\,774)$

i.e. $x = 0.372\,8$, correct to 4 significant figures

Practical applications of equations involving $\ln x$ and e^x are numerous and some typical examples are highlighted in Chapter 3.

Further problems on Napierian logarithms may be found in the following Section 2.5 (Problems 11 to 23), page 33.

2.5   Further problems

Exponential functions

In Problems 1 to 3 use the power series for e^x to determine the values of y, correct to 4 significant figures.

1. (a) $y = e^2$ (b) $y = e^{0.4}$ (c) $y = e^{0.1}$
    (a) [7.389]   (b) [1.492]   (c) [1.105]

2. (a) $y = e^{-0.3}$ (b) $y = e^{-0.1}$ (c) $y = e^{-2}$
    (a) [0.740 8]   (b) [0.904 8]   (c) [0.135 3]

3. (a) $y = 3\,e^4$ (b) $y = 0.86\,e^{-0.4}$ (c) $y = -5\,e^{-0.75}$
    (a) [163.8]   (b) [0.576 5]   (c) [−2.362]

4. Expand $e^{2x}(1 - 2x)$ to the term in x^4

$$\left[1 - 4x^2 - \frac{8x^3}{3} - 2x^4 \right]$$

5. Expand $(3\,e^{x^2})(x^{\frac{1}{2}})$ to six terms.

$$[3x^{\frac{1}{2}} + 3x^{\frac{5}{2}} + \tfrac{3}{2}x^{\frac{9}{2}} + \tfrac{1}{2}x^{\frac{13}{2}} + \tfrac{1}{8}x^{\frac{17}{2}} + \tfrac{1}{40}x^{\frac{21}{2}}]$$

In Problems 6 and 7 use a calculator to evaluate the functions given, correct to 4 significant figures.

6. (a) $e^{5.2}$ (b) $e^{-0.37}$ (c) $e^{0.86}$
    (a) [181.3]   (b) [0.690 7]   (c) [2.363]

7. (a) $e^{-0.58}$ (b) $e^{-0.17}$ (c) e^{12}
    (a) [0.559 9]   (b) [0.843 7]   (c) [162 800]

In Problems 8 and 9 evaluate correct to 5 significant figures.

8. (a) $\tfrac{1}{4}e^{2.963\,2}$ (b) $7.34\,e^{-1.125}$ (c) $\dfrac{5\,e^{2.749\,3}}{2\,e^{1.693\,2}}$
    (a) [4.840 0]   (b) [2.382 9]   (c) [7.187 8]

9. (a) $\dfrac{e^{3.791} - e^{-3.791}}{2}$ (b) $\dfrac{4.83}{e^{-2.763}}$ (c) $\dfrac{5(e^{-1.683} - 1)}{e^{2.483}}$
    (a) [22.139]   (b) [76.543]   (c) [−0.339 89]

10. The length of a bar, l, at temperature θ is given by $l = l_0 \, e^{\alpha\theta}$, where l_0 and α are constants. Evaluate l, correct to 4 significant figures, when $l_0 = 2.472$, $\theta = 316.8$ and $\alpha = 1.771 \times 10^{-4}$ [2.615]

Napierian logarithms

In Problems 11 to 16 determine the values of $\ln x$, correct to 4 decimal places, when x has the values shown.

11. (a) 2.614 (b) 6.775 (c) 9.213
    (a) [0.960 9] (b) [1.913 2] (c) [2.220 6]
12. (a) $3\frac{9}{17}$ (b) $\frac{127}{19}$ (c) $8\frac{4}{7}$
    (a) [1.261 1] (b) [1.899 7] (c) [2.148 4]
13. (a) 77.34 (b) 190 (c) 2 377
    (a) [4.348 2] (b) [5.247 0] (c) [7.773 6]
14. (a) 23.2 (b) 17 140 (c) 7 601
    (a) [3.144 2] (b) [9.749 2] (c) [8.936 0]
15. (a) 0.171 (b) 0.005 35 (c) 0.877 4
    (a) [−1.766 1] (b) [−5.230 7] (c) [−0.130 8]
16. (a) 3.74×10^{-3} (b) 7.818×10^{-2} (c) 9.671×10^{-4}
    (a) [−5.588 7] (b) [−2.548 7] (c) [−6.941 2]

In Problems 17 and 18 evaluate correct to 5 significant figures.

17. (a) $\frac{1}{8} \ln 4.729\,1$ (b) $\dfrac{\ln 71.49}{3.647}$ (c) $\dfrac{4.91 \ln 314.62}{e^{1.1762}}$
    (a) [0.194 22] (b) [1.170 7] (c) [8.700 7]
18. (a) $\dfrac{5.621 \ln e^{2.45}}{\lg 10^{1.62}}$ (b) $\dfrac{3\,e^{-0.1729}}{2 \ln 0.001\,78}$ (c) $\dfrac{\ln 2.931 - \ln 1.762}{4.32}$
    (a) [8.500 9] (b) [−0.199 31] (c) [0.117 80]

In Problems 19 to 23 solve the given equations, each correct to 4 significant figures.

19. $2 = 4\,e^{5t}$ [−0.138 6]
20. $5.36 = 2.81\,e^{-3.2x}$ [−0.201 8]
21. $21 = 33(1 - e^{-x/2})$ [2.023]

22. $6.48 = \ln\left(\dfrac{x}{3.9}\right)$ [2 543]

23. $2.48 = 3.62 \ln\left(\dfrac{1.46}{x}\right)$ [0.735 9]

Chapter 3

Curves of exponential growth and decay

3.1   Graphs of exponential functions

A graph of the curves $y = e^x$ and $y = e^{-x}$ over the range $x = -3$ to $x = 3$ is shown in Fig. 3.1. The values of e^x and e^{-x}, correct to 2 decimal places, are obtained by calculator and are shown below.

x	-3.0	-2.5	-2.0	-1.5	-1.0	-0.5	0	0.5	1.0	1.5	2.0	2.5	3.0
e^x	0.05	0.08	0.14	0.22	0.37	0.61	1	1.65	2.72	4.48	7.39	12.18	20.09
e^{-x}	20.09	12.18	7.39	4.48	2.72	1.65	1	0.61	0.37	0.22	0.14	0.08	0.05

A graph of $y = 1 - e^{-x}$ over the range $x = 0$ to $x = 3.5$ is shown in Fig. 3.2. A table of values is shown below.

x	0	0.5	1.0	1.5	2.0	2.5	3.0	3.5
e^{-x}	1	0.61	0.37	0.22	0.14	0.08	0.05	0.03
$1 - e^{-x}$	0	0.39	0.63	0.78	0.86	0.92	0.95	0.97

For graphs of the form $y = e^{kx}$, where k is any constant and can be positive or negative, 'k' has the effect of altering the scale of x. For graphs of the form $y = A e^{kx}$, where A is a constant, 'A' has the effect of altering the scale of y. Hence every curve of the form $y = A e^{kx}$ has the same general shape as shown in Fig. 3.1, and A and k are called **scale factors** of the graph. Their only function is to alter the values of x and y shown on the axes. Thus similar curves can be obtained for every function of the form $y = A e^{kx}$ by selecting appropriate scale factors. For example, the curve of $y = 2 e^{3x}$ becomes identical to the curve $y = e^x$ shown in Fig. 3.1 by making the y-axis markings $4, 8, 12, 16, \ldots$ instead of $2, 4, 6, 8, \ldots$ and the x-axis markings $\frac{1}{3}, \frac{2}{3}, 1$ instead of $1, 2, 3$. By similar reasoning, all curves of the form $y = a(1 - e^{-kx})$, where a and k are constants, have the same general shape as that shown in Fig. 3.2

   Curves of $y = A e^{kx}$, $y = A e^{-kx}$ and $y = a(1 - e^{-kx})$ have many applications since they define mathematically the laws of growth and decay which are discussed in Section 3.2

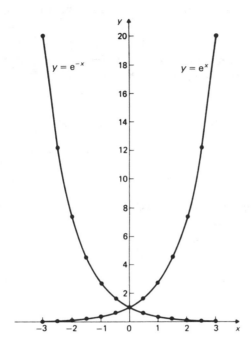

Fig. 3.1   Graph depicting $y = e^x$ and $y = e^{-x}$

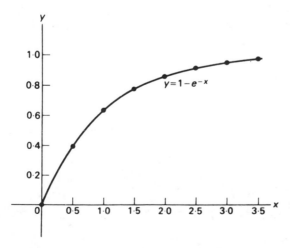

Fig. 3.2   Graph depicting $y = 1 - e^{-x}$

Worked problems on the graphs of exponential functions

Problem 1. Draw a graph of $y = 3\,e^{0.2x}$ over a range of $x = -3$ to $x = 3$ and hence determine the approximate value of y when $x = 1.7$ and the approximate value of x when $y = 3.3$

The values of y are calculated for integer values of x over the range required and are shown in the table below:

x	-3	-2	-1	0	1	2	3
$0.2x$	-0.6	-0.4	-0.2	0	0.2	0.4	0.6
$e^{0.2x}$	0.549	0.670	0.819	1	1.221	1.492	1.822
$3\,e^{0.2x}$	1.65	2.01	2.46	3	3.66	4.48	5.47

The values of the exponential functions are obtained by calculator. The points are plotted and the curve drawn as shown in Fig. 3.3.

From the graph, when $x = 1.7$, the corresponding value of y is **4.2** and when y is **3.3**, the corresponding value of x is **0.48**

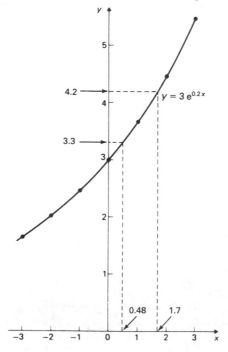

Fig. 3.3   Graph depicting $y = 3\,e^{0.2x}$

Problem 2. Draw a graph of $y = e^{-x^2}$ over a range $x = -2$ to $x = 2$
The values of the coordinates are calculated as shown below:

x	-2	-1.5	-1	0.5	0	0.5	1.0	1.5	2.0
$-x^2$	-4	-2.25	-1	-0.25	0	-0.25	-1	-2.25	-4
e^{-x^2}	0.02	0.11	0.37	0.78	1.0	0.78	0.37	0.11	0.02

The graph is plotted in Fig. 3.4.

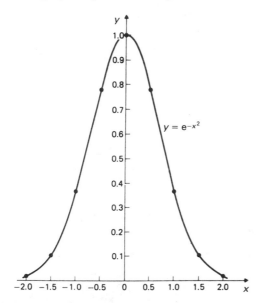

Fig. 3.4   Graph depicting $y = e^{-x^2}$

Problem 3. Draw a graph of $y = \frac{1}{5}(e^x - e^{-2x})$ over a range
$x = -1$ to $x = 4$. Determine from the graph the value of x when
$y = 6.6$

The values of the coordinates are calculated as shown below.
Since the values used to determine $\frac{1}{5}(e^x - e^{-2x})$ range from zero
to over 50, only 1 decimal place accuracy is taken.

x	-1	-0.5	0	0.5	1	2	3	4
e^x	0.4	0.6	1	1.6	2.7	7.4	20.1	54.6
$-2x$	2	1	0	-1	-2	-4	-6	-8
e^{-2x}	7.4	2.7	1	0.4	0.1	0.0	0.0	0.0
$e^x - e^{-2x}$	-7.0	-2.1	0	1.2	2.6	7.4	20.1	54.6
$\frac{1}{5}(e^x - e^{-2x})$	-1.4	-0.4	0	0.2	0.5	1.5	4.0	10.9

Using these values, the graph shown in Fig. 3.5 is drawn. From the graph, when $y = 6.6$, $x = \mathbf{3.5}$

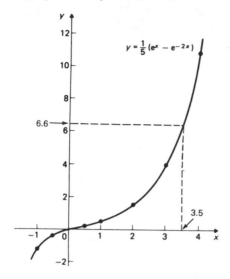

Fig. 3.5 Graph depicting $y = \frac{1}{5}(e^x - e^{-2x})$

Problem 4. Plot the curves $y = 2\,e^{-1.5x}$ and $y = 1.2(1 - e^{-2x})$ on the same axes from $x = 0$ to $x = 1$ and determine their point of intersection.

A table of values is drawn up as shown on page 39, values being taken correct to 2 decimal places.

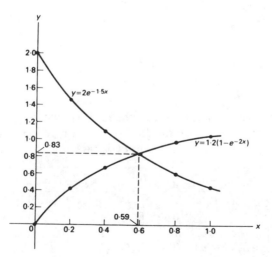

Fig. 3.6 Graph depicting $y = 2\,e^{-1.5x}$ and $y = 1.2(1 - e^{-2x})$

x	0	0.2	0.4	0.6	0.8	1.0
$2\,e^{-1.5x}$	2.00	1.48	1.10	0.81	0.60	0.45
$1.2(1 - e^{-2x})$	0	0.40	0.66	0.84	0.96	1.04

The curves are shown in Fig. 3.6 and are seen to intersect at
(0.59, 0.83). (Hence the solution of the simultaneous equations
$y = 2\,e^{-1.5x}$ and $y = 1.2(1 - e^{-2x})$ is $x = 0.59$ and $y = 0.83$)

*Further problems on graphs of exponential functions may be found in Section
3.3 (Problems 1 to 9), page 45.*

3.2   Laws of growth and decay

The laws of exponential growth or decay occur frequently in engineering
and science, and are always of the form $y = A\,e^{kx}$ and $y = A(1 - e^{kx})$, where
A and k are constants and can be either positive or negative. The natural
law $y = A\,e^{kx}$ relates quantities in which the rate of increase of y is
proportional to y itself for the growth law, or in which the rate of decrease
of y is proportional to y itself for the decay law. Some of the quantities
following exponential laws are given below.

(a)   Linear expansion

A rod of length l at temperature $\theta°C$ and having a positive coefficient of
linear expansion of α will get longer when heated. The natural growth law is

$$l = l_0\,e^{\alpha\theta}$$

where l_0 is the length of the rod at 0°C.

(b)   Change of electrical resistance with temperature

A resistor of resistance R_θ at temperature $\theta°C$ and having a positive
temperature coefficient of resistance of α increases in resistance when heated.
The natural growth law is

$$R_\theta = R_0\,e^{\alpha\theta}$$

where R_0 is the resistance at 0°C.

(c)   Tension in belts

A natural growth law governs the relationship between the tension T_1 in a
belt around a pulley wheel and its angle of lap α. It is of the form

$$T_1 = T_0\,e^{\mu\alpha}$$

where μ is the coefficient of friction between belt and pulley, and T_1 and T_0 are the tensions on the tight and slack sides of the belt respectively.

(d)   The growth of current in an inductive circuit

In a circuit of resistance R and inductance L having a final value of steady current I,

$$i = I(1 - e^{-Rt/L})$$

where i is the current flowing at time t. This is an equation which follows a growth law.

(e)   Biological growth

The rate of growth of bacteria is proportional to the amount present. When y is the number of bacteria present at time t and y_0 the number present at time $t = 0$ then

$$y = y_0 e^{kt}$$

where k is the growth constant.

(f)   Newton's law of cooling

The rate at which a body cools is proportional to the excess of its temperature above that of its surroundings. The law is

$$\theta = \theta_0 e^{-kt}$$

where the excess of temperature at time $t = 0$ is θ_0 and at time t is θ. The negative power of the exponent indicates a decay curve when k is positive.

(g)   Discharge of a capacitor

When a capacitor of capacitance C having an initial charge of Q is discharged through a resistor R, then

$$q = Q e^{-t/CR}$$

where q is the charge after time t.

(h)   Atmospheric pressure

The pressure p at height h above ground level is given by

$$p = p_0 e^{-h/c}$$

where p_0 is the pressure at ground level and c is a constant.

(i)   *The decay of current in an inductive circuit*

When a circuit having a resistance R, inductance L and initial current I is allowed to decay, it follows a natural law of the form

$$i = I e^{-Rt/L}$$

where i is the current flowing after time t.

(j)   *Radioactive decay*

The rate of disintegration of a radioactive nucleus having N_0 radioactive atoms present and a decay constant of λ is given by

$$N = N_0 e^{-\lambda t}$$

where N is the number of radioactive atoms present after time t.

These are just some of the relationships which exist which follow the natural laws of growth or decay.

Worked problems on the laws of growth and decay

Problem 1. A belt is in contact with a pulley for a sector of $\theta = 1.073$ radians and the coefficient of friction between these two surfaces is $\mu = 0.27$. Determine the tension on the taut side of the belt, T newtons, when the tension on the slack side is given by $T_0 = 23.8$ newtons, given that these quantities are related by the law $T = T_0 e^{\mu\theta}$. If we require the transmitted force $(T - T_0)$ to be increased to 25.0 newtons, assuming that T_0 remains at 23.8 newtons and θ at 1.073 radians, determine the coefficient of friction

$$T = T_0 e^{\mu\theta} = 23.8 e^{(0.27 \times 1.073)}$$
$$= 23.8 e^{0.290}$$

Hence $T = 23.8 \times 1.336\,4 = \textbf{31.81 newtons}$

For the transmitted force to be 25 newtons, T becomes $23.8 + 25$, or 48.8 newtons. Then

$$48.8 = 23.8 e^{\mu \times 1.073}$$

$$\frac{48.8}{23.8} = e^{1.073\mu} \text{ or } 2.050 = e^{1.073\mu}$$

Taking Napierian logarithms of each side of this equation gives

$$\ln 2.050 = 1.073\mu$$
$$0.7178 = 1.073\mu$$

i.e. **the coefficient of friction** $\mu = \dfrac{0.717\,8}{1.073} = \textbf{0.669\,0}$

Problem 2. The instantaneous current i amperes at time t seconds is given by $i = 6.0 \, e^{-t/CR}$ when a capacitor is being charged. The capacitance C is 8.3×10^{-6} farads and the resistance R has a value of 0.24×10^6 ohms. Determine the instantaneous current when t is 3.0 seconds. Also determine the time for the instantaneous current to fall to 4.2 amperes. Sketch a curve of current against time from $t = 0$ to $t = 6$ s

$$i = 6.0 \, e^{[-3/(8.3 \times 10^{-6} \times 0.24 \times 10^6)]}$$

$$= 6.0 \, e^{[-3/(8.3 \times 0.24)]}$$

$$= 6.0 \, e^{-1.5060}$$

Hence $i = 6 \times 0.221\,8 = 1.33$ amperes

That is, the current flowing when t is 3.0 seconds is 1.33 amperes

Is is usually easier to transpose and make t the subject of the formula before evaluation. Thus, since $i = 6.0 \, e^{-t/CR}$,

$$\frac{i}{6.0} = e^{-t/CR}$$

$$e^{t/CR} = \frac{6.0}{i}$$

$$\ln e^{t/CR} = \ln\left(\frac{6.0}{i}\right)$$

$$\frac{t}{CR} = \ln\left(\frac{6.0}{i}\right) \text{ or } t = CR \ln\left(\frac{6.0}{i}\right)$$

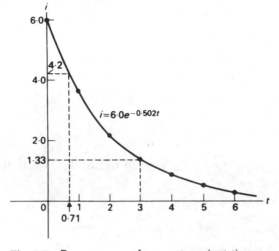

Fig. 3.7   Decay curve of current against time

Substituting the values of C, R and i gives

$$t = 8.3 \times 10^{-6} \times 0.24 \times 10^6 \ln \frac{6.0}{4.2}$$

$$= 1.992 \ln 1.428\ 6$$

$$= 1.992 \times 0.356\ 7 = 0.71 \text{ seconds}$$

i.e. **the time for the current to fall to 4.2 amperes is 0.71 seconds**

Since $i = 6.0\ e^{-t/CR}$ and $C = 8.3 \times 10^{-6}$ and $R = 0.24 \times 10^6$ then $i = 6.0\ e^{-0.502t}$. At $t = 0$, $i = 6.0\ e^0 = 6.0$ (since $e^0 = 1$) and when $t = \infty$, $i = 6.0\ e^{(-0.502)\infty} = 6.0(0) = 0$.

A decay curve representing $i = 6.0\ e^{-0.502t}$ is shown in Fig. 3.7.

Problem 3. The temperature θ_2 degrees Celsius of a winding which is being heated electrically, at time t seconds, is given by $\theta_2 = \theta_1(1 - e^{-t/T})$, where θ_1 is the temperature at time $t = 0$ seconds and T seconds is a constant. Calculate (a) θ_1 in degrees Celsius when θ_2 is 45°C, t is 28 s and T is 73 s, and (b) the time t seconds for θ_2 to be half the value of θ_1

(a) Transposing the formula to make θ_1 the subject gives

$$\theta_1 = \frac{\theta_2}{1 - e^{-t/T}}$$

Substituting the values of θ_2, t and T gives

$$\theta_1 = \frac{45}{1 - e^{-\frac{28}{73}}} = \frac{45}{1 - e^{-0.383\,6}}$$

i.e. $$\theta_1 = \frac{45}{1 - 0.681\ 4} = \frac{45}{0.318\ 6}$$

$$\theta_1 = 141.24°C$$

That is, the initial temperature is 141°C, correct to the nearest degree.

(b) Transposing to make t the subject of the formula:

$$\frac{\theta_2}{\theta_1} = 1 - e^{-t/T}$$

$$e^{-t/T} = 1 - \frac{\theta_2}{\theta_1}$$

Hence $$-\frac{t}{T} = \ln\left(1 - \frac{\theta_2}{\theta_1}\right)$$

i.e. $$t = -T \ln\left(1 - \frac{\theta_2}{\theta_1}\right)$$

Substituting the values of T, θ_1 and θ_2 gives

$$t = -73 \ln(1 - \tfrac{1}{2})$$
$$= -73 \ln 0.5$$
$$= -73 \times (-0.693\ 1) = 50.6 \text{ seconds}$$

i.e. **the time for the temperature to fall to half its original value is 50.6 seconds.**

Problem 4. The current i flowing in a coil is given by $i = \dfrac{E}{R}(1 - e^{-Rt/L})$ amperes, where R is 10 ohms, L is 2.5 henrys, E is 20 volts and t is the time in seconds. Draw a graph of current against time from $t = 0$ to $t = 1$ second. The time constant of the circuit is given by $\tau = \dfrac{L}{R}$. Show that in a time τ s the current rises to 63% of its final value.

Since $R = 10$, $L = 2.5$ and $E = 20$ then

$$i = \frac{20}{10}(1 - e^{-10t/2.5}), \text{ i.e. } i = 2(1 - e^{-4t})$$

A table of values is drawn up as shown below:

t	0	0.05	0.10	0.15	0.20	0.30	0.40	0.60	0.80	1.00
i	0	0.36	0.66	0.90	1.10	1.40	1.60	1.82	1.92	1.96

A graph of current i against time t is shown in Fig. 3.8

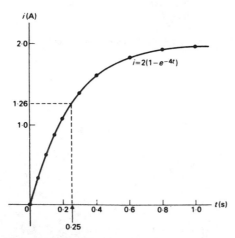

Fig. 3.8

When $t = \infty$, $i = 2(1 - e^{-4\infty}) = 2(1 - 0) = 2$ A

Thus the curve is tending towards a final value of $i = 2$ A $\left(\text{i.e. } \dfrac{E}{R}\right)$

Time constant $\tau = \dfrac{L}{R} = \dfrac{2.5}{10} = 0.25$ s. From Fig. 3.8, when

$t = 0.25$ s, $i = 1.26$ A, which is the same as 63% of 2 A

Hence in time $\tau = \dfrac{L}{R}$ the current rises to 63% of its final value

---

Further problems on the laws of growth and decay may be found in the following Section 3.3 (Problems 10 to 21), page 46.

3.3   Further problems

Graphs of exponential functions

In Problems 1 to 8 draw the graphs of the exponential functions given and use the graphs to determine the approximate values of x and y required.

1. $y = 5\,e^{0.4x}$ over a range $x = -3$ to $x = 3$ and determine the value of y when $x = 2.7$ and the value of x when $y = 10$
   [14.7, 1.7]

2. $y = 0.35\,e^{2.5x}$ over a range $x = -2$ to $x = 2$ and determine the value of y when $x = 1.8$ and the value of x when $y = 40$
   [31.5, 1.9]

3. $y = \frac{1}{3}\,e^{-2x}$ over a range $x = -2$ to $x = 2$ and determine the value of y when $x = -1.75$ and the value of x when $y = 5$
   [11, -1.35]

4. $y = 0.46\,e^{-0.27x}$ over a range $x = -10$ to $x = 10$ and determine the value of y when $x = -8.5$ and the value of x when $y = 6.1$
   [4.6, -9.6]

5. $y = 4\,e^{-2x^2}$ over a range $x = -1.5$ to $x = 1.5$ and determine the value of y when $x = -1.2$ and the value of x when $y = 2.9$
   [0.22, ± 0.4]

6. $y = 100\,e^{x^2/3}$ over a range $x = -3$ to $x = 3$ and determine the value of y when $x = \frac{7}{3}$ and the value of x when $y = 250$
   [614, ± 1.66]

7. $y = \frac{1}{2}(e^x - e^{-x})$ over a range $x = -3$ to $x = 3$ and determine the value of y when $x = -2.3$ and the value of x when $y = 5$
   [-4.94, 2.31]

8. $y = 3(e^{2x} - 4\,e^{-x})$ over a range $x = 2$ to $x = -2$ and determine the value of y when $x = -1.6$ and the value of x when $y = 35$
   [-59, 1.28]

9. Plot the curves $y = 3.2 e^{-1.4x}$ and $y = 1.7(1 - e^{-2x})$ on the same axes from $x = 0$ to $x = 1$ and determine their point of intersection.   [(0.67, 1.25)]

Laws of growth and decay

10. The instantaneous voltage v in a capacitive circuit is related to time t by the equation $v = Ve^{-t/CR}$, where V, C and R are constants. Determine v when $t = 27 \times 10^{-3}$, $C = 8.0 \times 10^{-6}$, $R = 57 \times 10^3$ and $V = 100$. Also determine R when $v = 83$, $t = 9.0 \times 10^{-3}$, $C = 8.0 \times 10^{-6}$ and $V = 100$. Sketch a curve of v against t [94.3, 6.04×10^3]

11. The length of a bar, l, at temperature θ is given by $l = l_0 e^{\alpha\theta}$, where l_0 and α are constants. Determine α when $l = 2.738$, $l_0 = 2.631$ and $\theta = 315.7$. Also determine l_0 when l and θ are as for the first part of the problem but $\alpha = 1.771 \times 10^{-4}$ [1.263×10^{-4}, 2.589]

12. Two quantities x and y are found to be related by the equation $y = a e^{-kx}$, where a and k are constants.
    (a) Determine y when $a = 1.671 \times 10^4$, $k = -4.60$ and $x = 1.537$
    (b) Determine x when $y = 76.31$, $a = 15.3$ and $k = 4.77$
    (a) [1.966×10^7]   (b) [$-0.336\,9$]

13. Quantities p and q are related by the equation $p = 7.413(1 - e^{kq/t})$, where k and t are constants. Determine p when $k = 3.7 \times 10^{-2}$, $q = 712.8$ and $t = 5.747$. Also determine t when $p = -98.3$ and q and k are as for the first part of the problem.   [-722, 9.92]

14. When quantities I and C are related by the equation $I = BT^2 e^{-C/T}$, and B and T are constants, determine I when $B = 14.3$, $T = 1.27$ and $C = 8.15$. Also determine C when $I = 7.47 \times 10^{-2}$ and B and T are as for the first part of the problem.   [0.037 66, 7.280]

15. In an experiment involving Newton's law of cooling the temperature θ after a time t of 73.0 seconds is found to be 51.8°C. Using the relationship $\theta = \theta_0 e^{-kt}$, determine k when $\theta_0 = 15.0°C$ [$-0.016\,98$]

16. The pressure p at height h above ground level is given by $p = p_0 e^{-h/c}$, where p_0 is the pressure at ground level and c is a constant. When p_0 is 1.013×10^5 pascals and the pressure at a height of 1 570 metres is 9.871×10^4 pascals, determine the value of c [60 620]

17. The current i amperes flowing in a capacitor at time t seconds is given by $i = 7.51(1 - e^{-t/CR})$, where the circuit resistance R is 27.4 kilohms and the capacitance C is 14.71 microfarads. Determine (a) the time for the current to reach 6.37 amperes and (b) the current flow after 0.458 seconds
    (a) [0.759 8 s]   (b) [5.099 A]

18. The voltage drop, V volts, across an inductor of L henrys at time t seconds is given by $V = 125\,e^{-Rt/L}$. Determine (a) the time for the voltage to reach 98.0 volts and (b) the voltage when t is 14.7 microseconds, given that the circuit resistance R is 128 ohms and its inductance is 10.3 millihenrys
    (a) [0.019 58 ms]   (b) [104.1 V]

19. The resistance R_t of an electrical conductor at temperature t degrees Celsius is given by $R_t = R_0\,e^{\alpha t}$, where α is a constant and R_0 is 3.41 kilohms. Calculate the value of α when R_t is 3.72 kilohms and $t = 1\,710$ degrees Celsius. For this conductor, at what temperature will the resistance be 3.50 kilohms?   [5.088×10^{-5}, 512°C]

20. The amount A after n years of a sum invested, P, is given by the compound interest law $A = P\,e^{rn/100}$ when the interest rate r is added continuously. Determine the amount after 10 years for a sum of £1 000 invested if the interest rate is 4% per annum   [£1 491.82]

21. The amount of product (x in mol cm^{-3}) formed in a chemical reaction starting with 3 mol cm^{-3} of reactant is given by $x = 3(1 - e^{-4t})$, where $t =$ time to form x in minutes. Plot a graph at 30-second intervals up to 3 minutes of this equation and determine x after 1 minute
    [2.945 mol cm^{-3}]

Chapter 4

Determination of laws and use of log-linear graph paper

4.1 Reduction of non-linear laws to a linear form

In experimental work, when two variables are believed to be connected, a set of corresponding measurements can be made. Such results can then be used to discover if there is a mathematical law relating the two variables. If such a law is found it can be used to predict further values. Usually the results obtained from experiments are plotted as a graph and an attempt is made to deduce the law from the graph. Now the relationship between two variables believed to be of the linear form $y = mx + c$ can be proved by plotting the measured values x and y and seeing if a straight-line graph results. If a straight line does fit the plotted points then the slope m and y-axis intercept c can be found, which establishes the law relating x and y for all values **in the given range**. However, frequently the relationship between variables x and y is not a linear one, i.e. when x is plotted against y a curve results. In such cases the non-linear equation is modified to the linear form $y = mx + c$ so that the constants can be found and the law relating the variables determined. This process is called the **determination of laws**. Some common forms of non-linear equations are given below. These are rearranged into a linear form by making a direct comparison with the straight-line form $y = mx + c$, that is, arranging in the form

a variable = (a constant × a variable) + a constant.

In the conversion of non-linear laws to a linear form it is useful to isolate the constant term first.

(i) $y = ax^2 + b$

$$\boxed{y} = a \boxed{x^2} + b$$
compares with $\boxed{y} = m \boxed{x} + c$

Hence plot y against x^2 to produce a straight-line graph of slope a and y-axis intercept b (see Worked Problem 1).

(ii) $y = \dfrac{a}{x} + b$

$$\boxed{y} = a \boxed{\dfrac{1}{x}} + b$$
compares with $\boxed{y} = m \boxed{x} + c$

Hence plot y against $\dfrac{1}{x}$ to produce a straight-line graph of slope a and y-axis intercept b (see Worked Problem 2).

(iii) $y = \dfrac{a}{x^2} + b$

$$\boxed{y} = a \boxed{\dfrac{1}{x^2}} + b$$

compares with $\boxed{y} = m \boxed{x} + c$

Hence plot y against $\dfrac{1}{x^2}$ to produce a straight-line graph of slope a and y-axis intercept b

(iv) $y = ax^2 + bx$

In this case there is no constant term as it stands, but by dividing throughout by x a constant term b is produced. Dividing both sides of the equation by x gives:

$$\boxed{\dfrac{y}{x}} = a \boxed{x} + b$$

compares with $\boxed{y} = m \boxed{x} + c$

Hence plot $\dfrac{y}{x}$ against x to produce a straight-line graph of slope a and $\dfrac{y}{x}$ − axis intercept b (see Worked Problem 3).

(v) $xy = ax + by$

Dividing both sides of the equation by y gives:

$$\boxed{x} = a \boxed{\dfrac{x}{y}} + b$$

compares with $\boxed{y} = m \boxed{x} + c$

Hence plot x against $\dfrac{x}{y}$ to produce a straight-line graph of slope a and x-axis intercept b

(vi) $y = ax^n$

Taking logarithms (usually to a base 10) of each side of the equation gives:

$\lg y = \lg(ax^n)$

$\lg y = \lg a + \lg x^n$

$\lg y = \lg a + n \lg x$

or $\boxed{\lg y} = n \boxed{\lg x} + \lg a$

compares with $\boxed{y} = m \boxed{x} + c$

Hence plot lg y against lg x to produce a straight-line graph of slope n and lg y-axis intercept lg a (a is obtained by taking antilogarithms of lg a) (see Worked Problems 4 and 5).

(vii) **$y = ab^x$**

Taking logarithms of each side of the equation gives:

lg $y =$ lg ab^x

lg $y =$ lg $a +$ lg b^x

lg $y =$ lg $a + x$ lg b

i.e. $\boxed{\lg y} = (\lg b)\,\boxed{x} + \lg a$

compares with $\boxed{y} = m\,\boxed{x} + c$

Hence plot lg y against x to produce a straight-line graph of slope lg b and lg y-axis intercept lg a (a and b are obtained by taking anti-logarithms of lg a and lg b respectively) (see Worked Problem 6).

Worked problems on determination of laws

Problem 1. The following experimental values of x and y are believed to be related by the law $y = ax^2 + b$. By plotting a suitable graph test if this is so and find the approximate values of a and b

x	0	1	2	3	4	5	6
y	3.0	4.8	12.0	23.3	38.1	59.5	83.5

If y is plotted against x the non-linear curve shown in Fig. 4.1 is produced. It is not possible to determine the values of a and b from such a curve.

Comparing $\boxed{y} = a\,\boxed{x^2} + b$ with the straight-line form $\boxed{y} = m\,\boxed{x} + c$

shows that y is to be plotted vertically against x^2 horizontally to produce a straight line of slope a and y-axis intercept b. Thus an extension to the above table of values needs to be produced. This is shown below.

y	3.0	4.8	12.0	23.3	38.1	59.5	83.5
x	0	1.0	2.0	3.0	4.0	5.0	6.0
x^2	0	1.0	4.0	9.0	16.0	25.0	36.0

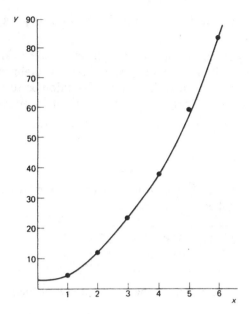

Fig. 4.1 Graph of y/x

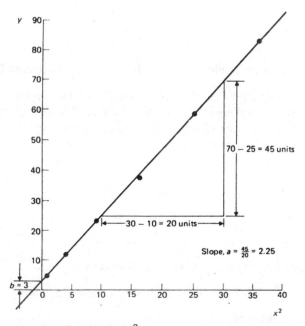

Fig. 4.2 Graph of y/x^2

The best straight line that fits the points is shown in the graph of Fig. 4.2

Note that not every point lies on the straight line drawn; rarely in experimental results will all points lie exactly on a straight line. In such cases there is likely to be a difference of opinion as to the exact position of the line. Hence any results are only approximate.

From the graph the slope a is found to be **2.25** and the y-axis intercept b is found to be **3**

Hence the law relating the variables x and y is $y = 2.25x^2 + 3$

Problem 2. In an experiment the following values of resistance R and voltage V were taken:

R ohms	45.3	49.8	52.4	57.6	62.3
V millivolts	113	102	96	86	79

It is thought that R and V are connected by a law of the form

$R = \dfrac{d}{V} + e$ where d and e are constants. Verify the law and find approximate values of d and e

$$R = \frac{d}{V} + e \qquad \text{i.e.} \quad R = d\,\frac{1}{V} + e$$

$$\text{Comparing with} \quad y = m\,x + c$$

shows that R is to be plotted vertically against $\dfrac{1}{V}$ horizontally to produce a straight line of slope d and R-axis intercept e. Another table of values is drawn up with V in this case changed from millivolts to volts so that when taking reciprocals of V, more manageable numbers result.

R	45.3	49.8	52.4	57.6	62.3
V	0.113	0.102	0.096	0.086	0.079
$\dfrac{1}{V}$	8.85	9.80	10.42	11.63	12.66

A graph of R against $\dfrac{1}{V}$ is shown plotted in Fig. 4.3.

A straight line fits the points which verifies that the law $R = \dfrac{d}{V} + e$ is obeyed.

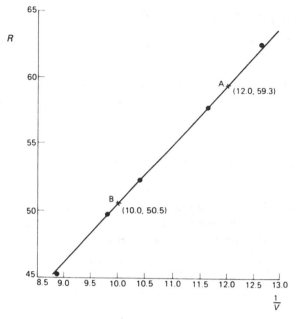

Fig. 4.3   Graph of $R \Big/ \dfrac{1}{V}$

It is not practical in this case to commence the scaling of each axis at zero. Thus it is not possible to find the R-axis intercept (i.e. at $\dfrac{1}{V} = 0$) from the graph. A simultaneous equation approach is therefore necessary. Any two points such as A and B may be used.

At A, $R = 59.3$ and $\dfrac{1}{V} = 12.0$

At B, $R = 50.5$ and $\dfrac{1}{V} = 10.0$

Hence, since $R = \dfrac{1}{V} d + e$:

$$59.3 = 12.0d + e \tag{1}$$

and $50.5 = 10.0d + e \tag{2}$

Subtracting equation (2) from equation (1) gives:

$8.8 = 2..0d$

$d = 4.4$

Substituting $d = 4.4$ in equation (2) gives:

$50.5 = (10.0)(4.4) + e$

$e = 6.5$

Checking in equation (1): R.H.S. $= (12.0)(4.4) + 6.5$

$$= 52.8 + 6.5$$

$$= 59.3 = \text{L.H.S.}$$

Hence the law connecting R and V is $R = \dfrac{4.4}{V} + 6.5$

Problem 3. The following table gives corresponding values of P and V which are believed to be related by a law of the form $P = aV^2 + bV$ where a and b are constants.

V	0.5	2.6	5.3	7.7	9.2	11.4	12.7
P	4.5	38.5	121.4	231.8	318.3	469.7	565.2

Verify the law and find the values of a and b. Hence find: (a) the value of P when V is 10.6, and (b) the positive value of V when $P = 150$

Dividing both sides of the equation $P = aV^2 + bV$ by V gives:

$$\boxed{\frac{P}{V}} = a \boxed{V} + b$$

Comparing with $\boxed{y} = m \boxed{x} + c$

shows that $\dfrac{P}{V}$ is to be plotted vertically against V horizontally to produce a straight line of slope a and $\dfrac{P}{V}$-axis intercept b.

Another table of values is drawn up.

V	0.5	2.6	5.3	7.7	9.2	11.4	12.7
P	4.5	38.5	121.4	231.8	318.3	469.7	565.2
$\dfrac{P}{V}$	9.0	14.8	22.9	30.1	34.6	41.2	44.5

A graph of $\dfrac{P}{V}$ against V is shown in Fig. 4.4 where a straight line fits the points. Thus the law is verified.

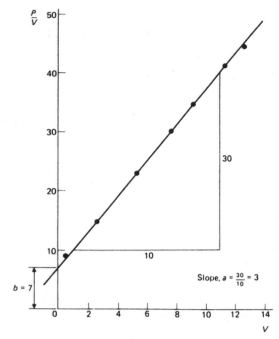

Fig. 4.4   Graph of $\dfrac{P}{V}\Big/V$

The $\dfrac{P}{V}$-axis intercept $b = 7$

The slope, or gradient, $a = 3$

Hence the law relating P and V is: $\boldsymbol{P = 3V^2 + 7V}$

(a)   When $V = 10.6$, $P = 3(10.6)^2 + 7(10.6)$

$$= 337.1 + 74.2 = \mathbf{411.3}$$

(b)   When $P = 150$, $150 = 3V^2 + 7V$

i.e. $3V^2 + 7V - 150 = 0$

Thus: $V = \dfrac{-7 \pm \sqrt{[(7)^2 - 4(3)(-150)]}}{2(3)}$

$$= \dfrac{-7 \pm \sqrt{[49 + 1\,800]}}{6} = \dfrac{-7 \pm 43}{6}$$

$$= 6 \text{ or } -8\dfrac{1}{3}$$

Thus for the law $P = 3V^2 + 7V$, $V = \mathbf{6}$ when $P = 150$

Problem 4. The power dissipated by a resistor was measured for various values of current flowing in the resistor and the results are shown:

Current (I amperes)   1.0 1.5 2.5 4.0 5.5   7.0
Power (P watts)   50 112 310 800 1 510 2 450

Prove that the law relating current and power is of the form $P = RI^n$, where R and n are constants, and determine the law.

To express the law $P = RI^n$ in a linear form, logarithms are taken of each side of the equation. This gives:

$$\lg P = \lg(RI^n)$$

or $\lg P = \lg R + \lg I^n$

that is, $\lg P = n \lg I + \lg R$

This is now in the form $y = mx + c$ and by plotting $\lg P$ vertically (since it corresponds to y) and $\lg I$ horizontally (since it corresponds to x), if a straight-line graph is produced then the law is verified.

I	1.0	1.5	2.5	4.0	5.5	7.0
$\lg I$	0	0.176	0.398	0.602	0.740	0.845
P	50	112	310	800	1 510	2 450
$\lg P$	1.699	2.049	2.491	2.903	3.179	3.389

The graph of $\lg I$ against $\lg P$ is shown in Fig. 4.5, and since a straight-line graph is produced the law $P = RI^n$ is verified.
    By selecting two points on the graph, say T having coordinates (0.8, 3.3) and Q having coordinates (0.1, 1.9), we can determine the gradient and hence obtain the value of n.

Hence $n = \dfrac{3.3 - 1.9}{0.8 - 0.1} = \dfrac{1.4}{0.7} = 2$

The vertical-axis intercept value when $\lg I$ is equal to zero is 1.7, this being the value of $\lg R$. By finding the antilogarithm of 1.7, i.e. 50.1, the value of R is ascertained.

Thus the required law is $P = 50.1\,I^2$

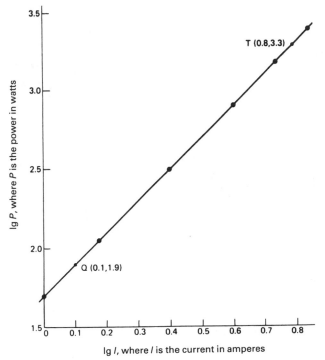

Fig. 4.5   Variation of lg I/lg P

Problem 5. Two quantities Q and H are believed to be related by the equation $Q = kH^n$. The experimental values obtained for Q and H are as shown:

Q	0.16	0.20	0.27	0.34	0.40	0.47	0.55
H	1.14	1.78	3.24	5.14	7.11	9.82	13.44

Determine the law connecting Q and H, and the value of Q when H is 6.00

Since $Q = kH^n$, then $\lg Q = n \lg H + \lg k$ and to determine the law connecting Q and H, $\lg Q$ is plotted against $\lg H$

$\lg Q$	−0.796	−0.699	−0.569	−0.469	−0.398	−0.328	−0.260
$\lg H$	0.057	0.250	0.511	0.711	0.852	0.992	1.128

The graph produced by plotting $\lg Q$ against $\lg H$ is shown in Fig. 4.6. Selecting two points which lie on the graph, say R (1.1, −0.275) and S (0.1, −0.775), gives a gradient of

$$\frac{-0.775 - (-0.275)}{0.1 - 1.1} \quad \text{or} \quad \frac{-0.5}{-1} \quad \text{i.e.} \quad \frac{1}{2}$$

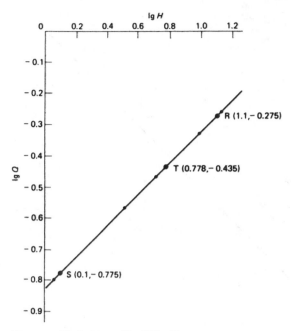

Fig. 4.6   Variation of lg Q/lg H

The vertical-axis intercept value lg $k = -0.825$ and by finding the antilogarithm, $k \approx 0.150$. Hence the law connecting Q and H is

$$Q = 0.150H^{\frac{1}{2}} \text{ or } Q = 0.150\sqrt{H}$$

When H is 6.00, lg H is 0.778. From the graph, the corresponding value of lg Q is -0.435, shown as point T, and finding the antilogarithm gives $Q = 0.367$ **when** $H = 6.00$

Problem 6. Values of x and y are believed to be related by a law of the form $y = ab^x$ where a and b are constants. The values of y and corresponding values of x are shown:

y	4.5	7.4	11.2	15.8	39.0	68.0	271.5
x	0.6	1.3	1.9	2.4	3.7	4.5	6.5

Verify that the law relating y and x is as stated and determine the approximate values of a and b.

Hence determine: (*a*) the value of y when x is 3.2, and (*b*) the value of x when y is 126.7

$y = ab^x$

Taking logarithms to base 10 of both sides of the equation gives:

$\lg y = \lg(ab^x) = \lg a + \lg b^x = \lg a + x \lg b$

i.e. $\lg y = (\lg b)x + \lg a$

Comparing with $y = m\ x + c$

shows that $\lg y$ is plotted vertically against x horizontally to produce a straight-line graph of slope $\lg b$ and $\lg y$-axis intercept $\lg a$. Another table of values is drawn up.

y	4.5	7.4	11.2	15.8	39.0	68.0	271.5
$\lg y$	0.65	0.87	1.05	1.20	1.59	1.83	2.43
x	0.6	1.3	1.9	2.4	3.7	4.5	6.5

A graph of $\lg y/x$ is shown in Fig. 4.7. A straight line fits the points, which verifies the law.

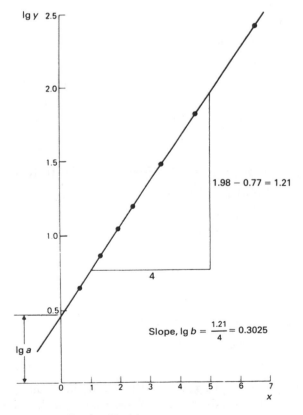

Fig. 4.7   Graph of $\lg y/x$

The intercept on the lg y-axis is 0.47, i.e. lg $a = 0.47$

Taking antilogarithms, $a = 2.951 = 3.0$, correct to 2 significant figures.

Slope, or gradient, lg $b = \dfrac{1.21}{4} = 0.302\,5$

Taking antilogarithms, $b = 2.007 = 2.0$ correct to 2 significant figures.

Hence the law relating x and y is $y = (3.0)(2.0)^x$

From the graph, when $x = 3.2$,   lg $y \approx 1.44$, giving $y \approx 27.5$

and when $y = 126.7$, lg $y = 2.10$, giving $x \approx 5.4$

Alternatively: (a) When $x = 3.2$, $y = (3.0)(2.0)^{3.2} = \mathbf{27.6}$

(b) When $y = 126.7$, $126.7 = (3.0)(2.0)^x$

i.e. $\dfrac{126.7}{3.0} = (2.0)^x$

$42.23 = (2.0)^x$

Taking logarithms: lg $42.23 = x$ lg 2.0

$x = \dfrac{\text{lg } 42.23}{\text{lg } 2.0} = \dfrac{1.625\,6}{0.301\,0} = \mathbf{5.40}$

Further problems on reduction of non-linear laws to a linear form (i.e. determination of laws) may be found in Section 4.4 (Problems 1 to 34), page 69.

4.2   Reducing equations of the form $y = ab^x$ to linear form using log-linear graph paper

As shown in the previous section, taking logarithms to a base of 10 of each side of the equation $y = ab^x$ gives

lg $y = $ lg(ab^x)

$= $ lg $a + $ lg b^x, by the laws of logarithms

$= x$ lg $b + $ lg a, where a and b are constants.

Comparing this equation with the straight-line equation $Y = mX + c$ gives:

$$\boxed{\text{lg } y} = \boxed{(\text{lg } b)}\ \boxed{x} + \text{lg } a$$
$$\boxed{Y} = \boxed{m}\ \boxed{X} + c$$

This shows that by plotting lg y against x, the slope of the resultant straight-line graph is lg b and the y-axis intercept value is lg a. In this case, graph paper which has a linear scale on one axis (x) and a logarithmic scale on the other axis (lg y) can be used. This type of graph paper is called **log-linear graph paper**.

Examination of the scale markings on a logarithmic axis shows that the scales do not have equal divisions. They are marked from 1 to 9 and this pattern of markings can be repeated several times since

$$\lg 1 - \lg 0.1 = 0 - (-1) = 1, \lg 10 - \lg 1 = 1 - 0 = 1$$

and

$$\lg 100 - \lg 10 = 2 - 1 = 1$$

and so on.

The number of times the pattern of markings is repeated signifies the number of cycles, the distance each cycle occupies on a logarithmic scale being the same. Thus, one cycle can be used to signify values from 0.1 to 1, or from 1 to 10, or from 10 to 100 and so on. Paper having three cycles on the logarithmic scale is called 'log 3 cycle × linear' graph paper (see Fig. 4.8).

To depict a set of numbers from 0.6 to 174, say, on the logarithmic axis of log-linear graph paper, 4 cycles will be required (0.1 to 1, 1 to 10, 10 to 100 and 100 to 1 000) and the start of each cycle is marked 0.1, 1, 10 and 100 or 10^{-1}, 10^0, 10^1 and 10^2. The divisions within a cycle are proportional to the logarithms of the numbers 1 to 10 and the distance from, say, the 1 to 2 marks is 0.301 0 (i.e. $\lg 2.000 0 - \lg 1.000 0$) of the total distance in the cycle. Similarly, the distance from the 9 to 10 marks is 0.045 76 of the total distance in the cycle, i.e. the distance between marks decreases logarithmically within the cycle.

The method of determining the constants a and b in the equation $y = ab^x$ is shown in the following worked problem.

Worked problem using log-linear graph paper to reduce an equation of the form $y = ab^x$ to linear form

Problem 1. Quantities x and y are believed to be related by a law of the form $y = ab^x$. The values of x and corresponding values of y are as shown.

x	0	0.5	1.0	1.5	2.0	2.5	3.0
y	1	3.2	10	31.6	100	316.2	1 000

Verify the law is as stated, find the approximate values of a and b and comment on the significance of the graph drawn.

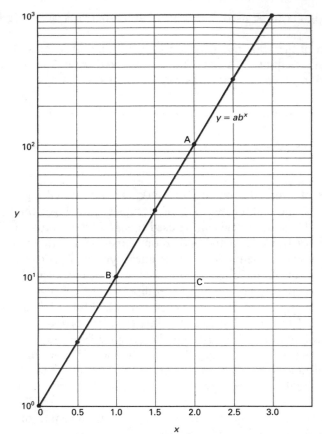

Fig. 4.8 Graph to verify a law of the form $y = ab^x$

Since $y = ab^x$, then $\lg y = \lg b \cdot x + \lg a$. Comparing this equation with the straight-line equation $Y = mX + c$ gives:

$$\boxed{\lg y} = \lg b\,\boxed{x} + \lg a$$
$$\boxed{Y} = m\,\boxed{X} + c$$

Thus $\lg y$ is plotted against x to verify that the law is as stated. Using log-linear graph paper, values of x are selected on the linear scale over a range 0 to 3. Values of y have a range from 1 to 1 000 and 3 cycles are needed to span this range. The graph is shown in Fig. 4.8.

A straight line can be drawn through the points so the law is verified. Since there is a mixture of linear values on the x-axis and logarithmic values on the y-axis, direct measurement of the slope of the graph is not possible to determine $\lg b$. Selecting any two points on the graph, say A and B, having co-ordinates $(2, 10^2)$

and $(1, 10^1)$ gives:

$$\text{slope} = \frac{AC}{BC} = \frac{\lg 10^2 - \lg 10^1}{2 - 1} = \frac{2 - 1}{2 - 1} = 1$$

Hence, slope $= \lg b = 1$, giving $b = 10$

Also, when $x = 0$, $y = 10^0 = 1$

i.e. $a = 1$

That is, the constants a and b have values of 1 and 10 respectively.

The significance of the graph is that the relationship is $y = 10^x$ and by the definition of a logarithm, when $y = 10^x$, $x = \lg y$. Hence the graph may be used to determine the approximate value of any logarithm between $y = 1$ and $y = 1\,000$. Also the approximate value of any antilogarithm between $x = 0$ and $x = 3$ can be found by finding the corresponding value of y. For example, when $y = 500$, from the graph, $x = 2.7$, i.e. the value of $\lg 500$ is approximately 2.7

By drawing a graph of $y = a^x$, where a is any positive number, the values of logarithms to a base of 'a' can be obtained, and this is one of the principal uses of equations of the form $y = ab^x$

Further problems on reducing equations of the form $y = ab^x$ to linear form may be found in Section 4.4 (Problems 35 and 36), page 74.

4.3   Reducing equations of the form $y = a\,e^{kx}$ to linear form using log-linear graph paper

As shown in Section 4.1 taking logarithms to a base of e of each side of the equation $y = a\,e^{kx}$ gives:

$$\ln y = \ln a\,e^{kx} \text{ or } \ln y = \ln a + \ln e^{kx}$$

Thus $\ln y = \ln a + kx \ln e$

However, by the basic definition of a logarithm,

when $y = a^x$ (3)

then $x = \log_a y$ (4)

If $y = a$, then from (3), $a = a^x$, or $x = 1$

From (4), $\log_a a = 1$

It follows that $\log_e e = 1$ or $\ln e = 1$

Hence, since $\ln y = kx \ln e + \ln a$, then $\ln y = kx + \ln a$, where a and k are constants.

Comparing this equation with the straight-line equation $Y = mX + c$

gives:

$$\boxed{\ln y} = k \boxed{x} + \ln a$$
$$\boxed{Y} = m \boxed{X} + c$$

This shows that by plotting ln y against x we will obtain a straight-line graph (and verify a relationship of the form $y = a\,e^{kx}$ where it exists).

The same log-linear graph paper can be used as for logarithms to a base of 10. Napierian logarithms and logarithms to a base of 10 are related by:

$$\ln x = 2.3026 \lg x$$

i.e.   ln x = (a constant)(lg x)

Thus, when using logarithmic graph paper to depict Napierian logarithms, the distances along an axis representing ln 100 to ln 10, ln 10 to ln 1 and so on are uniform, as they were for logarithms to a base of 10 and the distances within cycles alter logarithmically as they did for logarithms to a base of 10. Thus the effect of using Napierian logarithms instead of logarithms to a base of 10 is to introduce a scale factor, which is automatically allowed for in subsequent calculations and measurements, as shown in the worked problems. When log-linear graph paper is used, it is not necessary to determine the values of ln y, since values of y are plotted directly on the logarithmic axis.

In Section 3.1, Chapter 3, graphs of $y = e^x$ and $y = e^{-x}$ are plotted for values of x, from -3.0 to $+3.0$ with the resulting curves shown in Fig. 3.1, page 35. Graphs of $y = e^x$ and $y = e^{-x}$ may be plotted on log-linear graph paper as follows.

Taking Napierian logarithms of both sides of $y = e^x$ gives ln $y = x$. From the table of values shown on page 34, values of y range from 0.05 to 20.09 which requires 4 cycles on a logarithmic scale (i.e. 0.01 to 0.1, 0.1 to 1, 1 to 10 and 10 to 100). Thus 'log 4 cycle × linear' graph paper is required as shown in Fig. 4.9. When the graph is plotted a straight line results.

$$\text{Gradient of straight line} = \frac{AB}{BC} = \frac{\ln 4.48 - \ln 0.22}{1.5 - (-1.5)} = 1$$

The y-axis intercept value at $x = 0$ is $y = 1$.

When $y = e^{-x}$ is plotted on log-linear graph paper a straight line also results and has a gradient of -1 as shown in Fig. 4.9

The graphs of the form $y = a\,e^{kx}$ in the worked problems following show that the slope of the graph, k is given by $\dfrac{\ln y}{x}$ and is obtained by selecting two points on the curve and determining the values of y from the vertical axis and hence ln y and the values of x from the horizontal axis. When the straight-line graph cuts the ordinate $x = 0$, the intercept value gives the constant 'a' (as shown in Worked Problem 1). However, when the range of

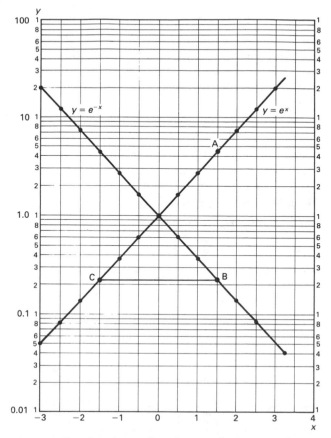

Fig. 4.9   Graphs of $y = e^x$ and $y = e^{-x}$

values is such that the straight-line graph does not cut the ordinate $x = 0$ within the scale selected, a point is selected on the straight-line graph and values of x, y and k are substituted in the equation $y = a\,e^{kx}$, to determine the value of 'a' (as shown in Worked Problem 2).

Worked problems to reduce equations of the form $y = a\,e^{kx}$ to linear form using log-linear graph paper

Problem 1. It is believed that x and y are related by a law of the form $y = a\,e^{kx}$ where a and k are constants. Values of x and y are measured and the results are as shown:

x	−0.9	0.25	0.9	2.1	2.8	3.7	4.8
y	2.5	6.0	10.0	25.0	42.5	85.0	198.0

Verify that the law stated does relate these quantities and determine the approximate values of a and k

Taking Napierian logarithms of each side of the equation $y = a\,e^{kx}$ gives:

$\ln y = kx + \ln a$

The values of y vary from 2.5 to 198, hence 'log 3 cycle × linear' graph paper is required. The graph is shown in Fig. 4.10

Since the points can be joined by a straight line, **the law $y = a\,e^{kx}$ for the values given is verified**. Selecting any two points on the line, say A having coordinates $(3, 50)$ and B having coordinates $(0, 5)$, the slope k is determined from:

$$\text{slope} = k = \frac{\ln y}{x} = \frac{\ln 50 - \ln 5}{3 - 0} = \frac{2.302\,6}{3} = 0.768$$

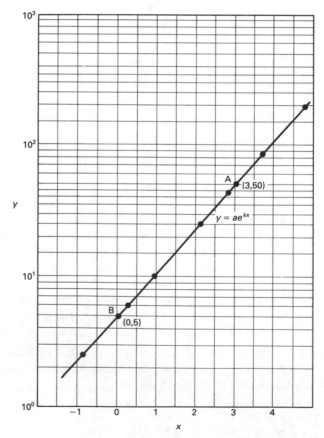

Fig. 4.10   Graph to verify a law of the form $y = a\,e^{kx}$ (Problem 1)

The y-axis intercept value at $x = 0$ is $y = 5$, hence $a = 5$.
Therefore the law is $y = 5\,e^{0.768x}$.

Alternatively, the values of a and k can be determined by solving simultaneous equations. Substituting in the equation $y = a\,e^{kx}$, the coordinate values of x and y from points A and B, gives:

$$50 = a\,e^{3k} \tag{5}$$

$$5 = a\,e^{0k} \tag{6}$$

Since $e^{0k} = e^0 = 1$, $a = 5$ from equation (6)

Substituting $a = 5$ in equation (5) gives:

$$50 = 5\,e^{3k}$$

or $e^{3k} = 10$

$$3k = \ln 10 = 2.302\,6$$

$$k = 0.768, \text{ as previously obtained.}$$

Problem 2. The current i (in milliamperes) flowing at an 8.3 microfarad capacitor, which is being discharged, varies with time t (in milliseconds), as shown:

| i milliamperes | 50.0 | 17.0 | 5.8 | 1.7 | 0.58 | 0.24 |
| t milliseconds | 200 | 255 | 310 | 375 | 425 | 475 |

Show that these results are connected by the law of the form $i = I\,e^{t/T}$ (where I and T are constants and I is the initial current flow in milliamperes) and determine the approximate values of the constants I and T.

Expressing $i = I\,e^{t/T}$ in the straight-line form of $Y = mX + c$ gives:

$$\ln i = \frac{1}{T}t + \ln I$$

Using 'log 3 cycle × linear' graph paper, the points are plotted on the graph shown in Fig. 4.11.

Since these points can be joined by a straight line, **the law** $i = I\,e^{t/T}$ **is verified**. Selecting any two points on the line, say A (400, 1) and B (282, 10), the slope $\dfrac{1}{T}$ is determined from

$$\frac{\ln i}{t} = \frac{\ln 1 - \ln 10}{400 - 282}$$

i.e. $\dfrac{1}{T} = \dfrac{-2.302\,6}{118} = -0.019\,5$

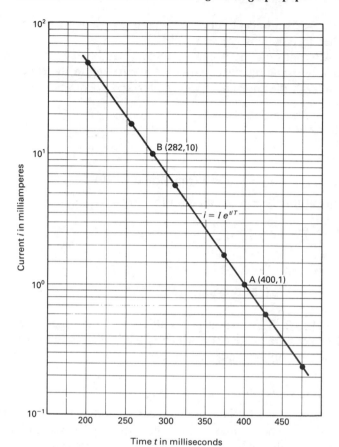

Fig. 4.11 Variation of current with time (Problem 2)

Hence $T = \dfrac{1}{-0.019\,5} = -51.3$, correct to 3 significant figures.

Since the line does not cross the i-axis at $t = 0$, the value of I is found by selecting a point on the line and using the coordinates of this point, together with the value of T in the equation $i = I\,e^{t/T}$. Using the coordinates of point A gives:

$$1 = I\,e^{(-400/51.3)}$$

$$I = e^{(400/51.3)}$$

$$= 2\,430\ \text{mA}$$

Hence the law is $i = 2\,430\,e^{-t/51.3}$, i.e. the values of I and T are 2 430 mA and −51.3 respectively

Further problems on reducing equations of the form $y = a\,e^{kx}$ to linear form may be found in the following Section 4.4 (Problems 37 to 42), page 74.

4.4 Further problems

Determination of laws

In Problems 1 to 12, x and y are variables and all other letters denote constants. For the stated laws to be verified it is necessary to modify them into a straight-line form. In order to plot a straight-line graph state: (*a*) what should be plotted on the vertical axis, (*b*) what should be plotted on the horizontal axis, (*c*) the slope or gradient and (*d*) the vertical-axis intercept.

1. $y = ax + b$ (*a*) $[y]$ (*b*) $[x]$ (*c*) $[a]$ (*d*) $[b]$
2. $y = cx^2 + d$ (*a*) $[y]$ (*b*) $[x^2]$ (*c*) $[c]$ (*d*) $[d]$
3. $y - f = e\sqrt{x}$ (*a*) $[y]$ (*b*) $[\sqrt{x}]$ (*c*) $[e]$ (*d*) $[f]$
4. $y = \dfrac{g}{x} + h$ (*a*) $[y]$ (*b*) $\left[\dfrac{1}{x}\right]$ (*c*) $[g]$ (*d*) $[h]$
5. $y = \sqrt{\left(\dfrac{j}{x}\right)} + (k)$ (*a*) $[y]$ (*b*) $\left[\dfrac{1}{\sqrt{x}}\right]$ (*c*) $[\sqrt{j}]$ (*d*) $[k]$
6. $y = lx^2 + mx$ (*a*) $\left[\dfrac{y}{x}\right]$ (*b*) $[x]$ (*c*) $[l]$ (*d*) $[m]$
7. $y - p = \dfrac{n}{x^2}$ (*a*) $[y]$ (*b*) $\left[\dfrac{1}{x^2}\right]$ (*c*) $[n]$ (*d*) $[p]$
8. $y = \dfrac{q}{x} + rx$ (*a*) $\left[\dfrac{y}{x}\right]$ (*b*) $\left[\dfrac{1}{x^2}\right]$ (*c*) $[q]$ (*d*) $[r]$
9. $y = sx^t$ (*a*) $[\lg y]$ (*b*) $[\lg x]$ (*c*) $[t]$ (*d*) $[\lg s]$
10. $y = uv^x$ (*a*) $[\lg y]$ (*b*) $[x]$ (*c*) $[\lg v]$ (*d*) $[\ln u]$
11. $y = \dfrac{a}{x - b}$ (*a*) $\left[\dfrac{1}{y}\right]$ (*b*) $[x]$ (*c*) $\left[\dfrac{1}{a}\right]$ (*d*) $\left[-\dfrac{b}{a}\right]$
12. $y = \dfrac{1}{wx + z}$ (*a*) $\left[\dfrac{1}{y}\right]$ (*b*) $[x]$ (*c*) $[w]$ (*d*) $[z]$

13. The following experimental values of x and y are believed to be related by the law $y = mx^2 + c$ where m and c are constants. By plotting a suitable graph verify this law and find the approximate values of m and c

x	2.3	4.1	6.0	8.4	9.9	11.3
y	13.9	31.2	60.2	111.8	153.0	197.5

$[m = 1.5, c = 6.0]$

14. Show that the following values of p and q obey a law of the
     form $p = m\sqrt{q} + n$, where m and n are constants.
     p 5.6   8.0   10.0 12.8 14.9 16.7 18.8
     q 0.64 3.61   9.0   18.5 30.3 39.7 54.8
     Determine approximate values of m and n
     $[m = 2.0, n = 4.1]$

15. Experimental values of load W newtons and distance
     l metres are shown in the following table:
     W 33.2   30.4   27.4   23.2   18.6   14.0   9.4   5.5
     l   0.741 0.364 0.233 0.163 0.116 0.093 0.076 0.067
     Verify that W and l are related by a law of the form

     $W = \dfrac{a}{l} + b$ and find the approximate values of a and b

     $[a = -2.05, b = 36]$

16. The solubility S of potassium chlorate is shown by the
     following table:

     $t(°C)$ 10   20 30 40 50 60 80 100
     S   4.5 7   10 14 19 25 39   57

     The relation between S and t is thought to be of the form
     $S = 3 + at + bt^2$. Plot a graph in order to test the
     supposition, and, if correct, use your graph to find probable
     values of a and b, explaining your methods.
     $[a = 0.12, b = 0.004]$

17. The pressure p and volume v of a gas at constant
     temperature are related by the law $pv = c$, where c is a
     constant. Show that the given values of p and v follow this
     law and determine the value of c

     Pressure, p (bar) 10.6 8.0 6.4 5.3 4.6 4.0
     Volume, v (m^3)   1.5 2.0 2.5 3.0 3.5 4.0 $[c = 15.85]$

18. Measurements of the resistance (R ohms) of varying
     diameters (d mm) of wire were made in an experiment with
     the following results:
     R 1.44 1.13 0.88 0.73 0.66
     d 1.13 1.38 1.78 2.26 2.76

     It is suspected that R is related to d by the law $R = \dfrac{c}{d^2} + a$

     Verify this and find approximate values for c and a. Estimate
     also the cross-sectional area of wire needed for a resistance
     reading of 0.52 ohms $[c = 1.2, a = 0.5, 47.1 \text{ mm}^2]$

19. Show that the following values of s and t follow a law of the
     type $s = at^3 + b$ (where a and b are constants) and find
     approximate values for a and b

s   49.7 47.4 41.4 29.5 10.0
t   1.0   2.0   3.0   4.0   5.0   [a = −0.32, b = 50]

20.   The periodic time, T, of oscillation of a pendulum is believed
      to be related to its length, l, by a law of the form $T = kl^n$
      where k and n are constants. Values of T were measured for
      various lengths of the pendulum and the results are as shown:

      Periodic time, T(s) 1.0 1.2 1.4 1.6 1.8 2.0   2.4
      Length, l (m)   2.4 3.5 4.8 6.3 7.9 9.8 14.1

      Prove the law is true. Determine the values of k and n and
      hence state the law. $[T = 0.64\sqrt{l}]$

21.   Current I and resistance R were measured experimentally
      and the results relating these quantities were:

      I   3.7 5.9 7.4 9.1 11.6 13.8
      R   6.2 3.9 3.1 2.5   2.0   1.7

      Show that I and R are related by a law of the form $R = aI^n$
      and determine the approximate values of a and n   [23, −1]

22.   The pH of a one-twentieth molar solution of potassium
      hydrogen phthalate varies with temperature (t) as shown:

      pH   4.011 4.001 4.001 4.011 4.031 4.061
      t°C   0   10   20   30   40   50

      Show that these results obey an equation of the form
      $pH = a + (t − 15)^2/b$ and determine the constants a and
      b   [a = 4.0, b = 20 000]

23.   The luminosity, I, of a lamp varies with the applied voltage,
      V, and it is anticipated that the relationship between these
      two quantities is of the form $I = kV^n$. Experimental values
      of I and V are:

      I candela 2.5 4.9 8.1 12.1 16.9 22.5
      V volts   50   70   90   110   130   150

      Verify the law and state it. Find the luminosity when
      100 volts is applied to the lamp. $[I = 0.001 V^2; 10\text{ candela}]$

24.   A physical quantity L, and another, C, were found to vary as
      shown when measured experimentally:

      L   300   350   400   450   500   550
      C   3464   3742   4000   4243   4472   4690

      Show that the law relating L and C is of the form $C = fL^n$
      where f and n are constants and determine the values of f
      and n $\left[200, \dfrac{1}{2}\right]$

25. The following table gives corresponding values of two
quantities x and y which are believed to be related by a law
of the form $y = ax^2 + bx$ where a and b are constants.

y 28.6 49.0 63.5 81.4 115.7 136.3
x   2.8   4.6   5.8   7.2   9.7   11.1

Verify the law and estimate the values of a and b. Hence
find: (i) the value of y when x is 6.6, and (ii) the value of x
when y is 102.5   [$a = 0.25$, $b = 9.5$, (i) 73.3 (ii) 8.8]

26. Quantitites p and q are believed to be related by a law of the
form $q = ab^p$. The value of p and corresponding values of q
are as shown:

p 0   0.5   1.0   1.5   2.0   2.5   3.0
q 1.0 3.2   10.0   31.6   100.0   316.2   1 000.0

Verify the law and find approximate values of a and
b [$a = 1$, $b = 10$]

27. Variation of the admittance Y of an electrical circuit with
the applied voltage V is as shown:

Voltage, V volts   0.37 0.51 0.74 0.98 1.13 1.34
Admittance, Y siemens 2.24 1.62 1.12 0.85 0.73 0.62

Show that the law connecting V and Y is of the form
$V = RY^n$ where R and n are constants and determine the
approximate values of R and n. From the graph, determine
the value of admittance when the voltage is 0.86 volts
[0.83, -1, 0.965 siemens]

28. Two variables x and y are believed to be related by the law
$xy = ax + by$ where a and b are constants. Results obtained
by experiment are as follows:

x   1.5 3.0 4.5 6.0 7.5 9.0
y 20.0 6.7 5.5 5.0 4.8 4.6

Verify the law and find approximate values of a and b.
Hence find (i) the value of y when x is 5.2; and (ii) the value
of x when y is 7.0   [$a = 4$, $b = 1.2$, (i) 5.2 (ii) 2.8]

29. In an experiment on moments, a bar was loaded with a
mass, W, at a distance x from the fulcrum. The results of the
experiment were:

x cm 28   30   32   34   36   38
W kg 22.1 20.7 19.4 18.2 17.2 16.3

Verify that a law of the form $W = ax^n$ is obeyed where a and
n are constants and determine the law.

$$\left[W = \frac{620}{x} \text{ or } W = 620x^{-1} \right]$$

30. Experimental values of the variation of current I amperes and the radii of copper strands r millimetres were measured and are as shown.

r (mm)	0.1	0.3	0.5	0.7	0.9	1.1	1.3
I (A)	0.000 17	0.001 5	0.004 3	0.008 3	0.014	0.021	0.029

    Prove that the law relating I and r is of the form $I = mr^n$ where m and n are constants and determine the law. What will be the radius of the strands when I is $0.006\,0$ A? $[I = 0.017r^2; r = 0.594$ mm$]$

31. The values of m and corresponding values of n are as follows:

m	0.2	0.7	1.3	1.9	2.4	3.6
n	6.0	14.2	39.4	109.7	257.2	1 990

    The law relating these quantities is of the form $n = ab^m$. Determine the approximate values of a and b $[a = 4.3, b = 5.5]$

32. The power dissipated by a resistor was measured for various values of current flowing in the resistor and the results are as shown below:

Current, I amperes	1.3	2.4	3.7	4.9	5.8
Power, P watts	37	127	301	528	740

    Prove that the law relating current and power is of the form $P = RI^n$, where R and n are constants, and determine the law. $[P = 22I^2]$

33. The vapour pressure (p) of triethylamine at various temperatures (T) are given below.

p (Pa)	0.439	10.120	93.330
T (K)	100	120	140

    Show that these results are related by an equation of the form

    $$\lg p = A - \frac{B}{T - 3.12}.$$ Find A and B $[A = 7.6, B = 770]$

34. The values of resistance R and voltage V were measured in an experiment and the results are shown below.

R ohms	59.5	74.2	84.1	94.8	152.2
V millivolts	135	107	94	83	51

    Resistance and voltage are thought to be connected by a law of the form $R = \dfrac{a}{V} + b$, where a and b are constants. Verify the law and find the approximate values of a and b. Determine the voltage when the resistance is 90.0 ohms $[a = 7.6, b = 3.2;$ 87.6 millivolts$]$

Reduction of $y = ab^x$ into linear form, using log-linear graph paper

35. Values of p and q are believed to be related by a law of the form $p = ab^q$ where a and b are constants. The values of p and corresponding values of q are:

     p 4.5 7.4 11.2 15.8 39.0 68.0 271.5
     q 0.6 1.3   1.9   2.4   3.7   4.5   6.5

     Verify that the law relating p and q is correct and determine the approximate values of a and b [3, 2]

36. The values of k and corresponding values of l are:

     l 0.2   0.7   1.3   1.9   2.4   3.6
     k 6.0 14.2 39.4 109.7 257.2 1 990.0

     The law relating these quantities is of the form $k = mn^l$. Determine the approximate values of m and n [4.3, 5.5]

Reduction of $y = a\,e^{kx}$ into linear form, using log-linear graph paper

37. Determine the law of the form $y = a\,e^{kx}$ which relates the following values:

     y 0.015 0.08   0.17   0.30   0.96   3.0
     x −6.0   8.5   15.0   20.0   30.0 40.0

     $[y = 0.03\,e^{0.115x}]$

38. The voltage drop across an inductor, v volts, is believed to be related to time, t milliseconds, by the relationship $v = V\,e^{t/T}$, where V and T are constants. The variation of voltage with time for this inductor is:

     v volts   700   400   190   100   50   17
     t milliseconds 27.3 31.7 37.5 42.5 47.8 56.3

     Show that the law relating these quantities is as stated and determine the approximate values of V and T
     [24.1 kV, −7.75 ms]

39. The tension in two sides of a belt, T and T_0 newtons, passing around a pulley wheel and in contact with the pulley over an angle of θ radians, is given by $T = T_0\,e^{\mu\theta}$ where T_0 and μ are constants. Determine the approximate values of T_0 and the coefficient of friction μ, from the following observations:

     T newtons 67.3   76.4   90.0   107.4   117.6   125.5
     θ radians   1.13   1.54   2.07   2.64   2.93   3.14

     [47.4, 0.31]

40. A liquid which is cooling is believed to follow a law of the form $\theta = \theta_0 \, e^{kt}$ where θ_0 and k are constants and θ is the temperature of the body at time t. Measurements are made of the temperature and time and the results are:

$\theta°C$	83	58	41.5	32	26
t minutes	16.7	25	32.5	37	43.5

    Show that these quantities are related by this law and determine the approximate values of θ_0 and k
    [174°C, −0.044]

41. The mass (m) of a given substance is believed to dissolve in one litre of water at temperature ($t°C$) according to the law $m = a \, e^{kt}$. m was measured at various temperatures and the following results obtained:

$t°C$	10	20	30	40	50
m kg	35.5	39.1	43.2	47.7	52.8

    Show that the law is true and find approximate values for a and k [$a = 32.1$, $k = 0.009\,90$]

42. The following results were obtained when measuring the growth of duckweed:

Weeks (t)	0	1	2	3	4	5
No. of fronds (n)	20	30	52	77	135	211

    It is thought the measurements are connected by the equation $n = A \, e^{kt}$. Verify that this is so and obtain approximate values of A and k [$A = 18.4$, $k = 0.49$]

Chapter 5

Curve sketching

5.1 Introduction

When a mathematical equation is known, coordinates may be calculated for a limited range of values, and the equation may be represented pictorially as a graph, within this range of calculated values. Sometimes it is useful to show all the characteristic features of an equation, and in this case a sketch depicting the equation can be drawn, in which all the important features are shown, but the accurate plotting of points is less important. This technique is called 'curve sketching' and involves the use of differential calculus, and hence is better left until a knowledge of differentiation and its applications has been covered. However, at this stage, certain basic curves such as the circle, parabola, ellipse and hyperbola can be readily drawn and are dealt with in this chapter. The quadratic graph is introduced in Section 5.2, and in Section 5.3 simple equations of the circle, parabola, ellipse and hyperbola, together with the characteristic curves associated with these equations are considered.

5.2 The quadratic graph

A general quadratic equation is given by $y = ax^2 + bx + c$, where a, b and c are constants, and $a \neq 0$.

(i) $y = ax^2$

The simplest quadratic equation is $y = x^2$. In order to plot a graph of $y = x^2$ a table of values is drawn up.

x	-3	-2	-1	0	1	2	3
$y = x^2$	9	4	1	0	1	4	9

Figure 5.1(a) shows a graph of $y = x^2$ which is symmetrical about the y-axis. (Note that the coordinates on the graph are joined by a smooth curve and not from point to point.) A curve of this shape is called a **parabola**. At the origin, i.e. at $(0, 0)$, the lowest point on the curve is reached. This is called a **turning-point**. The values of y on either side of this point are greater than at

the turning-point. Such a point is called a **minimum value**, and can be thought of as being the 'bottom of a valley'.

The curve depicting $y = ax^2$ could have been sketched by considering the following easily ascertained values:

    (*a*)   when x is 0, y is 0;
    (*b*)   when x is large and positive, y is large and positive;
and (*c*)   when x is large and negative, y is large and positive.

It can be seen that the curve shown in Fig. 5.1(*a*) meets these three criteria.

Figure 5.1(*b*) shows the effect of the constant *a* where $a = 1, 2, 3$ and $\frac{1}{2}$, i.e. it shows graphs of $y = x^2$, $y = 2x^2$, $y = 3x^2$ and $y = \frac{1}{2}x^2$. As the value of

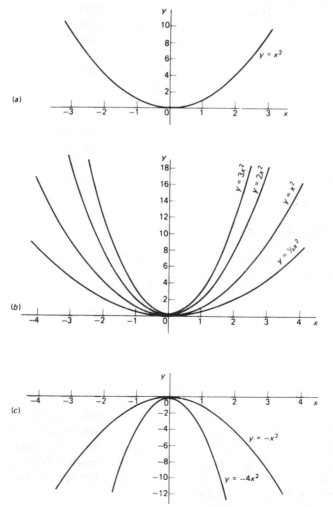

Fig. 5.1

a increases the curves become steeper; as a decreases the curves become less steep. All the curves of the form $y = ax^2$ are symmetrical about the y-axis, and the y-axis is called the axis of symmetry.

In addition to points (a) to (c) given above, the inclusion of one other pair of points will give the steepness of the curve, e.g. if $y = x^2$ and $x = \pm 2$ then $y = 4$.

(ii) $y = -ax^2$

Figure 5.1(c) shows graphs of $y = -x^2$ and $y = -4x^2$. This shows the effect of a being negative. The numerical values of y will be the same as previously obtained but will all be negative, giving the inverted parabolas shown. At the origin $(0, 0)$ a turning-point again exists. This time the values of y on either side of this point are less than at the turning-point. Such a point is called a **maximum value** and can be thought of as being the 'crest of a wave'.

(iii) $y = ax^2 + c$

A table of values for $y = x^2$ is shown on page 76. If a table of values is drawn up for, say, $y = x^2 + 3$ then all the values of y in the above table will be increased by 3. Similarly, if $y = x^2 - 2$ then all the values of y will be decreased by 2. Thus when the constant c is a positive value the parabola is raised by c units; when c is a negative value then the parabola is lowered by c units.

Figure 5.2(a) shows graphs of $y = x^2$, $y = 2x^2 + 3$ and $y = 3x^2 - 6$, and Fig. 5.2(b) shows graphs of $y = -x^2 + 5$ and $y = -3x^2 - 5$.

(iv) $y = ax^2 + bx + c$

A graph of $y = p(x + q)^2 + r$, where p, q and r are constants, is the same general shape as $y = ax^2$, i.e. a parabola.

When the constant r is a positive value the parabola is raised by r units from the zero position; when r is a negative value the parabola is lowered by r units (as in case (iii) above). When the constant q is a positive value the parabola is moved q units to the left from its zero position; when q is a negative value it is moved q units to the right.

For example, the graph of $y = 2(x + 3)^2 + 4$ is shown in Fig. 5.3. This is a parabola whose turning-point (a minimum value) is situated at $(-3, 4)$. Similarly, the graph of $y = -3(x - 2)^2 - 4$ is shown in Fig. 5.3. This is also a parabola, whose turning-point (a maximum value) is situated at $(2, -4)$. Of course, a quadratic expression is not normally expressed in the form $y = p(x + q)^2 + r$. If, however, the general expression $y = ax^2 + bx + c$ can be changed into the form $y = p(x + q)^2 + r$ then the curve can be readily sketched.

For example, let $y = x^2 + 8x + 7$. To obtain the q term in $(x + q)^2$ half the coefficient of the x term is taken, since when expanding $(x + q)^2$ the coefficient of x is $2q$. In this case $q = \frac{8}{2}$ or 4.

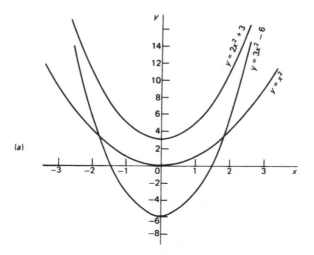

(a)

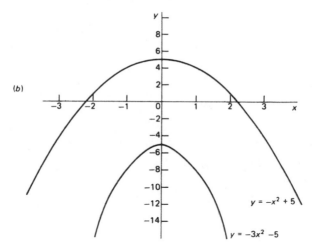

(b)

Fig. 5.2

However, $(x + 4)^2 = x^2 + 8x + 16$, which is 16 more than is needed. The 16 is therefore subtracted. Hence

$$y = x^2 + 8x + 7$$
$$= [(x + 4)^2 - 16] + 7$$
$$y = (x + 4)^2 - 9$$

which is a parabola which has a minimum value at $(-4, -9)$, as shown in Fig. 5.4(a).

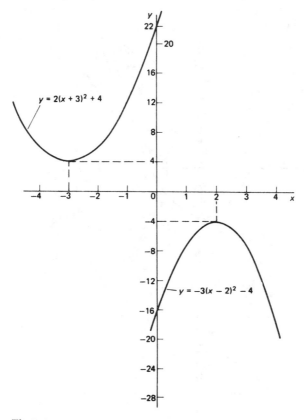

Fig. 5.3

Similarly, if $y = -4x^2 - 6x + 3$

then $\qquad y = -4\left(x^2 + \dfrac{6}{4}x\right) + 3$

$$= -4\left\{\left[x + \dfrac{3}{4}\right]^2 - \dfrac{9}{16}\right\} + 3$$

$$= -4\left[x + \dfrac{3}{4}\right]^2 + 2\dfrac{1}{4} + 3$$

i.e. $\qquad y = -4\left[x + \dfrac{3}{4}\right]^2 + 5\dfrac{1}{4}$

Comparing this with $y = p(x + q)^2 + r$ shows the parabola to be inverted (since p is negative), moved horizontally $\frac{3}{4}$ unit to the left from the zero position (since q is $+\frac{3}{4}$) and raised by $5\frac{1}{4}$ units from the zero position (since r is $+5\frac{1}{4}$). The turning-point on the curve (a maximum value) thus occurs at $(-\frac{3}{4}, 5\frac{1}{4})$, as shown in Fig. 5.4($b$).

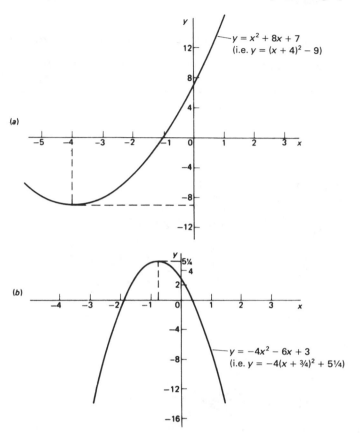

Fig. 5.4

Worked problems on sketching and plotting quadratic graphs

Problem 1. Sketch on the same axes the following curves:

(a) $y = 2x^2$ (b) $\dfrac{y}{4} = x^2$ (c) $2\sqrt{y} = x$ (d) $-y = 3x^2$

(a) $y = 2x^2$ is a parabola with its turning-point
(a minimum value) at $(0, 0)$. When $x = \pm 2$,
$y = 2(\pm 2)^2 = 8$

(b) $\dfrac{y}{4} = x^2$ or $y = 4x^2$ is a parabola with its turning-point

(a minimum value) at $(0, 0)$. This curve is steeper than
$y = 2x^2$ by a factor of 2. When $x = \pm 2$,
$y = 4(\pm 2)^2 = 16$

(c) $2\sqrt{y} = x$ or $y = \dfrac{x^2}{4}$ is a parabola with its turning-point

(a minimum value) at $(0, 0)$. This curve is less steep than

$y = 2x^2$ by a factor of $\frac{1}{8}$. When $x = \pm 2$, $y = \dfrac{(\pm 2)^2}{4} = 1$

(d) $-y = 3x^2$ or $y = -3x^2$ is an inverted parabola with its turning-point (a maximum value) at $(0, 0)$. When $x = \pm 2$, $y = -3(\pm 2)^2 = -12$

Figure 5.5 shows the four curves.

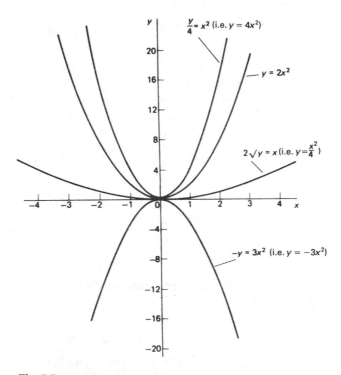

Fig. 5.5

Problem 2. Sketch the following curves and determine the coordinates of their turning-points:

(a) $y = 4x^2 + 5$ (b) $y = 4x^2 + 24x + 37$
(c) $y = -3x^2 - 2$ (d) $y = -2x^2 + 16x - 29$

(a) $y = 4x^2 + 5$ is a parabola with its turning-point (a minimum value) at $(0, 5)$. When $x = \pm 1$, $y = 4(\pm 1)^2 + 5 = 9$

(b) $y = 4x^2 + 24x + 37$
$$= 4(x^2 + 6x) + 37$$
$$= 4[(x + 3)^2 - 9] + 37$$
$$= 4(x + 3)^2 - 36 + 37$$
i.e. $y = 4(x + 3)^2 + 1$
Hence $y = 4x^2 + 24x + 37$ is a parabola with its
turning-point (a minimum value) at $(-3, 1)$.
When $x = -2$, $y = 5$; when $x = -4$, $y = 5$

(c) $y = -3x^2 - 2$ is an inverted parabola with its turning-point
(a maximum value) at $(0, -2)$. When $x = \pm 2$, $y = -14$

(d) $y = -2x^2 + 16x - 29$
$$= -2(x^2 - 8x) - 29$$
$$= -2[(x - 4)^2 - 16] - 29$$
$$= -2(x - 4)^2 + 32 - 29$$

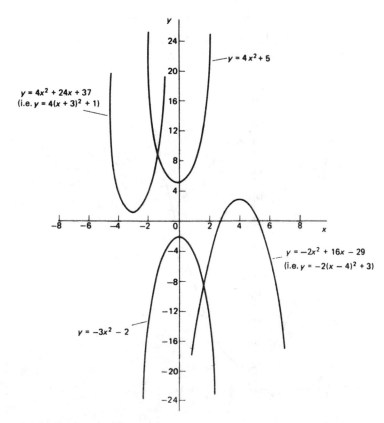

Fig. 5.6

i.e. $y = -2(x - 4)^2 + 3$

Hence $y = -2x^2 + 16x - 29$ is an inverted parabola with its turning-point (a maximum value) at **(4, 3)**.

When $x = 3$, $y = 1$; when $x = 5$, $y = 1$

Each of the curves is shown in Fig. 5.6

Problem 3. Plot the quadratic curve $y = 3x^2 - 2x + 4$ from $x = -3$ to $x = +4$ and find the coordinates of its turning-point.

To plot the quadratic curve a table is drawn up.

x	-3	-2	-1	0	1	2	3	4
$3x^2$	27	12	3	0	3	12	27	48
$-2x$	6	4	2	0	-2	-4	-6	-8
$+4$	4	4	4	4	4	4	4	4
$y = 3x^2 - 2x + 4$	37	20	9	4	5	12	25	44

The curve is plotted in Fig. 5.7. The turning-point (a minimum value) occurs at (0.3, 3.6)

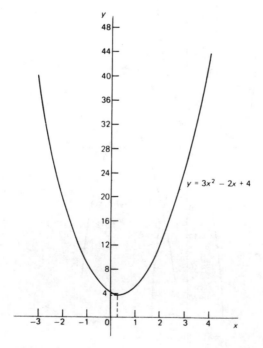

$y = 3x^2 - 2x + 4$

Fig. 5.7

The coordinates of the turning-point may be checked as follows:
Since $y = 3x^2 - 2x + 4$

then $y = 3\left(x^2 - \dfrac{2}{3}x\right) + 4$

$$= 3\left[\left(x - \dfrac{1}{3}\right)^2 - \dfrac{1}{9}\right] + 4$$

$$= 3\left(x - \dfrac{1}{3}\right)^2 - \dfrac{1}{3} + 4$$

i.e. $y = 3\left(x - \dfrac{1}{3}\right)^2 + 3\dfrac{2}{3}$

Hence the turning-point is at $\left(\dfrac{1}{3}, 3\dfrac{2}{3}\right)$

Further problems on the quadratic graph may be found in Section 5.4 (Problems 1 to 23), page 95.

5.3   Simple curves of the circle, parabola, ellipse and hyperbola

(i) The circle

The simplest form of the equation of a circle is that of a circle of radius a, having its centre at the origin of a rectangular coordinate system $(0, 0)$. The equation is

$$x^2 + y^2 = a^2 \text{ or } y = \sqrt{(a^2 - x^2)}$$

The coordinates of the equation $x^2 + y^2 = 9$ may be determined as shown below:

x	-3.0	-2.5	-2.0	-1.5	-1.0	-0.5
x^2	9.0	6.25	4.0	2.25	1.0	0.25
$9 - x^2$	0	2.75	5.0	6.75	8.0	8.75
$y = \sqrt{(9 - x^2)}$	0	± 1.66	± 2.24	± 2.60	± 2.83	± 2.96

x	0	0.5	1.0	1.5	2.0	2.5	3.0
x^2	0	0.25	1.0	2.25	4.0	6.25	9.0
$9 - x^2$	9.0	8.75	8.0	6.75	5.0	2.75	0
$y = \sqrt{(9 - x^2)}$	± 3	± 2.96	± 2.83	± 2.60	± 2.24	± 1.66	0

A graph of these values is shown in Fig. 5.8. The graph produced confirms that $x^2 + y^2 = 9$ is a circle, centre at the origin and of radius 3 units.

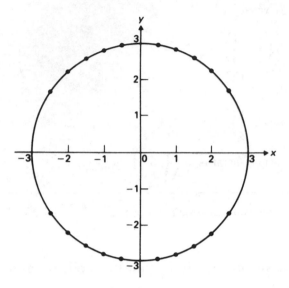

Fig. 5.8   Circle, $x^2 + y^2 = 9$

(ii) The parabola

Parabolas having an equation of the form $y = ax^2 + bx + c$ are introduced in Section 5.2, these parabolas having their axes of symmetry parallel to or coinciding with the y-axis. The simplest form of the equation of a parabola is

$$y = a\sqrt{x} \text{ or } y = ax^{\frac{1}{2}}$$

Parabolas having equations of this form have their axes of symmetry coincident with the x-axis and their vertices at the origin (0, 0). The coordinates of the equation $y = 2\sqrt{x}$, for example, may be determined as shown below for positive values of x (negative values of x give a complex result).

x	0	0.5	1.0	1.5	2.0	2.5	3.0
$\sqrt{x}$	0	± 0.71	± 1.0	± 1.22	± 1.41	± 1.58	± 1.73
$y = 2\sqrt{x}$	0	± 1.42	± 2.00	± 2.44	± 2.82	± 3.16	± 3.46

A graph of these values is shown in Fig. 5.9, the resulting curve being a parabola with its axis of symmetry coinciding with the x-axis and its vertex at the origin (0, 0).

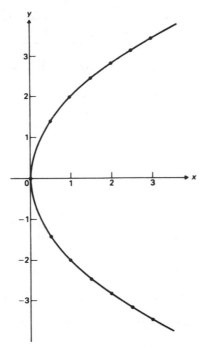

Fig. 5.9   Parabola, $y = 2x^{\frac{1}{2}}$

(iii) The ellipse

An ellipse has two axes at right angles to one another, corresponding to the maximum and minimum lengths which can be obtained within the ellipse. These axes are called the major axis for the greatest length and the minor axis for the least length.

The equation of an ellipse having its centre at the origin of a rectangular coordinate system and its axes, coinciding with the axes of the coordinate system, is

$$\frac{x^2}{a^2} + \frac{y^2}{b^2} = 1 \text{ or } y = b\sqrt{\left(1 - \frac{x^2}{a^2}\right)}$$

When $a = b$ this equation becomes $x^2 + y^2 = a^2$, i.e. a circle, centre at the origin of the coordinate system and of radius a (see (i) above).

The coordinates of the equation

$$\frac{x^2}{4} + \frac{y^2}{9} = 1, \text{ i.e. } y = 3\sqrt{\left(1 - \frac{x^2}{4}\right)}$$

may be determined as shown below.

x	-2.0	-1.5	-1.0	-0.5	0	0.5	1.0	1.5	2.0
$\dfrac{x^2}{4}$	1.0	0.563	0.25	0.063	0	0.063	0.25	0.563	1.0
$1 - \dfrac{x^2}{4}$	0	0.437	0.75	0.937	1.0	0.937	0.75	0.437	0
$\sqrt{\left(1 - \dfrac{x^2}{4}\right)}$	0	±0.661	±0.866	±0.968	±1.0	±0.968	±0.866	±0.661	0
$y = 3\sqrt{\left(1 - \dfrac{x^2}{4}\right)}$	0	±1.98	±2.60	±2.90	±3	±2.90	±2.60	±1.98	0

A graph of these values is shown in Fig. 5.10, the resulting curve being an ellipse with its centre at the origin. The width of the ellipse along the x-axis is 4 units and the total height along the y-axis is 6 units, i.e. the major axis is from $+3$ to -3 along the y-axis. The width AB in Fig. 5.10 is called the **minor axis** and is given by $2a$ in the general equation $\dfrac{x^2}{a^2} + \dfrac{y^2}{b^2} = 1$. The height CD in Fig. 5.10 is called the **major (or longer) axis** and is given by $2b$ in the equation $\dfrac{x^2}{a^2} + \dfrac{y^2}{b^2} = 1$.

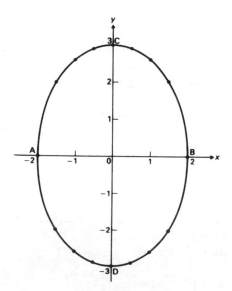

Fig. 5.10   Ellipse, $\dfrac{x^2}{4} + \dfrac{y^2}{9} = 1$

(iv) The hyperbola

The equation of a hyperbola is of the form

$$\frac{x^2}{a^2} - \frac{y^2}{b^2} = 1 \text{ or } y = b\sqrt{\left(\frac{x^2}{a^2} - 1\right)}$$

The coordinates of the equation

$$\frac{x^2}{4} - \frac{y^2}{9} = 1, \text{ i.e. } y = 3\sqrt{\left(\frac{x^2}{4} - 1\right)}$$

may be determined as shown below.

x	-10	-8	-6	-4	-2
$\dfrac{x^2}{4}$	25	16	9	4	1
$\dfrac{x^2}{4} - 1$	24	15	8	3	0
$\sqrt{\left(\dfrac{x^2}{4} - 1\right)}$	± 4.90	± 3.87	± 2.83	± 1.73	0
$y = 3\sqrt{\left(\dfrac{x^2}{4} - 1\right)}$	± 14.70	± 11.61	± 8.49	± 5.19	0

x	0	2	4	6	8	10
$\dfrac{x^2}{4}$	0	1	4	9	16	25
$\dfrac{x^2}{4} - 1$	-1	0	3	8	15	24
$\sqrt{\left(\dfrac{x^2}{4} - 1\right)}$	Complex	0	± 1.73	± 2.83	± 3.87	± 4.90
$y = 3\sqrt{\left(\dfrac{x^2}{4} - 1\right)}$	Complex	0	± 5.19	± 8.49	± 11.61	± 14.70

(For complex numbers and their representation see Chapter 9.)

The graph of these values is shown in Fig. 5.11, the resulting curve being a hyperbola which is symmetrical about both the x- and y-axes. The distance

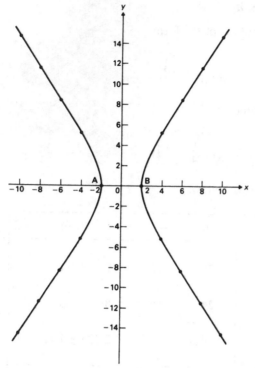

Fig. 5.11   Hyperbola, $\dfrac{x^2}{4} - \dfrac{y^2}{9} = 1$

AB in Fig. 5.11 is given by $2a$ in the general equation

$$\frac{x^2}{a^2} - \frac{y^2}{b^2} = 1$$

The distance OB is the value of x when y is equal to zero, i.e. when $\dfrac{x^2}{a^2} = 1$ or when $x = a$. Thus, due to symmetry, distance AB is $2a$ as stated.

The simplest equation of a hyperbola is of the form $y = \dfrac{a}{x}$, this being the equation of a rectangular hyperbola which is symmetrical about both the x- and y-axes, and lies entirely in the first and third quadrants. The coordinates of the equation $y = \dfrac{5}{x}$ may be determined as shown on page 91.

A graph of these values is shown in Fig. 5.12, the resulting curve being a rectangular hyperbola lying in the first and third quadrants and being symmetrical about both the x- and y-axes.

x	-7	-6	-5	-4	-3	-2	-1
$\dfrac{1}{x}$	-0.14	-0.17	-0.20	-0.25	-0.33	-0.50	-1
$y = \dfrac{5}{x}$	-0.70	-0.85	-1	-1.25	-1.65	-2.5	-5

x	0	1	2	3	4	5	6	7
$\dfrac{1}{x}$	∞	1	0.50	0.33	0.25	0.20	0.17	0.14
$y = \dfrac{5}{x}$	∞	5	2.5	1.65	1.25	1	0.85	0.70

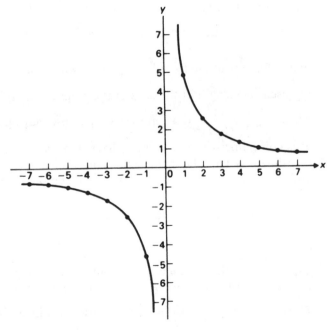

Fig. 5.12 Rectangular hyperbola, $y = \dfrac{5}{x}$

Worked problems on simple curves of the circle, parabola, ellipse and hyperbola

---

Problem 1. Sketch curves depicting the following equations, showing the significant values where possible:
(a) $x = \sqrt{(5 - y^2)}$ (b) $y^2 = 9x$ (c) $5x^2 = 35 - 7y^2$
(d) $3y^2 = 5x^2 - 15$ (e) $xy = 7$

(a)   Squaring both sides of the equation and transposing gives $x^2 + y^2 = 5$. Comparing this with the standard equation of a circle, centre origin and radius a, i.e. $x^2 + y^2 = a^2$, shows that $x^2 + y^2 = 5$ represents a circle, centre origin and radius $\sqrt{5}$. A sketch of this circle is shown in Fig. 5.13(a).

(b)   One form of the equation of a parabola is $y = a\sqrt{x}$. Squaring both sides of this equation gives $y^2 = a^2x$. The equation $y^2 = 9x$ is of this form and thus represents a parabola which is symmetrical about the x-axis and having its vertex at the origin $(0, 0)$. Also, when $x = 1$, $y = \pm 3$. A sketch of this parabola is shown in Fig. 5.13(b).

(c)   By dividing throughout by 35 and transposing the equation $5x^2 = 35 - 7y^2$ can be written as $\dfrac{x^2}{7} + \dfrac{y^2}{5} = 1$. The equation of an ellipse is of the form $\dfrac{x^2}{a^2} + \dfrac{y^2}{b^2} = 1$, where $2a$ and $2b$ represent the length of the axes of the ellipse.
Thus $\dfrac{x^2}{(\sqrt{7})^2} + \dfrac{y^2}{(\sqrt{5})^2} = 1$ represents an ellipse, having its axes coinciding with the x- and y-axes of a rectangular coordinate system, the major axis being $2\sqrt{7}$ units long and the minor axis $2\sqrt{5}$ units long, as shown in Fig. 5.13(c).

(d)   Dividing $3y^2 = 5x^2 - 15$ throughout by 15 and transposing gives $\dfrac{x^2}{3} - \dfrac{y^2}{5} = 1$. The equation $\dfrac{x^2}{a^2} - \dfrac{y^2}{b^2} = 1$ represents a hyperbola which is symmetrical about both the x- and y-axes, the distance between the vertices being given by $2a$.
Thus a sketch of $\dfrac{x^2}{3} - \dfrac{y^2}{5} = 1$ is as shown in Fig. 5.13(d), having a distance of $2\sqrt{3}$ between its vertices.

(e)   The equation $y = \dfrac{a}{x}$ represents a rectangular hyperbola lying entirely within the first and third quadrants. Transposing $xy = 7$ gives $y = \dfrac{7}{x}$, and therefore represents the rectangular hyperbola shown in Fig. 5.13(e).

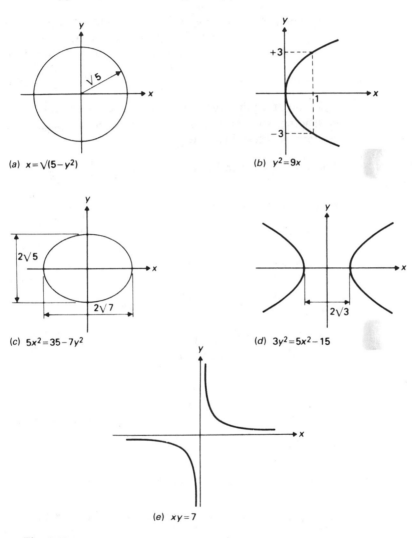

(a) $x = \sqrt{(5 - y^2)}$

(b) $y^2 = 9x$

(c) $5x^2 = 35 - 7y^2$

(d) $3y^2 = 5x^2 - 15$

(e) $xy = 7$

Fig. 5.13

Problem 2. Describe the shape of the curves represented by the following equations:

(a) $x = 3\sqrt{\left[1 - \left(\dfrac{y}{3}\right)^2\right]}$ (b) $\dfrac{y^2}{5} = 2x$ (c) $y = 6\left(1 - \dfrac{x^2}{16}\right)^{\frac{1}{2}}$

(d) $\quad x = 5\sqrt{\left[1 + \left(\dfrac{y}{2}\right)^2\right]}$ (e) $\dfrac{y}{4} = \dfrac{13}{3x}$

(a)   Squaring the equation gives $x^2 = 9\left[1 - \left(\dfrac{y}{3}\right)^2\right]$ and

transposing gives $x^2 + y^2 = 9$. Comparing this equation with $x^2 + y^2 = a^2$ shows that $x^2 + y^2 = 9$ is the equation of a circle having centre at the origin $(0, 0)$ and of radius 3 units.

(b)   Transposing $\dfrac{y^2}{5} = 2x$ gives $y = \sqrt{(10)}\sqrt{x}$. Thus $\dfrac{y^2}{5} = 2x$ is the

equation of a parabola having its axis of symmetry coinciding with the x-axis and its vertex at the origin of a rectangular coordinate system.

(c) $y = 6\left(1 - \dfrac{x^2}{16}\right)^{\frac{1}{2}}$ can be transposed to $\dfrac{y}{6} = \left(1 - \dfrac{x}{16}\right)^{\frac{1}{2}}$ and

squaring both sides gives $\dfrac{y^2}{36} = 1 - \dfrac{x^2}{16}$, i.e. $\dfrac{x^2}{16} + \dfrac{y^2}{36} = 1$

This is the equation of an ellipse, centre at the origin of a rectangular coordinate system, the major axis coinciding with the y-axis and being $2\sqrt{36}$, i.e. 12 units long. The minor axis coincides with the x-axis and is $2\sqrt{16}$, i.e. 8 units long.

(d)   Since $\quad x = 5\sqrt{\left[1 + \left(\dfrac{y}{2}\right)^2\right]}$

$$x^2 = 25\left[1 + \left(\dfrac{y}{2}\right)^2\right]$$

i.e. $\dfrac{x^2}{25} - \dfrac{y^2}{4} = 1$

This is a hyperbola which is symmetrical about both the x- and y-axes, the vertices being $2\sqrt{25}$, i.e. 10 units apart. (With reference to Section 5.3(iv), a is equal to ± 5.)

(e)   The equation $\dfrac{y}{4} = \dfrac{13}{3x}$ is of the form $y = \dfrac{a}{x}$, where

$a = \dfrac{52}{3} = 17.3$. This represents a rectangular hyperbola,

symmetrical about both the x- and y-axis, and lying entirely in the first and third quadrants, similar in shape to the curves shown in Fig. 5.13(e).

Further problems on simple curves of the circle, parabola, ellipse and hyperbola may be found in the following Section 5.4 (Problems 24 to 39), page 96.

5.4 Further problems

The quadratic graph

In Problems 1 to 10 sketch the quadratic graphs and determine the coordinates of the turning points.

1. $y = 5x^2$ $[(0, 0)]$
2. $\dfrac{y}{3} = x^2$ $[(0, 0)]$
3. $y = 2x^2 + 4$ $[(0, 4)]$
4. $y = 4x^2 - 1$ $[(0, -1)]$
5. $y + 2x^2 = 6$ $[(0, 6)]$
6. $y = -(5x^2 + 7)$ $[(0, -7)]$
7. $y = 2x^2 + 5x + 2$ $[(-1\frac{1}{4}, -1\frac{1}{8})]$
8. $y - 19 = 3x^2 - 12x$ $[(2, 7)]$
9. $y + 5 = 2x(2 - x)$ $[(1, -3)]$
10. $y + 3x^2 = 1 - 6x$ $[(-1, 4)]$

In Problems 11 to 19 plot the graphs and determine the coordinates of the turning points, stating whether they are maximum or minimum values.

11. $y = 3x^2 + 4x + 8$ $[(-\frac{2}{3}, 6\frac{2}{3}); \text{minimum}]$
12. $y = x^2 - 3x + 6$ $[(1\frac{1}{2}, 3\frac{3}{4}); \text{minimum}]$
13. $y + \frac{97}{24} = 7x - 6x^2$ $[(\frac{7}{12}, -2); \text{maximum}]$
14. $3x^2 + 4x + y = 5$ $[(-\frac{2}{3}, 6\frac{1}{3}); \text{maximum}]$
15. $y + \frac{2}{3} = x(3x + 16)$ $[(-2\frac{2}{3}, -22); \text{minimum}]$
16. $\frac{3}{16} + y = x(3 - 4x)$ $[(\frac{3}{8}, \frac{3}{8}); \text{maximum}]$
17. $y + 7x = x^2 + 2\frac{1}{4}$ $[(3\frac{1}{2}, -10); \text{minimum}]$
18. $15x + y = 18x^2 + 7\frac{1}{8}$ $[(\frac{5}{12}, 4); \text{minimum}]$
19. $20x^2 + 1 = 4x - y$ $[(\frac{1}{10}, -\frac{4}{5}); \text{maximum}]$
20. Find the point of intersection of the two curves $y = 3x^2$ and $y + 15x = 3x^2 + 5$ by plotting the two curves. $[(\frac{1}{3}, \frac{1}{3})]$
21. Determine the coordinates of the points of intersection of the two curves $y + 2 = 5x^2$ and $y = 2x^2 - 6x + 7$ by plotting the two curves. $[(1, 3); (-3, 43)]$
22. Plot the graph of $y = 5x^2 - 3x + 4$. From the graph find (a) the value of y when $x = 2.4$ and (b) the values of x when $y = 11.6$ (a) $[25.6]$ (b) $[1.57 \text{ and } -0.97]$
23. Draw up a table of values and plot the graph $y = -3x^2 + 4x - 6$, and from the graph find (a) the value of y when $x = 1.7$ and (b) the values of x when $y = -8.8$ (a) $[-7.87]$ (b) $[1.84 \text{ and } -0.51]$

Simple curves of the circle, parabola, ellipse and hyperbola

In Problems 24 to 31 sketch the curves depicting the equations given, showing significant values where possible.

24. $x = 7\sqrt{\left[1 - \left(\dfrac{y}{7}\right)^2\right]}$ [circle, centre $(0, 0)$, radius 7 units]

25. $\dfrac{y^2}{3x} = 4$ [parabola, symmetrical about x-axis, vertex at $(0, 0)$]

26. $\sqrt{x} = \dfrac{y}{5}$ [parabola, symmetrical about x-axis, vertex at $(0, 0)$]

27. $y^2 = \dfrac{x^2 - 12}{4}$ [hyperbola, symmetrical about x- and y-axes, distance between vertices $4\sqrt{3}$ units along x-axis]

28. $\dfrac{y^2}{5} = 5 - x^2$ [ellipse, centre $(0, 0)$, major axis 10 units along y-axis, minor axis $2\sqrt{5}$ units along x-axis]

29. $x = \frac{1}{3}\sqrt{(99 - 11y^2)}$ [ellipse, centre $(0, 0)$, major axis $2\sqrt{11}$ units along x-axis, minor axis 6 units along y-axis]

30. $x = 2\sqrt{(1 + y^2)}$ [hyperbola, symmetrical about x- and y-axes, distance between vertices 4 units along x-axis]

31. $x^2 y^2 = 5$ [rectangular hyperbola, lying in first and third quadrants only]

In Problems 32 to 39 describe the shape of the curves represented by the equations given.

32. $y = \sqrt{[2(1 - x^2)]}$ [ellipse, centre $(0, 0)$, major axis $2\sqrt{2}$ units along y-axis, minor axis 2 units along x-axis]

33. $y = \sqrt{[2(1 + x^2)]}$ [hyperbola, symmetrical about x- and y-axes, vertices 2 units apart along x-axis]

34. $y = \sqrt{(2 - x^2)}$ [circle, centre $(0, 0)$, radius $\sqrt{2}$ units]

35. $y = 2x^{-1}$ [rectangular hyperbola, lying in first and third quadrants, symmetrical about x- and y-axes]

36. $y = (2x)^{\frac{1}{2}}$ [parabola, vertex at $(0, 0)$, symmetrical about the x-axis]

37. $2y^2 - 6 = -3x^2$ [ellipse, centre $(0,0)$, major axis
$2\sqrt{3}$ units along the y-axis, minor axis
$2\sqrt{2}$ units along the x-axis]

38. $y^2 = 13 - x^2$ [circle, centre $(0,0)$, radius $\sqrt{13}$]

39. $4x^2 - 3y^2 = 12$ [hyperbola, symmetrical about
x- and y-axes, vertices on x-axis
distance $2\sqrt{3}$ apart]

Chapter 6

Partial fractions

6.1 Introduction

Consider the following addition of algebraic fractions:

$$\frac{1}{x+1} + \frac{1}{x+2} = \frac{(x+2) + (x+1)}{(x+1)(x+2)} = \frac{2x+3}{x^2 + 3x + 2}$$

If we start with the expression

$$\frac{2x+3}{x^2 + 3x + 2}$$

and find the fractions whose sum gives this result, the two fractions obtained $\left(\text{i.e. } \dfrac{1}{x+1} \text{ and } \dfrac{1}{x+2}\right)$ are called the **partial fractions** of $\dfrac{2x+3}{x^2 + 3x + 2}$

This process of expressing a fraction in terms of simpler fractions, called resolving into partial fractions, is used as a preliminary to integrating certain functions (see Chapter 24) and the techniques used are explained by example later in this chapter. However, before attempting to resolve an algebraic expression into partial fractions, the following points must be considered and appreciated:

(a) **The denominator of the algebraic expression** must **factorise**. In the above example the denominator $x^2 + 3x + 2$ factorises as $(x+1)(x+2)$.

(b) In the above example, the numerator, $2x + 3$, is said to be of degree one since the highest powered x term is x^1. The denominator, $x^2 + 3x + 2$, is said to be of degree two since the highest powered x term is x^2. **In order to resolve an algebraic expression into partial fractions, the numerator** must **be at least one degree less than the denominator.** When the degree of the numerator is equal to or higher than the degree of the denominator, the denominator must be divided into the numerator until the remainder is of lower degree than the denominator. For example, $\dfrac{x^2 + x - 5}{x^2 - 2x - 3}$ cannot be resolved into partial fractions as it stands, since the numerator and denominator are of the same degree.

Dividing $x^2 + x - 5$ by $x^2 - 2x - 3$ gives:

$$
\begin{array}{r}
1 \\
x^2 - 2x - 3 \overline{)\, x^2 + x - 5} \\
x^2 - 2x - 3 \\
\hline
3x - 2
\end{array}
$$

Thus: $\dfrac{x^2 + x - 5}{x^2 - 2x - 3} = 1 + \dfrac{3x - 2}{x^2 - 2x - 3}$

Since $x^2 - 2x - 3$ factorises as $(x + 1)(x - 3)$ then $\dfrac{3x - 2}{x^2 - 2x - 3}$ may be resolved into partial fractions (see Section 6.2).

(c) Given an identity such as:

$$5x^2 - 3x + 2 \equiv Ax^2 + Bx + C$$

(note: $\equiv$ means 'identically equal to'), then $A = 5$, $B = -3$ and $C = 2$, since the identity is true for all values of x and the coefficients of x^n (where $n = 0, 1, 2 \ldots$) on the left-hand side of the identity are equal to the coefficients of x^n on the right-hand side of the identity. Similarly, if $ax^3 + bx^2 - cx + d \equiv 2x^3 + 5x - 7$ then $a = 2$, $b = 0$, $c = -5$ and $d = -7$.

6.2   Denominator containing linear factors

The corresponding partial fractions of an algebraic expression $\dfrac{f(x)}{(x - a)(x - b)}$ are of the form $\dfrac{A}{(x - a)} + \dfrac{B}{(x - b)}$, where $f(x)$ is a polynomial of degree less than 2. Similarly, the corresponding partial fractions of $\dfrac{f(x)}{(x + a)(x - b)(x - c)}$ are of the form $\dfrac{A}{(x + a)} + \dfrac{B}{(x - b)} + \dfrac{C}{(x - c)}$, where $f(x)$ is a polynomial of degree less than 3, and so on.

Problem 1. Resolve $\dfrac{x - 8}{x^2 - x - 2}$ into partial fractions.

The denominator factorises as $(x + 1)(x - 2)$ and the numerator is of less degree than the denominator. Thus $\dfrac{x - 8}{x^2 - x - 2}$ may be resolved into partial fractions.

Let $\dfrac{x - 8}{x^2 - x - 2} = \dfrac{x - 8}{(x + 1)(x - 2)} \equiv \dfrac{A}{x + 1} + \dfrac{B}{x - 2}$

where A and B are constants to be determined.

Adding the two fractions on the right-hand side gives:

$$\frac{x - 8}{(x + 1)(x - 2)} \equiv \frac{A(x - 2) + B(x + 1)}{(x + 1)(x - 2)}$$

Since the denominators are the same on each side of the identity then the numerators must be equal to each other.

Hence $x - 8 \equiv A(x - 2) + B(x + 1)$

There are two methods whereby A and B may be determined using the properties of identities introduced in Section 6.1

Method 1

Since an identity is true for all real values of x, substitute into the identity a value of x to reduce one of the unknown constants to zero.

Let $x = 2$.   Then $2 - 8 = A(0) + B(3)$

i.e. $-6 = 3B$

$B = -2$

Let $x = -1$.   Then $-1 - 8 = A(-3) + B(0)$

$-9 = -3A$

$A = 3$

Method 2

Since the coefficients of x^n ($n = 0, 1, 2, \ldots$) on the left-hand side of an identity equal the coefficients of x^n on the right-hand side, equate the respective coefficients on each side of the identity.

Since $x - 8 \equiv A(x - 2) + B(x + 1)$

then $x - 8 \equiv Ax - 2A + Bx + B = (A + B)x + (-2A + B)$

Thus $1 = A + B$ (by equating the coefficients of x)   (1)

and $-8 = -2A + B$ (by equating the constants)   (2)

Solving the two simultaneous equations gives $A = 3$ and $B = -2$ as before.

Thus $\dfrac{x - 8}{x^2 - x - 2} \equiv \dfrac{3}{x + 1} + \dfrac{-2}{x - 2} \equiv \dfrac{3}{x + 1} - \dfrac{2}{x - 2}$

It is usually quicker and easier to adopt the first method as far as possible although with other types of partial fractions a combination of the two methods is necessary.

Problem 2. Express $\dfrac{6x^2 + 7x - 25}{(x-1)(x+2)(x-3)}$ in partial fractions.

Let $\dfrac{6x^2 + 7x - 25}{(x-1)(x+2)(x-3)}$

$$\equiv \frac{A}{(x-1)} + \frac{B}{(x+2)} + \frac{C}{(x-3)}$$

$$\equiv \frac{A(x+2)(x-3) + B(x-1)(x-3) + C(x-1)(x+2)}{(x-1)(x+2)(x-3)}$$

Equating the numerators gives:

$$6x^2 + 7x - 25 \equiv A(x+2)(x-3) + B(x-1)(x-3)$$
$$+ C(x-1)(x+2)$$

Let $x = 1$.

Then $6 + 7 - 25 = A(3)(-2) + B(0)(-2) + C(0)(3)$

i.e. $-12 = -6A$

$$A = 2$$

Let $x = -2$.

Then $6(-2)^2 + 7(-2) - 25 = A(0)(-5) + B(-3)(-5) + C(-3)(0)$

i.e. $-15 = 15B$

$$B = -1$$

Let $x = 3$.

Then $6(3)^2 + 7(3) - 25 = A(5)(0) + B(2)(0) + C(2)(5)$

i.e. $50 = 10C$

$$C = 5$$

Thus: $\dfrac{6x^2 + 7x - 25}{(x-1)(x+2)(x-3)} \equiv \dfrac{2}{(x-1)} - \dfrac{1}{(x+2)} + \dfrac{5}{(x-3)}$

Problem 3. Convert $\dfrac{x^3 - x^2 - 5x}{x^2 - 3x + 2}$ into partial fractions.

The numerator is of higher degree than the denominator, thus dividing gives:

$$\begin{array}{r} x + 2 \\ x^2 - 3x + 2 \overline{\smash{\big)}\ x^3 -\ x^2 - 5x} \\ \underline{x^3 - 3x^2 + 2x} \\ 2x^2 - 7x \\ \underline{2x^2 - 6x + 4} \\ -x - 4 \end{array}$$

Thus

$$\frac{x^3 - x^2 - 5x}{x^2 - 3x + 2} = x + 2 + \frac{-x - 4}{x^2 - 3x + 2} = x + 2 - \frac{x + 4}{(x - 1)(x - 2)}$$

Let $\dfrac{x + 4}{(x - 1)(x - 2)} \equiv \dfrac{A}{x - 1} + \dfrac{B}{x - 2} \equiv \dfrac{A(x - 2) + B(x - 1)}{(x - 1)(x - 2)}$

Equating the numerators gives:

$$x + 4 \equiv A(x - 2) + B(x - 1)$$

Let $x = 1$.   Then $5 = -A$

$$A = -5$$

Let $x = 2$.   Then $6 = B$

Thus $\dfrac{x + 4}{(x - 1)(x - 2)} \equiv \dfrac{-5}{(x - 1)} + \dfrac{6}{(x - 2)}$

Thus $\dfrac{x^3 - x^2 - 5x}{x^2 - 3x + 2} \equiv x + 2 - \left[\dfrac{-5}{(x - 1)} + \dfrac{6}{(x - 2)} \right]$

$$\equiv x + 2 + \frac{5}{(x - 1)} - \frac{6}{(x - 2)}$$

6.3   Denominator containing repeated linear factors

When the denominator of an algebraic expression has a factor $(x - a)^n$ then the corresponding partial fractions are $\dfrac{A}{(x - a)} + \dfrac{B}{(x - a)^2} + \cdots + \dfrac{C}{(x - a)^n}$, since $(x - a)^n$ is assumed to hide the factors $(x - a)^{n-1}, (x - a)^{n-2}, \ldots, (x - a)$

Problem 1. Express $\dfrac{x + 5}{(x + 3)^2}$ in partial fractions.

Let $\dfrac{x + 5}{(x + 3)^2} \equiv \dfrac{A}{(x + 3)} + \dfrac{B}{(x + 3)^2} \equiv \dfrac{A(x + 3) + B}{(x + 3)^2}$

Equating the numerators gives:

$$x + 5 \equiv A(x + 3) + B$$

Let $x = -3$.   Then $-3 + 5 = A(0) + B$

i.e. $B = 2$

Equating the coefficient of x gives $A = 1$

(Checking by equating constant terms gives: $5 = 3A + B$ which is true when $A = 1$ and $B = 2$)

Thus $\dfrac{x + 5}{(x + 3)^2} \equiv \dfrac{1}{(x + 3)} + \dfrac{2}{(x + 3)^2}$

Problem 2. Resolve $\dfrac{5x^2 - 19x + 3}{(x - 2)^2(x + 1)}$ into partial fractions.

The given denominator is a combination of the linear factor type and repeated linear factor type.

Let $\dfrac{5x^2 - 19x + 3}{(x - 2)^2(x + 1)} \equiv \dfrac{A}{(x - 2)} + \dfrac{B}{(x - 2)^2} + \dfrac{C}{(x + 1)}$

$$\equiv \dfrac{A(x - 2)(x + 1) + B(x + 1) + C(x - 2)^2}{(x - 2)^2(x + 1)}$$

Equating the numerators gives:

$$5x^2 - 19x + 3 \equiv A(x - 2)(x + 1) + B(x + 1) + C(x - 2)^2$$

Let $x = 2$.

Then $5(2)^2 - 19(2) + 3 = A(0)(3) + B(3) + C(0)^2$

i.e $-15 = 3B$

$$B = -5$$

Let $x = -1$.

Then $5(-1)^2 - 19(-1) + 3 = A(-3)(0) + B(0) + C(-3)^2$

i.e. $27 = 9C$

$$C = 3$$

$5x^2 - 19x + 3 \equiv A(x - 2)(x + 1) + B(x + 1) + C(x - 2)^2$
$\equiv A(x^2 - x - 2) + B(x + 1) + C(x^2 - 4x + 4)$
$\equiv (A + C)x^2 + (-A + B - 4C)x + (-2A + B + 4C)$

Equating the coefficients of x^2 gives:

$5 = A + C$

Since $C = 3$ then $A = 2$

[Check:

Equating the coefficients of x gives: $\qquad -19 = -A + B - 4C$

If $A = 2$, $B = -5$ and $C = 3$ then $-A + B - 4C = -19 =$ L.H.S.

Equating the constant terms gives: $\qquad 3 = -2A + B + 4C$

If $A = 2$, $B = -5$ and $C = 3$ then $-2A + B + 4C = 3 =$ L.H.S.]

Thus $\dfrac{5x^2 - 19x + 3}{(x - 2)^2(x + 1)} \equiv \dfrac{2}{(x - 2)} - \dfrac{5}{(x - 2)^2} + \dfrac{3}{(x + 1)}$

Problem 3. Convert $\dfrac{2x^2 - 13x + 13}{(x + 4)^3}$ into partial fractions.

Let $\dfrac{2x^2 - 13x + 13}{(x - 4)^3} \equiv \dfrac{A}{(x - 4)} + \dfrac{B}{(x - 4)^2} + \dfrac{C}{(x - 4)^3}$

$\qquad\qquad \equiv \dfrac{A(x - 4)^2 + B(x - 4) + C}{(x - 4)^3}$

Equating the numerators gives:

$2x^2 - 13x + 13 \equiv A(x - 4)^2 + B(x - 4) + C$

Let $x = 4$. Then $2(4)^2 - 13(4) + 13 = A(0)^2 + B(0) + C$

i.e. $\qquad\qquad\qquad\qquad\qquad -7 = C$

Also $2x^2 - 13x + 13 \equiv A(x^2 - 8x + 16) + B(x - 4) + C$

$\qquad\qquad\qquad \equiv Ax^2 + (-8A + B)x + (16A - 4B + C)$

Equating the coefficients of x^2 gives: $\quad \mathbf{2 = A}$

Equating the coefficients of x gives: $-13 = -8A + B$

from which $\qquad\qquad\qquad\qquad \mathbf{B = 3}$

[Check: Equating the constant terms gives: $13 = 16A - 4B + C$

$\qquad$ If $A = 2$, $B = 3$ and $C = -7$

$\qquad$ then R.H.S. $= 16(2) - 4(3) - 7 = 13 =$ L.H.S.]

Thus $\dfrac{2x^2 - 13x + 13}{(x - 4)^3} \equiv \dfrac{2}{(x - 4)} + \dfrac{3}{(x - 4)^2} - \dfrac{7}{(x - 4)^3}$

6.4   Denominator containing a quadratic factor

When the denominator contains a quadratic factor of the form $px^2 + qx + r$ (where p, q and r are constants), which does not factorise without containing surds or imaginary values, then the corresponding partial fraction is of the form $\dfrac{Ax + B}{px^2 + qx + r}$, i.e. the numerator is assumed to be a polynomial of degree one less than the denominator. Hence the corresponding partial fractions of an algebraic expression $\dfrac{f(x)}{(px^2 + qx + r)(x - a)}$ are of the form

$$\frac{Ax + B}{px^2 + qx + r} + \frac{C}{x - a}$$

Problem 1. Resolve $\dfrac{8x^2 - 3x + 19}{(x^2 + 3)(x - 1)}$ into partial fractions.

Let $\dfrac{8x^2 - 3x + 19}{(x^2 + 3)(x - 1)} \equiv \dfrac{Ax + B}{(x^2 + 3)} + \dfrac{C}{(x - 1)}$

$$\equiv \frac{(Ax + B)(x - 1) + C(x^2 + 3)}{(x^2 + 3)(x - 1)}$$

Equating the numerators gives:

$$8x^2 - 3x + 19 \equiv (Ax + B)(x - 1) + C(x^2 + 3)$$

Let $x = 1$. Then $8(1)^2 - 3(1) + 19 = (A + B)(0) + C(4)$

i.e. $24 = 4C$

$$C = 6$$

$$8x^2 - 3x + 19 \equiv (Ax + B)(x - 1) + C(x^2 + 3)$$
$$\equiv Ax^2 - Ax + Bx - B + Cx^2 + 3C$$
$$\equiv (A + C)x^2 + (-A + B)x + (-B + 3C)$$

Equating the coefficients of the x^2 terms gives:

$$8 = A + C$$

Since $C = 6$, $A = 2$

Equating the coefficients of the x terms gives:

$$-3 = -A + B$$

Since $A = 2$, $B = -1$

[Check: Equating the constant terms gives: $19 = -B + 3C$
R.H.S. $= -B + 3C = -(-1) + 3(6) = 19 =$ L.H.S.]

Hence $\dfrac{8x^2 - 3x + 19}{(x^2 + 3)(x - 1)} \equiv \dfrac{2x - 1}{(x^2 + 3)} + \dfrac{6}{(x - 1)}$

Problem 2. Resolve $\dfrac{2 + x + 6x^2 - 2x^3}{x^2(x^2 + 1)}$ into partial fractions.

Terms such as x^2 may be treated as $(x + 0)^2$, i.e. it is a repeated linear factor type.

Let $\dfrac{2 + x + 6x^2 - 2x^3}{x^2(x^2 + 1)} \equiv \dfrac{A}{x} + \dfrac{B}{x^2} + \dfrac{Cx + D}{(x^2 + 1)}$

$$\equiv \frac{Ax(x^2 + 1) + B(x^2 + 1) + (Cx + D)x^2}{x^2(x^2 + 1)}$$

Equating the numerators gives:

$$2 + x + 6x^2 - 2x^3 = Ax(x^2 + 1) + B(x^2 + 1) + (Cx + D)x^2$$

$$\equiv (A + C)x^3 + (B + D)x^2 + Ax + B$$

Let $x = 0$. Then $2 = B$

Equating the coefficients of x^3 gives:

$$-2 = A + C \tag{3}$$

Equating the coefficients of x^2 gives:

$$6 = B + D$$

Since $B = 2,$ $D = 4$

Equating the coefficients of x gives:

$$1 = A$$

From equation (3), $C = -3$

[Check: Equating the constant terms: $2 = B$ as before]

Hence $\dfrac{2 + x + 6x^2 - 2x^3}{x^2(x^2 + 1)} \equiv \dfrac{1}{x} + \dfrac{2}{x^2} + \dfrac{4 - 3x}{x^2 + 1}$

6.5 Summary

Denominator containing	Expression	Form of partial fractions
Linear factors	$\dfrac{f(x)}{(x+a)(x+b)(x+c)}$	$\dfrac{A}{(x+a)} + \dfrac{B}{(x+b)} + \dfrac{C}{(x+c)}$
Repeated linear factors	$\dfrac{f(x)}{(x-a)^3}$	$\dfrac{A}{(x-a)} + \dfrac{B}{(x-a)^2} + \dfrac{C}{(x-a)^3}$
Quadratic factors	$\dfrac{f(x)}{(ax^2+bx+c)(x-d)}$	$\dfrac{Ax+B}{(ax^2+bx+c)} + \dfrac{C}{(x-d)}$
General example:	$\dfrac{f(x)}{(x^2+a)(x+b)^2(x+c)}$	$\dfrac{Ax+B}{(x^2+a)} + \dfrac{C}{(x+b)} + \dfrac{D}{(x+b)^2} + \dfrac{E}{(x+c)}$

In each of the above cases $f(x)$ must be of less degree than the relevant denominator. If it is not, then the denominator must be divided into the numerator. For every possible factor of the denominator there is a corresponding partial fraction.

6.6 Further problems

Resolve the following into partial fractions:

1. $\dfrac{8}{x^2-4}$ $\left[\dfrac{2}{(x-2)} - \dfrac{2}{(x+2)}\right]$

2. $\dfrac{3x+5}{x^2+2x-3}$ $\left[\dfrac{2}{(x-1)} + \dfrac{1}{(x+3)}\right]$

3. $\dfrac{y-13}{y^2-y-6}$ $\left[\dfrac{3}{(y+2)} - \dfrac{2}{(y-3)}\right]$

4. $\dfrac{17x^2-21x-6}{x(x+1)(x-3)}$ $\left[\dfrac{2}{x} + \dfrac{8}{(x+1)} + \dfrac{7}{(x-3)}\right]$

5. $\dfrac{6x^2+7x-49}{(x-4)(x+1)(2x-3)}$ $\left[\dfrac{3}{(x-4)} - \dfrac{2}{(x+1)} + \dfrac{4}{(2x-3)}\right]$

6. $\dfrac{x^2+2}{(x+4)(x-2)}$ $\left[1 - \dfrac{3}{(x+4)} + \dfrac{1}{(x-2)}\right]$

7. $\dfrac{2x^2+4x+19}{2(x-3)(x+4)}$ $\left[1 + \dfrac{7}{2(x-3)} - \dfrac{5}{2(x+4)}\right]$

8. $\dfrac{2x^3 + 7x^2 - 2x - 27}{(x-1)(x+4)}$ $\left[2x + 1 - \dfrac{4}{(x-1)} + \dfrac{7}{(x+4)}\right]$

9. $\dfrac{2t-1}{(t+1)^2}$ $\left[\dfrac{2}{(t+1)} - \dfrac{3}{(t+1)^2}\right]$

10. $\dfrac{8x^2 + 12x - 3}{(x+2)^3}$ $\left[\dfrac{8}{(x+2)} - \dfrac{20}{(x+2)^2} + \dfrac{5}{(x+2)^3}\right]$

11. $\dfrac{6x+1}{(2x+1)^2}$ $\left[\dfrac{3}{(2x+1)} - \dfrac{2}{(2x+1)^2}\right]$

12. $\dfrac{1}{x^2(x+2)}$ $\left[\dfrac{1}{2x^2} - \dfrac{1}{4x} + \dfrac{1}{4(x+2)}\right]$

13. $\dfrac{9x^2 - 73x + 150}{(x-7)(x-3)^2}$ $\left[\dfrac{5}{(x-7)} + \dfrac{4}{(x-3)} - \dfrac{3}{(x-3)^2}\right]$

14. $\dfrac{-(9x + 4x + 4)}{x^2(x^2 - 4)}$ $\left[\dfrac{1}{x} + \dfrac{1}{x^2} + \dfrac{2}{(x+2)} - \dfrac{3}{(x-2)}\right]$

15. $\dfrac{-(a^2 + 5a + 13)}{(a^2 + 5)(a - 2)}$ $\left[\dfrac{2a-1}{(a^2+5)} - \dfrac{3}{(a-2)}\right]$

16. $\dfrac{3-x}{(x^2+3)(x+3)}$ $\left[\dfrac{1-x}{2(x^2+3)} + \dfrac{1}{2(x+3)}\right]$

17. $\dfrac{12 - 2x - 5x^2}{(x^2 + x + 1)(3 - x)}$ $\left[\dfrac{2x+5}{(x^2+x+1)} - \dfrac{3}{(3-x)}\right]$

18. $\dfrac{x^3 + 7x^2 + 8x + 10}{x(x^2 + 2x + 5)}$ $\left[1 + \dfrac{2}{x} + \dfrac{3x-1}{(x^2+2x+5)}\right]$

19. $\dfrac{5x^3 - 3x^2 + 41x - 64}{(x^2 + 6)(x - 1)^2}$ $\left[\dfrac{2-3x}{(x^2+6)} + \dfrac{8}{(x-1)} - \dfrac{3}{(x-1)^2}\right]$

20. $\dfrac{6x^3 + 5x^2 + 4x + 3}{(x^2 + x + 1)(x^2 - 1)}$ $\left[\dfrac{2x-1}{(x^2+x+1)} + \dfrac{3}{(x-1)} + \dfrac{1}{(x+1)}\right]$

Chapter 7

The notation of second order matrices and determinants

7.1   Introduction

Matrices are used in engineering and science for solving linear simultaneous equations.

Terms used in connection with matrices
Consider the linear simultaneous equations:

$$2x + 3y = 4 \tag{1}$$
$$\text{and } 5x - 6y = 7 \tag{2}$$

In **matrix** notation, the coefficients of x and y are written as $\begin{pmatrix} 2 & 3 \\ 5 & -6 \end{pmatrix}$, that is, occupying the same relative positions as in equations (1) and (2) above. The grouping of the coefficients of x and y in this way is called an **array** and the coefficients forming the array are called the **elements** of the matrix.

If there are m rows across an array and n columns down an array, then the matrix is said to be of order $m \times n$, called 'm by n'. Thus for the equations

$$2x + 3y - 4z = 5 \tag{3}$$
$$6x - 7y + 8z = 9 \tag{4}$$

the matrix of the coefficients of x, y and z is $\begin{pmatrix} 2 & 3 & -4 \\ 6 & -7 & 8 \end{pmatrix}$ and is a '2 by 3' matrix. A matrix having a single row is called a **row** matrix and one having a single column is called a **column** matrix. For example, in equation (3) above, the coefficients of x, y and z form a row matrix of $(2 \quad 3 \quad -4)$ and the coefficients of x in equations (3) and (4) form a column matrix of $\begin{pmatrix} 2 \\ 6 \end{pmatrix}$.

A matrix having the same number of rows as columns is called a **square** matrix. Thus the matrix for the coefficients of x and y in equations (1) and (2) above, i.e. $\begin{pmatrix} 2 & 3 \\ 5 & -6 \end{pmatrix}$, is a square matrix, and is called a second order matrix. Matrices are generally denoted by capital letters and if the matrix representing the coefficients of x and y in equations (1) and (2) above is A, then $A = \begin{pmatrix} 2 & 3 \\ 5 & -6 \end{pmatrix}$

7.2   Addition, subtraction and multiplication of second order matrices

In arithmetic, once the basic procedures associated with addition, subtraction, multiplication and division have been mastered, simple problems may be solved. With matrices, the various rules governing them have to be understood before they can be used to solve practical problems. In this section the basic laws of addition, subtraction and multiplication are introduced.

A matrix does not have a single numerical value and cannot be simplified to a particular answer. The main advantage of using matrices is that by applying the laws of matrices, given in this section, they can be simplified, and by comparing one matrix with another similar matrix, values of unknown elements can be determined. It will be seen in Chapter 8 that matrices can be used in this way for solving simultaneous equations. Matrices can be added, subtracted and multiplied and suitable definitions for these operations are formulated, so that they obey most of the laws which govern the algebra of numbers.

Addition

Only matrices of the same order may be added. Thus a 2 by 2 matrix can be added to a 2 by 2 matrix by adding corresponding elements, but a 3 by 2 matrix cannot be added to a 2 by 2 matrix, since some elements in one matrix do not have corresponding elements in the other. The sum of two matrices is the matrix obtained by adding the elements occupying corresponding positions in the matrix, and results in two matrices being simplified to a single matrix.

For example, the matrices

$$\begin{pmatrix} 1 & 3 \\ 2 & -4 \end{pmatrix} \text{ and } \begin{pmatrix} 2 & 5 \\ 6 & -7 \end{pmatrix}$$

are added as follows:

$$\begin{pmatrix} 1 & 3 \\ 2 & -4 \end{pmatrix} + \begin{pmatrix} 2 & 5 \\ 6 & -7 \end{pmatrix} = \begin{pmatrix} 1+2 & 3+5 \\ 2+6 & (-4)+(-7) \end{pmatrix}$$

$$= \begin{pmatrix} 3 & 8 \\ 8 & -11 \end{pmatrix}$$

Subtraction

Only matrices of the same order can be subtracted and the difference between two matrices, say $A - B$, is the matrix obtained by subtracting the elements of matrix B from those occupying the corresponding positions in matrix A.

For example:

$$\begin{pmatrix} 1 & 3 \\ 2 & -4 \end{pmatrix} - \begin{pmatrix} 2 & 5 \\ 6 & -7 \end{pmatrix} = \begin{pmatrix} 1-2 & 3-5 \\ 2-6 & (-4)-(-7) \end{pmatrix}$$

$$= \begin{pmatrix} -1 & -2 \\ -4 & 3 \end{pmatrix}$$

By adding the single matrix obtained by adding A and B to, say, matrix C, the single matrix representing $A + B + C$ is obtained. By taking, say, matrix D from this single matrix, $A + B + C - D$ is obtained. Thus the laws of addition and subtraction can be applied to more than two matrices, providing that they are all of the same order.

Multiplication

(a) Scalar multiplication

When a matrix is multiplied by a number, the resultant matrix is one of the same order having each element multiplied by the number.

Thus, if matrix $A = \begin{pmatrix} 1 & 3 \\ 2 & -4 \end{pmatrix}$, then $2A = 2\begin{pmatrix} 1 & 3 \\ 2 & -4 \end{pmatrix}$

$$= \begin{pmatrix} 2 \times 1 & 2 \times 3 \\ 2 \times 2 & 2 \times (-4) \end{pmatrix}$$

$$= \begin{pmatrix} 2 & 6 \\ 4 & -8 \end{pmatrix}$$

(b) Multiplication of matrices

Two matrices can only be multiplied together when the number of columns in the first one is equal to the number of rows in the second one. This is because the process of matrix multiplication depends on finding the sum of the products of the rows in one matrix with the columns in the other. Thus it is possible to multiply a 2 by 2 matrix by a column matrix having two elements or by another 2 by 2 matrix, but it is not possible to multiply it by a row matrix. Thus if:

$$A = \begin{pmatrix} 2 & 3 \\ 5 & 6 \end{pmatrix}, B = \begin{pmatrix} 1 \\ 8 \end{pmatrix} \text{ and } C = (4 \quad 9)$$

it is possible to find $A \times B$, since the number of columns in A is equal to the number of rows in B, but it is not possible to find $A \times C$ since there are two columns in A but only one row in C. If a 2 by 2 matrix A is multiplied by a column matrix, B, having two elements, the resulting matrix is a two-element column matrix. The top element is the sum of the products obtained by taking the elements of the top row of A with B. The bottom element is the sum of the products obtained by taking the bottom elements of A with B.

If $\quad A = \begin{pmatrix} a & b \\ c & d \end{pmatrix}$ and $B = \begin{pmatrix} p \\ q \end{pmatrix}$

then $A \times B = \begin{pmatrix} a & b \\ c & d \end{pmatrix} \times \begin{pmatrix} p \\ q \end{pmatrix} = \begin{pmatrix} ap + bq \\ cp + dq \end{pmatrix}$

For example, multiplying the matrices $\begin{pmatrix} 2 & -5 \\ 4 & 3 \end{pmatrix}$ and $\begin{pmatrix} 1 \\ 6 \end{pmatrix}$ gives

$$\begin{pmatrix} 2 & -5 \\ 4 & 3 \end{pmatrix} \times \begin{pmatrix} 1 \\ 6 \end{pmatrix} = \begin{pmatrix} 2 \times 1 + (-5) \times 6 \\ 4 \times 1 + 3 \times 6 \end{pmatrix} = \begin{pmatrix} 2 - 30 \\ 4 + 18 \end{pmatrix}$$

$$= \begin{pmatrix} -28 \\ 22 \end{pmatrix}$$

If a 2 by 2 matrix, say A, is multiplied by a 2 by 2 matrix, say B, the resulting matrix is a 2 by 2 matrix, say C. The top elements of C are the sum of the products obtained by taking the top row of A with the columns of B. The bottom elements of C are the sum of the products obtained by taking the bottom row of A with the columns of B.

For example:

$$\begin{pmatrix} 1 & 3 \\ -2 & 4 \end{pmatrix} \times \begin{pmatrix} 5 & 0 \\ 7 & -6 \end{pmatrix} = \begin{pmatrix} 1 \times 5 + 3 \times 7 & 1 \times 0 + 3 \times (-6) \\ -2 \times 5 + 4 \times 7 & -2 \times 0 + 4 \times (-6) \end{pmatrix}$$

$$= \begin{pmatrix} 26 & -18 \\ 18 & -24 \end{pmatrix}$$

In general, when a matrix of dimension (m by n) is multiplied by a matrix of dimension (n by q), the resulting matrix is one of dimension (m by q).

Although the laws of matrices are so formulated that they follow most of the laws which govern the algebra of numbers, frequently in the multiplication of matrices

$$A \times B \neq B \times A$$

It is shown above that

$$A \times B = \begin{pmatrix} 1 & 3 \\ -2 & 4 \end{pmatrix} \times \begin{pmatrix} 5 & 0 \\ 7 & -6 \end{pmatrix}$$

$$= \begin{pmatrix} 26 & -18 \\ 18 & -24 \end{pmatrix}$$

However, $B \times A = \begin{pmatrix} 5 & 0 \\ 7 & -6 \end{pmatrix} \times \begin{pmatrix} 1 & 3 \\ -2 & 4 \end{pmatrix}$

$$= \begin{pmatrix} 5 \times 1 + 0 \times (-2) & 5 \times 3 + 0 \times 4 \\ 7 \times 1 + (-6) \times (-2) & 7 \times 3 + (-6) \times 4 \end{pmatrix}$$

i.e. $B \times A = \begin{pmatrix} 5 & 15 \\ 19 & -3 \end{pmatrix}$

That is, $A \times B \neq B \times A$ in this case. The results are said to be *non-commutative* (i.e. they are not in agreement).

Worked problems on the addition, subtraction and multiplication of second order matrices

---

Problem 1. If $A = \begin{pmatrix} 1 & 4 \\ -3 & 2 \end{pmatrix}$, $B = \begin{pmatrix} 5 & -1 \\ 0 & 1 \end{pmatrix}$ and $C = \begin{pmatrix} 3 & -4 \\ 7 & 2 \end{pmatrix}$
determine the single matrix for (a) $A + C$, (b) $A - C$ and (c) $A + B - C$

(a) $A + C = \begin{pmatrix} 1 & 4 \\ -3 & 2 \end{pmatrix} + \begin{pmatrix} 3 & -4 \\ 7 & 2 \end{pmatrix} = \begin{pmatrix} 1 + 3 & 4 + (-4) \\ (-3) + 7 & 2 + 2 \end{pmatrix}$

$$= \begin{pmatrix} 4 & 0 \\ 4 & 4 \end{pmatrix}$$

(b) $A - C = \begin{pmatrix} 1 & 4 \\ -3 & 2 \end{pmatrix} - \begin{pmatrix} 3 & -4 \\ 7 & 2 \end{pmatrix} = \begin{pmatrix} 1 - 3 & 4 - (-4) \\ (-3) - 7 & 2 - 2 \end{pmatrix}$

$$= \begin{pmatrix} -2 & 8 \\ -10 & 0 \end{pmatrix}$$

(c) From part (b), $A - C = \begin{pmatrix} -2 & 8 \\ -10 & 0 \end{pmatrix}$

Hence $A + B - C = \begin{pmatrix} -2 & 8 \\ -10 & 0 \end{pmatrix} + \begin{pmatrix} 5 & -1 \\ 0 & 1 \end{pmatrix}$

$$= \begin{pmatrix} -2 + 5 & 8 + (-1) \\ -10 + 0 & 0 + 1 \end{pmatrix}$$

$$= \begin{pmatrix} 3 & 7 \\ -10 & 1 \end{pmatrix}$$

Problem 2. Determine the single matrix for (a) $A \cdot C$ and (b) $A \cdot B$, where $A = \begin{pmatrix} 2 & 4 \\ 1 & -3 \end{pmatrix}$, $B = \begin{pmatrix} 3 & -7 \\ 4 & -5 \end{pmatrix}$ and $C = \begin{pmatrix} 2 \\ -5 \end{pmatrix}$

(a) $A \cdot C = \begin{pmatrix} 2 & 4 \\ 1 & -3 \end{pmatrix} \times \begin{pmatrix} 2 \\ -5 \end{pmatrix} = \begin{pmatrix} 2 \times 2 + & 4 \times (-5) \\ 1 \times 2 + (-3) \times (-5) \end{pmatrix}$

$$= \begin{pmatrix} 4 - 20 \\ 2 + 15 \end{pmatrix} = \begin{pmatrix} -16 \\ 17 \end{pmatrix}$$

(b) $A \cdot B = \begin{pmatrix} 2 & 4 \\ 1 & -3 \end{pmatrix} \times \begin{pmatrix} 3 & -7 \\ 4 & -5 \end{pmatrix}$

$$= \begin{pmatrix} [2 \times 3 + 4 \times 4] & [2 \times (-7) + 4 \times (-5)] \\ [1 \times 3 + (-3) \times 4] & [1 \times (-7) + (-3) \times (-5)] \end{pmatrix}$$

$$= \begin{pmatrix} 6 + 16 & -14 + (-20) \\ 3 + (-12) & -7 + 15 \end{pmatrix}$$

$$= \begin{pmatrix} 22 & -34 \\ -9 & 8 \end{pmatrix}$$

Further problems on the addition, subtraction and multiplication of matrices may be found in Section 7.5 (Problems 1 to 10), page 116.

7.3   The unit matrix

A unit matrix is one in which the values of the elements in the leading diagonal ($\backslash$) are 1, the remaining elements being 0. Thus, a 2 by 2 unit matrix is $\begin{pmatrix} 1 & 0 \\ 0 & 1 \end{pmatrix}$, and is usually denoted by the symbol I.

If A is a square matrix and I the unit matrix, then $A \times I = I \times A$, that is, this is one case in matrices where the law $A \times B = B \times A$ of the algebra of numbers is true. The unit matrix is analogous to the number 1 in ordinary algebra.

7.4   Second order determinants

The solution of the linear simultaneous equations:

$$a_1 x + b_1 y + c_1 = 0 \tag{5}$$

$$a_2 x + b_2 y + c_2 = 0 \tag{6}$$

may be found by the elimination method of solving simultaneous equations.

To eliminate y:

$$\text{Equation (1)} \times b_2: a_1b_2x + b_1b_2y + c_1b_2 = 0$$
$$\text{Equation (2)} \times b_1: a_2b_1x + b_1b_2y + c_2b_1 = 0$$

Subtracting: $\quad (a_1b_2 - a_2b_1)x + (c_1b_2 - c_2b_1) = 0$

Thus, $\quad x = \dfrac{-(c_1b_2 - c_2b_1)}{a_1b_2 - a_2b_1}$

i.e. $\quad x = \dfrac{(b_1c_2 - b_2c_1)}{a_1b_2 - a_2b_1}$ $\qquad\qquad$ (7)

Similarly, to eliminate x:

$$\text{Equation (1)} \times a_2: a_1a_2x + a_2b_1y + a_2c_1 = 0$$
$$\text{Equation (2)} \times a_1: a_1a_2x + a_1b_2y + a_1c_2 = 0$$

Subtracting: $\quad (a_2b_1 - a_1b_2)y + (a_2c_1 - a_1c_2) = 0$

Thus, $\quad y = \dfrac{-(a_2c_1 - a_1c_2)}{(a_2b_1 - a_1b_2)} = \dfrac{(a_1c_2 - a_2c_1)}{(a_2b_1 - a_1b_2)} = \dfrac{(a_1c_2 - a_2c_1)}{-(a_1b_2 - a_2b_1)}$

i.e. $f - y = \dfrac{(a_1c_2 - a_2c_1)}{(a_1b_2 - a_2b_1)}$ $\qquad\qquad$ (8)

Equations (7) and (8) can be written in the form

$$\frac{x}{b_1c_2 - b_2c_1} = \frac{-y}{a_1c_2 - a_2c_1} = \frac{1}{a_1b_2 - a_2b_1} \qquad\qquad (9)$$

The denominators of equation (9) are all of the general form

$$pq - rs$$

Although as stated in Section 7.2 a matrix does not have a single numerical value and cannot be simplified to a particular answer, coefficients written in this form may be expressed as a special matrix, denoted by an array within vertical lines rather than brackets. In this case:

$$\begin{vmatrix} a & b \\ c & d \end{vmatrix} = ad - bc, \text{ and is called a second order } \mathbf{determinant}$$

It is shown in Chapter 8 that determinants can be used to solve linear simultaneous equations such as those given in equations (5) and (6) above.

Worked problem on second order determinants

Problem 1. Evaluate the determinants:

(a) $\begin{vmatrix} 3 & -1 \\ 4 & 2 \end{vmatrix}$ and (b) $\begin{vmatrix} a & -2b \\ 2a & -3b \end{vmatrix}$

By the definition of a determinant, $\begin{vmatrix} a & b \\ c & d \end{vmatrix} = ad - bc$, hence

(a) $\begin{vmatrix} 3 & -1 \\ 4 & 2 \end{vmatrix} = (3 \times 2) - ((-1) \times 4) = 6 + 4 = \mathbf{10}$

(b) $\begin{vmatrix} a & -2b \\ 2a & -3b \end{vmatrix} = (a \times (-3b)) - ((-2b) \times 2a) = -3ab + 4ab$
$$= \mathbf{ab}$$

Further problems on second order determinants may be found in Section 7.5 following (Problems 11 to 15), page 118.

7.5   Further problems

Addition, subtraction and multiplication of second order matrices

In Problems 1 to 5, matrices A, B, C and D are given by

$$A = \begin{pmatrix} 1 & 4 \\ -3 & 2 \end{pmatrix}, B = \begin{pmatrix} 2 & 7 \\ -1 & 0 \end{pmatrix}$$

$$C = \begin{pmatrix} 5 & -1 \\ 0 & 1 \end{pmatrix} \text{ and } D = \begin{pmatrix} 3 & -4 \\ 7 & 2 \end{pmatrix}$$

Determine the single matrix for the expressions given.

1. (a) $A + B$ $\left[\begin{pmatrix} 3 & 11 \\ -4 & 2 \end{pmatrix} \right]$

    (b) $A + C$ $\left[\begin{pmatrix} 6 & 3 \\ -3 & 3 \end{pmatrix} \right]$

2. (a) $C + D$ $\left[\begin{pmatrix} 8 & -5 \\ 7 & 3 \end{pmatrix} \right]$

    (b) $B + D$ $\left[\begin{pmatrix} 5 & 3 \\ 6 & 2 \end{pmatrix} \right]$

3. (a) $B - A$ $\left[\begin{pmatrix} 1 & 3 \\ 2 & -2 \end{pmatrix} \right]$

   (b) $D - B$ $\left[\begin{pmatrix} 1 & -11 \\ 8 & 2 \end{pmatrix} \right]$

4. (a) $C - A$ $\left[\begin{pmatrix} 4 & -5 \\ 3 & -1 \end{pmatrix} \right]$

   (b) $B - C$ $\left[\begin{pmatrix} -3 & 8 \\ -1 & -1 \end{pmatrix} \right]$

5. (a) $A + B + C$ $\left[\begin{pmatrix} 8 & 10 \\ -4 & 3 \end{pmatrix} \right]$

   (b) $A - B + C - D$ $\left[\begin{pmatrix} 1 & 0 \\ -9 & 1 \end{pmatrix} \right]$

In Problems 6 to 10, matrices A, B, C, D and E are given by

$$A = \begin{pmatrix} 3 & 1 \\ -2 & 4 \end{pmatrix}, B = \begin{pmatrix} 2 & -5 \\ 0 & 1 \end{pmatrix}, C = \begin{pmatrix} -1 & 6 \\ 3 & 0 \end{pmatrix},$$

$$D = \begin{pmatrix} 2 \\ 3 \end{pmatrix} \text{ and } E = \begin{pmatrix} -1 \\ 4 \end{pmatrix}$$

Determine the single matrix for the expressions given.

6. (a) $A \cdot D$ $\left[\begin{pmatrix} 9 \\ 8 \end{pmatrix} \right]$

   (b) $B \cdot E$ $\left[\begin{pmatrix} -22 \\ 4 \end{pmatrix} \right]$

7. (a) $C \cdot D$ $\left[\begin{pmatrix} 16 \\ 6 \end{pmatrix} \right]$

   (b) $A \cdot E$ $\left[\begin{pmatrix} 1 \\ 18 \end{pmatrix} \right]$

8. (a) $A \cdot C$ $\left[\begin{pmatrix} 0 & 18 \\ 14 & -12 \end{pmatrix} \right]$

   (b) $C \cdot B$ $\left[\begin{pmatrix} -2 & 12 \\ 6 & -15 \end{pmatrix} \right]$

9. (a) $A \cdot B$ $\left[\begin{pmatrix} 6 & -14 \\ -4 & 14 \end{pmatrix}\right.$

$\left.\begin{pmatrix} 16 & -18 \\ -2 & 4 \end{pmatrix}\right]$ (Note that $A \cdot B \neq B \cdot A$)

(b) $B \cdot A$

10. (a) $B \cdot C$ $\left[\begin{pmatrix} -17 & 12 \\ 3 & 0 \end{pmatrix}\right.$

$\left.\begin{pmatrix} -15 & 23 \\ 9 & 3 \end{pmatrix}\right]$

(b) $C \cdot A$

Second order determinants

In Problems 11 to 15 evaluate the determinants given.

11. (a) $\begin{vmatrix} 2 & 3 \\ 4 & 5 \end{vmatrix}$ (b) $\begin{vmatrix} -1 & -1 \\ 7 & 2 \end{vmatrix}$ (a) [−2] (b) [5]

12. (a) $\begin{vmatrix} 3 & -1 \\ 4 & 7 \end{vmatrix}$ (b) $\begin{vmatrix} 5 & -2 \\ 3 & 1 \end{vmatrix}$ (a) [25] (b) [11]

13. (a) $\begin{vmatrix} -2 & 4 \\ 3 & 1 \end{vmatrix}$ (b) $\begin{vmatrix} 1 & -4 \\ 5 & 1 \end{vmatrix}$ (a) [−14] (b) [21]

14. (a) $\begin{vmatrix} x & 2x \\ -3x & 5 \end{vmatrix}$ (b) $\begin{vmatrix} y^2 & 3y \\ 4y^2 & -2y \end{vmatrix}$

(a) $[5x + 6x^2]$ (b) $[-14y^3]$

15. (a) $\begin{vmatrix} c & -2b \\ 4c & -3b \end{vmatrix}$ (b) $\begin{vmatrix} a & c \\ 2a & 4c \end{vmatrix}$ (a) [5bc] (b) [2ac]

Chapter 8

The solution of simultaneous equations having two unknowns using determinants and matrices

8.1 The solution of simultaneous equations having two unknowns using determinants

When introducing determinants in Section 7.4, the simultaneous equations

$$a_1 x + b_1 y + c_1 = 0 \tag{1}$$
$$a_2 x + b_2 y + c_2 = 0 \tag{2}$$

are solved using the elimination method of solving simultaneous equations, and it is shown that

$$\frac{x}{b_1 c_2 - b_2 c_1} = \frac{-y}{a_1 c_2 - a_2 c_1} = \frac{1}{a_1 b_2 - a_2 b_1} \tag{3}$$

It is also stated that the denominators of equation (3) are all of the general form

$$pq - rs$$

This algebraic expression is denoted by a special matrix having its array within vertical lines and is called a **determinant**. Thus

$$\begin{vmatrix} a & b \\ c & d \end{vmatrix} = ad - bc$$

and is called a second order determinant.

The denominators of equation (3) can be written in determinant form, giving

$$\frac{x}{\begin{vmatrix} b_1 & c_1 \\ b_2 & c_2 \end{vmatrix}} = \frac{-y}{\begin{vmatrix} a_1 & c_1 \\ a_2 & c_2 \end{vmatrix}} = \frac{1}{\begin{vmatrix} a_1 & b_1 \\ a_2 & b_2 \end{vmatrix}} \tag{4}$$

This expression is used to solve simultaneous equations by determinants and can be remembered by the 'cover-up' rule. In this rule:

(i) the equations are written in the form $a_1 x + b_1 y + c_1 = 0$;
(ii) if equation (4) is written in the form

$$\frac{x}{|D_1|} = \frac{-y}{|D_2|} = \frac{1}{|D|}, \text{ then}$$

(iii) $|D_1|$ is obtained by covering up the x column and writing down the remaining coefficients in determinant form in positions corresponding to the positions they occupy in the equations.

(iv) $|D_2|$ is obtained by covering up the y column and treating the coefficients as in (iii) above.

(v) $|D|$ is obtained by covering up the constants column and treating the coefficients as in (iii) above.

For example, to solve the simultaneous equations

$$2x + 3y = 11 \tag{5}$$
$$4x + 2y = 10 \tag{6}$$

by using determinants:

(i)   the equations are written as

$$2x + 3y - 11 = 0$$
$$4x + 2y - 10 = 0$$

(ii) $\dfrac{x}{|D_1|} = \dfrac{-y}{|D_2|} = \dfrac{1}{|D|}$

(iii) $|D_1| = \begin{vmatrix} 3 & -11 \\ 2 & -10 \end{vmatrix}$, obtained by covering up the x column in (i) above.

(iv) $|D_2| = \begin{vmatrix} 2 & -11 \\ 4 & -10 \end{vmatrix}$, obtained by covering up the y column in (i) above.

(v) $|D| = \begin{vmatrix} 2 & 3 \\ 4 & 2 \end{vmatrix}$, obtained by covering up the constants column in (i) above.

Thus

$$\frac{x}{\begin{vmatrix} 3 & -11 \\ 2 & -10 \end{vmatrix}} = \frac{-y}{\begin{vmatrix} 2 & -11 \\ 4 & -10 \end{vmatrix}} = \frac{1}{\begin{vmatrix} 2 & 3 \\ 4 & 2 \end{vmatrix}}$$

i.e.

$$\frac{x}{3 \times (-10) - 2 \times (-11)} = \frac{-y}{2 \times (-10) - 4 \times (-11)} = \frac{1}{2 \times 2 - 4 \times 3}$$

$$\frac{x}{-8} = \frac{-y}{24} = \frac{1}{-8}$$

giving $x = \dfrac{-8}{-8} = 1$ and $-y = \dfrac{24}{-8}$, i.e. $y = 3$

Checking in the original equations:

L.H.S. of equation (5) = $2 \times 1 + 3 \times 3 = 11$ = R.H.S.

L.H.S. of equation (6) = $4 \times 1 + 2 \times 3 = 10$ = R.H.S.

Hence, $x = 1$, $y = 3$ is the correct solution.

If, in a determinant of the form $D = \begin{vmatrix} a_1 & b_1 \\ a_2 & b_2 \end{vmatrix}$, $a_1 b_2 = a_2 b_1$, then $a_1 b_2 - a_2 b_1 = 0$, i.e. $D = 0$. This means that in the simultaneous equations on which the determinant is based (say of the form $ax + by = c$), a_1 and b_1 are each multiplied by the same constant to give a_2 and b_2. So essentially there is only one equation with two unknown quantities which is not capable of solution. An example is the determinant $\begin{vmatrix} 2 & 1 \\ 10 & 5 \end{vmatrix}$, which arises when trying to solve the equations, say

$$2x + y = 3$$

and $\quad 10x + 5y = 15$

and in this case $D = 2 \times 5 - 10 \times 1 = 0$, i.e. the equations cannot be solved. The matrix of the coefficients of x and y, i.e. $\begin{pmatrix} 2 & 1 \\ 10 & 5 \end{pmatrix}$, is called a **singular matrix** when the determinant of the matrix is equal to 0.

Worked problems on the solution of simultaneous equations having two unknowns using determinants

Problem 1. Solve the simultaneous equations:

$$\tfrac{3}{2}p - 2q = \tfrac{1}{2} \tag{7}$$
$$p + \tfrac{3}{2}q = 6 \tag{8}$$

by using determinants.

(i) Writing the equations in the form $ax + by + c = 0$ gives
$$\tfrac{3}{2}p - 2q - \tfrac{1}{2} = 0$$
$$p + \tfrac{3}{2}q - 6 = 0$$

(ii) $\dfrac{p}{|D_1|} = \dfrac{-q}{|D_2|} = \dfrac{1}{|D|}$ (note the signs are $+$, $-$, $+$)

(iii) Covering up the p column gives $|D_1| = \begin{vmatrix} -2 & -\tfrac{1}{2} \\ \tfrac{3}{2} & -6 \end{vmatrix}$

(iv) Covering up the q column gives $|D_2| = \begin{vmatrix} \tfrac{3}{2} & -\tfrac{1}{2} \\ 1 & -6 \end{vmatrix}$

(v) Covering up the constants column gives $|D| = \begin{vmatrix} \tfrac{3}{2} & -2 \\ 1 & \tfrac{3}{2} \end{vmatrix}$

Thus

$$\frac{p}{\begin{vmatrix} -2 & -\frac{1}{2} \\ \frac{3}{2} & -6 \end{vmatrix}} = \frac{-q}{\begin{vmatrix} \frac{3}{2} & -\frac{1}{2} \\ 1 & -6 \end{vmatrix}} = \frac{1}{\begin{vmatrix} \frac{3}{2} & -2 \\ 1 & \frac{3}{2} \end{vmatrix}}$$

i.e. $\dfrac{p}{(-2) \times (-6) - (\frac{3}{2}) \times (-\frac{1}{2})} = \dfrac{-q}{(\frac{3}{2}) \times (-6) - (1) \times (-\frac{1}{2})}$

$$= \frac{1}{(\frac{3}{2}) \times (\frac{3}{2}) - (1) \times (-2)}$$

so $\qquad\qquad \dfrac{p}{12\frac{3}{4}} = \dfrac{-q}{-8\frac{1}{2}} = \dfrac{1}{4\frac{1}{4}}$

Hence $p = \dfrac{12\frac{3}{4}}{4\frac{1}{4}} = 3$ and $q = \dfrac{8\frac{1}{2}}{4\frac{1}{4}} = 2$

Checking in the original equations:

L.H.S. of equation (7) $= \frac{3}{2} \times 3 - 2 \times 2 = \frac{1}{2} =$ R.H.S.

L.H.S. of equation (8) $= 3 + \frac{3}{2} \times 2 = 6 =$ R.H.S.

Hence $p = 3$, $q = 2$ is the correct solution.

Problem 2. Use determinants to solve the simultaneous equations:

$$-0.5f + 0.4g = 0.7 \tag{9}$$
$$1.2f - 0.3g = 3.6 \tag{10}$$

Writing the equations in the form $ax + by + c = 0$ gives

$-0.5f + 0.4g - 0.7 = 0$
$1.2f - 0.3g - 3.6 = 0$ $\quad$ Hence $\quad \dfrac{f}{|D_1|} = \dfrac{-g}{|D_2|} = \dfrac{1}{|D|}$

Covering up the f column gives

$$|D_1| = \begin{vmatrix} 0.4 & -0.7 \\ -0.3 & -3.6 \end{vmatrix}$$

$\qquad = (0.4) \times (-3.6) - (-0.3) \times (-0.7) = -1.44 - 0.21 = -1.65$

Covering up the g column gives

$$|D_2| = \begin{vmatrix} -0.5 & -0.7 \\ 1.2 & -3.6 \end{vmatrix}$$

$\qquad = (-0.5) \times (-3.6) - (1.2) \times (-0.7) = 1.8 - (-0.84) = 2.64$

Covering up the constants column gives

$$|D| = \begin{vmatrix} -0.5 & 0.4 \\ 1.2 & -0.3 \end{vmatrix}$$

$\qquad = (-0.5) \times (-0.3) - (1.2) \times (0.4) = 0.15 - 0.48 = -0.33$

Hence $\dfrac{f}{-1.65} = \dfrac{-g}{2.64} = \dfrac{1}{-0.33}$

i.e. $f = \dfrac{-1.65}{-0.33} = 5$ and $g = \dfrac{-2.64}{-0.33} = 8$

Checking:

L.H.S. of equation (9) $= -0.5 \times 5 + 0.4 \times 8 = 0.7 =$ R.H.S.

L.H.S. of equation (10) $= 1.2 \times 5 - 0.3 \times 8 = 3.6 =$ R.H.S.

Thus $f = 5$, $g = 8$ is the correct solution.

Problem 3. Use determinants to solve the simultaneous equations:

$$\frac{10}{a} - \frac{4}{b} = 3 \tag{11}$$

$$\frac{6}{a} + \frac{8}{b} = 7 \tag{12}$$

Let $x = \dfrac{1}{a}$ and $y = \dfrac{1}{b}$, then

$10x - 4y = 3$

$6x + 8y = 7$

Writing in the $ax + by + c = 0$ form gives

$10x - 4y - 3 = 0$

$6x + 8y - 7 = 0$

Applying the cover-up rule gives

$$\frac{x}{\begin{vmatrix} -4 & -3 \\ 8 & -7 \end{vmatrix}} = \frac{-y}{\begin{vmatrix} 10 & -3 \\ 6 & -7 \end{vmatrix}} = \frac{1}{\begin{vmatrix} 10 & -4 \\ 6 & 8 \end{vmatrix}}$$

i.e. $\dfrac{x}{28 + 24} = \dfrac{-y}{-70 + 18} = \dfrac{1}{80 + 24}$ and $x = \frac{52}{104} = \frac{1}{2}$, $y = \frac{52}{104} = \frac{1}{2}$

Since $x = \dfrac{1}{a}$, $a = 2$, and since $y = \dfrac{1}{b}$, $b = 2$

Checking:

L.H.S. of equation (11) $= \frac{10}{2} - \frac{4}{2} = 3 =$ R.H.S.

L.H.S. of equation (12) $= \frac{6}{2} + \frac{8}{2} = 7 =$ R.H.S.

Hence $a = 2$, $b = 2$ is the correct solution.

Problem 4. The forces acting on a bolt are resolved horizontally and vertically, giving the simultaneous equations shown below. Use determinants to find the values of F_1 and F_2, correct to three significant figures:

$$3.4F_1 - 0.83F_2 = 3.9 \tag{13}$$

$$0.7F_1 + 1.47F_2 = -2.05 \tag{14}$$

Writing the equations in the $ax + by + c = 0$ form gives

$$3.4F_1 - 0.83F_2 - 3.9 = 0$$

$$0.7F_1 + 1.47F_2 + 2.05 = 0$$

Applying the cover-up rule gives

$$\frac{F_1}{\begin{vmatrix} -0.83 & -3.9 \\ 1.47 & 2.05 \end{vmatrix}} = \frac{-F_2}{\begin{vmatrix} 3.4 & -3.9 \\ 0.7 & 2.05 \end{vmatrix}} = \frac{1}{\begin{vmatrix} 3.4 & -0.83 \\ 0.7 & 1.47 \end{vmatrix}}$$

Hence

$$\frac{F_1}{(-0.83) \times 2.05 - 1.47 \times (-3.9)} = \frac{-F_2}{3.4 \times 2.05 - 0.7 \times (-3.9)}$$

$$= \frac{1}{3.4 \times 1.47 - 0.7 \times (-0.83)}$$

that is, $\dfrac{F_1}{-1.701\,5 + 5.733} = \dfrac{-F_2}{6.970 + 2.730} = \dfrac{1}{4.998 + 0.581}$

i.e. $\dfrac{F_1}{4.032} = \dfrac{-F_2}{9.700} = \dfrac{1}{5.579}$

Thus $F_1 = \dfrac{4.032}{5.579} = 0.723$, correct to three significant figures,

and $F_2 = \dfrac{-9.700}{5.579} = -1.74$, correct to three significant figures.

Checking:

L.H.S. of equation (13) = $3.4 \times 0.723 - 0.83 \times (-1.74)$
    = 3.90, correct to three significant figures.
L.H.S. of equation (14) = $0.7 \times 0.723 + 1.47 \times (-1.74)$
    = -2.05, correct to three significant figures.

Hence $F_1 = 0.723$, $F_2 = -1.74$ is the correct solution.

Further problems on solving simultaneous equations having two unknowns using determinants may be found in Section 8.4 (Problems 1 to 15), page 131.

8.2   The inverse of a matrix

The inverse or reciprocal of matrix A is the matrix A^{-1}, such that $A \cdot A^{-1} = I = A^{-1} \cdot A$, where I is the unit matrix, introduced in Chapter 7, Section 7.3.   The process of inverting a matrix makes division possible. If three matrices, A, B and X, are such that

$$A \cdot X = B$$

then $$X = \frac{B}{A} = A^{-1} \cdot B$$

Let the inverse of matrix A be $A^{-1} = \begin{pmatrix} a & b \\ c & d \end{pmatrix}$, and let matrix A be, say, $\begin{pmatrix} 2 & 3 \\ -1 & 1 \end{pmatrix}$. By the definition of the inverse of a matrix, $\begin{pmatrix} 2 & 3 \\ -1 & 1 \end{pmatrix} \times \begin{pmatrix} a & b \\ c & d \end{pmatrix} = \begin{pmatrix} 1 & 0 \\ 0 & 1 \end{pmatrix}$, the unit matrix. Multiplying the matrices on the left-hand side gives $\begin{pmatrix} 2a + 3c & 2b + 3d \\ -a + c & -b + d \end{pmatrix} = \begin{pmatrix} 1 & 0 \\ 0 & 1 \end{pmatrix}$ (15)

Since these two matrices are equal to one another, the corresponding elements are equal to one another, hence

$$-a + c = 0, \text{ that is, } a = c$$
$$2b + 3d = 0, \text{ that is, } b = -\tfrac{3}{2}d$$

Substituting for a and b in equation (15) above gives

$$\begin{pmatrix} 5c & 0 \\ 0 & \dfrac{5d}{2} \end{pmatrix} = \begin{pmatrix} 1 & 0 \\ 0 & 1 \end{pmatrix}$$

Thus, $5c = 1$, that is, $c = \tfrac{1}{5}$

$$\frac{5d}{2} = 1, \text{ that is, } d = \tfrac{2}{5}$$

Since $a = c$, $a = \tfrac{1}{5}$, and since $b = -\tfrac{3}{2}d$, $b = -\tfrac{3}{5}$. Thus, the inverse of matrix

$$\begin{pmatrix} 2 & 3 \\ -1 & 1 \end{pmatrix} \text{ is } \begin{pmatrix} \tfrac{1}{5} & -\tfrac{3}{5} \\ \tfrac{1}{5} & \tfrac{2}{5} \end{pmatrix}$$

There is an alternative method of finding the inverse of a matrix. If the inverses of many matrices are determined and the inverses of the matrices are compared with the matrices, a relationship is seen to exist between matrices and their inverses. This relationship for a matrix of the form $\begin{pmatrix} a & b \\ c & d \end{pmatrix}$

is that in the inverse:

  (i)   the position of the a and d elements are interchanged,
  (ii) the sign of both the b and c elements is changed, and
  (iii) the matrix is multiplied by $\dfrac{1}{ad-bc}$, i.e. the reciprocal of the determinant of the matrix.

Thus, the inverse of matrix $\begin{pmatrix} a & b \\ c & d \end{pmatrix}$ is $\dfrac{1}{ad-bc}\begin{pmatrix} d & -b \\ -c & a \end{pmatrix}$. For the matrix $\begin{pmatrix} 2 & 3 \\ -1 & 1 \end{pmatrix}$ considered previously the inverse is

$$\frac{1}{2\times 1 - 3\times(-1)}\begin{pmatrix} 1 & -3 \\ 1 & 2 \end{pmatrix} = \tfrac{1}{5}\begin{pmatrix} 1 & -3 \\ 1 & 2 \end{pmatrix} = \begin{pmatrix} \tfrac{1}{5} & -\tfrac{3}{5} \\ \tfrac{1}{5} & \tfrac{2}{5} \end{pmatrix}$$

as shown previously.

Worked problem on the inverse of a matrix

Problem 1. Determine the inverse of the matrix $A = \begin{pmatrix} 5 & -3 \\ -2 & 1 \end{pmatrix}$

Let the inverse matrix be $A^{-1} = \begin{pmatrix} a & b \\ c & d \end{pmatrix}$

Since $A \cdot A^{-1} = I$, the unit matrix, then

$$\begin{pmatrix} 5 & -3 \\ -2 & 1 \end{pmatrix} \times \begin{pmatrix} a & b \\ c & d \end{pmatrix} = \begin{pmatrix} 1 & 0 \\ 0 & 1 \end{pmatrix}$$

Applying the multiplication law to the left-hand side gives

$$\begin{pmatrix} 5a - 3c & 5b - 3d \\ -2a + c & -2b + d \end{pmatrix} = \begin{pmatrix} 1 & 0 \\ 0 & 1 \end{pmatrix} \tag{16}$$

Equating corresponding elements gives

$$-2a + c = 0, \text{ i.e. } a = \frac{c}{2}, \text{ and } 5b - 3d = 0, \text{ i.e. } b = \tfrac{3}{5}d$$

Substituting in equation (16): $\begin{pmatrix} -\dfrac{c}{2} & 0 \\ 0 & -\dfrac{d}{5} \end{pmatrix} = \begin{pmatrix} 1 & 0 \\ 0 & 1 \end{pmatrix}$

i.e. $-\dfrac{c}{2} = 1$, $c = -2$ and $-\dfrac{d}{5} = 1$, $d = -5$

Since $a = \dfrac{c}{2}$, $a = -1$ and since $b = \tfrac{3}{5}d$, $b = -3$

Thus the inverse matrix of $\begin{pmatrix} 5 & -3 \\ -2 & 1 \end{pmatrix}$ is $\begin{pmatrix} -1 & -3 \\ -2 & -5 \end{pmatrix}$

The solution may be checked, using $A \cdot A^{-1} = I$. Thus

$$\begin{pmatrix} 5 & -3 \\ -2 & 1 \end{pmatrix} \times \begin{pmatrix} -1 & -3 \\ -2 & -5 \end{pmatrix} = \begin{pmatrix} -5+6 & -15+15 \\ 2-2 & 6-5 \end{pmatrix}$$

$$= \begin{pmatrix} 1 & 0 \\ 0 & 1 \end{pmatrix}, \text{ the inverse matrix.}$$

Hence the solution $\begin{pmatrix} -1 & -3 \\ -2 & -5 \end{pmatrix}$ is correct.

Alternatively, the relationship that the inverse of matrix

$\begin{pmatrix} a & b \\ c & d \end{pmatrix}$ is $\dfrac{1}{ad-bc} \begin{pmatrix} d & -b \\ -c & a \end{pmatrix}$ could have been applied.

The inverse of $\begin{pmatrix} 5 & -3 \\ -2 & 1 \end{pmatrix} = \dfrac{1}{5 \times 1 - (-2) \times (-3)} \begin{pmatrix} 1 & 3 \\ 2 & 5 \end{pmatrix}$

$$= \dfrac{1}{-1} \begin{pmatrix} 1 & 3 \\ 2 & 5 \end{pmatrix}$$

Applying the law for scalar multiplication gives $\begin{pmatrix} -1 & -3 \\ -2 & -5 \end{pmatrix}$,

as obtained previously. The alternative method of applying a formula is the easiest method of determining the inverse of a matrix.

Further problems on the inverse of a matrix may be found in Section 8.4 (Problems 16 to 20), page 133.

8.3 The solution of simultaneous equations having two unknowns using matrices

Matrices may be used to solve linear simultaneous equations. For equations having two unknown quantities there is no advantage in using a matrix method. However, for equations having three or more unknown quantities, solution by a matrix method can usually be performed more quickly and accurately.

Two linear simultaneous equations, such as

$$2x + 3y = 4 \tag{17}$$
$$x - 5y = 6 \tag{18}$$

may be written in matrix form as

$$\begin{pmatrix} 2 & 3 \\ 1 & -5 \end{pmatrix} \begin{pmatrix} x \\ y \end{pmatrix} = \begin{pmatrix} 4 \\ 6 \end{pmatrix} \tag{19}$$

The inverse of the matrix $\begin{pmatrix} 2 & 3 \\ 1 & -5 \end{pmatrix}$ is obtained as shown in Section 8.2

and is $\begin{pmatrix} \frac{5}{13} & \frac{3}{13} \\ \frac{1}{13} & -\frac{2}{13} \end{pmatrix}$. Multiplying both sides of equation (19) by this inversed matrix gives

$$\begin{pmatrix} 1 & 0 \\ 0 & 1 \end{pmatrix}\begin{pmatrix} x \\ y \end{pmatrix} = \begin{pmatrix} \frac{5}{13} & \frac{3}{13} \\ \frac{1}{13} & -\frac{2}{13} \end{pmatrix}\begin{pmatrix} 4 \\ 6 \end{pmatrix} = \begin{pmatrix} \frac{20}{13} + \frac{18}{13} \\ \frac{4}{13} - \frac{12}{13} \end{pmatrix}$$

$$= \begin{pmatrix} \frac{38}{13} \\ -\frac{8}{13} \end{pmatrix}$$

i.e.
$$\begin{pmatrix} x \\ y \end{pmatrix} = \begin{pmatrix} \frac{38}{13} \\ -\frac{8}{13} \end{pmatrix}$$

Equating corresponding elements gives

$$x = \tfrac{38}{13} \text{ and } y = -\tfrac{8}{13}$$

Check: L.H.S. of equation (17) is $2 \times \frac{38}{13} + 3 \times (-\frac{8}{13}) = 4 =$ R.H.S.

L.H.S. of equation (18) is $\frac{38}{13} - 5(-\frac{8}{13}) = 6 =$ R.H.S.

Hence $x = \frac{38}{13}$, $y = -\frac{8}{13}$ is the correct solution.

Summary

To solve linear simultaneous equations with two unknown quantities by using matrices:

(i)   write the equations in the standard form

$$ax + by = c$$
$$dx + ey = f$$

(ii)   write this in matrix form, i.e.

$$\begin{pmatrix} a & b \\ d & e \end{pmatrix}\begin{pmatrix} x \\ y \end{pmatrix} = \begin{pmatrix} c \\ f \end{pmatrix}$$

(iii)   determine the inverse of matrix $\begin{pmatrix} a & b \\ d & e \end{pmatrix}$

(iv)   multiply each side of (ii) by the inversed matrix, and express in the form

$$\begin{pmatrix} x \\ y \end{pmatrix} = \begin{pmatrix} g \\ h \end{pmatrix}$$

(v)   solve for x and y by equating corresponding elements, and

(vi)   check the solution in the original equations.

Worked problems on solving simultaneous equations having two unknowns using matrices

Problem 1. Use matrices to solve the simultaneous equations:

$$4a - 3b = 18 \tag{20}$$

$$a + 2b = -1 \tag{21}$$

With reference to the above summary:

(i)   The equations are in the standard form.

(ii)   The matrices are $\begin{pmatrix} 4 & -3 \\ 1 & 2 \end{pmatrix}\begin{pmatrix} a \\ b \end{pmatrix} = \begin{pmatrix} 18 \\ -1 \end{pmatrix}$

(iii)   The inverse matrix of

$$\begin{pmatrix} 4 & -3 \\ 1 & 2 \end{pmatrix} \text{ is } \frac{1}{4 \times 2 - (-3) \times 1}\begin{pmatrix} 2 & 3 \\ -1 & 4 \end{pmatrix}$$

i.e. $\frac{1}{11}\begin{pmatrix} 2 & 3 \\ -1 & 4 \end{pmatrix}$, that is, $\begin{pmatrix} \frac{2}{11} & \frac{3}{11} \\ -\frac{1}{11} & \frac{4}{11} \end{pmatrix}$

(iv)   Multiplying each side of (ii) by this inversed matrix gives

$$\begin{pmatrix} a \\ b \end{pmatrix} = \begin{pmatrix} \frac{2}{11} & \frac{3}{11} \\ -\frac{1}{11} & \frac{4}{11} \end{pmatrix}\begin{pmatrix} 18 \\ -1 \end{pmatrix}$$

$$\begin{pmatrix} a \\ b \end{pmatrix} = \begin{pmatrix} \frac{36}{11} + (-\frac{3}{11}) \\ -\frac{18}{11} + (-\frac{4}{11}) \end{pmatrix} = \begin{pmatrix} 3 \\ -2 \end{pmatrix}$$

(v)   Thus $a = 3$, $b = -2$

(iv)   Checking:

L.H.S. of equation (20) is $4 \times 3 - 3(-2) = 18 = $ R.H.S.

L.H.S. of equation (21) is $3 + 2(-2) = -1 = $ R.H.S.

Hence $a = 3$, $b = -2$ is the correct solution.

Problem 2. Solve the simultaneous equations:

$$\frac{3}{x} - \frac{2}{y} = 0 \tag{22}$$

$$\frac{1}{x} + \frac{4}{y} = 14 \tag{23}$$

by using matrices.

With reference to the summary:

(i) The equations may be expressed in standard form by letting $\dfrac{1}{x}$ be p and $\dfrac{1}{y}$ be q. Thus equations (22) and (23) become

$$3p - 2q = 0$$
$$p + 4q = 14$$

(ii) The matrices are $\begin{pmatrix} 3 & -2 \\ 1 & 4 \end{pmatrix}\begin{pmatrix} p \\ q \end{pmatrix} = \begin{pmatrix} 0 \\ 14 \end{pmatrix}$

(iii) The inverse of $\begin{pmatrix} 3 & -2 \\ 1 & 4 \end{pmatrix}$ is $\frac{1}{14}\begin{pmatrix} 4 & 2 \\ -1 & 3 \end{pmatrix}$

(iv) Multiplying each side of (ii) by (iii) gives

$$\begin{pmatrix} p \\ q \end{pmatrix} = \tfrac{1}{14}\begin{pmatrix} 4 & 2 \\ -1 & 3 \end{pmatrix}\begin{pmatrix} 0 \\ 14 \end{pmatrix}$$

i.e. $\begin{pmatrix} p \\ q \end{pmatrix} = \tfrac{1}{14}\begin{pmatrix} 4 \times 0 + 2 \times 14 \\ -1 \times 0 + 3 \times 14 \end{pmatrix}$

$$= \tfrac{1}{14}\begin{pmatrix} 28 \\ 42 \end{pmatrix} = \begin{pmatrix} \frac{1}{14} \times 28 \\ \frac{1}{14} \times 42 \end{pmatrix} = \begin{pmatrix} 2 \\ 3 \end{pmatrix}$$

(v) Thus $p = 2$ and $q = 3$, i.e. $x = \frac{1}{2}$, $y = \frac{1}{3}$

(vi) Checking:

L.H.S. of equation (22) is $\frac{3}{1/2} - \frac{2}{1/3} = 6 - 6 = 0 = $ R.H.S.

L.H.S. of equation (23) is $\frac{1}{1/2} + \frac{4}{1/3} = 2 + 12 = 14 = $ R.H.S.

Hence $x = \frac{1}{2}$, $y = \frac{1}{3}$ is the correct solution.

Problem 3. A force system is analysed, and by resolving the forces horizontally and vertically the following equations are obtained:

$$6F_1 - F_2 = 5 \tag{24}$$
$$5F_1 + 2F_2 = 7 \tag{25}$$

Use matrices to solve for F_1 and F_2

The matrices are $\begin{pmatrix} 6 & -1 \\ 5 & 2 \end{pmatrix}\begin{pmatrix} F_1 \\ F_2 \end{pmatrix} = \begin{pmatrix} 5 \\ 7 \end{pmatrix}$

The inverse of $\begin{pmatrix} 6 & -1 \\ 5 & 2 \end{pmatrix}$ is $\frac{1}{17}\begin{pmatrix} 2 & 1 \\ -5 & 6 \end{pmatrix}$

Hence $\begin{pmatrix} F_1 \\ F_2 \end{pmatrix} = \tfrac{1}{17}\begin{pmatrix} 2 & 1 \\ -5 & 6 \end{pmatrix}\begin{pmatrix} 5 \\ 7 \end{pmatrix} = \tfrac{1}{17}\begin{pmatrix} 10 + 7 \\ -25 + 42 \end{pmatrix} = \begin{pmatrix} 1 \\ 1 \end{pmatrix}$

Thus $F_1 = 1$, $F_2 = 1$

Checking: L.H.S. of equation (24) is $6 - 1 = 5 = $ R.H.S.

L.H.S. of equation (25) is $5 + 2 = 7 = $ R.H.S.

Thus, $F_1 = 1$ and $F_2 = 1$ is the correct solution.

Any technical problems, such as the equations formed by the resolution of vector quantities or the equations relating load and effort in machines, which were previously solved by using simultaneous equations, may be solved either by using determinants or by using matrices.

Further problems on the solution of simultaneous equations having two unknowns using matrices may be found in Section 8.4 following (Problems 21 to 30), page 133.

8.4 Further problems

The solution of simultaneous equations having two unknowns using determinants

In Problems 1 to 11 use determinants to solve the simultaneous equations given.

1. $4v_1 - 3v_2 = 18$
   $v_1 + 2v_2 = -1$ $\qquad [v_1 = 3, v_2 = -2]$

2. $3m - 2n = -4.5$
   $4m + 3n = 2.5$ $\qquad [m = -\frac{1}{2}, n = 1\frac{1}{2}]$

3. $\dfrac{a}{3} + \dfrac{b}{4} = 8$

   $\dfrac{a}{6} - \dfrac{b}{8} = 1$ $\qquad [a = 15, b = 12]$

4. $s + t = 17$

   $\dfrac{s}{5} - \dfrac{t}{7} = 1$ $\qquad [s = 10, t = 7]$

5. $\dfrac{c}{5} + \dfrac{d}{3} = \dfrac{43}{30}$

   $\dfrac{c}{9} - \dfrac{d}{6} = -\dfrac{1}{12}$ $\qquad [c = 3, d = 2\frac{1}{2}]$

6. $0.5i_1 - 1.2i_2 = -13$
   $0.8i_1 + 0.3i_2 = 12.5$ $\qquad [i_1 = 10, i_2 = 15]$

7. $1.25L_1 - 0.75L_2 = 1$
   $0.25L_1 + 1.25L_2 = 17$ $\qquad [L_1 = 8.0, L_2 = 12.0]$

8. $\dfrac{1}{2a} + \dfrac{3}{5b} = 7$

$\dfrac{4}{a} + \dfrac{1}{2b} = 13$ $[a = \frac{1}{2}, b = \frac{1}{10}]$

9. $\dfrac{3}{v_1} - \dfrac{2}{v_2} = \dfrac{1}{2}$

$\dfrac{5}{v_1} + \dfrac{3}{v_2} = \dfrac{29}{12}$ $[v_1 = 3, v_2 = 4]$

10. $\dfrac{4}{p_1 - p_2} = \dfrac{16}{21}$

$\dfrac{3}{p_1 + p_2} = \dfrac{4}{9}$ $[p_1 = 6, p_2 = \frac{3}{4}]$

11. $\dfrac{2x + 1}{5} - \dfrac{1 - 4y}{2} = \dfrac{5}{2}$

$\dfrac{1 - 3x}{7} + \dfrac{2y - 3}{5} + \dfrac{32}{35} = 0$ $[x = 2, y = 1]$

12. A vector system to determine the shortest distance between two moving bodies is analysed, producing the following equations:

$11S_1 - 10S_2 = 30$
$21S_2 - 20S_1 = -40$

Use determinants to find the values of S_1 and S_2
$[S_1 = 7.42, S_2 = 5.16]$

13. The power in a mechanical device is given by $p = aN + \dfrac{b}{N}$

where a and b are constants. Use determinants to find the value of a and b if $p = 13$ when $N = 3$ and $p = 12$ when $N = 2$
$[a = 3, b = 12]$

14. The law connecting friction F and load L for an experiment to find the friction force between two surfaces is of the type $F = aL + b$, where a and b are constants.
When $F = 6.0$, $L = 7.5$ and when $F = 2.7$, $L = 2.0$
Find the values of a and b by using determinants.
$[a = 0.60, b = 1.5]$

15. The length L metres of an alloy at temperature $t\,°C$ is given by $L = L_0(1 + \alpha t)$, where L_0 and α are constants.
Use determinants to find the values of L_0 and α if $L = 20$ m when t is $52\,°C$ and $L = 21$ m when $t = 100\,°C$
$[L_0 = 18.92$ m, $\alpha = 0.001\,1]$

The inverse of a matrix

In Problems 16 to 20 find the inverse of the matrices given.

16. $\begin{pmatrix} 2 & -1 \\ -5 & -1 \end{pmatrix}$ $\left[\begin{pmatrix} \frac{1}{7} & -\frac{1}{7} \\ -\frac{5}{7} & -\frac{2}{7} \end{pmatrix} \right]$

17. $\begin{pmatrix} 1 & -3 \\ 1 & 7 \end{pmatrix}$ $\left[\begin{pmatrix} \frac{7}{10} & \frac{3}{10} \\ -\frac{1}{10} & \frac{1}{10} \end{pmatrix} \right]$

18. $\begin{pmatrix} 3 & 5 \\ -2 & 1 \end{pmatrix}$ $\left[\begin{pmatrix} \frac{1}{13} & -\frac{5}{13} \\ \frac{2}{13} & \frac{3}{13} \end{pmatrix} \right]$

19. $\begin{pmatrix} -2 & -1 \\ 4 & 3 \end{pmatrix}$ $\left[\begin{pmatrix} -\frac{3}{2} & -\frac{1}{2} \\ 2 & 1 \end{pmatrix} \right]$

20. $\begin{pmatrix} -4 & -3 \\ 5 & 3 \end{pmatrix}$ $\left[\begin{pmatrix} 1 & 1 \\ -\frac{5}{3} & -\frac{4}{3} \end{pmatrix} \right]$

The solution of simultaneous equations having two unknowns using matrices

In Problems 21 to 26 solve the simultaneous equations given by using matrices.

21. $p + 3q = 11$
$p + 2q = 8$ $[p = 2, q = 3]$

22. $3a + 4b - 5 = 0$
$12 = 5b - 2a$ $[a = -1, b = 2]$

23. $\dfrac{m}{3} + \dfrac{n}{4} = 6$

$\dfrac{m}{6} - \dfrac{n}{8} = 0$ $[m = 9, n = 12]$

24. $4a - 6b + 2.5 = 0$
$7a - 5b + 0.25 = 0$ $[a = \frac{1}{2}, b = \frac{3}{4}]$

25. $\dfrac{x}{8} + \dfrac{5}{2} = y$

$13 - \dfrac{y}{3} - 3x = 0$ $[x = 4, y = 3]$

26. $\dfrac{a - 1}{3} + \dfrac{b + 2}{2} = 3$

$\dfrac{1 - a}{6} + \dfrac{4 - b}{2} = \dfrac{1}{2}$ $[a = 4, b = 2]$

27. When determining the relative velocity of a system, the following equations are produced:

$$3.0 = 0.10v_1 + (v_1 - v_2)$$
$$-2.0 = 0.05v_2 - (v_1 - v_2)$$

Use matrices to find the values of v_1 and v_2
$[v_1 = 7.42, v_2 = 5.16]$

28. Applying Newton's laws of motion to a mechanical system gives the following equations:

$$14 = 0.2u + 2u + 8(u - v)$$
$$0 = -8(u - v) + 2v + 10v$$

Use matrices to find the values of u and v
$[u = 2.0, v = 0.8]$

29. Equations connecting the lens system in a position transducer are:

$$\frac{4}{u_1} + \frac{6}{v_1} + \frac{9}{v_2} = 6$$

$$\frac{15}{u_1} + \frac{11}{v_1} + \frac{2}{v_2} = 8\frac{1}{12}$$

If $v_1 = v_2$, use matrices to find the values of u_1, v_1 and v_2
$[u_1 = 4, v_1 = v_2 = 3]$

30. When an effort E is applied to the gearbox on a diesel motor it is found that a resistance R can be overcome and that E and R are connected by a formula $E = a + bR$, where a and b are constants. An effort of 3.5 newtons overcomes a resistance of 5 newtons and an effort of 5.3 newtons overcomes a resistance of 8 newtons. Use matrices to find the values of a and b $[a = 0.50, b = 0.60]$

Chapter 9

Complex numbers

9.1 Introduction

The solution of the quadratic equation $ax^2 + bx + c = 0$ may be obtained by using the quadratic formula:

$$x = \frac{-b \pm \sqrt{(b^2 - 4ac)}}{2a}$$

Hence the solutions of $2x^2 + x - 3 = 0$ are

$$x = \frac{-1 \pm \sqrt{[1^2 - 4(2)(-3)]}}{2(2)}$$

$$= \frac{-1 \pm \sqrt{25}}{4} = -\tfrac{1}{4} \pm \tfrac{5}{4}$$

i.e. $x = 1$ or $x = -1\tfrac{1}{2}$

However, a problem exists if $(b^2 - 4ac)$ in the quadratic formula results in a negative number, since we cannot obtain in real terms the square root of a negative number. The only numbers we have met to date have been real numbers. These are either integers (such as $+1$, $+5$, -7, etc.) or rational numbers (such as $\tfrac{4}{1}$ or 4.000, $-\tfrac{7}{9}$ or $-0.7\dot{7}$, $\tfrac{1}{2}$ or $0.500\,0$, etc.) or irrational numbers (such as $\pi = 3.141\,592\ldots$, $\sqrt{3} = 1.732\ldots$, etc.).

If $x^2 - 2x + 5 = 0$ then

$$x = \frac{2 \pm \sqrt{[(-2)^2 - 4(1)(5)]}}{2(1)} = \frac{2 \pm \sqrt{-16}}{2}$$

In order to deal with such a problem as determining $\sqrt{-16}$ the concept of complex numbers has been evolved.

9.2 Definition of a complex number

(a) Imaginary numbers

Let b and c be real numbers. An imaginary number is written in the form jb or jc (i.e. $j \times b$ or $j \times c$) where operator j is defined by the following two rules:

(i) For addition $jb + jc = j(b + c)$ (1)
(ii) For multiplication $jb \times jc = -bc$ (2)

i.e. $j^2 bc = -bc$

Thus $j^2 = -1$

or $j = \sqrt{-1}$

It is immaterial whether the operator j is placed in front of or after the number, i.e. $j4 = 4j$, and so on. An imaginary number $j3$ means $(\sqrt{-1}) \times 3$ and $-j2$ means $-(\sqrt{-1}) \times 2$ or $(-2) \times (\sqrt{-1})$.

Similarly, from equation (1) $j2 + j5 = j7$

and $j5 - j2 = j3$

(b) Complex numbers

From Section 9.1

$$x^2 - 2x + 5 = 0 \text{ and } x = \frac{2 \pm \sqrt{-16}}{2}$$

$\sqrt{-16}$ can be split into $(\sqrt{-1}) \times \sqrt{16}$, i.e. $j\sqrt{16}$ or $\pm j4$

Hence $x = \frac{2 \pm j4}{2} = 1 \pm j2$

i.e. $x = 1 + j2 \text{ or } x = 1 - j2$

The solutions of the quadratic equation $x^2 - 2x + 5 = 0$ are of the form $a + jb$, where 'a' is a real number and 'jb' an imaginary number. Numbers in the form $a + jb$ are called complex numbers. Hence $1 + j2$, $5 - j7$, $-2 + j3$ and $-\pi + j\sqrt{2}$ are all examples of complex numbers.

In algebra, if a quantity x is added to a quantity $3y$ the result is written as $x + 3y$ since x and y are separate quantities. Similarly, it is important to appreciate that real and imaginary numbers are different types of numbers and must be kept separate.

When imaginary numbers were first introduced the symbol i (i.e. the first letter of the word imaginary) was used to indicate $\sqrt{-1}$ and this symbol is still used in pure mathematics. However, in engineering, the symbol i indicates electric current, and to avoid any possible confusion the next letter in the alphabet, i.e. j, is used to represent $\sqrt{-1}$.

9.3   The Argand diagram

A complex number can be represented pictorially on rectangular or cartesian axes. The horizontal (or x) axis is used to represent the real axis and the vertical (or y) axis is used to represent the imaginary axis. Such a diagram is called an **Argand diagram** (named after an eighteenth-century French mathematician) and is shown in Fig. 9.1(a).

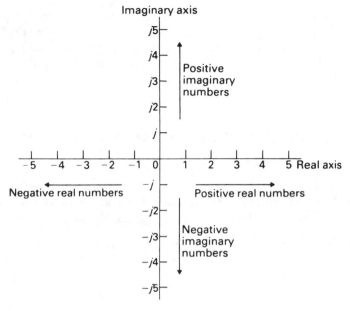

(a)

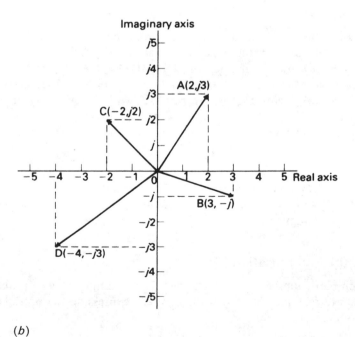

(b)

Fig. 9.1   The Argand diagram

In Fig. 9.1(*b*) the point A represents the complex number $2 + j3$ and is obtained by plotting the coordinates $(2, j3)$ as in graphical work. Often an arrow is drawn from the origin to the point as shown. Figure 9.1(*b*) also shows the Argand points B, C and D, representing the complex numbers $3 - j$, $-2 + j2$ and $-4 - j3$ respectively.

A complex number of the form $a + jb$ is called a **cartesian complex number**.

9.4   Operations involving cartesian complex numbers

(a) Addition and subtraction

Let two complex numbers be represented by $Z_1 = a + jb$ and $Z_2 = c + jd$. Two complex numbers are added/subtracted by adding/subtracting separately the two real parts and the two imaginary parts. Hence

$$Z_1 + Z_2 = (a + jb) + (c + jd)$$
$$= (a + c) + j(b + d)$$
and $$Z_1 - Z_2 = (a + jb) - (c + jd)$$
$$= (a - c) + j(b - d)$$

Thus, if $Z_1 = 3 + j2$ and $Z_2 = 2 - j4$ then

$$Z_1 + Z_2 = (3 + j2) + (2 - j4) = 3 + j2 + 2 - j4 = 5 - j2$$
and $$Z_1 - Z_2 = (3 + j2) - (2 - j4) = 3 + j2 - 2 + j4 = 1 + j6$$

The addition and subtraction of complex numbers may be achieved graphically as shown in the Argand diagram in Fig. 9.2

In Fig. 9.2(*a*), by vector addition, $\mathbf{OP} + \mathbf{OQ} = \mathbf{OR_1}$. R_1 is found to be the Argand point $(5, -j2)$, i.e. $(3 + j2) + (2 - j4) = 5 - j2$ (as above).

In Fig. 9.2(*b*) vector $\mathbf{OQ}$ is reversed (shown as $\mathbf{OQ'}$) since it is being subtracted.

(Note that $\mathbf{OQ} = 2 - j4$

Thus $-\mathbf{OQ} < > (2 - j4)$

$= -2 + j4$,   shown as $\mathbf{OQ'}$)

Thus $\mathbf{OP} - \mathbf{OQ} = \mathbf{OP} + \mathbf{OQ'} = \mathbf{OR_2}$. R_2 is found to be the Argand point $(1, j6)$, i.e. $(3 + j2) - (2 - j4) = 1 + j6$ (as above). Vector or phasor addition is covered in Chapter 17.

(b) Multiplication

Two complex numbers are multiplied by assuming that all quantities involved are real and then, by using $j^2 = -1$, expressing the product in the form $a + jb$. Hence

$$(a + jb)(c + jd) = ac + a(jd) + (jb)c + (jb)(jd)$$
$$= ac + jad + jbc + j^2bd$$

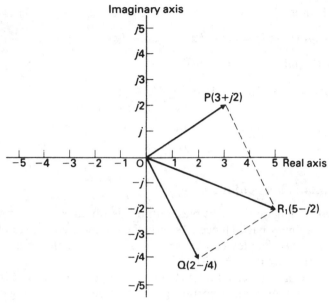

(a)

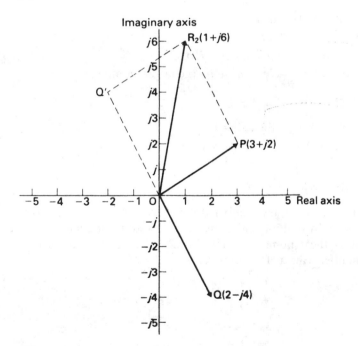

(b)

Fig. 9.2   (a) **OP + OQ = OR₁** and (b) **OP − OQ = OR₂**

But $j^2 = -1$, thus

$$(a + jb)(c + jd) = (ac - bd) + j(ad + bc)$$

If $Z_1 = 3 + j2$ and $Z_2 = 2 - j4$ then

$$\begin{aligned}
Z_1 Z_2 &= (3 + j2)(2 - j4) = 6 - j12 + j4 - j^2 8 \\
&= (6 - (-1)8) + j(-12 + 4) \\
&= 14 + j(-8) \\
&= 14 - j8
\end{aligned}$$

(c) Complex conjugate

The complex conjugate of a complex number is obtained by changing the sign of the imaginary part. Hence $a + jb$ is the complex conjugate of $a - jb$ and $-2 + j$ is the complex conjugate of $-2 - j$. (Note that only the sign of the imaginary part is changed.)

The product of a complex number Z and its complex conjugate $\bar{Z}$ is always a real number, and this is an important property used when dividing complex numbers. Thus, if

$$Z = a + jb \text{ then } \bar{Z} = a - jb$$

and

$$\begin{aligned}
Z\bar{Z} &= (a + jb)(a - jb) \\
&= -a^2 - jab + jab - j^2 b^2 \\
&= a^2 - -b^2 = a^2 + b^2 \text{ (i.e. a real number)}
\end{aligned}$$

Similarly, if

$$Z = 1 + j2 \text{ then } \bar{Z} = 1 - j2$$

and

$$Z\bar{Z} = (1 + j2)(1 - j2) = 1^2 + 2^2 = 5$$

(d) Division

Expressing one complex number divided by another, in the form $a + jb$, is accomplished by multiplying the numerator and denominator by the complex conjugate of the denominator. This has the effect of making the denominator a real number. Hence, for example,

$$\begin{aligned}
\frac{2 + j4}{3 - j4} &= \frac{2 + j4}{3 - j4} \times \frac{3 + j4}{3 + j4} \\
&= \frac{6 + j8 + j12 + j^2 16}{3^2 + 4^2} \\
&= \frac{-10 + j20}{25} = \frac{-10}{25} + j\frac{20}{25} \text{ or } -0.4 + j0.8
\end{aligned}$$

(e) Complex equations

If two complex numbers are equal then their real parts are equal and their imaginary parts are equal. Hence, if

$$a + jb = c + jd$$

then $\qquad a = c$ and $b = d$

This is a useful property since equations having two unknown quantities can be solved from one equation (see Worked Problem 4 following).

Worked problems on operations involving cartesian complex numbers

Problem 1. Solve the following quadratic equations:
(a) $x^2 + 9 = 0$ (b) $4y^2 - 3y + 5 = 0$

(a) $x^2 + 9 = 0$

$$x^2 = -9$$
$$x = \sqrt{-9} = (\sqrt{-1})(\sqrt{9}) = (\sqrt{-1})(\pm 3) = (j)(\pm 3) = \pm j3$$

(b) $4y^2 - 3y + 5 = 0$

Using the quadratic formula:

$$y = \frac{-(-3) \pm \sqrt{[(-3)^2 - 4(4)(5)]}}{2(4)} = \frac{3 \pm \sqrt{(9 - 80)}}{8}$$

$$= \frac{3 \pm \sqrt{-71}}{8} = \frac{3 \pm (\sqrt{-1})(\sqrt{71})}{8} = \frac{3 \pm j\sqrt{71}}{8}$$

$$= \frac{3}{8} \pm j \frac{\sqrt{71}}{8} = \mathbf{0.375 \pm j1.053}, \text{ correct to 3 decimal places}$$

Problem 2. Evaluate (a) j^3 (b) j^4 (c) j^5 (d) j^6 (e) j^7 (f) $\dfrac{3}{j^9}$

(a) $j = \sqrt{-1}, \quad j^2 = -1, \quad j^3 = j \times j^2 = (j)(-1) = -j$

(b) $j^4 = j \times j^3 = (j)(-j) = -j^2 = +1$

(c) $j^5 = j \times j^4 = (j) \times (1) = j$

(d) $j^6 = j \times j^5 = (j) \times (j) = j^2 = -1$

(e) $j^7 = j \times j^6 = (j) \times (-1) = -j$

(f) $j^9 = j^2 \times j^7 = (-1) \times (-j) = j$

Hence $\dfrac{3}{j^9} = \dfrac{3}{j} = \dfrac{3}{j} \times \dfrac{(-j)}{(-j)} = \dfrac{-j3}{-j^2} = \dfrac{-j3}{+1} = -j3$

Problem 3. If $Z_1 = 1 + j2$, $Z_2 = 2 - j3$, $Z_3 = -4 + j$ and $Z_4 = -3 - j2$ evaluate in $a + jb$ form the following:
(a) $Z_1 + Z_2 - Z_3$ (b) $Z_1 Z_3$ (c) $Z_2 Z_3 Z_4$
(d) $\dfrac{Z_2}{Z_4}$ (e) $\dfrac{Z_1 - Z_4}{Z_2 + Z_3}$

(a)
$$Z_1 + Z_2 - Z_3 = (1 + j2) + (2 - j3) - (-4 + j)$$
$$= 1 + j2 + 2 - j3 + 4 - j$$
$$= 7 - j2$$

(a)
$$Z_1 Z_3 = (1 + j2)(-4 + j)$$
$$= -4 + j - j8 + j^2 2$$
$$= -6 - j7$$

(c)
$$Z_2 Z_3 Z_4 = (2 - j3)(-4 + j)(-3 - j2)$$
$$= (-8 + j2 + j12 - j^2 3)(-3 - j2)$$
$$= (-5 + j14)(-3 - j2)$$
$$= 15 + j10 - j42 - j^2 28$$
$$= 43 - j32$$

(d)
$$\frac{Z_2}{Z_4} = \frac{2 - j3}{-3 - j2}$$
$$= \frac{2 - j3}{-3 - j2} \times \frac{-3 + j2}{-3 + j2}$$
$$= \frac{-6 + j4 + j9 - j^2 6}{3^2 + 2^2}$$
$$= \frac{0 + j13}{13}$$
$$= 0 + j1 \text{ or } j$$

(e)
$$\frac{Z_1 - Z_4}{Z_2 + Z_3} = \frac{(1 + j2) - (-3 - j2)}{(2 - j3) + (-4 + j)}$$
$$= \frac{4 + j4}{-2 - j2}$$
$$= \frac{4(1 + j)}{-2(1 + j)}$$
$$= -2$$

Problem 4. Solve the following complex equations:
(a) $3(a + jb) = 9 - j2$
(b) $(2 + j)(-2 + j) = x + jy$
(c) $(a - j2b) + (b - j3a) = 5 + j2$

(a) $3(a + jb) = 9 - j2$
$3a + j3b = 9 - j2$
Equating real parts gives $3a = 9$, i.e. $a = 3$
Equating imaginary parts gives $3b = -2$, i.e. $b = -\frac{2}{3}$

(b) $(2 + j)(-2 + j) = x + jy$
$-4 + j2 - j2 + j^2 = x + jy$
$- 5 + j0 = x + jy$
Equating real and imaginary parts gives $x = -5$, $y = 0$

(c) $(a - j2b) + (b - j3a) = 5 + j2$
$(a + b) + j(-2b - 3a) = 5 + j2$

| Hence | $a + b = 5$ | (1) |
| and | $-2b - 3a = 2$ | (2) |

We have two simultaneous equations to solve:
Multiplying equation (1) by 2 gives $2a + 2b = 10$ (3)
Adding equations (2) and (3) gives $-a = 12$, i.e. $a = -12$
From equation (1) $b = 17$

Further problems on operations involving cartesian complex numbers may be found in Section 9.9 (Problems 1 to 23), page 159.

9.5   The polar form of a complex number

Let a complex number Z, in cartesian form, be $x + jy$. This is shown in the Argand diagram of Fig. 9.3(a). Let r be the distance OZ and θ the angle OZ makes with the positive real axis. From the trigonometry

$$\cos \theta = \frac{x}{r}, \text{ i.e. } x = r \cos \theta$$

and $$\sin \theta = \frac{y}{r}, \text{ i.e. } y = r \sin \theta$$

$$Z = x + jy$$
$$= r \cos \theta + jr \sin \theta$$
$$= r(\cos \theta + j \sin \theta)$$

This latter form is usually abbreviated to $Z = r \angle \theta$ and is called the polar form of a complex number. The complex number is now specified in terms of r and θ instead of x and y.

r is called the **modulus** (or magnitude) of Z and is written as mod Z or $|Z|$.
r is determined from Pythagoras's theorem on triangle OAZ, i.e.

$$|Z| = r = \sqrt{(x^2 + y^2)}$$

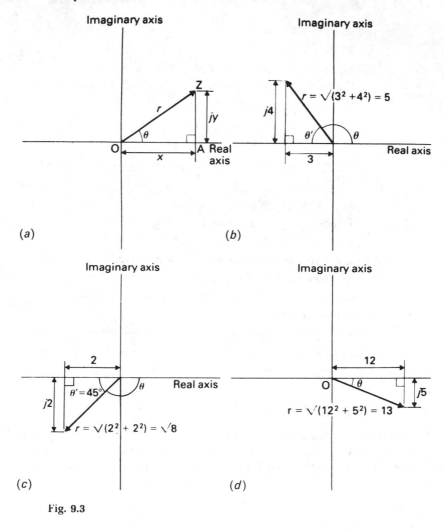

Fig. 9.3

The modulus is represented on the Argand diagram by the distance OZ. θ is called the **argument** (or amplitude) of Z and is written as arg Z. θ is also deduced from triangle OAZ, giving

$$\mathbf{arg}\ Z = \theta = \mathbf{arctan}\ \frac{y}{x}$$

In Fig. 9.3(a) the Argand point Z is shown in the first quadrant. However, the above results apply to any point in the Argand diagram.

By convention, the principal value of θ is used, i.e. the numerically least value such that $-\pi \leqslant \theta \leqslant \pi$. For example, in Fig. 9.3(b), $\theta' = \arctan \frac{4}{3} = 53.13°$ or $53°\ 8'$. Hence $\theta = 180° - 53°\ 8' = 126°\ 52'$. Therefore $-3 + j4 = 5\ \angle\ 126°\ 52'$.

Similarly, in Fig. 9.3(c) θ is $180° - 45°$, i.e. $135°$ measured in the negative direction. Hence $-2 - j2 = \sqrt{8} \angle -135°$. (This is the same as $\sqrt{8} \angle 225°$. However, the principal value of $\sqrt{8} \angle -135°$ is normally used.)

In Fig. 9.3(d), $\theta = \arctan \frac{5}{12} = 22.62°$ or $22° \, 37'$. Hence

$$12 - j5 = 13 \angle -22° \, 37'$$

Whenever changing from a cartesian form of complex number to a polar form, or vice versa, a sketch is invaluable for deciding the quadrant in which the complex number occurs. There are always two possible values of

$$\theta = \arctan \frac{y}{x}$$

only one of which is correct for a particular complex number.

9.6 Multiplication and division using complex numbers in polar form

An important use of the polar form of a complex number is in multiplication and division, which is achieved more easily than with cartesian form.

(a) Multiplication

Let $\qquad Z_1 = r_1 \angle \theta_1$ and $Z_2 = r_2 \angle \theta_2$

Then $\quad Z_1 Z_2 = [r_1 \angle \theta_1][r_2 \angle \theta_2]$

$$= [r_1(\cos \theta_1 + j \sin \theta_1)] \times [r_2(\cos \theta_2 + j \sin \theta_2)]$$

$$= r_1 r_2(\cos \theta_1 \cos \theta_2 + j \sin \theta_2 \cos \theta_1$$
$$+ j \cos \theta_2 \sin \theta_1 + j^2 \sin \theta_1 \sin \theta_2)$$

$$= r_1 r_2[(\cos \theta_1 \cos \theta_2 - \sin \theta_1 \sin \theta_2)$$
$$+ j(\sin \theta_1 \cos \theta_2 + \cos \theta_1 \sin \theta_2)]$$

$$= r_1 r_2[\cos(\theta_1 + \theta_2) + j \sin(\theta_1 + \theta_2)] \quad \text{(from compound angle formulae, Chapter 16)}$$

$$= r_1 r_2 \angle (\theta_1 + \theta_2)$$

Hence to obtain the product of complex numbers in polar form their moduli are multiplied together and their arguments are added. This result is true for all polar complex numbers. Thus

$$3 \angle 25° \times 2 \angle 32° = 6 \angle 57°$$
$$4 \angle 11° \times 5 \angle -18° = 20 \angle -7°$$

$$2 \angle \frac{\pi}{3} \times 7 \angle \frac{\pi}{6} = 14 \angle \frac{\pi}{2} \text{ and so on}$$

(b) Division

Let $Z_1 = r_1 \angle \theta_1$ and $Z_2 = r_2 \angle \theta_2$

Then $\dfrac{Z_1}{Z_2} = \dfrac{r_1 \angle \theta_1}{r_2 \angle \theta_2} = \dfrac{r_1(\cos \theta_1 + j \sin \theta_1)}{r_2(\cos \theta_2 + j \sin \theta_2)}$

$$= \frac{r_1(\cos \theta_1 + j \sin \theta_1)}{r_2(\cos \theta_2 + j \sin \theta_2)} \times \frac{(\cos \theta_2 - j \sin \theta_2)}{(\cos \theta_2 - j \sin \theta_2)}$$

$$= \frac{r_1(\cos \theta_1 \cos \theta_2 - j \sin \theta_2 \cos \theta_1 + j \sin \theta_1 \cos \theta_2 - j^2 \sin \theta_1 \sin \theta_2)}{r_2(\cos^2 \theta_2 + \sin^2 \theta_2)}$$

$$= \frac{r_1[(\cos \theta_1 \cos \theta_2 + \sin \theta_1 \sin \theta_2) + j(\sin \theta_1 \cos \theta_2 - \sin \theta_2 \cos \theta_1)]}{r_2(1)}$$

$$= \frac{r_1}{r_2}[\cos(\theta_1 - \theta_2) + j \sin(\theta_1 - \theta_2)]$$

               (from compound angle formulae, Chapter 16)

$$= \frac{r_1}{r_2} \angle (\theta_1 - \theta_2)$$

Hence to obtain the ratio of two complex numbers in polar form their moduli are divided and their arguments subtracted. This result is true for all polar complex numbers. Thus

$$\frac{8 \angle 58°}{2 \angle 11°} = 4 \angle 47°$$

$$\frac{9 \angle 136°}{3 \angle -60°} = 3 \angle [136° - (-60°)] = 3 \angle 196° = 3 \angle -164°$$

$$\frac{10 \times \dfrac{\pi}{2}}{5 \angle -\dfrac{\pi}{4}} = 2 \angle \frac{3\pi}{4} \text{ and so on}$$

Worked problems on the polar form of complex numbers

Problem 1. Determine the modulus and argument of the complex number $Z = -5 + j8$ and express Z in polar form.

A sketch, shown in Fig. 9.4(a), indicates that the complex number $-5 + j8$ lies in the second quadrant of the Argand diagram.

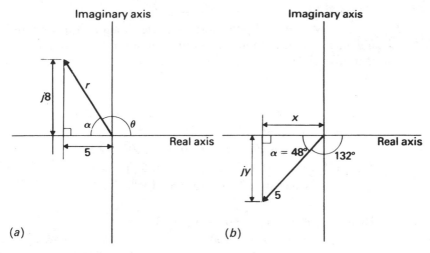

Fig. 9.4

Modulus: $|Z| = r = \sqrt{[(-5)^2 + (8)^2]}$
$\qquad\qquad = \textbf{9.434}$, correct to 4 significant figures
$\quad \alpha = \arctan \tfrac{8}{5} = \textbf{58}°$
Argument, $\arg Z = \theta = 180° - 58° = 122°$
In polar form $-5 + j8$ is written as $\textbf{9.434} \angle \textbf{122}°$

Problem 2. Convert $5 \angle -132°$ into $a + jb$ form correct to 4 significant figures.

A sketch, shown in Fig. 9.4(b), indicates that the polar complex number $5 \angle -132°$ lies in the third quadrant of the Argand diagram.
  Using trigonometrical ratios, $x = 5 \cos 48° = 3.346$
$\qquad\qquad\qquad\qquad$ and $y = 5 \sin 48° = 3.716$
Hence $\textbf{5} \angle \textbf{-132}° = \textbf{-3.346} - \textbf{j3.716}$

Problem 3. Evaluate (a) $\dfrac{15 \angle 30° \times 8 \angle 45°}{4 \angle 60°}$
$\qquad\qquad$ and (b) $4 \angle 30° + 3 \angle -60° - 5 \angle -135°$
giving answers in polar and in cartesian forms, correct to 3 significant figures.

(a) $\dfrac{15 \angle 30° \times 8 \angle 45°}{4 \angle 60°} = \dfrac{15 \times 8}{4} \angle (30° + 45° - 60°)$
$\qquad\qquad\qquad\qquad = \textbf{30.0} \angle \textbf{15}°$, in polar form
$\qquad\qquad\qquad\qquad = 30.0(\cos 15° + j \sin 15°)$
$\qquad\qquad\qquad\qquad = \textbf{29.0} + \textbf{j7.76}$, in cartesian form

(b) $4 \angle 30° + 3 \angle -60° - 5 \angle -135°$

The advantages of polar form are seen when multiplying and dividing complex numbers. Addition or subtraction in polar form is not possible. Each polar complex number has to be converted into cartesian form first. Thus

$$4 \angle 30° = 4(\cos 30° + j \sin 30°)$$
$$= 4 \cos 30° + j4 \sin 30° = 3.464 + j2.000$$
$$3 \angle -60° = 3[\cos(-60°) + j \sin(-60°)]$$
$$= 3 \cos 60° - j3 \sin 60° = 1.500 - j2.598$$
$$5 \angle -135° = 5[\cos(-135°) + j \sin(-135°)]$$
$$= 5 \cos(-135°) + j5 \sin(-135°)$$
$$= -3.536 - j3.536$$

Hence $4 \angle 30° + 3 \angle -60° - 5 \angle -135°$

$$= (3.464 + j2.000)$$
$$+ (1.500 - j2.598) - (-3.536 - j3.536)$$
$$= \mathbf{8.50 + j2.94}, \text{ in cartesian form}$$
$$= \sqrt{[(8.50)^2 + (2.94)^2]} \angle \arctan \frac{2.94}{8.50}$$

and since the complex number is in the first quadrant
$$= \mathbf{8.99 \angle 19° 5'}, \text{ in polar form}$$

Further problems on the polar form of complex numbers may be found in Section 9.9 (Problems 24 to 38), page 161.

9.7   De Moivre's theorem – powers and roots of complex numbers

From Section 9.6

$$r \angle \theta \times r \angle \theta = r^2 \angle 2\theta$$
$$r \angle \theta \times r \angle \theta \times r \angle \theta = r^3 \angle 3\theta \text{ and so on}$$

Such results are generally stated in De Moivre's theorem, which may be stated as

$$[r \angle \theta]^n = r^n \angle n\theta \quad (= r^n(\cos n\theta + j \sin n\theta))$$

This result is true for all positive, negative or fractional values of n. De Moivre's theorem is thus useful in determining powers and roots of complex numbers. For example,

$$[2 \angle 15°]^6 = 2^6 \angle (6 \times 15°) = 64 \angle 90° = 0 + j64$$

A square root of a complex number is determined as follows:

$$\sqrt{[r \angle \theta]} = [r \angle \theta]^{\frac{1}{2}} = r^{\frac{1}{2}} \angle \tfrac{1}{2}\theta$$

However, it is important to realise that a real number has two square roots, equal in size but opposite in sign. On an Argand diagram the roots are 180° apart (see Worked Problem 3).

Worked problems on powers and roots of complex numbers

Problem 1. Determine the square of the complex numbers $3 - j4$ (*a*) in cartesian form and (*b*) in polar form, using De Moivre's theorem. Compare the results obtained and show the roots on an Argand diagram.

(*a*)   In cartesian form

$$(3 - j4)^2 = (3 - j4)(3 - j4)$$
$$= 9 - j12 - j12 + j^2 16$$
$$= -7 - j24$$

(*b*)   In polar form

$$(3 - j4) = \sqrt{[(3)^2 + (4)^2]} \angle \arctan \tfrac{4}{3} \text{ (see Fig. 9.5)}$$
$$= 5\angle -53° \, 8'$$
$$[5\angle -53° \, 8']^2 = 5^2 \angle (2 \times -53° \, 8')$$
$$= 25 \angle -106° \, 16'$$

The complex number $(3 - j4)$, together with its square, i.e. $(3 - j4)^2$, is shown in Fig. 9.5

$25 \angle -106° \, 16'$ in cartesian form is
$25 \cos(-106° \, 16') + j25 \sin(-106° \, 16')$, i.e. $-7 - j24$,
as in part (*a*).

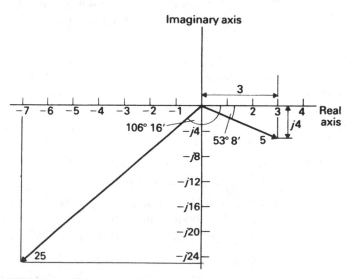

Fig. 9.5

Problem 2. Determine $(-2 + j3)^5$ in polar and cartesian form.

$Z = -2 + j3$ is situated in the second quadrant of the Argand diagram. Thus $r = \sqrt{[(2)^2 + (3)^2]} = \sqrt{13}$

and $\qquad \alpha = \arctan \frac{3}{2} = 56.31°$ or $56° \ 19'$

Hence the argument $\theta = 180° - 56° \ 19' = 123° \ 41'$

Thus $2 - j3$ in polar form is $\sqrt{13} \angle 123° \ 41'$

$$
\begin{aligned}
(-2 + j3)^5 &= [\sqrt{13} \angle 123° \ 41']^5 \\
&= (\sqrt{13})^5 \angle (5 \times 123° \ 41'), \text{ from De Moivre's theorem} \\
&= 13^{\frac{5}{2}} \angle 618° \ 25' \\
&= 13^{\frac{5}{2}} \angle 258° \ 25' \\
&\quad (\text{since } 618° \ 25' = 618° \ 25' - 360°) \\
&= 13^{\frac{5}{2}} \angle -101° \ 35' \\
&= \mathbf{609.3 \angle -101° \ 35'}
\end{aligned}
$$

In cartesian form

$$
609.3 \angle -101° \ 35' = 609.3 \cos(-101° \ 35') + j609.3 \sin(-101° \ 35')
$$
$$
= \mathbf{-122.3 - j596.9}
$$

Problem 3. Determine the two square roots of the complex number $12 + j5$ in cartesian and polar form, correct to 3 significant figures. Show the roots on an Argand diagram.

In polar form $12 + j5 = \sqrt{(12^2 + 5^2)} \angle \arctan \frac{5}{12}$, since $12 + j5$ is in the first quadrant of the Argand diagram, i.e.

$12 + j5 = 13 \angle 22° \ 37'$

Since we are finding the square roots of $13 \angle 22° \ 37'$ there will be two solutions. To obtain the second solution it is helpful to express $13 \angle 22° \ 37'$ also as $13 \angle (360° + 22° \ 37')$, i.e. $13 \angle 382° \ 37'$ (we have merely rotated one revolution to obtain this result). The reason for doing this is that when we divide the angles by 2 we still obtain angles less than $360°$, as shown below. Hence

$$
\begin{aligned}
\sqrt{(12 + j5)} &= \sqrt{[13 \angle 22° \ 37']} \text{ or } \sqrt{[13 \angle 382° \ 37']} \\
&= [13 \angle 22° \ 37']^{\frac{1}{2}} \text{ or } [13 \angle 382° \ 37']^{\frac{1}{2}} \\
&= 13^{\frac{1}{2}} \angle (\tfrac{1}{2} \times 22° \ 37') \text{ or } 13^{\frac{1}{2}} \angle (\tfrac{1}{2} \times 382° \ 37'), \\
&\quad \text{from De Moivre's theorem} \\
&= \sqrt{13} \angle 11° \ 19' \text{ or } \sqrt{13} \angle 191° \ 19' \\
&= 3.61 \angle 11° \ 19' \text{ or } 3.61 \angle -168° \ 41'
\end{aligned}
$$

These two solutions of $\sqrt{(12 + j5)}$ are shown in the Argand diagram of Fig. 9.6

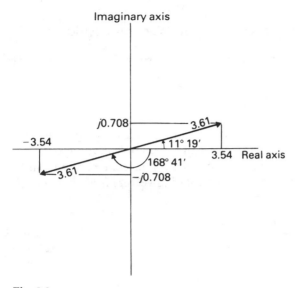

Fig. 9.6

$3.61 \angle 11° 19'$ is in the first quadrant of the Argand diagram. Thus
$3.61 \angle 11° 19' = 3.61(\cos 11° 19' + j \sin 11° 19')$
$\qquad\qquad\quad = 3.54 + j0.708$
$3.61 \angle -168° 41'$ is in the third quadrant of the Argand diagram. Thus
$3.61 \angle -168° 41' = 3.61[\cos(-168° 41') + j \sin(-168° 41')]$
$\qquad\qquad\qquad = -3.54 - j0.708$
Thus in cartesian form the two roots are $\pm(3.54 + j0.708)$

From the Argand diagram the roots are seen to be 180° apart,
i.e. they lie on a straight line. This is always true when finding
square roots of complex numbers.

Further problems on powers and roots of complex numbers may be found in
Section 9.9 (Problems 39 to 48), page 162.

9.8   Applications of complex numbers

Complex numbers are widely used in the analysis of electrical circuits
supplied by an alternating voltage and are also used for solving problems
involving coplanar vectors.

(a) Phasor applications to series connected circuits

In an electrical circuit containing resistance only (R ohms) the current
(I amperes) is in phase with the applied voltage (V volts) and the ratio of

voltage to current is given by $\dfrac{V}{I} = R$. When an electrical circuit contains inductance only (L henrys) the ratio of voltage to current is called the inductive reactance (X_L ohms) and

$$\frac{V}{I} = X_L = 2\pi f L \text{ ohms} \tag{3}$$

where f is the frequency in hertz of the applied voltage. The current in a purely inductive circuit lags 90° behind the applied voltage.

When both resistance and inductance are present, as shown in Fig. 9.7(a), the phasor diagram (a phasor is a rotating vector) is as shown in Fig. 9.7(b).

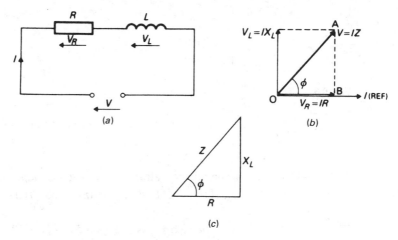

Fig. 9.7

The ratio of $\dfrac{V}{I}$ in any a.c. electrical circuit containing both resistance and reactance is called the **impedance** Z of the circuit and is measured in ohms. With reference to triangle OAB in Fig. 9.7(b), the impedance triangle shown in Fig. 9.7(c) is produced by dividing each quantity by current I. From the impedance triangle it can be seen that the modulus of impedance

$$|Z| = \sqrt{(R^2 + X_L^2)} \text{ ohms} \tag{4}$$

and circuit phase angle

$$\phi = \arctan \frac{X_L}{R} \tag{5}$$

If the impedance triangle is superimposed on an Argand diagram, as shown in Fig. 9.8, then impedance

$$Z = (R + jX_L) \text{ ohms} \tag{6}$$

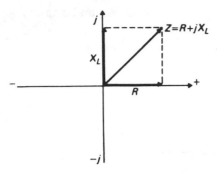

Fig. 9.8

When an electrical circuit contains capacitance (C farads) only, the ratio of voltage to current is called the capacitive reactance (X_C ohms) and

$$\frac{V}{I} = X_C = \frac{1}{2\pi f C} \text{ ohms} \tag{7}$$

The current in a purely capacitive circuit leads the applied voltage by 90°. When both resistance and capacitance are present, as shown in Fig. 9.9(a), the phasor diagram is as shown in Fig. 9.9(b) and the impedance triangle is obtained as for the circuit containing inductance and is shown in Fig. 9.9(c). From the impedance triangle it can be seen that modulus of impedance

$$|Z| = \sqrt{(R^2 + X_C^2)} \text{ ohms} \tag{8}$$

and circuit phase angle

$$\phi = \arctan \frac{X_C}{R} \tag{9}$$

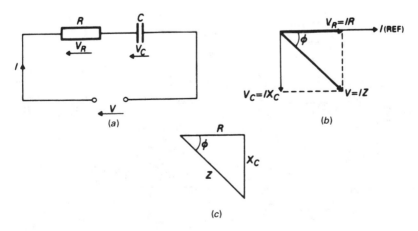

Fig. 9.9

If the impedance triangle is superimposed on an Argand diagram, as shown in Fig. 9.10, then impedance

$$Z = R - jX_C \qquad (10)$$

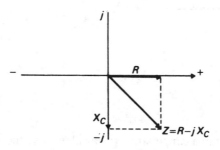

Fig. 9.10

For an electrical circuit containing resistance, inductance and capacitance connected in series the circuit diagram is as shown in Fig. 9.11(*a*), the phasor diagram is as shown in Fig. 9.11(*b*) and the impedance triangle is as shown in Fig. 9.11(*c*). From the impedance triangle it can be seen that modulus of impedance

$$|Z| = \sqrt{[R^2 + (X_L - X_C)^2]}\ \text{ohms} \qquad (11)$$

and circuit phase angle

$$\phi = \arctan \frac{X_L - X_C}{R} \qquad (12)$$

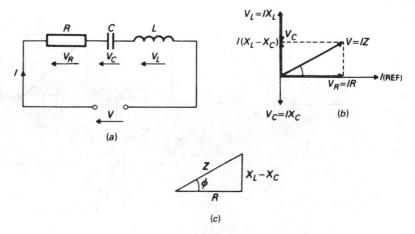

Fig. 9.11

When superimposing the impedance triangle on an Argand diagram, as shown in Fig. 9.12, the resultant impedance is given by

$$Z = R + j(X_L - X_C) \text{ ohms} \tag{13}$$

(see Worked Problems 1 to 4)

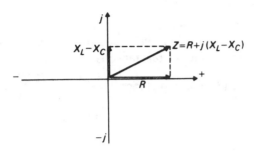

Fig. 9.12

(b) Vector applications

Problems involving coplanar vector quantities, such as those dealing with displacement, velocity, acceleration, force, moment and momentum, may be resolved using complex number theory. Examples of the application of complex number theory to such problems are shown in Worked Problems 5 to 7.

Worked problems on the applications of complex numbers

Problem 1. Determine the resistance and the series connected inductance or capacitance for each of the following impedances:
(a) $12 + j5$, (b) $-j40$ and (c) $30 \angle 60°$
Assume a frequency of 50 Hz

(a)   From equation (6), $Z = R + jX_L$
    Hence $12 + j5$ shows that the **resistance is 12 ohms** and the inductive reactance is 5 ohms.
    From equation (3), since $X_L = 2\pi f L$, $L = \dfrac{X_L}{2\pi f}$

$$= \frac{5}{2\pi(50)} = 0.016 \text{ H}$$

   i.e. **the inductance is 16 mH**

(b)   From equation (10), $Z = R - jX_C$
    Hence $0 - j40$ shows that the **resistance is zero** and the capacitive reactance is 40 ohms.

From equation (7) $X_C = \dfrac{1}{2\pi f C}$, $C = \dfrac{1}{2\pi f X_C}$

$$= \dfrac{1}{2\pi(50)(40)} \, F = \dfrac{10^6}{2\pi(50)(40)} \, \mu F$$

$$= 79.6 \, \mu F$$

i.e. **the capacitance is 79.6 μF**

(c) $30 \angle 60° = 30(\cos 60° + j \sin 60°) = 15 + j25.98$
From equation (6), $15 + j25.98$ shows that the **resistance is 15 ohms** and the inductive reactance is 25.98 ohms.

From equation (3), the inductance $L = \dfrac{X_L}{2\pi f} = \dfrac{25.98}{2\pi(50)}$

$$= 0.0827 \, H$$

i.e. **the inductance is 82.7 mH**

Problem 2. The impedance of an electrical circuit is $30 - j50$ ohms. Determine (a) the resistance, (b) the capacitance, (c) the modulus of the impedance and (d) the current flowing, when the circuit is connected to a 240 V, 50 Hz supply.

(a)   Since $Z = R - jX_C$, the **resistance is 30 ohms**

(b)   Since $Z = R - jX_C$, the capacitance reactance is 50 ohms
From equation (7) capacitance

$$C = \dfrac{1}{2\pi f X_C} = \dfrac{10^6}{2\pi(50)(50)} \, \mu F = \textbf{63.7 μF}$$

(c)   The modulus of the impedance
$$|Z| = \sqrt{[R^2 + (X_C)^2]} = \sqrt{[30^2 + (-50)^2]} = \textbf{58.31 ohms}$$

(d)   From equation (9) the circuit phase angle

$$\phi = \arctan \dfrac{X_C}{R} = \arctan \dfrac{50}{30} = 59.04° \text{ or } 59° \, 2'$$

Since $Z = R - jX_C$, this angle is in the fourth quadrant, i.e. $-59° \, 2'$. Thus an alternative way of expressing the impedance is
$$Z = 58.31 \angle -59° \, 2'$$
The current flowing
$$I = \dfrac{V}{Z} = \dfrac{240 \angle 0°}{58.31 \angle -59° \, 2'} = 4.116 \angle 59° \, 2'$$
(Since the voltage is given as 240 volts, it is $240 + j0$ volts in rectangular form and $240 \angle 0°$ volts in polar form.)

Problem 3. A series connected electrical circuit has a resistance of 32 ohms and an inductance of 0.15 H. It is connected to a 200 V, 50 Hz supply. Determine (a) the inductive reactance, (b) the impedance in rectangular and polar forms, (c) the current and the circuit phase angle, (d) the voltage drop across the resistor and (e) the voltage drop across the inductor.

(a) From equation (3), inductive reactance

$$X_L = 2\pi fL = 2\pi(50)(0.15) = \textbf{47.1 ohms}$$

(b) From equation (6), impedance

$Z = R + jX_L$, i.e. $Z = 32 + j47.1$

Thus from equation (4)

$|Z| = \sqrt{(32^2 + 47.1^2)} = 57$

and from equation (5) circuit phase angle

$$\phi = \arctan\frac{47.1}{32} = 55.81° \text{ or } 55° \ 48'$$

Thus $Z = \textbf{57} \ \angle \ \textbf{55°} \ \textbf{48' ohms}$, in polar form

(c) Current $I = \dfrac{V}{Z} = \dfrac{200 \ \angle \ 0°}{57 \ \angle \ 55° \ 48'} = 3.51 \ \angle -55° \ 48'$

i.e. **the current is 3.51 A lagging V by 55° 48'**

(d) The voltage drop across the 32 ohm resistor

$V_R = IR = (3.51 \ \angle -55° \ 48')(32)$

$\qquad = \textbf{112.3} \ \angle -\textbf{55°} \ \textbf{48' volts}$

(e) The voltage drop across the 0.15 H inductor

$V_L = IX_L = (3.51 \ \angle -55° \ 48')(47.1 \ \angle \ 90°)$

$\qquad = \textbf{165.3} \ \angle \ \textbf{34°} \ \textbf{12' volts}$

Problem 4. A 240 V, 50 Hz voltage is applied across a series connected circuit having a resistance of 12 Ω, an inductance of 0.10 H and a capacitance of 120 μF. Determine the current flowing in the circuit.

From equation (3), inductive reactance

$X_L = 2\pi fL = 2\pi(50)0.10 = 31.4 \ \Omega$

From equation (7), capacitive reactance

$$X_C = \frac{1}{2\pi fC} = \frac{10^6}{2\pi(50)(120)} = 26.5 \ \Omega$$

From equation (13), the impedance

$Z = R + j(X_L - X_C) = 12 + j(31.4 - 26.5) = 12 + j4.9$

In polar form

$$Z = \sqrt{(12^2 + 4.9^2)} \angle \arctan \frac{4.9}{12} = 13 \angle 22° \, 13' \text{ ohms}$$

Hence current

$$I = \frac{V}{Z} = \frac{240 \angle 0°}{13 \angle 22° \, 13'} = 18.5 \angle -22° \, 13' \text{ amperes}$$

i.e. **the current flowing is 18.5 amperes, lagging the voltage by 22° 13′**

Problem 5. Coplanar forces of $(7 + jx)$ newtons, $(3 - j5)$ newtons and $(y + j3)$ newtons act on a body which is in equilibrium. Determine the values of the three forces, both in rectangular and polar coordinate forms.

Since the body is in equilibrium, the algebraic sum of the forces is zero, i.e.

$$(7 + jx) + (3 - j5) + (y + j3) = 0 \qquad (14)$$

Also, since the resultant force is zero then the horizontal and vertical components of the force are zero. It follows that the real and imaginary parts of equation (14) are zero. Thus

$$7 + 3 + y = 0, \quad \text{i.e. } y = -10$$

and $j(x - 5 + 3) = j(0)$, i.e. $x = 2$

Thus the three forces are $(7 + j2)$, $(3 - j5)$ and $(-10 + j3)$

In polar form

$$(7 + j2) = \sqrt{(7^2 + 2^2)} \angle \arctan\left(\frac{2}{7}\right) = 7.280 \angle 15° \, 57'$$

$$(3 - j5) = \sqrt{(3^2 + 5^2)} \angle \arctan\left(-\frac{5}{3}\right) = 5.831 \angle -59° \, 2'$$

$$(-10 + j3) = \sqrt{(10^2 + 3^2)} \angle \arctan\left(-\frac{3}{10}\right) = 10.440 \angle 163° \, 18'$$

Problem 6. The velocity of ship A is $(4 - j5)$ knots and that of ship B is $(-6 - j6)$ knots. Find the magnitude and direction of the velocity of ship B relative to ship A.

Let v_A be the velocity of ship A, i.e. $v_A = (4 - j5)$, and let v_B be the velocity of ship B, i.e. $v_B = (-6 - j6)$

The velocity of B relative to A is $v_B - v_A$

$$v_B - v_A = (-6 - j6) - (4 - j5) = -10 - j$$

$$= \sqrt{(10^2 + 1^2)} \angle \arctan\left(\frac{-1}{-10}\right) = 10.05 \angle 185° \, 43'$$

Thus the magnitude of the velocity of B relative to A is **10.05 knots** and the direction of B relative to A is **185° 43′**

Problem 7. A projectile is given an initial velocity of $u = (21 + j8)$ m/s. Determine (a) the magnitude and direction of its velocity and (b) the distance and direction relative to the origin after 1.5 seconds. (Take g as 9.81 m/s^2, velocity as $v = u - jgt$ and displacement as $s = ut - j\frac{1}{2}gt^2$, where t is the time in seconds.)

(a) The velocity at time t seconds is $v = u - jgt$ (since g acts vertically downwards), hence

$$v = (21 + j8) - jgt$$
$$= 21 + j(8 - 9.81t)$$

When $t = 1.5$ s $v = 21 + j[8 - (9.81)(1.5)]$
$$= 21 - j6.715$$

In polar form $v = \sqrt{(21^2 + 6.715^2)} \angle \arctan\left(\dfrac{-6.715}{21}\right)$
$$= 22.05 \angle -17° \, 44'$$

Thus the magnitude of the velocity after 1.5 seconds is **22.05 m/s** and the direction is **−17° 44'**

(b) The displacement $s = ut - j\frac{1}{2}gt^2$
$$= (21 + j8)t - j(\tfrac{1}{2})(9.81)t^2$$
$$= 21t + j(8t - (\tfrac{1}{2})(9.81)t^2)$$

When $t = 1.5$ s $\quad s = 21(1.5) + j[(8)(1.5) - (\tfrac{1}{2})(9.81)(1.5)^2]$
$$= 31.5 + j0.963\,8$$

In polar form $\quad s = \sqrt{[(31.5)^2 + (0.963\,8)^2]} \angle \arctan\left(\dfrac{0.963\,8}{31.5}\right)$
$$= 31.51 \angle 1° \, 45'$$

Thus the distance from the origin is **31.51 m** and the direction is **1° 45'**

Further problems on the applications of complex numbers may be found in the following Section 9.9 (Problems 49 to 59), page 163.

9.9 Further problems

Operations on cartesian complex numbers

In Problems 1 to 5 solve the quadratic equations.

1. $x^2 + 16 = 0$ $\qquad [x = \pm j4]$
2. $x^2 - 2x + 2 = 0$ $\qquad [x = 1 \pm j]$
3. $2x^2 + 3x + 4 = 0$ $\qquad \left[x = -\dfrac{3}{4} \pm j\dfrac{\sqrt{23}}{4}\right]$
4. $5y^2 + 2y = -3$ $\qquad \left[y = -\dfrac{1}{5} \pm j\dfrac{\sqrt{14}}{5}\right]$
5. $4t^2 = t - 1$ $\qquad \left[t = \dfrac{1}{8} \pm j\dfrac{\sqrt{15}}{8}\right]$

6. Show on an Argand diagram the following complex numbers:
   (a) $3 + j6$ (b) $2 - j3$ (c) $-3 + j4$ (d) $-1 - j5$
7. Write down the complex conjugates of the following complex
   numbers: (a) $4 + j$ (b) $3 - j2$ (c) $-5 - j$
   (a) $[4 - j]$ (b) $[3 + j2]$ (c) $[-5 + j]$

In Problems 8 to 12 evaluate in $a + jb$ form, assuming that
$Z_1 = 2 + j3$, $Z_2 = 3 - j4$, $Z_3 = -1 + j2$ and $Z_4 = -2 - j5$

8. (a) $Z_1 - Z_2$ (b) $Z_2 + Z_3 - Z_4$
   (a) $[-1 + j7]$ (b) $[4 + j3]$

9. (a) $Z_1 Z_2$ (b) $Z_3 Z_4$
   (a) $[18 + j]$ (b) $[12 + j]$

10. (a) $Z_1 Z_3 Z_4$ (b) $Z_2 Z_3 + Z_4$
    (a) $[21 + j38]$ (b) $[3 + j5]$

11. (a) $\dfrac{Z_1}{Z_2}$ (b) $\dfrac{Z_1 + Z_2}{Z_3 + Z_4}$

    (a) $\left[-\dfrac{6}{25} + j\dfrac{17}{25} \right]$ (b) $\left[-\dfrac{2}{3} + j \right]$

12. (a) $\dfrac{Z_1 Z_2}{Z_1 + Z_2}$ (b) $Z_1 + \dfrac{Z_2}{Z_3} + Z_4$

    (a) $\left[\dfrac{89}{26} + j\dfrac{23}{26} \right]$ (b) $\left[-\dfrac{11}{5} - j\dfrac{12}{5} \right]$

13. Evaluate $\left[\dfrac{(1 + j)^2 - (1 - j)^2}{j} \right]$ [4]

14. If $Z_1 = 4 - j3$ and $Z_2 = 2 + j$ evaluate x and y, given
    $$x + jy = \frac{1}{Z_1 - Z_2} + \frac{1}{Z_1 Z_2}$$ $[x = 0.188, \; y = 0.216]$

15. Evaluate (a) j^8 (b) j^{11} (c) $\dfrac{3}{j^3}$ (d) $\dfrac{5}{j^6}$
    (a) $[1]$ (b) $[-j]$ (c) $[j3]$ (d) $[-5]$

16. Evaluate (a) $(1 + j)^4$ (b) $\dfrac{2 - j}{2 + j}$ (c) $\dfrac{1}{2 + j3}$
    (a) $[-4]$ (b) $\left[\dfrac{3}{5} - j\dfrac{4}{5} \right]$ (c) $\left[\dfrac{2}{13} - j\dfrac{3}{13} \right]$

17. If $Z = \dfrac{1 + j3}{1 - j2}$ evaluate Z^2 in $a + jb$ form $[0 - j2]$

18. Evaluate (a) j^{33} (b) $\dfrac{1}{(2 - j2)^4}$ (c) $\dfrac{1 + j3}{2 + j4} + \dfrac{3 - j2}{5 - j}$
    (a) $[j]$ (b) $\left[-\dfrac{1}{64} \right]$ (c) $[1.354 - j0.169]$

In Problems 19 to 23 solve the given complex equations.

19. $4(a + jb) = 7 - j3$ $\left[a = \dfrac{7}{4}, b = -\dfrac{3}{4} \right]$

20. $(3 + j4)(2 - j3) = x + jy$ $[x = 18, y = -1]$

21. $(1 + j)(2 - j) = j(p + jq)$ $[p = 1, q = -3]$

22. $(a - j3b) + (b - j2a) = 4 + j6$ $[a = 18, b = -14]$

23. $5 + j2 = \sqrt{(e + jf)}$ $[e = 21, f = 20]$

Polar form of complex numbers

In Problems 24 to 26 determine the modulus and the argument of each of the complex numbers given.

24. (a) $3 + j4$ (b) $2 - j5$
  (a) $[5, 53° \, 8']$ (b) $[5.385, -68° \, 12']$

25. (a) $-4 + j$ (b) $-5 - j3$
  (a) $[4.123, 165° \, 58']$ (b) $[5.831, -149° \, 2']$

26. (a) $(2 + j)^2$ (b) $j(3 - j)$
  (a) $[5, 53° \, 8']$ (b) $[3.162, 71° \, 34']$

In Problems 27 to 29 express the given cartesian complex numbers in polar form, leaving answers in surd form.

27. (a) $6 + j5$ (b) $3 - j2$ (c) -3 (a) $[\sqrt{61} \angle 39° \, 48']$
  (b) $[\sqrt{13} \angle -33° \, 41']$ (c) $[3 \angle 180°$ or $3 \angle \pi]$

28. (a) $-5 + j$ (b) $-4 - j3$ (c) $-j2$
  (a) $[\sqrt{26} \angle 168° \, 41']$ (b) $[5 \angle -143° \, 8']$
  (c) $\left[2 \angle -90°$ or $2 \angle -\dfrac{\pi}{2} \right]$

29. (a) $(-1 + j)^3$ (b) $-j(1 - j)$ (c) $j^3(2 - j3)$
  (a) $[\sqrt{8} \angle 45°]$ (b) $[\sqrt{2} \angle -135°]$ (c) $[\sqrt{13} \angle -146° \, 19']$

In Problems 30 to 32 convert the given polar complex numbers into $(a + jb)$ form, giving answers correct to 4 significant figures.

30. (a) $6 \angle 30°$ (b) $4 \angle 60°$ (c) $3 \angle 45°$ (a) $[5.196 + j3.000]$
  (b) $[2.000 + j3.464]$ (c) $[2.121 + j2.121]$

31. (a) $2 \angle \dfrac{\pi}{2}$ (b) $3 \angle \pi$ (c) $5 \angle \dfrac{5\pi}{6}$ (a) $[0 + j2.000]$
  (b) $[-3.000 + j0]$ (c) $[-4.330 + j2.500]$

32. (a) $8 \angle 150°$ (b) $4.2 \angle -120°$ (c) $3.6 \angle -25°$
  (a) $[-6.928 + j4.000]$ (b) $[-2.100 - j3.637]$
  (c) $[3.263 - j1.521]$

33. Using an Argand diagram, evaluate in polar form
  (a) $2 \angle 30° + 3 \angle 40°$ (b) $5.5 \angle 120° - 2.5 \angle -50°$
  (a) $[4.982 \angle 36°]$ (b) $[7.974 \angle 123° \, 7']$

In Problems 34 to 36 evaluate in polar form.

34. (a) $2 \angle 40° \times 5 \angle 20°$ (b) $2.6 \angle 72° \times 4.3 \angle 45°$
(a) $[10 \angle 60°]$ (b) $[11.18 \angle 117°]$

35. (a) $5.8 \angle 35° \div 2 \angle -10°$ (b) $4 \angle 30° \times 3 \angle 70° \div 2 \angle -15°$
(a) $[2.9 \angle 45°]$ (b) $[6 \angle 115°]$

36. (a) $\dfrac{4.1 \angle 20° \times 3.2 \angle -62°}{1.2 \angle 150°}$ (b) $6 \angle 25° + 3 \angle -36° - 4 \angle 72°$
(a) $[10.93 \angle 168°]$ (b) $[7.289 \angle 24° 35']$

37. Solve the complex equations, giving answers correct to 4 significant figures:

(a) $\dfrac{12 \angle \dfrac{\pi}{2} \times 3 \angle \dfrac{3\pi}{4}}{2 \angle -\dfrac{\pi}{3}} = x + jy$

(b) $15 \angle \dfrac{\pi}{3} + 12 \angle \dfrac{\pi}{2} - 6 \angle -\dfrac{\pi}{3} = r \angle \theta$

(a) $[x = 4.659, y = -17.39]$ (b) $[r = 30.52, \theta = 81° 31']$

38. Three vectors are represented by P, $2 \angle 30°$; Q, $3 \angle 90°$; and R, $4 \angle -60°$. Determine in polar form the vectors represented by (a) $P + Q + R$ and (b) $P - Q - R$
(a) $[3.770 \angle 8° 10']$ (b) $[1.488 \angle 100° 22']$

Powers and roots of complex numbers

In Problems 39 to 42 evaluate in cartesian and in polar form.

39. (a) $(2 + j3)^2$ (b) $(4 - j5)^2$
(a) $[-5 + j12; 13 \angle 112° 37']$ (b) $[-9 - j40; 41 \angle 102° 41']$

40. (a) $(-3 + j2)^5$ (b) $(-2 - j)^3$
(a) $[597 + j122; 609.3 \angle 11° 33']$
(b) $[-2 - j11; 11.18 \angle -100° 18']$

41. (a) $(4 \angle 32°)^4$ (b) $(2 \angle 125°)^5$
(a) $[-157.6 + j201.7; 256 \angle 128°]$
(b) $[-2.789 - j31.88; 32 \angle -95°]$

42. (a) $\left(3 \angle -\dfrac{\pi}{3}\right)^3$ (b) $(1.5 \angle -160°)^4$
(a) $[-27 + j0; 27 \angle -\pi]$ (b) $[0.879\,2 + j4.986; 5.063 \angle 80°]$

In Problems 43 to 45 determine the two square roots of the given complex numbers in cartesian form and show the results on an Argand diagram.

43. (a) $2 + j$ (b) $3 - j2$
(a) $[\pm(1.455 + j0.344)]$ (b) $[\pm(1.817 - j0.550)]$

44. (a) $-3 + j4$ (b) $-1 - j3$
    (a) $[\pm(1 + j2)]$ (b) $[\pm(1.040 - j1.443)]$

45. (a) $5 \angle 36°$ (b) $14 \angle \dfrac{3\pi}{2}$
    (a) $[\pm(2.127 + j0.691)]$ (b) $[\pm(-2.646 + j2.646)]$

46. If $Z = 3 \angle 30°$ evaluate in polar form
    (a) Z^2 (b) $\sqrt{Z}$
    (a) $[9 \angle 60°]$ (b) $[\sqrt{3} \angle 15°$ and $\sqrt{3} \angle -165°]$

47. Convert $2 - j$ into polar form and hence evaluate $(2 - j)^7$ in polar form $\qquad [\sqrt{5} \angle -26° \, 34'; \, 279.5 \angle 174° \, 3']$

48. Simplify,
$$\frac{\left(\cos\dfrac{\pi}{9} - j\sin\dfrac{\pi}{9}\right)^4}{\left(\cos\dfrac{\pi}{9} + j\sin\dfrac{\pi}{9}\right)^5} \qquad [-1]$$

Application of complex numbers

49. Determine the resistance R and series inductance L (or capacitance C) for each of the following impedances, assuming the frequency to be 50 Hz:
    (a) $4 + j7$ (b) $3 - j2$ (c) $j10$
    (d) $-j200$ (e) $15 \angle \dfrac{\pi}{3}$ (f) $6 \angle -45°$
    (a) $[R = 4\,\Omega, L = 22.3\,\text{mH}]$ (b) $[R = 3\,\Omega, C = 1\,592\,\mu\text{F}]$
    (c) $[R = 0, L = 31.8\,\text{mH}]$ (d) $[R = 0, C = 15.92\,\mu\text{F}]$
    (e) $[R = 7.5\,\Omega, L = 41.3\,\text{mH}]$ (f) $[R = 4.243\,\Omega, C = 750.3\,\mu\text{F}]$

50. An alternating voltage of 100 V, 50 Hz is applied across an impedance of $(20 - j30)$ ohms. Calculate (a) the resistance, (b) the capacitance, (c) the current and (d) the phase angle between current and voltage.
    (a) $[20\,\Omega]$ (b) $[106.1\,\mu\text{F}]$ (c) $[2.774\,\text{A}]$ (d) $[56° \, 19']$

51. Two voltages are represented by $(15 + j10)$ and $(12 - j4)$ volts. Determine the magnitude of the resultant voltage when these voltages are added. $\qquad [27.66\,\text{V}]$

52. Two impedances, $Z_1 = (2 + j6)$ ohms and $Z_2 = (5 - j2)$ ohms, are connected in series to a supply voltage of 100 V. Determine the magnitude of the current and its phase angle relative to the voltage.
    $[12.40\,\text{A}; \, 29° \, 45' \text{ lagging}]$

53. A resistance of 45 ohms is connected in series with a capacitor of 42 μF. If the applied voltage is 250 V, 50 Hz determine (*a*) the capacitive reactance, (*b*) the impedance, (*c*) the current and its phase relative to the applied voltage, (*d*) the voltage across the resistance and (*e*) the voltage across the capacitance.   (*a*) [75.79 Ω]   (*b*) [88.14 Ω] (*c*) [2.836 A at 59° 18′ leading]   (*d*) [127.6 V] (*e*) [214.9 V]

54. Forces of $(-3 - j5)$ N, $(13 + j2)$ N, $(-8 + j4)$ N and $(x + jy)$ N are in equilibrium. Find x and y $[x = -2\,\text{N}, y = -1\,\text{N}]$

55. Find the resultant of forces $F_1 = (3 - j17)$ newtons, $F_2 = (10 - j2)$ newtons and $F_3 = (-8 + j2)$ newtons. Express the resultant force in polar form. $[17.72\,\text{N} \angle -73°\,37′]$

56. A body moves from $(0 + j0)$ to point A at $(9 - j10)$. It then moves to points B and C at $(22 + j4)$ and $(-7 - j)$ respectively. Determine its distance from the origin and its angle relative to the real positive axis. $[25, \angle -16°\,16′]$

57. An aircraft, A, flying at a constant height has a velocity of $(380 + j270)$ km/h. Another aircraft, B, at the same height, has a velocity of $(290 - j417)$ km/h. Determine (*a*) the velocity of A relative to B and (*b*) the velocity of B relative to A, expressing the results in polar form. (*a*) [692.9 km/h at 82° 32′]   (*b*) [692.9 km/h at −97° 28′]

58. A projectile is given an initial velocity, *u*, of $(24 + j27)$ m/s. A second projectile launched from the same place in the same vertical plane at the same time has a velocity of $(31 + j17)$ m/s. After launching, both projectiles move freely under gravity. Determine the magnitude and direction of the velocity of the second projectile relative to the first. $(v = u - jgt)$ [12.21 m at $\angle -55°\,0′$]

59. In the hydrogen atom the angular momentum, *p*, of the de Broglie wave is given by $p\Psi = -(jh/2\pi)(\pm jm\Psi)$. Find *p* $$\left[p = \pm\frac{mh}{2\pi}\right]$$

Chapter 10

Polar coordinates

10.1   Introduction

There are two ways in which the position of a point in a plane can be represented. These are:

(a)   by **cartesian coordinates**, i.e. (x, y), and
(b)   by **polar coordinates**, i.e. (r, θ), where r is a 'radius' from a fixed point and θ is an angle from a fixed axis.

Conversion from cartesian to polar coordinates

With reference to Fig. 10.1, and also to Section 9.5, if x and y are known values, then r and θ may be calculated from: $r = \sqrt{(x^2 + y^2)}$, by Pythagoras's theorem and $\theta = \arctan \dfrac{x}{y}$, from trigonometric ratios.

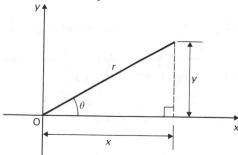

Fig. 10.1

Conversion from polar to cartesian coordinates

From the trigonometric ratios in Fig. 10.1, $\cos \theta = \dfrac{x}{y}$, from which $x = r \cos \theta$, and $\sin \theta = \dfrac{y}{r}$, from which $y = r \sin \theta$. Hence if r and θ are known values in Fig. 10.1, then:

$$x = r \cos \theta \quad \text{and} \quad y = r \sin \theta$$

Worked problems on converting cartesian to polar coordinates, and vice versa

Problem 1. Express in polar coordinates the position (5, 2).

A diagram representing the position using cartesian coordinates (5, 2) is shown in Fig. 10.2

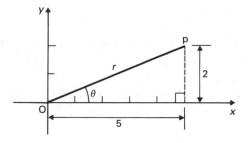

Fig. 10.2

From Pythagoras's theorem, $r = \sqrt{(5^2 + 2^2)} = \sqrt{29} = 5.385$

From trigonometric ratios, $\theta = \arctan \dfrac{2}{5} = 21.80°$ (or 21° 48′) or 0.381 radians

Hence the position of point P in polar coordinates is (5.385, 21.80°) or (5.385, 0.381 rad)

Problem 2. Express (−3, 4) in polar coordinates.

A sketch showing the position (−3, 4) is shown in Fig. 10.3 (to avoid errors it is suggested that a diagram should always be drawn).

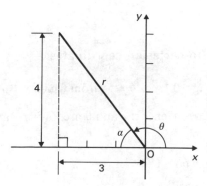

Fig. 10.3

From Fig. 10.3, $r = \sqrt{(3^2 + 4^2)} = 5$

and $\alpha = \arctan\dfrac{4}{3} = 53.13°$ (or $53°\ 8'$) or 0.927 rad

Hence $\theta = 180° - 53.13° = 126.87°$

or $\theta = \pi - 0.927 = 2.215$ rad

Thus $(-3, 4)$ in cartesian coordinates corresponds to $(5, 126.87°)$ or $(5, 2.215$ rad$)$ in polar coordinates.

(Note that angle θ must always be measured anticlockwise from the positive x-axis.)

Problem 3. Express $(-12, -5)$ in polar coordinates.

A sketch showing the position $(-12, -5)$ is shown in Fig. 10.4

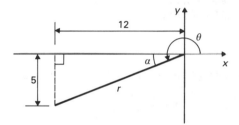

Fig. 10.4

From Fig. 10.4, $r = \sqrt{(12^2 + 5^2)} = 13$

and $\alpha = \arctan\dfrac{5}{12} = 22.62°$ (or $22°\ 37'$) or 0.395 rad

Hence $\theta = 180° + 22.62° = 202.62°$

and $\theta = \pi + 0.395 = 3.537$ rad

Thus $(-12, -5)$ in cartesian coordinates corresponds to $(13, 202.62°)$ or $(13, 3.537$ rad$)$ in polar coordinates.

Problem 4. Express $(3, -7)$ in polar coordinates.

A sketch showing the position $(3, -7)$ is shown in Fig. 10.5.

From Fig. 10.5, $r = \sqrt{(3^2 + 7^2)} = \sqrt{58} = 7.616$

and $\alpha = \arctan\dfrac{7}{3} = 66.80°$ (or $66°\ 48'$) or 1.166 rad

Hence $\theta = 360° - 66.80° = 293.20°$

or $\theta = 2\pi - 1.166 = 5.117$ rad

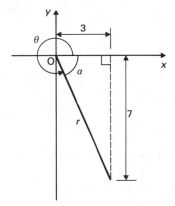

Fig. 10.5

Thus (3, −7) in cartesian coordinates corresponds to (7.616, 293.20°) or (7.616, 5.117 rad)

Problem 5. Express (4, 125°) in cartesian coordinates.

A sketch showing the position (4, 125°) is shown in Fig. 10.6

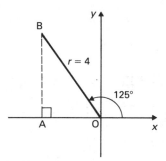

Fig. 10.6

From Fig. 10.6, $x = r\cos\theta = 4\cos 125° = -2.294$, which corresponds to length OA in Fig. 10.6, and
$y = 4\sin\theta = 4\sin 125° = 3.277$, which corresponds to length AB in Fig. 10.6.

Thus (4, 125°) in polar coordinates corresponds to (−2.294, 3.277) in cartesian coordinates.

(Note that when converting from polar to cartesian coordinates, it is not essential to draw a sketch. Use of $x = r\cos\theta$ and $y = r\sin\theta$ automatically produces the correct signs.)

Problem 6. Express (3.21, 5.37 rad) in cartesian coordinates.

A sketch showing the position (3.21, 5.37 rad) is shown in Fig. 10.7 (where 5.37 rad = 5.37 $\times \dfrac{180°}{\pi} = 307.68°$)

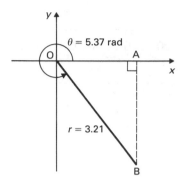

Fig. 10.7

From Fig. 10.7, $x = r \cos \theta = 3.21 \cos 5.37 = 1.962$, which corresponds to length OA in Fig. 10.7, and $y = r \sin \theta = 3.21 \sin 5.37 = -2.541$, which corresponds to length AB in Fig. 10.7

Thus (3.21, 5.37 rad) in polar coordinates corresponds to (1.962, −2.541) in cartesian coordinates.

Further problems on converting cartesian to polar coordinates, and vice versa, may be found in Section 10.3 (Problems 1 to 16), page 179.

10.2   Polar curves

With cartesian coordinates, the equation of a curve is expressed as a general relationship between x and y, i.e. $y = f(x)$. Typical examples include $y = 5x^2 - 3x + 1$, $y = \sin 2x$ and $y = 3 \cos^2 x$.

Similarly, with polar coordinates the equation of a curve is expressed in the form $r = f(\theta)$. Typical examples include $r = 3 \sin \theta$, $r = 4 \cos^2 \theta$ and $r = 3\theta$.

When a graph of $r = f(\theta)$ is required to be drawn, a table of values needs to be produced, thus giving the (r, θ) coordinates.

Worked problems on polar curves

Problem 1. Plot the polar graph of $r = 3 \sin \theta$ between $0°$ and $360°$ using $30°$ intervals.

A table of values at $30°$ intervals is produced as shown below.

θ	0	30°	60°	90°	120°	150°	180°
$r = 3 \sin \theta$	0	1.50	2.60	3.00	2.60	1.50	0

θ	210°	240°	270°	300°	330°	360°
$r = 3 \sin \theta$	-1.50	-2.60	-3.00	-2.60	-1.50	0

The polar graph $r = 3 \sin \theta$ is plotted as shown in Fig. 10.8. Initially the zero line OP is constructed, and the broken lines in Fig. 10.8 at $30°$ intervals are produced. The maximum value of r is 3.00, hence OP is scaled and circles drawn as shown with the largest radius of 3.00 units. The polar coordinates $(0, 0°)$, $(1.50, 30°)$, $(2.60, 60°)$, $(3.00, 90°)$, ... are plotted and are shown as points O, Q, R, S, T, U, O in Fig. 10.8. When the points are joined with a smooth curve, a circle is seen to have been produced. When plotting the next point $(-1.50, 210°)$, since r is negative, it is

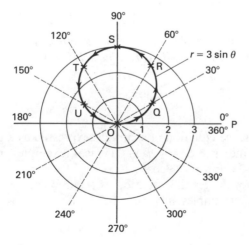

Fig. 10.8

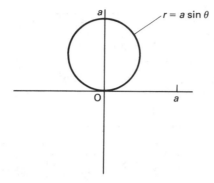

Fig. 10.9

plotted in the opposite direction to 210°, i.e. 1.50 units long on the 30° axis. Hence the point $(-1.50, 210°)$ is equivalent to the point $(1.50, 30°)$. Similarly, $(-2.60, 240°)$ is the same point as $(2.60, 60°)$, and so on. When all the coordinates are plotted, the graph of $r = 3 \sin \theta$ appears as a single circle, although, in fact, it is two circles, one on top of the other.

In general, a polar curve $r = a \sin \theta$ is as shown in Fig. 10.9. In a similar manner to that explained above, it may be shown that the polar curve $r = a \cos \theta$ is as shown in Fig. 10.10

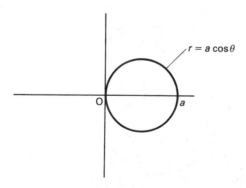

Fig. 10.10

Problem 2. Plot the polar graph of $r = 4 \cos^2 \theta$ between $\theta = 0$ and $\theta = 2\pi$ radians, using intervals of $\dfrac{\pi}{6}$ rad.

A table of values is produced as shown below.

θ	0	$\dfrac{\pi}{6}$	$\dfrac{\pi}{3}$	$\dfrac{\pi}{2}$	$\dfrac{2\pi}{3}$	$\dfrac{5\pi}{6}$	π
$\cos\theta$	1	0.866	0.50	0	-0.50	-0.866	-1
$r = 4\cos^2\theta$	4	3	1	0	1	3	4

θ	$\dfrac{7\pi}{6}$	$\dfrac{4\pi}{3}$	$\dfrac{3\pi}{2}$	$\dfrac{5\pi}{3}$	$\dfrac{11\pi}{6}$	2π
$\cos\theta$	-0.866	-0.50	0	0.50	0.866	1
$r = 4\cos^2\theta$	3	1	0	1	3	4

Initially, the zero line OP is constructed and the broken lines at intervals of $\dfrac{\pi}{6}$ rad (or 30°) are produced as shown in Fig. 10.11.

The maximum value of r is 4, hence OP is scaled, and circles produced as shown, with the largest at a radius of 4 units.

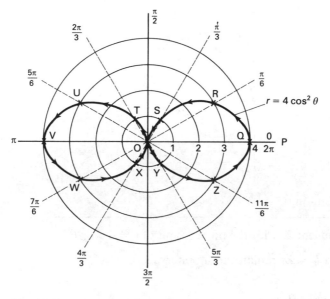

Fig. 10.11

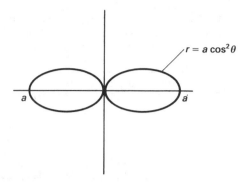

Fig. 10.12

The polar coordinates $(4, 0)$, $\left(3, \dfrac{\pi}{6}\right)$, $\left(1, \dfrac{\pi}{3}\right)$, ..., $(4, \pi)$ are plotted and shown as points Q, R, S, O, T, U and V respectively. Then $\left(3, \dfrac{7\pi}{6}\right)$, $\left(1, \dfrac{4\pi}{3}\right)$, ..., $(4, 2\pi)$ are plotted as shown by points W, X, O, Y, Z and Q respectively. Thus two distinct loops are produced as shown in Fig. 10.11

In general, a polar curve $y = a\cos^2\theta$ is as shown in Fig. 10.12. In a similar manner to that explained above, it may be shown that the polar curve $r = a\sin^2\theta$ is as shown in Fig. 10.13

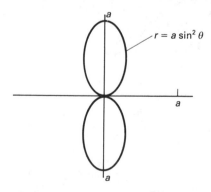

Fig. 10.13

Problem 3. Plot the graph of $r = 2\sin 2\theta$ between $\theta = 0°$ and $\theta = 360°$, using $15°$ intervals.

A table of values is produced as shown below.

θ	0	15°	30°	45°	60°	75°	90°	105°	120°
$r = 2 \sin 2\theta$	0	1.00	1.73	2.00	1.73	1.00	0	−1.00	−1.73

θ	135°	150°	165°	180°	195°	210°	225°	240°
$r = 2 \sin 2\theta$	−2.00	−1.73	−1.00	0	1.00	1.73	2.00	1.73

θ	255°	270°	285°	300°	315°	330°	345°	360°
$r = 2 \sin 2\theta$	1.00	0	−1.00	−1.73	−2.00	−1.73	−1.00	0

The polar graph of $r = 2 \sin 2\theta$ is plotted as shown in Fig. 10.14 and is seen to contain four similar loops displaced at 90° from each other.

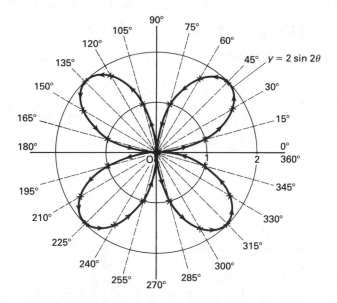

Fig. 10.14

In general, a polar curve $r = a \sin 2\theta$ is as shown in Fig. 10.15

In a similar manner, it may be shown that polar curves of $r = a \cos 2\theta$, $r = a \sin 3\theta$ and $r = a \cos 3\theta$ are as shown in Fig. 10.16

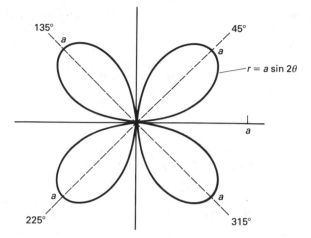

Fig. 10.15

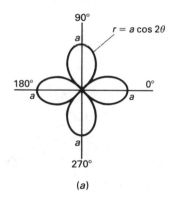

(a)

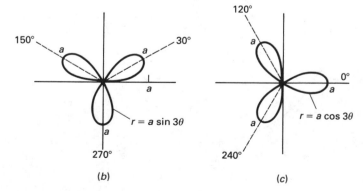

(b)   (c)

Fig. 10.16

Problem 4. Draw the polar curve $r = 3\theta$ between $\theta = 0$ and $\theta = \dfrac{5\pi}{2}$ rad at intervals of $\dfrac{\pi}{6}$

A table of values is produced as shown below.

θ	0	$\dfrac{\pi}{6}$	$\dfrac{\pi}{3}$	$\dfrac{\pi}{2}$	$\dfrac{2\pi}{3}$	$\dfrac{5\pi}{6}$	π	$\dfrac{7\pi}{6}$
$r = 3\theta$	0	1.57	3.14	4.17	6.28	7.85	9.42	11.00

θ	$\dfrac{4\pi}{3}$	$\dfrac{3\pi}{2}$	$\dfrac{5\pi}{3}$	$\dfrac{11\pi}{6}$	2π	$\dfrac{13\pi}{6}$	$\dfrac{7\pi}{3}$	$\dfrac{5\pi}{2}$
$r = 3\theta$	12.57	14.14	15.71	17.28	18.85	20.42	21.99	23.56

The polar curve of $r = 3\theta$ is shown in Fig. 10.17, and is seen to be an ever-increasing spiral.

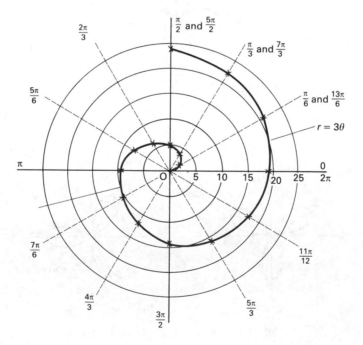

Fig. 10.17

Problem 5. Plot the polar curve $r = 2(1 + \cos \theta)$ from $\theta = 0°$ to $\theta = 360°$, using 30° intervals.

A table of values is produced as shown below.

θ	0	30°	60°	90°	120°	150°	180°
$r = 2(1 + \cos \theta)$	4.00	3.73	3.00	2.00	1.00	0.27	0

θ	210°	240°	270°	300°	330°	360°
$r = 2(1 + \cos \theta)$	0.27	1.00	2.00	3.00	3.73	4.00

The polar curve $r = 2(1 + \cos \theta)$ is shown in Fig. 10.18

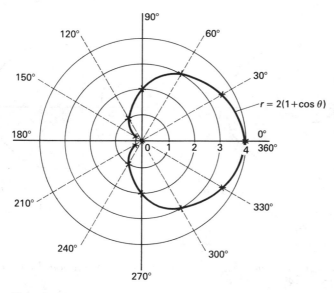

Fig. 10.18

In general, a polar curve $r = a(1 + \cos \theta)$ is as shown in Fig. 10.19 and the shape produced is called a **cardioid**.

It may also be shown that the polar curve $r = a + b \cos \theta$ varies in shape depending on the relative values of a and b.

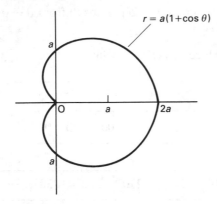

$r = a(1 + \cos \theta)$

a

O   a   2a

a

Fig. 10.19

(i)   When $a = b$, the polar curve shown in Fig. 10.19 results.
(ii) When $a < b$, the general shape shown in Fig. 10.20(a) results.
(iii) When $a > b$, the general shape shown in Fig. 10.20(b) results

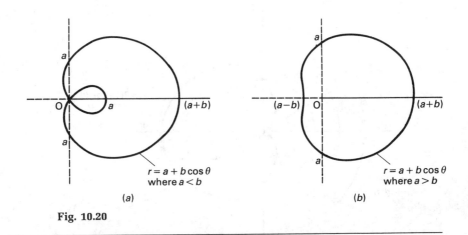

$r = a + b\cos\theta$
where $a < b$

$r = a + b\cos\theta$
where $a > b$

(a)   (b)

Fig. 10.20

Further problems on polar curves may be found in Section 10.3 following (Problems 17 to 28), page 179.

10.3   Further problems

Conversion of cartesian to polar coordinates, and vice versa

In Problems 1 to 8, express the given cartesian coordinates as polar coordinates, correct to 2 decimal places, in both degrees and radians.

1. (1, 6)   [(2.65, 80.54°) or (2.65, 1.41 rad)]
2. (5.91, 1.75)   [(6.16, 16.49°) or (6.16, 0.29 rad)]
3. (−4, 7)   [(8.06, 119.74°) or (8.06, 2.09 rad)]
4. (−6.29, 2.32)   [(6.70, 159.75°) or (6.70, 2.79 rad)]
5. (−9, −5)   [(10.30, 209.05°) or (10.30, 3.65 rad)]
6. (−2.16, −6.92)   [(7.25, 252.66°) or (7.25, 4.41 rad)]
7. (4, −2)   [(4.47, 333.43°) or (4.47, 5.82 rad)]
8. (7.83, −11.61)   [(14.00, 304.00°) or (14.00, 5.31 rad)]

In Problems 9 to 16, express the given polar coordinates as cartesian coordinates, correct to 3 decimal places.

9. (3, 62°)   [(1.408, 2.650)]
10. (2.31, 1.29 rad)   [(0.640, 2.220)]
11. (4, 152°)   [(−3.532, 1.878)]
12. (1.17, 2.43 rad)   [(−0.886, 0.764)]
13. (7.2, 212°)   [(−6.106, −3.815)]
14. (5.73, 3.97 rad)   [(−3.874, −4.222)]
15. (2.30, 297°)   [(1.044, −2.049)]
16. (5.32, 5.41 rad)   [(3.418, −4.077)]

Polar curves

In Problems 17 to 28, sketch the given polar curves.

17. $r = 2 \sin \theta$ [Similar to Fig. 10.9, when $a = 2$]
18. $r = 7 \cos \theta$ [Similar to Fig. 10.10, when $a = 7$]
19. $r = 3 \sin^2 \theta$ [Similar to Fig. 10.12, when $a = 3$]
20. $r = 5 \cos^2 \theta$ [Similar to Fig. 10.13, when $a = 5$]
21. $r = 6 \sin 2\theta$ [Similar to Fig. 10.15, when $a = 6$]
22. $r = 2 \cos 2\theta$ [Similar to Fig. 10.16(a), when $a = 2$]
23. $r = 4 \sin 3\theta$ [Similar to Fig. 10.16(b), when $a = 4$]
24. $r = \cos 3\theta$ [Similar to Fig. 10.16(c), when $a = 1$]
25. $r = 5\theta$ [Similar to Fig. 10.17]
26. $r = 7(1 + \cos \theta)$ [Similar to Fig. 10.19, when $a = 7$]
27. $r = 2 + 3 \cos \theta$
    [Similar to Fig. 10.20(a), when $a = 2$ and $b = 3$]
28. $r = 4 + \cos \theta$
    [Similar to Fig. 10.20(b), when $a = 4$ and $b = 1$]

Chapter 11

Arithmetic and geometric progressions

11.1 Arithmetic progressions

When a sequence has a constant difference between successive terms it is called an **arithmetic progression**, which is often abbreviated to A.P. Examples include

  (i)   1, 5, 9, 13, 17, ... where the **common difference** is 4,
  (ii) 9, 6, 3, 0, -3, ... where the **common difference** is -3 and
  (iii) $a, a + d, a + 2d, a + 3d, ...$ where the **common difference** is d.

nth term of an arithmetic progression

If the first term of an arithmetic progression is 'a' and the common difference is 'd', then:

the nth term is given by $a + (n - 1)d$

In example (i) above, the 8th term is given by

$$1 + (8 - 1)(4) = \mathbf{29}$$

and the 17th term is given by

$$1 + (17 - 1)(4) = \mathbf{65}$$

Similarly, in example (ii) above, the 10th term is given by

$$9 + (10 - 1)(-3) = \mathbf{-18}$$

and the 18th term is given by

$$9 + (18 - 1)(-3) = \mathbf{-42}$$

Sum of an arithmetic progression

The sum S of an arithmetic progression may be obtained by multiplying the average of all the terms by the number of terms. The average of all the terms is $\dfrac{a+1}{2}$, where 'a' is the first term and 'l' is the last term. For n terms, $l = a + (n - 1)d$.

Hence the sum of n terms,

$$S_n = n\left(\frac{a+l}{2}\right)$$

$$= \frac{n}{2}(a + l) = \frac{n}{2}[a + \{a + (n - 1)d\}]$$

i.e. $$S_n = \frac{n}{2}[2a + (n - 1)d]$$

For example, the sum of the first six terms of the series $1, 5, 9, 13, \ldots$ is given by

$$S_6 = \frac{6}{2}[2(1) + (6 - 1)4]$$

since $a = 1$ and $d = 4$,

i.e. $S_6 = 3(2 + 20) = \mathbf{66}$

and the sum of the first seven terms of the series $9, 6, 3, 0, -3, \ldots$ is given by

$$S_7 = \frac{7}{2}[2(9) + (7 - 1)(-3)]$$

$$= \frac{7}{2}[18 - 18] = \mathbf{0}$$

Arithmetic progressions – summary

If a = first term, d = common difference and n = number of terms then the arithmetic progression is: $a, a + d, a + 2d, \ldots$.

The nth term is: $a + (n - 1)d$

Sum of n terms, $S_n = \dfrac{n}{2}[2a + (n - 1)d] = \dfrac{n}{2}(a + l)$, where l = last term.

Worked problems on arithmetic progressions

Problem 1. Determine (a) the eighth, and (b) the nineteenth term of the series $3, 8, 13, \ldots$.

$3, 8, 13, \ldots$ is an arithmetic progression with a common difference d of 5

(a)   The nth term of an A.P. is given by $a + (n - 1)d$. Since the first term, $a = 3$, $d = 5$ and $n = 8$, then the eighth term is

$3 + (8 - 1)(5) = 3 + 35 = \mathbf{38}$

(b)   The nineteenth term is $3 + (19 - 1)(5) = 3 + 90 = \mathbf{93}$

Problem 2. The fifth term of an A.P. is 18 and the twelfth term is 46. Determine the 17th term.

The nth term of an A.P. is $a + (n - 1)d$

The 5th term is:

$$a + 4d = 18 \tag{1}$$

The 12th term is:

$$a + 11d = 46 \tag{2}$$

Equation (2) − equation (1) gives $7d = 28$

from which, $d = \dfrac{28}{7} = 4$

Substituting in equation (1) gives $a + 4(4) = 18$

from which, $a = 2$

Hence the 17th term is $a + (n - 1)d = 2 + (17 - 1)(4) = \mathbf{66}$

Problem 3. Determine the number of the term whose value is 40 in the series $1\frac{1}{2}, 4\frac{1}{4}, 7, 9\frac{3}{4}, \ldots$

$1\frac{1}{2}, 4\frac{1}{4}, 7, 9\frac{3}{4}, \ldots$ is an A.P. where $a = 1\frac{1}{2}$ and $d = 2\frac{3}{4}$

Hence if the nth term is 40,

then: $a + (n - 1)d = 40$

i.e. $1\frac{1}{2} + (n - 1)\left(2\frac{3}{4}\right) = 40$

$$2\frac{3}{4}(n - 1) = 40 - 1\frac{1}{2} = 38\frac{1}{2}$$

Therefore $(n - 1) = \dfrac{38\frac{1}{2}}{2\frac{3}{4}} = \dfrac{38.5}{2.75} = 14$

and $n = 14 + 1 = 15$

Hence 40 is the 15th term of the A.P.

Problem 4. Find the sum of the first eleven terms of the series 4, 10, 16, 22, ...

4, 10, 16, 22, . . . is an A.P. where $a = 4$ and $d = 6$

The sum of n terms of an A.P., $\qquad S_n = \dfrac{n}{2}[2a + (n - 1)d]$

Hence the sum of the first eleven terms, $S_n = \dfrac{11}{2}[2(4) + (11 - 1)(6)]$

$$= \dfrac{11}{2}[8 + 60] = \mathbf{374}$$

Problem 5. Determine the sum of the first 19 terms of the series 2.4, 3.1, 3.8, 4.5, ...

2.4, 3.1, 3.8, 4.5, . . . is an A.P. where $a = 2.4$ and $d = 0.7$

The sum of the first 19 terms, $S_{19} = \dfrac{n}{2}[2a + (n - 1)d]$

$$= \dfrac{19}{2}[2(2.4) + (19 - 1)0.7]$$

$$= \dfrac{19}{2}(4.8 + 12.6) = \dfrac{19}{2}(17.4)$$

$$= \mathbf{165.3}$$

Problem 6. The sum of 8 terms of an arithmetic progression is 42 and the common difference is 2. Determine the first term of the series.

$n = 8$, $d = 2$ and $S_8 = 42$

Since the sum of n terms of an A.P. is given by

$$S_n = \dfrac{n}{2}[2a + (n - 1)d] \text{ then } 42 = \dfrac{8}{2}[2a + (8 - 1)(2)] = 4(2a + 14)$$

Hence $\dfrac{42}{4} = 2a + 14$, i.e. $10.5 = 2a + 14$

$$10.5 - 14 = 2a$$

from which $\qquad a = \dfrac{-3.5}{2} = -1.75$

i.e. the first term is $\mathbf{-1.75}$

Problem 7. Three numbers are in arithmetic progression. Their sum is 21 and their product is 231. Determine the three numbers.

Let the three numbers be $(a - d)$, a and $(a + d)$

Then $(a - d) + a + (a + d) = 21$, i.e. $3a = 21$, from which, $a = 7$

Also $a(a - d)(a + d) = 231$

i.e. $a(a^2 - d^2) = 231$

Since $a = 7$, $7(7^2 - d^2) = 231$

$$7^2 - d^2 = \frac{231}{7} = 33$$

$$7^2 - 33 = d^2$$

from which, $d^2 = 49 - 33 = 16$ and $d = 4$

The three numbers are therefore $(7 - 4)$, 7 and $(7 + 4)$

i.e. **3, 7 and 11**

Problem 8. Find the sum of all the numbers between 3 and 198 which are exactly divisible by 3

The series $3, 6, 9, 12, \ldots, 198$ is an A.P. whose first term $a = 3$ and common difference $d = 3$

The last term is $a + (n - 1)d = 198$

i.e. $3 + (n - 1)(3) = 198$

from which, $(n - 1) = \dfrac{198 - 3}{3} = \dfrac{195}{3} = 65$

Hence $n = 65 + 1 = 66$

The sum of 66 terms is given by $S_{66} = \dfrac{n}{2}[2a + (n - 1)d]$

$$= \frac{66}{2}[2(3) + (66 - 1)(3)]$$

$$= 33[6 + 195] = 33(201)$$

$$= \mathbf{6633}$$

Problem 9. The first, eightieth and last term of an arithmetic progression are $2\frac{1}{2}$, 121 and 181 respectively. Determine (a) the number of terms of the series, (b) the sum of all the terms, and (c) the 70th term.

(a)   Let the A.P. be $a, a + d, a + 2d, \ldots, a + (n - 1)d$, where $a = 2\frac{1}{2}$

The 80th term is $a + (80 - 1)d = 121$

i.e. $2\frac{1}{2} + 79d = 121$

from which, $79d = 121 - 2\frac{1}{2} = 118\frac{1}{2}$

and $d = \dfrac{118.5}{79} = 1.5$

The last term is $a + (n - 1)d$,

i.e. $2\frac{1}{2} + (n - 1)(1\frac{1}{2}) = 181$

$$(n - 1) = \frac{181 - 2\frac{1}{2}}{1\frac{1}{2}} = \frac{178.5}{1.5} = 119$$

Hence the number of terms in the series, $n = 119 + 1 = 120$

(b)   Sum of all the terms, $S_{120} = \frac{n}{2}[2a + (n - 1)d]$

$$= \frac{120}{2}[2(2\frac{1}{2}) + (120 - 1)(1\frac{1}{2})]$$

$$= 60[5 + 119(1\frac{1}{2})] = 60(183.5)$$

$$= 11\,010$$

(c)   The 70th term is $a + (n - 1)d = 2\frac{1}{2} + (70 - 1)(1\frac{1}{2})$

$$= 2\frac{1}{2} + 69(1\frac{1}{2}) = 2\frac{1}{2} + 103\frac{1}{2} = \mathbf{106}$$

Further problems on arithmetic progressions may be found in Section 11.3 (Problems 1 to 16), page 191.

11.2   Geometric progressions

When a sequence has a constant ratio between successive terms it is called a **geometric progression**, which is often abbreviated to G.P. The constant is called the **common ratio**.

Examples include

(i)   1, 2, 4, 8, ... where the common ratio is 2,

(ii)   20, 10, 5, $2\frac{1}{2}$, ... where the common ratio is $\frac{1}{2}$, and

(iii) $a, ar, ar^2, ar^3, \ldots$ where the common ratio is r.

nth term of a geometric progression

If the first term of a geometric progression is 'a' and the common ratio is r, then:

the nth term is ar^{n-1}

which can be readily checked from the above examples.

For example, the 7th term of the G.P. 1, 2, 4, 8, ... is $(1)(2)^6 = \mathbf{64}$ since $a = 1$ and $r = 2$, and the 6th term of the G.P. 20, 10, 5, $2\frac{1}{2}$, ... is $(20)\left(\frac{1}{2}\right)^5 = \frac{20}{32} = \frac{5}{8}$ or **0.625**.

Sum of a geometric progression

Let a geometric progression be $a, ar, ar^2, ar^3, \ldots, ar^{n-1}$ then the sum of n terms,

$$S_n = a + ar + ar^2 + ar^3 + \cdots + ar^{n-1} \cdots \qquad (3)$$

Multiplying throughout by r gives

$$rS_n = ar + ar^2 + ar^3 + ar^4 + \cdots + ar^{n-1} + ar^n \cdots \qquad (4)$$

Subtracting equation (3) from equation (4) gives:

$$rS_n - S_n = ar^n - a$$

and $S_n(r - 1) = a(r^n - 1)$

Thus the sum of n terms, $S_n = \dfrac{a(r^n - 1)}{(r - 1)}$

which is valid when $r > 1$.

Subtracting equation (4) from equation (3) gives:

$$S_n - rS_n = a - ar^n$$

and $S_n(1 - r) = a(1 - r^n)$

Thus the sum of n terms, $S_n = \dfrac{a(1 - r^n)}{(1 - r)}$, which is valid when $r < 1$.

For example, the sum of the first 9 terms of the G.P. 1, 2, 4, 8, ... is given by $S_9 = \dfrac{a(r^n - 1)}{(r - 1)} = \dfrac{1(2^9 - 1)}{(2 - 1)}$, since $a = 1$ and $r = 2$,

i.e. $S_9 = \dfrac{1(512 - 1)}{1} = \mathbf{511}$,

and the sum of the first 5 terms of the G.P. 20, 10, 5, $2\frac{1}{2}$, ... is given by $S_5 = \dfrac{a(1 - r^n)}{(1 - r)} = \dfrac{20(1 - (\frac{1}{2})^5)}{(1 - \frac{1}{2})}$, since $a = 20$ and $r = \dfrac{1}{2}$,

i.e. $S_5 = \dfrac{20\left(1 - \dfrac{1}{32}\right)}{\dfrac{1}{2}} = \dfrac{20\left(\dfrac{31}{32}\right)}{\dfrac{1}{2}} = \mathbf{38.75}$

Sum to infinity of a geometric progression

When the common ratio r of a geometric progression is less than unity, the sum of n terms is given by $S_n = \dfrac{a(1 - r^n)}{(1 - r)}$, which may be written as

$$S_n = \frac{a}{(1-r)} - \frac{ar^n}{(1-r)}$$

Since $r < 1$, r^n becomes less as n increases, i.e. $r^n \to 0$ as $n \to \infty$.

Hence $\dfrac{ar^n}{(1-r)} \to 0$ as $n \to \infty$

Thus $\quad S_n \to \dfrac{a}{(1-r)}$ as $n \to \infty$

The quantity $\dfrac{a}{(1-r)}$ is called the **sum to infinity**, S_∞, and is the limiting value

of a sum of an infinite number of terms,

i.e. $\quad S_\infty = \dfrac{a}{(1-r)}$, which is valid when $-1 < r < 1$

For example, the sum to infinity of the G.P. 20, 10, 5, $2\frac{1}{2}$, ... is $S_\infty = \dfrac{20}{(1-\frac{1}{2})}$,

since $a = 20$ and $r = \dfrac{1}{2}$

i.e. $\quad S_\infty = \mathbf{40}$

Similarly, the sum to infinity of the G.P. $1, \frac{1}{2}, \frac{1}{4}, \ldots$ is $S_\infty = \dfrac{1}{(1-\frac{1}{2})} = \mathbf{2}$, since

$a = 1$ and $r = \dfrac{1}{2}$

Geometric progressions – summary

If $a =$ first term, $r =$ common ratio and $n =$ number of terms then the geometric progression is: $a, ar, ar^2, \ldots$. The nth term is ar^{n-1}

Sum of n terms, $S_n = \dfrac{a(r^n - 1)}{(r-1)}$ or $\dfrac{a(1-r^n)}{(1-r)}$

If $-1 < r < 1$, $S_\infty = \dfrac{a}{(1-r)}$

Worked problems on geometric progressions

Problem 1. Determine the ninth term of the series 2, 4, 8, 16, ...

2, 4, 8, 16, ... is a G.P. with a common ratio, r, of 2. The nth term of a G.P. is ar^{n-1}, where a is the first term. Hence the ninth term is $(2)(2)^{9-1} = (2)(2)^8 = 2(256) = \mathbf{512}$

Problem 2. Find the sum of the first eight terms of the series 2, 6, 18, 54, ...

2, 6, 18, 54, ... is a G.P. with a common ratio, $r = 3$

The sum of n terms, $S_n = \dfrac{a(r^n - 1)}{(r - 1)}$ when $r > 1$

Hence $S_8 = \dfrac{2(3^8 - 1)}{(3 - 1)} = \dfrac{2(6\,561 - 1)}{2} = \mathbf{6\,560}$

Problem 3. The first term of a G.P. is 10 and the fourth term is 33.75. Determine the 6th term and the 10th term, each correct to 5 significant figures.

The fourth term of a G.P. is given by $ar^3 = 33.75$, where the first term, $a = 10$

Hence $10r^3 = 33.75$, $r^3 = \dfrac{33.75}{10} = 3.375$

and $\qquad r = \sqrt[3]{(3.375)} = 1.5$

The 6th term is $ar^5 = 10(1.5)^5 = \mathbf{75.938}$

and the 10th term is $ar^9 = 10(1.5)^9 = \mathbf{384.43}$

Problem 4. Which term of the series 729, 243, 81, ... is $\dfrac{1}{27}$?

729, 243, 81, ... is a G.P. with a common ratio, $r = \dfrac{1}{3}$ and first term, $a = 729$.

The nth term of a G.P. is given by ar^{n-1}

Hence $\dfrac{1}{27} = 729\left(\dfrac{1}{3}\right)^{n-1}$

from which, $\left(\dfrac{1}{3}\right)^{n-1} = \dfrac{1}{(27)(729)} = \dfrac{1}{3^3 3^6} = \dfrac{1}{3^9} = \left(\dfrac{1}{3}\right)^{-9}$

Thus $n - 1 = 9$ and $n = 9 + 1 = 10$

Hence $\dfrac{1}{27}$ is the 10th term of the G.P.

Problem 5. Find the sum of the first 8 terms of the series 57.00, 34.20, 20.52, ... correct to 6 significant figures.

The common ratio $r = \dfrac{ar}{a} = \dfrac{34.20}{57.00} = 0.60$

$$\left(\text{Also } \frac{ar^2}{ar} = \frac{20.52}{34.20} = 0.60\right)$$

Hence $a = 57.00$, $r = 0.60$ and the sum of 8 terms,

$$S_8 = \frac{a(1 - r^n)}{(1 - r)} = \frac{57.00(1 - 0.60^8)}{(1 - 0.60)}$$

$$= \frac{57.00(1 - 0.016\,796\,1\cdots)}{0.40} = \mathbf{140.107}$$

Problem 6. Find the sum to infinity of the series $4, 1, \dfrac{1}{4}, \ldots$

$4, 1, \dfrac{1}{4}, \ldots$ is a G.P. of common ratio, $r = \dfrac{1}{4}$

The sum to infinity, $S_\infty = \dfrac{a}{(1 - r)} = \dfrac{4}{\left(1 - \dfrac{1}{4}\right)} = \dfrac{4}{3/4} = \dfrac{16}{3}$

$$= 5\frac{1}{3}$$

Problem 7. In a geometric progression the fifth term is 8 times the second term and the sum of the 6th and 8th terms is 160. Determine (a) the common ratio, (b) the first term, and (c) the sum of the fourth to tenth terms, inclusive.

(a) Let the G.P. be $a, ar, ar^2, ar^3, \ldots, ar^{n-1}$. The second term is ar and the fifth term is ar^4. The fifth term is 8 times the second term, hence $ar^4 = 8ar$, from which, $r^3 = 8$ and $r = \sqrt[3]{8} = 2$, i.e. **the common ratio is 2**

(b) The sum of the 6th and 8th terms is 160

Hence $ar^5 + ar^7 = 160$

From part (a), $r = 2$, hence $a(2)^5 + a(2)^7 = 160$

i.e. $32a + 128a = 160$

and $\qquad 160a = 160$, from which, $a = 1$

Hence the first term is 1

(c) The sum of the 4th to 10th terms, inclusive, is given by

$$S_{10} - S_3 = \frac{a(r^{10} - 1)}{(r - 1)} - \frac{a(r^3 - 1)}{(r - 1)}$$

$$= \frac{1(2^{10} - 1)}{(2 - 1)} - \frac{1(2^3 - 1)}{(2 - 1)}$$

$$= (2^{10} - 1) - (2^3 - 1) = 2^{10} - 2^3 = 1\,024 - 8 = \mathbf{1\,016}$$

Problem 8. If £250 is invested at compound interest of 8% per annum, determine (a) the value after 8 years, (b) the time, correct to the nearest year, it takes to reach more than £600.

(a)   Let the G.P. be a, ar, ar^2, ..., ar^{n-1}

The first term $a = $ £250 and the common ratio $r = 1.08$

Hence the second term is $ar = (250)(1.08) = $ £270, which is the value after 1 year, the third term is
$ar^2 = (250)(1.08)^2 = $ £291.60 which is the value after 2 years, and so on.

Thus the value after 8 years is $ar^8 = (250)(1.08)^8$

$$= \textbf{£462.73}$$

(b)   When £600 has been reached, $600 = ar^n$

i.e. $600 = (250)(1.08)^n$

$$\frac{600}{250} = (1.08)^n \text{ and } 2.4 = (1.08)^n$$

Taking logarithms to base 10 of both sides gives

$\lg 2.4 = \lg(1.08)^n = n \lg(1.08)$, by the laws of logarithms,

from which, $n = \dfrac{\lg 2.4}{\lg 1.08} = 11.38$

Hence it will take 12 years to reach more than £600

Problem 9. A hire tool company finds that their net return for hiring tools is decreasing by 15% per annum. If their net gain on a certain tool this year is £750, find the possible total of all future profits from this tool (assuming the tool lasts for ever).

The net gain forms a series £750, £750 × 0.85,
£750 × $(0.85)^2$, ... which is a G.P. with $a = $ £750 and $r = 0.85$

The sum to infinity, $S_\infty = \dfrac{a}{(1-r)} = \dfrac{750}{(1-0.85)} = $ £5 000

Hence the total future profit is £5 000

Problem 10. A drilling machine is to have 7 speeds ranging from 100 rev/min to 1 000 rev/min. If the speeds form a geometric progression, determine their value, each correct to the nearest whole number.

Let the G.P. of n terms be given by a, ar, ar^2, ..., ar^{n-1}. The first term $a = 100$ rev/min

The seventh term is given by ar^{7-1}, which is 1 000 rev/min,

i.e. $ar^6 = 1\,000$, from which, $r^6 = \dfrac{1\,000}{100} = 10$

Thus the common ratio, $r = \sqrt[6]{10} = 1.467\,799\,3$

The first term,	$a = 100$ rev/min	
the second term,	$ar = (100)(1.467\,799\,3)$	$= 146.78$
the third term,	$ar^2 = (100)(1.467\,799\,3)^2$	$= 215.44$
the fourth term,	$ar^3 = (100)(1.467\,799\,3)^3$	$= 316.23$
the fifth term,	$ar^4 = (100)(1.467\,799\,3)^4$	$= 464.16$
the sixth term,	$ar^5 = (100)(1.467\,799\,3)^5$	$= 681.29$

and the seventh term, $ar^6 = (100)(1.467\,799\,3)^6 = 1\,000$

Hence, correct to the nearest whole number, the 7 speeds of the drilling machine are

100, 147, 215, 316, 464, 681 and 1 000 rev/min

Further problems on geometric progressions may be found in Section 11.3 following (Problems 17 to 29), page 192.

11.3   Further problems

Arithmetic progressions

1. Find the 13th term of the series 6, 11, 16, 21, ...   [62]
2. Determine the 19th term of the series 15, 14.4, 13.8, 13.2, ...   [4.2]
3. The 8th term of a series is 23 and the 13th term is 38. Determine the 18th term.   [53]
4. Find the 14th term of the arithmetic progression of which the first term is $1\frac{1}{2}$ and the tenth term is $19\frac{1}{2}$ [$27\frac{1}{2}$]
5. Determine the number of the term which is 26.4 in the series 4, 5.6, 7.2, 8.8, ...   [15th]
6. Find the sum of the first 15 terms of the series 4, 7, 10, 13, ...   [46]
7. Determine the sum of the series 4.2, 5.7, 7.2, 8.7, ..., 31.2   [336.3]
8. The sum of 22 terms of an A.P. is 407 and the common difference is 2. Determine the first term of the series. [$-2\frac{1}{2}$]
9. Three numbers are in arithmetic progression. Their sum is 15 and their product is 45. Determine the three numbers. [1, 5, 9]

10. Find the sum of all the numbers between 4 and 176 which are exactly divisible by 5   [2 975]

11. Find the number of terms in the series 4, 7, 10, ... of which the sum is 4 186   [52]

12. Insert three terms between $2\frac{1}{2}$ and $21\frac{1}{2}$ to form an arithmetic progression.   [7, $11\frac{1}{2}$, 16]

13. The first, fiftieth and last terms of an A.P. are 6, 202 and 382. Find (a) the number of terms, (b) the sum of all the terms and (c) the 75th term.   [(a) 95   (b) 18 430   (c) 302]

14. When starting a job a man is paid a salary of £17 500 per annum and receives annual increments of £450. Determine his salary in the 8th year and calculate the total he will have earned in the first 11 years.   [£20 650; £217 250]

15. The fifth term of an A.P. is 23 and the eleventh term is 35. Find the first term, the common difference and the sum of the first ten terms.   [15, 2, 240]

16. An oil company bores a hole 90 m deep. Estimate the cost of boring if the cost is £50 for drilling the first metre, with an increase in cost of £3 per metre for succeeding metres. [£12 510]

Geometric progressions

17. Determine the 9th term of the series 3, 6, 12, 24, ...   [768]

18. Find the sum of the first 8 terms of the series $\frac{1}{2}$, $1\frac{1}{2}$, $4\frac{1}{2}$, $13\frac{1}{2}$, ...   [1 640]

19. The first term of a geometric progression is 5 and the fifth term is 160. Determine the 7th and 10th terms.   [320; 2 560]

20. Which term of the series 2, 6, 18, ... is 13 122?   [9th]

21. Determine the sum of the first seven terms of the series 1, $2\frac{1}{2}$, $6\frac{1}{4}$, ... correct to 4 significant figures.   [406.2]

22. Find the sum to infinity of 5, 1, $\frac{1}{5}$, ...   [6.25]

23. Determine the sum to infinity of 4, -1, $\frac{1}{4}$, ... $\left[5\frac{1}{3}\right]$

24. In a geometric progression the 8th term is 8 times the 5th term and the sum of the 6th and 7th term is 288. Determine (a) the common ratio, (b) the first term and (c) the sum of the 5th to 10th terms inclusive.   [(a) 2   (b) 3   (c) 6 096]

25. The value of a metal forming machine originally valued at £5 000 depreciates 12% per annum. Determine its value after 5 years, correct to the nearest pound. The machine is sold when its value is less than £800. After how many years is the machine sold?   [£2 998, 15 years]

26. If the population of Great Britain is 54 million and is decreasing at 2% per annum, determine the population in 7 years time.   [46.88 M]

27. 250 grammes of a radioactive substance disintegrates at a rate of 2.5% per annum. How much of the substance is left after 15 years?   [175.4 g]

28. A drilling machine is to have 6 speeds ranging from 150 rev/min to 800 rev/min. If the speeds form a G.P., determine their values, each correct to the nearest whole number.   [150, 210, 293, 410, 572, 800 rev/min]

29. If £500 is invested at compound interest at 6.5% p.a., determine (a) the value after 10 years, (b) the time, correct to the nearest year, it takes to reach £1 200
     [(a) £938.57 (b) 14 years]

Chapter 12

The solution of triangles

12.1 Solution of scalene triangles

A scalene triangle is one in which all its sides and angles are unequal. Solving a triangle means finding all unknown sides and angles.

Use is made of the sine and cosine rules when solving scalene triangles. With reference to triangle ABC shown in Fig. 12.1:

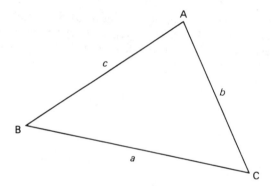

Fig. 12.1

the **sine rule** states: $\dfrac{a}{\sin A} = \dfrac{b}{\sin B} = \dfrac{c}{\sin C}$

and the **cosine rule** states: $a^2 = b^2 + c^2 - 2bc \cos A$

or $b^2 = a^2 + c^2 - 2ac \cos B$

or $c^2 = a^2 + b^2 - 2ab \cos C$

Worked problems on the solution of scalene triangles

Problem 1. In a triangle ABC, $C = 39° \, 14'$, side $c = 6.47$ cm and side $b = 8.23$ cm. Solve the triangle.

Using the sine rule: $\dfrac{8.23}{\sin B} = \dfrac{6.47}{\sin 39° \ 14'}$

$\sin B = \dfrac{8.23 \sin 39° \ 14'}{6.47}$

Hence $B = 53° \ 34'$ or $126° \ 26'$

If $B = 53° \ 34'$ then $A = 180° - 53° \ 34' - 39° \ 14' = 87° \ 12'$

If $B = 126° \ 26'$ then $A = 180° - 126° \ 26' - 39° \ 14' = 14° \ 20'$

Thus there are two possible solutions of triangle ABC. This occurrence is known as the ambiguous case.

Case 1. When $A = 87° \ 12'$ then

$\dfrac{a}{\sin 87° \ 12'} = \dfrac{6.47}{\sin 39° \ 14'}$

$a = \dfrac{6.47 \sin 87° \ 12'}{\sin 39° \ 14'} = \textbf{10.22 cm}$

Case 2. When $A = 14° \ 20'$ then

$\dfrac{a}{\sin 14° \ 20'} = \dfrac{6.47}{\sin 39° \ 14'}$

$a = \dfrac{6.47 \sin 14° \ 20'}{\sin 39° \ 14'} = \textbf{2.53 cm}$

Figure 12.2 shows the two possible solutions of triangle ABC.

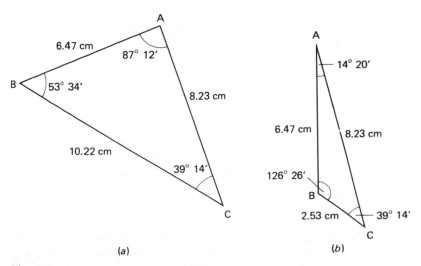

(a)   (b)

Fig. 12.2

Both sets of answers are feasible since in each case the lengths of the sides are related directly to the sizes of the corresponding angles (i.e. the longest side is opposite the largest angle and so on).

Problem 2. Solve triangle DEF given $F = 64° 55'$, side $d = 14.83$ cm and side $e = 10.72$ cm

Using the cosine rule:

$$f^2 = d^2 + e^2 - 2de \cos F$$
$$= (14.83)^2 + (10.72)^2 - [2(14.83)(10.72) \cos 64° 55']$$
$$= 200.1$$
$$f = 14.14 \text{ cm}$$

Using the sine rule:

$$\frac{14.14}{\sin 64° 55'} = \frac{10.72}{\sin E}$$

$$\sin E = \frac{10.72 \sin 64° 55'}{14.14}$$

Hence $E = 43° 22'$ or $136° 38'$

When $E = 43° 22'$, $D = 180° - 43° 22' - 64° 55' = 71° 43'$

When $E = 136° 38'$, $D = 180° - 136° 38' - 64° 55' = -21° 33'$

This latter solution is impossible and is neglected.

Hence in triangle DEF, $f = 14.14$ cm, $D = 71° 43'$ and $E = 43° 22'$

Problem 3. Find the three angles of a triangular template having sides of 6.42 cm, 8.31 cm and 9.78 cm

Let a triangle ABC represent the template with $a = 6.42$ cm, $b = 8.31$ cm and $c = 9.78$ cm.

The largest angle is C since 9.78 cm is the longest side.

Since $c^2 = a^2 + b^2 - 2ab \cos C$

$$\cos C = \frac{a^2 + b^2 - c^2}{2ab} = \frac{(6.42)^2 + (8.31)^2 - (9.78)^2}{2(6.42)(8.31)}$$

$\therefore$ $\cos C = 0.137\,1$

$C = \text{arcos } 0.137\,1 = 82° 7'$

Since $\cos C$ is positive the triangle is acute-angled. If the value of $\cos C$ had been negative C would have been an angle between 90° and 180° and the triangle would have been obtuse-angled.

By finding the largest angle first, when given the three sides of a triangle, an immediate indication of the type of triangle is given.

Using the sine rule: $\dfrac{9.78}{\sin 82° \, 7'} = \dfrac{8.31}{\sin B}$

$\sin B = \dfrac{8.31 \sin 82° \, 7'}{9.78}$

$B = 57° \, 19'$

$A = 189° - 82° \, 7' - 57° \, 19' = 40° \, 34'$

Hence **the three angles of the triangular template are 40° 34',
57° 19' and 82° 7'**

Problem 4. Two phasors are shown in Fig. 12.3. When
$i_1 = 14.6$ amperes and $i_2 = 23.8$ amperes, calculate the value of
their resultant (i.e. length AC) and the angle it makes with i_1

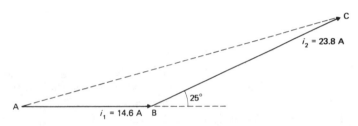

Fig. 12.3

In triangle ABC shown in Fig. 12.3:

$B = 180° - 25° = 155°$

Using the cosine rule:

$AC^2 = AB^2 + BC^2 - [2(AB)(BC) \cos 155°]$
$\quad\;\; = (14.6)^2 + (23.8)^2 - [2(14.6)(23.8)(\cos 155°)]$
$\quad\;\; = 1\,409$

Here the resultant $AC = \sqrt{1\,409} = 37.54$ amperes.

Using the sine rule: $\dfrac{23.8}{\sin A} = \dfrac{37.54}{\sin 155°}$

$\sin A = \dfrac{23.8 \sin 155°}{37.54}$

$A = 15° \, 32'$

Hence **the resultant AC makes an angle of 15° 32' with i_1**

Problem 5. A quadrilateral plot of ground ABDC has the dimensions: AB = 60 m, BD = 130 m, DC = 145 m and CA = 124 m. The angle BAC is 64°. Determine BC and the angle BDC.

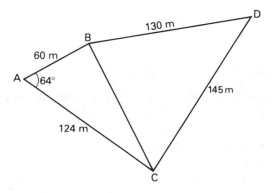

Fig. 12.4

The plot of ground is shown in Fig. 12.4

The quadrilateral ABDC is divided into two triangles by joining BC. In triangle ABC the given information is 2 sides and the included angle. Thus the cosine rule can be used.

$$\text{Hence } BC^2 = 60^2 + 124^2 - 2(60)(124)\cos 64°$$

$$= 3\,600 + 15\,376 - 6\,523$$

$$= 12\,453$$

$$BC = \sqrt{122\,453} \text{ m}$$

$$\mathbf{BC = 111.6\ m}$$

In triangle BCD the given information is three sides. Thus the cosine rule can be used again.

$$\cos \angle BDC = \frac{130^2 + 145^2 - 111.6^2}{2 \times 130 \times 145}$$

$$= \frac{16\,900 + 21\,025 - 12\,455}{37\,700}$$

$$= 0.675\,6$$

$$\textbf{Hence } \angle \textbf{BDC} = \textbf{47.50°} = \textbf{47°30′}$$

Further problems on the solution of scalene triangles may be found in the following Section 12.2 (Problems 1 to 18), page 199.

12.2 Further problems

1. Solve the following triangles ABC:
   (a) $A = 53°$, $B = 61°$, $a = 12.6$ cm
   (b) $B = 83° 16'$, $b = 16.48$ mm, $c = 12.92$ mm
   (c) $A = 37°$, $B = 73°$, $b = 4.3$ m
   (a) $[C = 66°$, $b = 13.80$ cm, $c = 14.41$ cm]
   (b) $[A = 45° 36'$, $c = 51° 8'$, $a = 11.86$ mm]
   (c) $[a = 2.71$ m, $c = 4.23$ m, $C = 70°]$

2. Solve the following triangles DEF:
   (a) $F = 41°$, $e = 9.30$ cm, $f = 7.21$ cm
   (b) $D = 58°$, $E = 69°$, $e = 7.4$ cm
   (c) $D = 114° 8'$, $F = 21° 5'$, $d = 14.6$ cm
   (a) $[D = 81° 12'$, $E = 57° 48'$, $d = 10.86$ cm
      or $D = 16°48'$, $E = 122° 12'$, $d = 3.177$ cm]
   (b) $[d = 6.72$ cm, $f = 6.33$ cm, $F = 53°]$
   (c) $[e = 11.27$ cm, $f = 5.755$ cm, $E = 44° 47']$

3. Solve the following triangles GHI:
   (a) $g = 8.52$ m, $i = 12.7$ m, $I = 24° 9'$
   (b) $h = 17.86$ m, $i = 12.67$ m, $H = 83° 46'$
   (c) $h = 45.3$ mm, $i = 35.7$ mm, $I = 36° 47'$
   (a) $[G = 15° 56'$, $H = 139° 55'$, $h = 20.03$ m]
   (b) $[G = 51° 23'$, $I = 44° 51'$, $g = 14.04$ m]
   (c) $[H = 49° 21'$, $G = 93° 52'$, $g = 59.57$ mm
      or $H = 130° 39'$, $G = 12° 34'$, $g = 12.97$ mm]

4. Solve the following triangles JKL:
   (a) $J = 71°$, $K = 36°$, $j = 23.7$ mm
   (b) $j = 37.2$ cm, $k = 31.6$ cm, $K = 37°$
   (c) $j = 19.53$ cm, $k = 15.96$ cm, $J = 102° 57'$
   (a) $[k = 14.73$ mm, $l = 23.97$ mm, $L = 73°]$
   (b) $[J = 45° 7'$, $L = 97° 53'$, $l = 52.01$ cm
      or $J = 134° 53'$, $L = 8° 7'$, $l = 7.414$ cm]
   (c) $[K = 52° 47'$, $L = 24° 16'$, $l = 8.237$ cm]

5. Solve the following triangles MNP:
   (a) $m = 1.46$ cm, $p = 1.62$ cm, $M = 26° 0'$
   (b) $N = 111° 49'$, $P = 15° 19'$, $p = 43.81$ cm
   (c) $m = 45.0$ mm, $p = 35.0$ mm, $M = 51° 19'$
   (a) $[P = 29° 6'$, $N = 124° 54'$, $n = 2.732$ cm
      or $P = 150° 54'$, $N = 3° 6'$, $n = 0.180\,1$ cm]
   (b) $[m = 132.2$ cm, $n = 154.0$ cm, $M = 52° 52']$
   (c) $[P = 37° 23'$, $N = 91° 18'$, $n = 57.63$ mm]

6. Solve the following triangles QRS:
   (a) $R = 68° 15'$, $S = 57° 47'$, $r = 1.47$ m
   (b) $r = 43.36$ cm, $s = 30.20$ cm, $S = 29° 32'$

(c) $r = 1.00$ m, $q = 24.0$ cm, $R = 53°$
(a) $[q = 1.280$ m, $s = 1.339$ m, $Q = 53° \ 58']$
(b) $[R = 45° \ 3', Q = 105° \ 25', q = 59.06$ cm
        or $R = 134° \ 57', Q = 15° \ 31', q = 16.39$ cm]
(c) $[Q = 11° \ 3', S = 115° \ 57', s = 112.7$ cm]

7.   Solve the following triangles TUV:
(a) $t = 62.0$ mm, $u = 41.2$ mm, $V = 62° \ 11'$
(b) $T = 102° \ 8', u = 32.8$ cm, $v = 43.7$ cm
(c) $t = 9$ cm, $u = 7$ cm, $v = 6$ cm
(a) $[v = 56.81$ mm, $T = 77° \ 55', U = 39° \ 54']$
(b) $[t = 59.90$ cm, $U = 32° \ 22', V = 45° \ 30']$
(c) $[T = 87° \ 16', U = 50° \ 59', V = 41° \ 45']$

8.   Solve the following triangles WXY:
(a) $w = 129$ mm, $x = 158$ mm, $y = 59$ mm
(b) $Y = 67°, w = 16.4$ cm, $x = 11.8$ cm
(c) $W = 123° \ 17', y = 72.0$ mm, $x = 43.0$ mm
(a) $[W = 50° \ 43', X = 108° \ 33', Y = 20° \ 44']$
(b) $[W = 70° \ 21', X = 42° \ 39', y = 16.03$ cm]
(c) $[X = 20° \ 37', Y = 36° \ 6', w = 102.1$ mm]

9.   Solve the following triangles ABC:
(a) $b = 11$ cm, $c = 15$ cm, $A = 55°$
(b) $a = 8.983$ m, $b = 12.460$ m, $c = 15.910$ m
(c) $A = 73°, b = 7.20$ cm, $c = 9.00$ cm
(a) $[a = 12.5$ cm, $B = 46° \ 8', C = 78° \ 52']$
(b) $[A = 34° \ 13', B = 51° \ 20', C = 94° \ 27']$
(c) $[a = 9.746$ cm, $B = 44° \ 57', C = 62° \ 3']$

10.   Solve the following triangles DEF:
(a) $d = 19.47$ cm, $f = 17.63$ cm, $E = 38° \ 29'$
(b) $d = 7.0$ m, $e = 5.0$ m, $f = 8.0$ m
(c) $d = 25$ mm, $e = 32$ mm, $f = 39$ mm
(a) $[e = 12.35$ cm, $D = 78° \ 50', F = 62° \ 41']$
(b) $[D = 60°, E = 38° \ 13', F = 81° \ 47']$
(c) $[D = 39° \ 43', E = 54° \ 52', F = 85° \ 25']$

11.   Two phasors are shown in Fig. 12.5. If $v_1 = 60.0$ volts and
      $v_2 = 90.0$ volts calculate the value of their resultant (i.e. the
      length AB) and the angle it makes with v_1
      [145.1 volts, $18° \ 4'$]

12.   Two ships P and Q leave port at the same time. P sails at a
      steady speed of 52.0 km/h S 32° W and B at 38.0 km/h
      S 24° E. Find their distance apart after 2 hours 30 minutes.
      [110.1 km]

13.   A man leaves a town walking at a steady speed of 6.00 km/h
      in a direction of S 30° W. Another man leaves the same

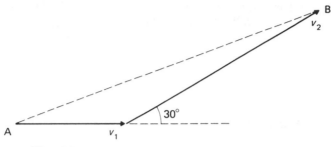

Fig. 12.5

town, cycling at a steady speed in a direction S 23° E. After 4.0 hours the two men are 100.00 km apart. Find the speed of the cyclist. [28.15 km/h]

14. A ship B sails from a port in a direction N 41° W at an average speed of 30.0 km/h. Another ship D sails at the same time as B but in a direction N 32° E at an average speed of 24.0 km/h. Calculate their distance apart after 3.0 hours. [97.44 km]

15. A room 9.00 m wide has a span roof which slopes at 32° on one side and 41° on the other. Find the length of the roof slopes. [6.174 m, 4.987 m]

16. A jib crane consists of a vertical post PQ 5.2 m in length, the inclined jib QR, 12.8 m in length, and a tie PR. Angle QPR is 122°. Calculate: (*a*) the length of the tie, and (*b*) the inclination of the jib to the vertical. (*a*) [9.26 m] (*b*) [37° 51′]

17. A reciprocating engine mechanism is shown in Fig. 12.6, where XY represents the rotating crank, YZ the connecting rod and Z the piston which moves vertically along the broken line XZ. If the rotating crank is 1.240 cm in length and the connecting rod is 6.480 cm, calculate for the position shown: (*a*) the inclination of the connecting rod to the vertical, and (*b*) the distance XZ (*a*) [10° 39′] (*b*) [6.689 cm]

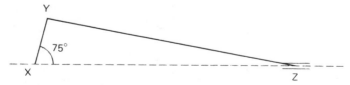

Fig. 12.6

18. A park is in the form of a quadrilateral ABCD as shown in Fig. 12.7 and its area is 2 791 m². Determine the length of (a) the perimeter fencing, and (b) the short-cut BD across the park.   (a) [219.4 m]   (b) [70.37 m]

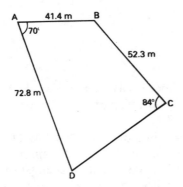

Fig. 12.7

Chapter 13

The solution of three-dimensional triangulation problems

13.1   The angle between a line and a plane

The angle between a line and a plane is defined as the angle between the line and its projection on the plane. In Fig. 13.1 the line AB is shown making an angle with the plane DEFG. If the line AC is constructed perpendicular to plane DEFG then the projection of the line AB on the plane DEFG is given by the length BC (where BC lies in the plane DEFG and is thus perpendicular to AC). In Fig. 13.1 angle θ is the angle between the line AB and the plane DEFG.

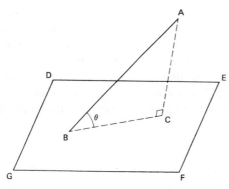

Fig. 13.1

13.2   The angle between two intersecting planes

Figure 13.2 shows two planes, ABCD and CDEF, having a common edge CD. To find the angle between these two intersecting planes it is necessary to establish a line that is perpendicular to the edge DC in each of the planes, i.e. GH is perpendicular to CD in plane ABCD and HJ is perpendicular to CD in plane CDEF.

The angle between the two intersecting planes is $\angle$ GHJ, shown as angle θ in Fig. 13.2

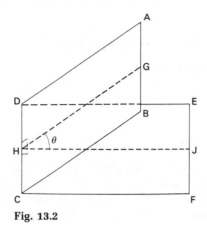

Fig. 13.2

13.3   Three-dimensional triangulation problems

In three-dimensional trigonometry it is important that the problem can be visualised and hence a clearly labelled sketch should always be made. It is often useful to redraw relevant constituent triangles, each of them fully dimensioned with the given information. Any triangle which has three facts given (either two sides and an angle, or two angles and a side, or three sides) may be solved using either (a) trigonometric ratios and Pythagoras's theorem for right-angled triangles or (b) the sine or cosine rule for triangles which are not right-angled. Provided the solution of relevant triangles can be obtained then angles between lines and planes or angles between intersecting planes may be found.

Worked problems on 'three-dimensional' trigonometry

Problem 1. The base of a right pyramid of vertex A is a rectangle BCDE. X and Y are the mid-points of sides BC and CD respectively. AB = 9.20 cm, BC = 7.60 cm and CD = 4.20 cm. Calculate (a) the perpendicular height of the pyramid, (b) the angle edge AE makes with the base, (c) the angle faces ABC and ACD make with the base, and (d) the angle the plane AXY makes with the base.
   The pyramid is shown in Fig. 13.3

(a)   Let the diagonals of the base, EC and BD, intersect at point F.
   Length of diagonal EC = $\sqrt{[(7.60)^2 + (4.20)^2]}$ = 8.683 cm

$$\text{Length EF} = \frac{1}{2}\,\text{EC} = \frac{8.683}{2} = 4.342 \text{ cm}$$

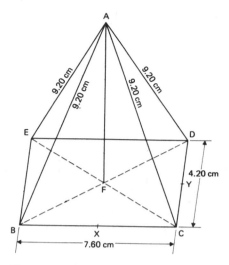

Fig. 13.3

Using the theorem of Pythagoras on the right-angled triangle AEF (see Fig. 13.4(a)) gives

$$AF = \sqrt{[(9.20)^2 - (4.342)^2]} = 8.111 \text{ cm}$$

Hence **the height of the pyramid is 8.111 cm**

(b)   The angle which a line makes with a plane is the angle which it makes with its projection on the plane. In the pyramid, EF is the projection of AE on to the base BCDE. From triangle AEF, shown in Fig. 13.4(a),

$$\cos E = \frac{EF}{EA} = \frac{4.342}{9.20}$$

Thus $\angle AEF = 61° 50'$

Hence **AE makes an angle of 61° 50′ with the base.**

(c)   Since point X is the mid-point of side BC, FX is perpendicular to BC. Also AX is the perpendicular height of the triangle ABC. Thus the angle between the two intersecting lines AX and XF is the angle between the face ABC and the base. In triangle AFX, shown in Fig. 13.4(b),

$$AF = 8.111 \text{ cm and } FX = \frac{4.20}{2} = 2.10 \text{ cm}$$

Hence $\angle AXF = \arctan \dfrac{8.111}{2.10} = 75° 29'$

Since Y is the mid-point of side CD, FY is perpendicular to CD. Also AY is the perpendicular height of triangle ACD. Thus the angle between the two intersecting lines AY and YF

is the angle between the face ACD and the base. In triangle
AFY, shown in Fig. 13.4(c),

$$AF = 8.111 \text{ cm and } FY = \frac{7.60}{2} = 3.80 \text{ cm}$$

Hence $\angle \mathbf{AYF} = \arctan \dfrac{8.111}{3.80} = \mathbf{64° \, 54'}$

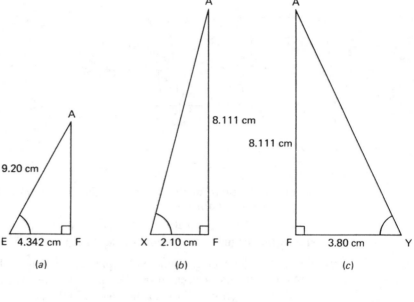

(a)   (b)   (c)

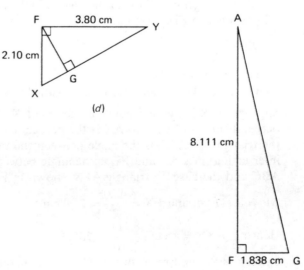

(d)

(e)

Fig. 13.4

(*d*) To find the angle between plane AXY and the base BCDE it is necessary to establish a perpendicular to the common edge XY in each of the planes.

Triangle FXY is shown in Fig. 13.4(*d*)

$$\angle \text{FXY} = \arctan \frac{3.80}{2.10} = 61° \, 4'$$

Let FG be perpendicular to XY. Then
FG = 2.10 sin 61° 4′ = 1.838 cm
From the triangle AFG, shown in Fig. 13.4(*e*),

$$\angle \text{AGF} = \arctan \frac{\text{AF}}{\text{FG}} = \arctan \frac{8.111}{1.838} = 77° \, 14'$$

Hence **the angle between the plane AXY and the base is 77° 14′**

Problem 2. A 4.00-cm cube has top ABCD and base EFGH with vertical edges AE, BF, CG and DH. I and J are the mid-points of sides EH and HG respectively. Calculate (*a*) the angle between DF and CF, (*b*) the angle between DI and GI, (*c*) the angle between the line CE and the plane EFGH, and (*d*) the angle between the plane BIJ and the plane EFGH.

The cube is shown in Fig. 13.5

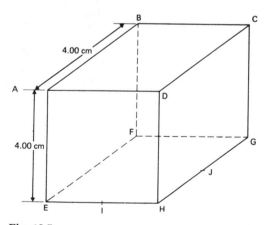

Fig. 13.5

(*a*) *To find the angle between DF and CF*
DF is one side of triangle DFH, shown in Fig. 13.6(*a*), where
DH = 4.00 cm and FH = $\sqrt{[(4.00)^2 + (4.00)^2]}$ = 5.657 cm.
Hence all the diagonals of the cube are 5.657 cm. Since ∠ DHF is 90°, DF = $\sqrt{[(4.00)^2 + (5.657)^2]}$, i.e. DF = 6.928 cm.

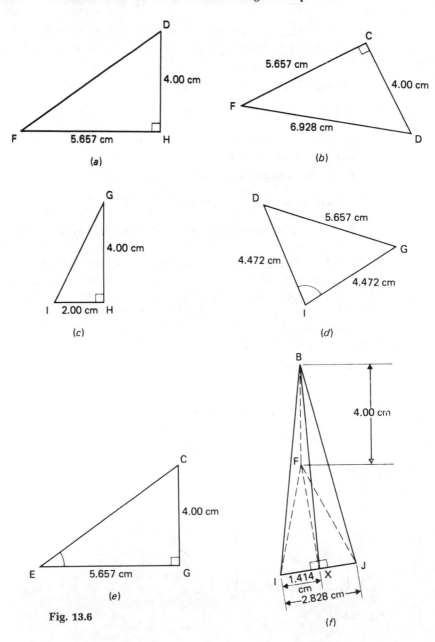

Fig. 13.6

In triangle CFD, shown in Fig. 13.6(b), FD = 6.928 cm, CF = 5.657 cm and CD = 4.00 cm

$$\angle\, \mathbf{CFD} = \arctan \frac{4.00}{5.657} = \mathbf{35°\ 16'}$$

(b) *To find the angle between DI and GI*
GI is one side of triangle GHI, shown in Fig. 13.6(c), where
GH = 4.00 cm, HI = 2.00 cm and $\angle$ GHI = 90°. Hence

$$GI = \sqrt{[(4.00)^2 + (2.00)^2]} = 4.472 \text{ cm}$$

Similarly,   DI = 4.472 cm (i.e. the same length as GI)

Diagonal DG = 5.657 cm

In triangle GID, shown in Fig. 13.6(d), $\angle$ DIG is given by the
cosine rule:

$$\cos DIG = \frac{(DI)^2 + (IG)^2 - (DG)^2}{2(DI)(IG)}$$

$$= \frac{(4.472)^2 + (4.472)^2 - (5.657)^2}{2(4.472)(4.472)} = 0.199\ 9$$

Hence $\angle$ **DIG = 78° 28′**

(c) *To find the angle between the line CE and the plane EFGH*
The projection of the line CE on to the plane EFGH is given
by the diagonal EG, where EG and CG are perpendicular to
each other.

In triangle CEG, shown in Fig. 13.6(e), GE = 5.657 cm and
$\angle$ CEG is the angle between line CE and plane EFGH. Hence

$$\angle \textbf{CEG} = \arctan \frac{CG}{EG} = \arctan \frac{4.00}{5.657} = \textbf{35° 16′}$$

(d) *To find the angle between plane BIJ and plane EFGH*
From Fig. 13.6(f) IJ is the common edge between planes
BIJ and EFGH:

$$IJ = \sqrt{[(2.00)^2 + (2.00)^2]} = 2.828 \text{ cm}$$

Let X be the mid-point of IJ; then IX = 1.414 cm,
FI = 4.472 cm (i.e. length FI is the same as length GI) and
FX = $\sqrt{[(4.472)^2 - (1.414)^2]}$ = 4.243 cm. FX is perpendicular to
IJ and BX is perpendicular to IJ. Thus the angle between planes
BIJ and EFGH is given by angle BXF. Hence

$$\angle \textbf{BXF} = \arctan \frac{4.00}{4.243} = \textbf{43° 19′}$$

Problem 3. A ship is observed from the top of a cliff 152 m high
in a direction S 28° 19′ W at an angle of depression of 8° 46′.
Six minutes later the same ship is seen in a direction W 17° 13′ N
at an angle of depression of 9° 52′. Calculate the speed of the
ship in km/h.

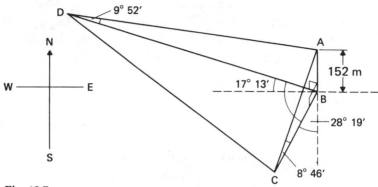

Fig. 13.7

In Fig. 13.7 A represents the top of the cliff AB, C the initial position of the ship and D the position of the ship at the second observation. To find the speed of the ship the distance CD is required. To find CD distances BC and BD need to be calculated:

$$\angle CBD = (90° - 20° \ 19') + 17° \ 13' = 78° \ 54'$$

From triangle ABC, BC = 152 cot 8° 46' = 985.7 m

From triangle ABD, BD = 152 cot 9° 52' = 873.9 m

Using the cosine rule on triangle BCD gives

$$CD^2 = (985.7)^2 + (873.9)^2$$
$$- [2(985.7)(873.9) \cos 78° \ 54']$$

Therefore   CD = 1 185 m or 1.185 km

$$\textbf{Speed of ship} = \frac{\text{distance travelled}}{\text{time}} = \frac{1.185 \text{ km}}{\frac{6}{60} \text{ h}}$$

$$= \textbf{11.85 km/h}$$

Problem 4. From a point P, at ground level, a mine shaft is constructed to a depth of 0.264 km. Tunnels are constructed from Q, 3.450 km due west of P, and R, 2.875 km S 40° E, to meet at the base S of the shaft. Assuming that P, Q and R are on horizontal ground and that the two tunnels descend uniformly, find (a) their angles of descent and (b) the angle between the directions of the two tunnels.

The mine is shown in Fig. 13.8
The constituent triangles are shown in Fig. 13.9

The angle of descent of tunnel QS, i.e. $\angle$ PQS of Fig. 13.9(a), is

given by arctan $\dfrac{0.264}{3.450}$, i.e. **4° 23'**

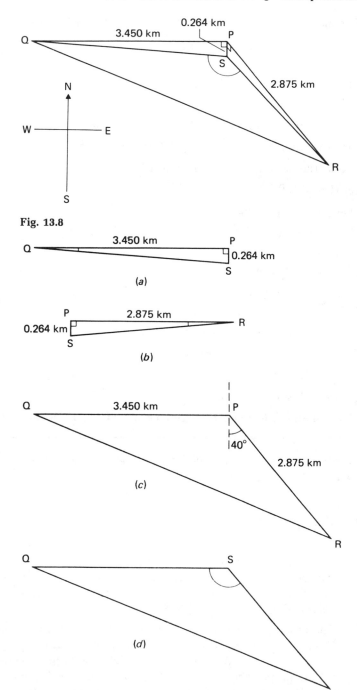

Fig. 13.8

(a)

(b)

(c)

(d)

Fig. 13.9

The angle of descent of tunnel RS, i.e. $\angle$ PRS of Fig. 13.9(b), is given by $\arctan\dfrac{0.264}{2.875}$, i.e. **5° 15′**

The angle between the directions of the two tunnels is given by $\angle$ RSQ of Fig. 13.9(d)

From triangle PQS, QS $= \sqrt{[(3.450)^2 + (0.264)^2]} = 3.460$ km

From triangle PRS, RS $= \sqrt{[(2.875)^2 + (0.264)^2]} = 2.887$ km

From triangle PQR, shown in Fig. 13.9(c),

$\angle$ QPR $= 130°$

Using the cosine rule:

$$QR^2 = (3.450)^2 + (2.875)^2 - [2(3.450)(2.875)\cos 130°]$$

Hence QR $= 5.738$ m

From triangle QRS

$$\cos RSQ = \frac{(SQ)^2 + (SR)^2 - (QR)^2}{2(SQ)(SR)}$$

$$= \frac{(3.460)^2 + (2.887)^2 - (5.738)^2}{2(3.460)(2.887)}$$

$$= -0.631\,6$$

Hence $\angle$ RSQ $= 129°\ 10′$

Hence **the angle between the directions of the two tunnels is 129° 10′**

Further problems on three-dimensional trigonometry may be found in the following Section 13.4 (Problems 1 to 14).

13.4   Further problems

1. A vertical aerial stands on horizontal ground. A surveyor standing due west of the aerial finds the angle of elevation of the top to be 51° 5′. He moves due south 20.0 m and finds the elevation of the top of the aerial to be 47° 19′. Calculate the height of the aerial.   [44.88 m]

2. The base of a right pyramid of vertex V is a rectangle ABCD. E and F are the mid-points of AB and BC respectively. VA = 15.00 cm, AB = 11.00 cm and BC = 8.00 cm. Calculate (a) the perpendicular height of the pyramid, (b) the angle the edge VA makes with the base, (c) the angles which the faces VAB and VBC make with the base, and (d) the angle the

plane VEF makes with the base ABCD.

(a) [13.37 cm]   (b) [63° 3′]   (c) [73° 21′, 67° 38′]   (d) [76° 24′]

3. An aeroplane is sighted due west from a radar station at an elevation of 38° and a height of 7 500 m and later at an elevation of 34° at height 5 000 m in direction W 48° S. If it is descending uniformly, find the angle of descent. Find the speed of the aeroplane (in m/s and in km/h) if the time between the two observations is 45.0 seconds.
[19° 9′, 169.4 m/s, 609.9 km/h]

4. From a point X, at ground level, a mine shaft is constructed to a depth of 300 m. Tunnels are constructed from Y, 5.0 km due south of X, and Z, 3.0 km N 30° W from X, to meet at the base A of the shaft. Find the angle between the directions of the two tunnels (i.e. ∠YAZ), assuming that X, Y and Z are on horizontal ground.   [148° 41′]

5. A tent is in the form of a regular octagonal pyramid with each side of the base 4 m in length. If the perpendicular height of the tent is 10 m, find (a) the angle between a sloping face and the base, and (b) the angle of inclination of an edge of the sloping face to the base.   (a) [64° 14′]   (b) [62° 24′]

6. The base of a roof, shown in Fig. 13.10, is a rectangle PQRS. The equal faces PSX and QRY each make an angle of 45° with the base; the equal faces PQYX and RSXY each make an angle of 50° with the base. Calculate (a) the perpendicular height of the roof, (b) the length XY, (c) the length of the sloping edge RY, and (d) the angle the edge SX makes with the base.

(a) [5.96 m]   (b) [4.08 m]   (c) [9.80 m]   (d) [37° 27′]

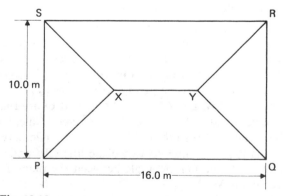

Fig. 13.10

7. A crane lifts a load from position X and deposits it at position Y on the same ground level, and in so doing the crane pivots through an angle of 65° and the jib moves from an angle of 68° to the vertical to an angle of 48° to the vertical. Find the distance XY if the jib is of length 8.0 m   [7.29 m]

8. An aircraft is sighted due north from a radar station at an elevation of 49° 4′ at a height of 10 000 m, and later at an elevation of 37° 19′ and height 6 250 m in direction E 24° S. If it is descending uniformly, find the angle of descent and the speed of the aircraft if the time between the two observations is 72 seconds   [14° 50′, 203.3 m/s or 732.0 km/h]

9. From a ship situated at a point S 17° W of a vertical lighthouse the angle of elevation of the top of the lighthouse is 13°. Later, when the ship is at a point S 2° E from the lighthouse, the angle of elevation is 9°. Find the height of the lighthouse if the distance travelled by the ship between the two observations is 146 m   [55.55 m]

10. A man standing W 30° S of a tower measures the angle of elevation of the top of the tower as 44° 19′. From a position E 28° S from the tower the elevation of the top is 36° 43′. Calculate the height of the tower if the distance between the two observations is 120 m   [57.86 m]

11. The access to an underground cave X may be made from three entrances – either in a vertical lift 0.42 km in length from point A at ground level or by a tunnel 6.00 km long from a point B at ground level due west of A, or by another tunnel 4.20 km from point C at ground level E 41° S of A. Find the angle between the directions of the tunnels (i.e. ∠ BXC), assuming that A, B and C are on horizontal ground.   [137° 55′]

12. A coastguard situated at the top of a cliff 200 m high observes a ship in a direction S 32° W at an angle of depression of 9° 14′. Five minutes later the same ship is seen in a direction W 25° N at an angle of depression of 10° 51′. Calculate the speed of the ship in km/h   [18.16 km/h]

13. A 3.00-cm cube has top ABCD and base WXYZ with vertical edges AW, BX, CY and DZ. P and Q are the mid-points of sides WZ and ZY respectively. Calculate (a) the angle between DX and CX, (b) the angle between DP and YP, (c) the angle between the line CW and the plane WXYZ, and (d) the angle between the plane BPQ and the plane WXYZ
(a) [35° 16′]   (b) [78° 28′]   (c) [35° 16′]   (d) [43° 19′]

14. A 15.30-cm-high right circular cone of vertex A has a base diameter of 9.80 cm. If point B is on the circumference of the base, find the inclination of AB to the base. C is also a point on the circumference of the base, distance 7.60 cm from B. Find the inclination of the plane ABC to the base.
[72° 15′, 78° 34′]

Chapter 14

Graphs of trigonometric functions

14.1 Graphs of trigonometric ratios

One method of plotting graphs of trigonometric ratios is to initially draw up a table of values. This is achieved by using a calculator. In the graphs plotted below, 15° intervals have been used and values in the tables are taken correct to 3 decimal places.

(i) $y = \sin A$

$A°$	0	15	30	45	60	75	90	105	120	135
$\sin A$	0	0.259	0.500	0.707	0.866	0.966	1.000	0.966	0.866	0.707

$A°$	150	165	180	195	210	225	240	255	270
$\sin A$	0.500	0.259	0	−0.259	−0.500	−0.707	−0.866	−0.966	−1.000

$A°$	285	300	315	330	345	360
$\sin A$	−0.966	−0.866	−0.707	−0.500	−0.259	0

(ii) $y = \sin 2A$

$A°$	0	15	30	45	60	75	90	105	120	135
$2A$	0	30	60	90	120	150	180	210	240	270
$\sin 2A$	0	0.500	0.866	1.000	0.866	0.500	0	−0.500	−0.866	−1.000

$A°$	150	165	180	195	210	225	240	255	270
$2A$	300	330	360	390	420	450	480	510	540
$\sin 2A$	−0.866	−0.500	0	0.500	0.866	1.000	0.866	0.500	0

$A°$	285	300	315	330	345	360
$2A$	570	600	630	660	690	720
$\sin 2A$	−0.500	−0.866	−1.000	−0.866	−0.500	0

(iii) $y = \sin \frac{1}{2}A$

$A°$	0	15	30	45	60	75	90	105	120	135
$\frac{1}{2}A$	0	$7\frac{1}{2}$	15	$22\frac{1}{2}$	30	$37\frac{1}{2}$	45	$52\frac{1}{2}$	60	$67\frac{1}{2}$
$\sin\frac{1}{2}A$	0	0.131	0.259	0.383	0.500	0.609	0.707	0.793	0.866	0.924

$A°$	150	165	180	195	210	225	240	255	270
$\frac{1}{2}A$	75	$82\frac{1}{2}$	90	$97\frac{1}{2}$	105	$112\frac{1}{2}$	120	$127\frac{1}{2}$	135
$\sin\frac{1}{2}A$	0.966	0.991	1.000	0.991	0.966	0.924	0.866	0.793	0.707

$A°$	285	300	315	330	345	360
$\frac{1}{2}A$	$142\frac{1}{2}$	150	$157\frac{1}{2}$	165	$172\frac{1}{2}$	180
$\sin\frac{1}{2}A$	0.609	0.500	0.383	0.259	0.131	0

Graphs of $y = \sin A$, $y = \sin 2A$ and $y = \sin \frac{1}{2}A$ are shown in Fig. 14.1

(iv) $y = \cos A$

$A°$	0	15	30	45	60	75	90	105	120	135
$\cos A$	1.000	0.966	0.866	0.707	0.500	0.259	0	-0.259	-0.500	-0.707

$A°$	150	165	180	195	210	225	240	255	270
$\cos A$	-0.866	-0.966	-1.000	-0.966	-0.866	-0.707	-0.500	-0.259	0

$A°$	285	300	315	330	345	360
$\cos A$	0.259	0.500	0.707	0.866	0.966	1.000

(v) $y = \cos 2A$

$A°$	0	15	30	45	60	75	90	105	120	135
$2A$	0	30	60	90	120	150	180	210	240	270
$\cos 2A$	1.000	0.866	0.500	0	-0.500	-0.866	-1.000	-0.866	-0.500	0

$A°$	150	165	180	195	210	225	240	255	270
$2A$	300	330	360	390	420	450	480	510	540
$\cos 2A$	0.500	0.866	1.000	0.866	0.500	0	-0.500	-0.866	-1.000

$A°$	285	300	315	330	345	360
$2A$	570	600	630	660	690	720
$\cos 2A$	-0.866	-0.500	0	0.500	0.866	1.000

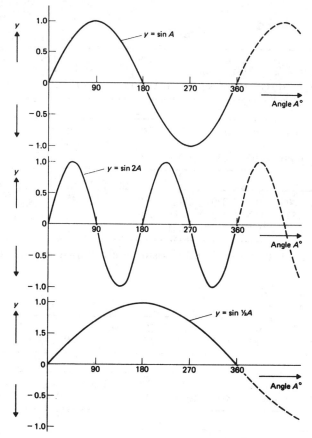

Fig. 14.1 Graphs of sin A, sin $2A$ and sin $\frac{1}{2}A$

(vi) $y = \cos \frac{1}{2}A$

$A°$	0	15	30	45	60	75	90	105	120	135
$\frac{1}{2}A$	0	$7\frac{1}{2}$	15	$22\frac{1}{2}$	30	$37\frac{1}{2}$	45	$52\frac{1}{2}$	60	$67\frac{1}{2}$
$\cos \frac{1}{2}A$	1.000	0.991	0.966	0.924	0.866	0.793	0.707	0.609	0.500	0.383

$A°$	150	165	180	195	210	225	240	255	270
$\frac{1}{2}A$	75	$82\frac{1}{2}$	90	$97\frac{1}{2}$	105	$112\frac{1}{2}$	120	$127\frac{1}{2}$	135
$\cos \frac{1}{2}A$	0.259	0.131	0	-0.131	-0.259	-0.383	-0.500	-0.609	-0.707

$A°$	285	300	315	330	345	360
$\frac{1}{2}A$	$142\frac{1}{2}$	150	$157\frac{1}{2}$	165	$172\frac{1}{2}$	180
$\cos \frac{1}{2}A$	-0.793	-0.866	-0.924	-0.966	-0.991	-1.000

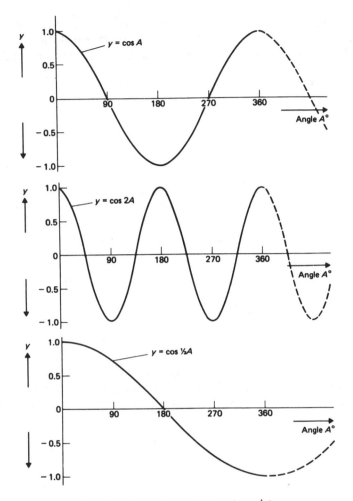

Fig. 14.2 Graphs of cos A, cos 2A and cos ½A

Graphs of $y = \cos A$, $y = \cos 2A$ and $y = \cos \frac{1}{2}A$ are plotted in Fig. 14.2

Period
Each of the graphs shown in Figs 14.1 and 14.2 will repeat themselves and such functions are known as **periodic functions**.

Since $y = \sin A$ and $y = \cos A$ repeat themselves every time the angle increases by 360° or 2π radians then 360° or 2π radians is called the period.

Since $y = \sin 2A$ and $y = \cos 2A$ repeat themselves every time the angle increases by 180° or π radians then 180° or π radians is the period of these waveforms.

Since $y = \sin \frac{1}{2}A$ and $y = \cos \frac{1}{2}A$ repeat themselves every $720°$ or 4π radians then $720°$ or 4π radians is the period of these waveforms.

Generally, if $y = \sin pA$ or $y = \cos pA$, where p is a constant, then the period of the waveform will be $\dfrac{360°}{p}$ or $\dfrac{2\pi}{p}$ radians.

Hence if $y = \sin A$ or $y = \cos A$,   period $= \dfrac{360°}{1} = 360°$ or 2π radians

$\qquad y = \sin 2A$ or $y = \cos 2A$, period $= \dfrac{360°}{2} = 180°$ or π radians

$\qquad y = \sin\frac{1}{2}A$ or $y = \cos\frac{1}{2}A$, period $= \dfrac{360°}{\frac{1}{2}} = 720°$ or 4π radians

$\qquad y = \sin 6A$ or $y = \cos 6A$, period $= \dfrac{360°}{6} = 60°$ or $\dfrac{\pi}{3}$ radians

$\qquad y = \sin \frac{1}{5}A$ or $y = \cos \frac{1}{5}A$, period $= \dfrac{360°}{\frac{1}{5}} = 1\,800°$ or 10π radians

and so on.

Leading and lagging angles

A sine or cosine curve may not always start at $0°$. To show this, a periodic function is represented by $y = \sin(A \pm \alpha)$ or $y = \cos(A \pm \alpha)$, where α is a phase difference compared with $y = \sin A$ or $y = \cos A$. α is called the **phase angle**.

(i) $y = \sin(A + 60°)$

$A°$	0	15	30	45	60	75	90	105
$(A + 60°)$	60	75	90	105	120	135	150	165
$\sin(A + 60°)$	0.866	0.966	1.000	0.966	0.866	0.707	0.500	0.259

$A°$	120	135	150	165	180	195	210	225
$(A + 60°)$	180	195	210	225	240	255	270	285
$\sin(A + 60°)$	0	−0.259	−0.500	−0.707	−0.866	−0.966	−1.000	−0.966

$A°$	240	255	270	285	300	315	330	345	360
$(A + 60)°$	300	315	330	345	360	375	390	405	420
$\sin(A + 60°)$	−0.866	−0.707	−0.500	−0.259	0	0.259	0.500	0.707	0.866

A graph of $y = \sin(A + 60°)$ is shown in Fig. 14.3

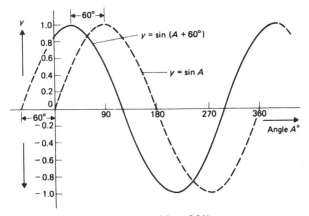

Fig. 14.3 Graph of $y = \sin(A + 60°)$

(ii) $y = \cos(A - 45°)$

$A°$	0	15	30	45	60	75	90	105
$(A - 45)°$	-45	-30	-15	0	15	30	45	60
$\cos(A - 45°)$	0.707	0.866	0.966	1.000	0.966	0.866	0.707	0.500

$A°$	120	135	150	165	180	195	210	225
$(A - 45)°$	75	90	105	120	135	150	165	180
$\cos(A - 45°)$	0.259	0	-0.259	-0.500	-0.707	-0.866	-0.966	-1.000

$A°$	240	255	270	285	300	315	330	345	360
$(A - 45)°$	195	210	225	240	255	270	285	300	315
$\cos(A - 45°)$	-0.966	-0.866	-0.707	-0.500	-0.259	0	0.259	0.500	0.707

A graph of $y = \cos(A - 45°)$ is shown in Fig. 14.4

If the graph of $y = \sin A$ is assumed to commence at $0°$ then the graph of $y = \sin(A + 60°)$ from Fig. 14.3 is seen to start $60°$ earlier. Thus the graph of $y = \sin(A + 60°)$ is said to **lead** the graph of $y = \sin A$ by $60°$.

Similarly, if the graph of $y = \cos A$ is assumed to commence at $0°$ then the graph of $y = \cos(A - 45°)$ from Fig. 14.4 is seen to start $45°$ later. Thus the graph of $y = \cos(A - 45°)$ is said to **lag** the graph of $y = \cos A$ by $45°$.

In each of the above two examples the angle of lead or lag may also be seen on the graphs by comparing the positions of the maximum values.

Generally, a graph of $y = \sin(A + \alpha)$ leads $y = \sin A$ by angle α. A graph of $y = \sin(A - \alpha)$ lags $y = \sin A$ by angle α.

It may be seen from Figs 14.1 and 14.2 that a cosine curve is the same shape as a sine curve, except that it starts $90°$ earlier, i.e. leads by $90°$.

Hence $\cos A = \sin(A + 90°)$.

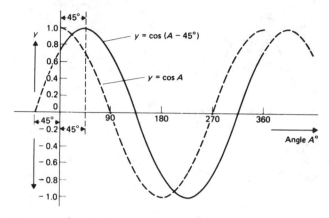

Fig. 14.4 Graph of $y = \cos(A - 45°)$

Since sine and cosine curves are of the same shape both are referred to generally as 'sine waves'.

Note that $y = \sin(2t + 60°)$ leads $y = \sin 2t$ by $\dfrac{60°}{2}$, i.e. 30°.

Similarly, $y = \cos(3t - 45°)$ lags $y = \cos 3t$ by $\dfrac{45°}{3}$, i.e. 15°, and so on.

Generally, $y = \sin(pt + \alpha°)$ leads $y = \sin pt$ by $\dfrac{\alpha°}{p}$ and $y = \sin(pt - \alpha°)$ lags $y = \sin pt$ by $\dfrac{\alpha°}{p}$

Graphs of $\sin^2 A$ and $\cos^2 A$

(i) $y = \sin^2 A$

$A°$	0	15	30	45	60	75	90	105	120	135
$\sin A$	0	0.259	0.500	0.707	0.866	0.966	1.000	0.966	0.866	0.707
$(\sin A)^2$	0	0.067	0.250	0.500	0.750	0.933	1.000	0.933	0.750	0.500

$A°$	150	165	180	195	210	225	240	255	270
$\sin A$	0.500	0.259	0	−0.259	−0.500	−0.707	−0.866	−0.966	−1.000
$(\sin A)^2$	0.250	0.067	0	0.067	0.250	0.500	0.750	0.933	1.000

$A°$	285	300	315	330	345	360
$\sin A$	−0.966	−0.866	−0.707	−0.500	−0.259	0
$(\sin A)^2$	0.933	0.750	0.500	0.250	0.067	0

A graph of $y = \sin^2 A$ is shown in Fig. 14.5

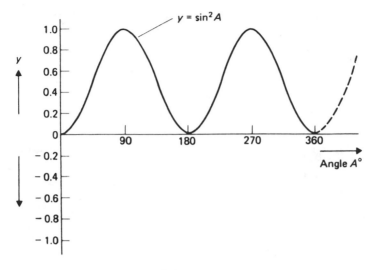

Fig. 14.5 Graph of $y = \sin^2 A$

(ii) $y = \cos^2 A$

$A°$	0	15	30	45	60	75	90	105	120
$\cos A$	1.000	0.966	0.866	0.707	0.500	0.259	0	−0.259	−0.500
$(\cos A)^2$	1.000	0.933	0.750	0.500	0.250	0.067	0	0.067	0.250

$A°$	135	150	165	180	195	210	225	240
$\cos A$	−0.707	−0.866	−0.966	−1.000	−0.966	−0.866	−0.707	−0.500
$(\cos A)^2$	0.500	0.750	0.933	1.000	0.933	0.750	0.500	0.250

$A°$	255	270	285	300	315	330	345	360
$\cos A$	−0.259	0	0.259	0.500	0.707	0.866	0.966	1.000
$(\cos A)^2$	0.067	0	0.067	0.250	0.500	0.750	0.933	1.000

A graph of $y = \cos^2 A$ is shown in Fig. 14.6

The graphs of $y = \sin^2 A$ and $y = \cos^2 A$ shown in Figs 14.5 and 14.6 are periodic functions of period 180° or π radians. Both graphs display only positive values.

A graph of $y = \sin^2 2A$ would have a period of $\dfrac{180°}{2}$, i.e. 90° or $\dfrac{\pi}{2}$ radians. Similarly, a graph of $y = \cos^2 5A$ would have a period of $\dfrac{180}{5}$, i.e. 36° or $\dfrac{\pi}{5}$ radians and so on.

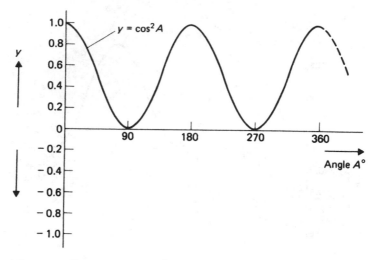

Fig. 14.6   Graph of $y = \cos^2 A$

Amplitude

Amplitude is the name given to the maximum value of a sine wave. For each of the sine waves shown in Figs 14.1 to 14.4 the maximum value is $+1$ (and hence -1 in the negative direction). Thus for $y = \sin A$, $y = \cos A$ and so on, the amplitude is 1.

However, if $y = 3 \sin A$, then each of the values in the table is multiplied by 3 and the maximum value, and thus the amplitude, is 3. Similarly, if $y = 7 \cos 2A$ then the amplitude is 7 and if $y = 4 \sin(A - 30°)$ the amplitude is 4. In each of these examples, the period of the graph is unaffected by the amplitude.

Sketching graphs of trigonometric ratios

It is often useful to be able to sketch waveforms. This can be achieved reasonably accurately without the time consuming process of drawing up a table. If the amplitude and period of a function are known then a graph of that function may be sketched.

(A 'sketch' is assumed to mean 'the general outline showing important points on the axes'.)

Worked problems on sketching graphs of trigonometrical ratios

Problem 1. Sketch $y = 2 \sin 3A$ from $A = 0$ to $A = 2\pi$ radians.

Amplitude = 2, hence the maximum value is +2 and the minimum value is −2

Period = $\dfrac{2\pi}{3}$ radians = 120°

Therefore there are 3 cycles (sine waves) which are completed at 120°, and then at intervals of 120°

The first maximum value occurs at $\dfrac{1}{4}$ (120°) and then at intervals of 120°

The first minimum value occurs at $\dfrac{3}{4}$ (120°) and then at intervals of 120°

A sketch of $y = 2 \sin 3A$ is shown in Fig. 14.7

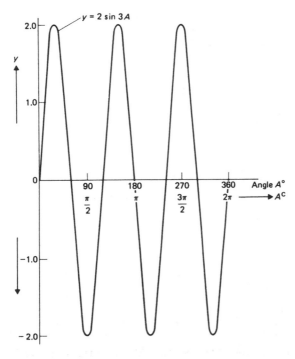

Fig. 14.7 A sketch of $y = 2 \sin 3A$

Problem 2. Sketch $y = 4 \cos 2t$

Amplitude = 4, hence the maximum value is +4 and the minimum value is −4

$$\text{Period} = \frac{360°}{2} = 180° \text{ (or } \pi \text{ radians)}$$

Two cycles are completed between 0° and 360°, at 180° and 360°

Maximum values occur at 0° and 180° (cos $2t = 1$)

Minimum values occur at 90° and 270° (cos $2t = -1$)

A sketch of $y = 4 \cos 2t$ is shown in Fig. 14.8

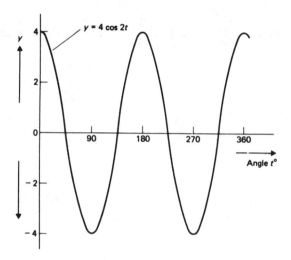

Fig. 14.8   A sketch of $y = \cos 2t$

Problem 3. Sketch $y = 3 \sin \dfrac{2}{5} A$ over one cycle (i.e. one complete sine wave)

Amplitude $= 3$, hence the maximum value is $+3$ and the minimum value is -3

$$\text{Period} = \frac{360°}{\frac{2}{5}} = 360° \times \frac{5}{2} = 900°,$$

i.e. one cycle is completed between 0° and 900° (and half a cycle is completed after 450°)

   The maximum value is at $\dfrac{450°}{2} = 225°$ and the minimum value is at $225° + 450° = 675°$

   A sketch of $y = 3 \sin \frac{2}{5} A$ is shown in Fig. 14.9

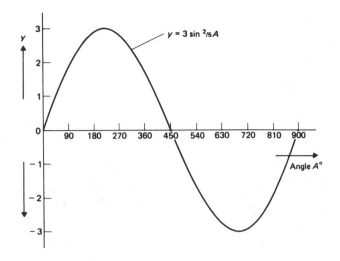

Fig. 14.9   A sketch of $y = 3 \sin \frac{2}{5}A$

Problem 4. Sketch the following graphs from $t = 0$ to $t = 2\pi$ seconds showing relevant details:

(a) $y = 2 \sin 2t$ (b) $y = 5 \cos 3t$

(c) $y = 4 \sin\left(t - \dfrac{\pi}{4}\right)$ (d) $y = 3 \cos\left(t + \dfrac{\pi}{3}\right)$

(e) $y = 4 \sin^2 t$ (f) $y = 3 \cos^2 2t$

(a) $y = 2 \sin 2t$ has an amplitude of 2 and a period of $\dfrac{2\pi}{2}$, i.e. π radians, as shown in Fig. 14.10(a)

(b) $y = 5 \cos 3t$ has an amplitude of 5 and a period of $\dfrac{2\pi}{3}$ radians, as shown in Fig. 14.10(b)

(c) $y = 4 \sin\left(t - \dfrac{\pi}{4}\right)$ has an amplitude of 4, a period of $\dfrac{2\pi}{1}$, i.e. 2π radians, and lags $y = 4 \sin t$ by $\dfrac{\pi}{4}$ radians, as shown in Fig. 14.10(c)

(d) $y = 3 \cos\left(t + \dfrac{\pi}{3}\right)$ has an amplitude of 3, a period of 2π radians and leads $y = 3 \cos t$ by $\dfrac{\pi}{3}$ radians, as shown in Fig. 14.10(d)

(e) $y = 4 \sin^2 t$ has an amplitude of 4 and a period of $\dfrac{\pi}{1}$, i.e. π radians, as shown in Fig. 14.10(e)

(f) $y = 3 \cos^2 2t$ has an amplitude of 3 and a period of $\dfrac{\pi}{2}$ radians, as shown in Fig. 14.10(f)

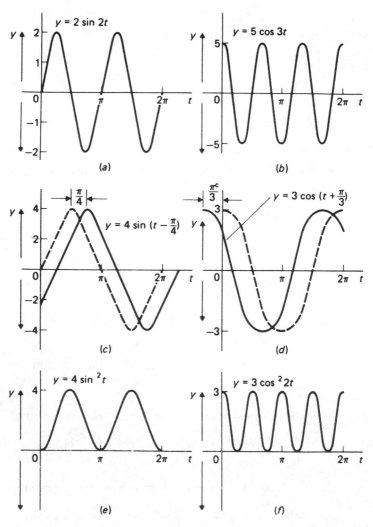

Fig. 14.10

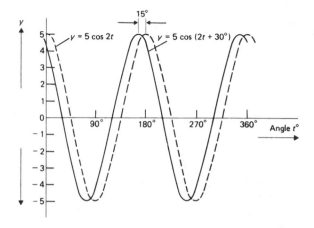

Fig. 14.11   A sketch of $y = 5\cos(2t + 30°)$

Problem 5. Sketch $y = 5\cos(2t + 30°)$

Amplitude $= 5$

$$\text{Period} = \frac{360°}{2} = 180°$$

The graph of $y = 5\cos(2t + 30°)$ **leads** the graph of $y = 5\cos 2t$ by $\dfrac{30°}{2}$, i.e. $15°$ (i.e. the waveform starts $15°$ earlier).

   $y = 5\cos 2t$ is sketched as shown in Fig. 14.11 and then $y = 5\cos(2t + 30°)$ is sketched, leading $5\cos 2t$ by $15°$.

Problem 6. Sketch $y = 3\sin(2t - 45°)$

Amplitude $= 3$

$$\text{Period} = \frac{360°}{2} = 180°$$

The graph of $y = 3\sin(2t - 45°)$ lags the graph of $y = 3\sin 2t$ by $\dfrac{45°}{2}$, i.e. $22\frac{1}{2}°$ (i.e. the waveform starts $22\frac{1}{2}°$ later).

   $y = 3\sin 2t$ is sketched as shown in Fig. 14.12, and then $y = 3\sin(2t - 45°)$ is sketched, lagging $3\sin 2t$ by $22\frac{1}{2}°$.

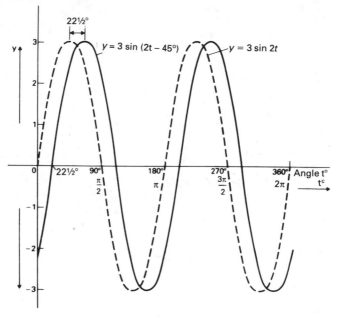

Fig. 14.12 A sketch of $y = 3\sin(2t - 45°)$

Further problems on sketching trigonometric ratios may be found in Section 14.5 (Problems 1 to 5), page 235.

14.2   Phasors, periodic time and frequency

In Fig. 14.13 let OA represent a vector that is free to rotate anticlockwise about O at a velocity of ω radians per second. A rotating vector is known as a **phasor**.

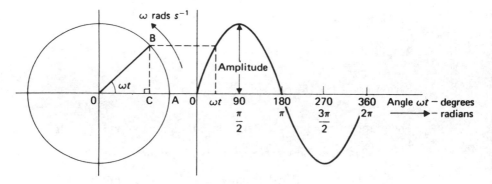

Fig. 14.13 Production of sine curve by rotating vector

After a time t seconds the vector OA will turn through an angle ωt radians (shown as angle AOB in Fig. 14.13). If the line BC is constructed perpendicular to OA as shown then:

$$\sin \omega t = \frac{BC}{OB}$$

i.e. $BC = OB \sin \omega t$

If all such vertical components are projected on to a graph of y against angle ωt in radians a sine curve results of amplitude OA. This method of producing a sine wave provides an alternative to that discussed in Section 14.1.

Periodic time

Let T seconds be the time for the rotating vector OA to make one revolution (i.e. 2π radians, which is one period).

Then: $2\pi = \omega T$

or $T = \dfrac{2\pi}{\omega}$ seconds

Time T is known as the **periodic time**. Thus the base of a sine curve (i.e. the horizontal axis) has a time scale as well as an angular scale.

Frequency

The number of complete waveforms (or cycles) occurring per second is called the frequency, f. (The electrical unit of frequency is the hertz (Hz).)

$$\text{Frequency} = \frac{\text{Number of cycles}}{\text{second}} = \frac{1}{T} = \frac{\omega}{2\pi} \text{ Hz}$$

i.e. $f = \dfrac{\omega}{2\pi} \text{ Hz}$

Angular velocity $\omega = 2\pi f$

14.3   General equation of a periodic function

Given the general sinusoidal periodic function

$$y = A \sin(\omega t \pm \alpha)$$

the following information can be obtained.

1. Amplitude $= A$
2. Angular velocity $= \omega$ radians per second
3. Periodic time, $T = \dfrac{2\pi}{\omega}$ seconds
4. Frequency, $f = \dfrac{\omega}{2\pi}$ Hz
5. $\alpha =$ angle of lead or lag (compared with $y = A \sin \omega t$)

Worked problems on periodic functions of the form $y = A \sin(\omega t \pm \alpha)$

Problem 1. An alternating voltage is given by $v = 50 \sin(200\pi t - 0.25)$ volts. Find the amplitude, periodic time, frequency and phase angle (with reference to $50 \sin 200\pi t$) of the oscillation.

$v = 50 \sin(200\pi t - 0.25)$

This compares with the general form $y = A \sin(\omega t \pm \alpha)$

Hence, **amplitude = 50 volts**

Angular velocity $\omega = 200\pi$

Therefore **periodic time,** $T = \dfrac{2\pi}{\omega} = \dfrac{2\pi}{200\pi}$

$$= \dfrac{1}{100} = \textbf{0.01 s}$$

Frequency, $f = \dfrac{1}{T} = \dfrac{1}{0.01} = \textbf{100 Hz}$

Phase angle $\alpha = 0.25$ radians (or 14° 19′) lagging $\sin 200\pi t$

Problem 2. The current in an alternating current circuit at any time t seconds is given by $i = 75.0 \sin(100\pi t + 0.320)$ amperes. Find
(a) the amplitude, periodic time, frequency and phase angle (with reference to $75.0 \sin 100\pi t$),
(b) the value of the current when $t = 0$,
(c) the value of the current when $t = 0.006$ s,
(d) the time when the current first reaches 50.0 amperes, and
(e) the time when the current is maximum.

Sketch one cycle of the oscillation.

(a) **Amplitude = 75.0 amperes**

Periodic time, $T = \dfrac{2\pi}{\omega} = \dfrac{2\pi}{100\pi}$ (since $\omega = 100\pi$)

$$= \dfrac{1}{50} \text{ or } \mathbf{0.02 \text{ s}}$$

Frequency, $f = \dfrac{1}{T} = \dfrac{1}{0.02} = \mathbf{50 \text{ Hz}}$

Phase angle $\alpha = 0.320$ radians (or $18° \ 20'$) leading $75.0 \sin 100\pi t$

(b)   When $t = 0$, $i = 75.0 \sin(0 + 0.320)$

$$= 75.0 \sin 18° \ 20'$$

$$= 75.0 \ (0.314 \ 6)$$

Hence $i = \mathbf{23.60 \text{ amperes}}$

(c)   When $t = 0.006s$, $i = 75.0 \sin[100\pi(0.006) + 0.320]$

$$= 75.0 \sin(0.6\pi + 0.320)$$

$$= 75.0 \sin(2.205)$$

$$= 75.0 \sin(126.34)°$$

$$= 75.0(0.805 \ 5)$$

Hence $i = \mathbf{60.41 \text{ amperes}}$

(d)   When $i = 50.0$ amperes, $50.0 = 75.0 \sin(100\pi t + 0.320)$

$$\dfrac{50.0}{75.0} = \sin(100\pi t + 0.320)$$

$$0.666 \ 7 = \sin(100\pi t + 0.320)$$

$$100\pi t + 0.320 = \arcsin 0.666 \ 7$$

$$= 41° \ 49' \text{ or } 0.729 \ 8 \text{ radians}$$

$$100\pi t + 0.320 = 0.729 \ 8$$

$$100\pi t = 0.729 \ 8 - 0.320$$

$$= 0.409 \ 8$$

Hence $t = \dfrac{0.409 \ 8}{100\pi} = \mathbf{0.001 \ 30 \text{ s}}$

(e)   When the current is a maximum,

$i = $ amplitude $= 75.0$ amperes

$$75.0 = 75.0 \sin(100\pi t + 0.320)$$

$$1 = \sin(100\pi t + 0.320)$$

$$100\pi t + 0.320 = \arcsin 1 = 90° \text{ or } \dfrac{\pi}{2} \text{ radians}$$

$$100\pi t + 0.320 = 1.570 \ 8$$

$$100\pi t = 1.570\,8 - 0.320$$

$$= 1.250\,8$$

Hence $t = \dfrac{1.250\,8}{100\pi}\,\text{s} = \mathbf{0.004\,0\ s}$

A sketch of $i = 75.0\sin(100\pi t + 0.320)$ amperes is shown in Fig. 14.14.

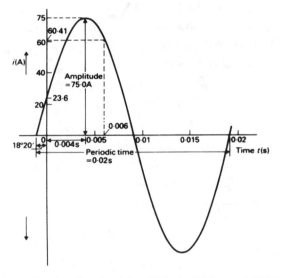

Fig. 14.14   Graph of $i = 75.0\sin(100\pi t + 0.320)$ amperes

---

Further problems on periodic functions of the form $y = A\sin(\omega t \pm \alpha)$ may be found in Section 14.5 (Problems 6 to 15), page 237.

14.4   Trigonometric approximations for small angles

If angle x is a **small angle** (i.e. less than about $5°$) and is expressed in **radians** then the following trigonometric approximations may be shown to be true:

(a) $\sin x \approx x$

(b) $\tan x \approx x$

(c) $\cos x \approx 1 - \dfrac{x^2}{2}$

For example, let $x = 1°$ which is $\left(1 \times \dfrac{\pi}{180}\right)$ radians, i.e. $0.017\,45$ radians, correct to 5 decimal places (since 2π radians $= 360°$). By calculator,

$\sin 1° = 0.01745$ and $\tan 1° = 0.01746$, showing that $\sin x = x \approx \tan x$ when $x = 0.01745$ radians.

Also, $\cos 1° = 0.99985$

When $x = 1°$, i.e. 0.01745 radians, $1 - \dfrac{x^2}{2} = 1 - \dfrac{0.01745^2}{2} = 0.99985$ correct

to 5 decimal places, showing that $\cos x = 1 - \dfrac{x^2}{2}$ when $x = 0.01745$ rads.

Similarly, let $x = 5°$ which is $\left(5 \times \dfrac{\pi}{180}\right)$ radians, i.e. 0.08727 radians,

correct to 5 decimal places. By calculator,

$\qquad \sin 5° = 0.08716$, thus $\sin x \approx x$,

$\qquad \tan 5° = 0.08749$, thus $\tan x \approx x$,

and $\qquad \cos 5° = 0.99619$

Since $x = 0.08727$ radians, $1 - \dfrac{x^2}{2} = 1 - \dfrac{0.08727^2}{2} = 0.99619$, showing that

$\cos x = 1 - \dfrac{x^2}{2}$ when $x = 0.08727$ radians.

If $\sin x \approx x$ for small angles then $\dfrac{\sin x}{x} \approx 1$ and this relationship can be used

when determining the differential coefficients of $\sin x$ and $\cos x$.

14.5 Further problems

Graphs of trigonometrical ratios

In Problems 1 to 4, find the amplitude and period of the wave (in degrees) and sketch the curve between $0°$ and $360°$.

1. (a) $y = 2 \sin 3A$ (b) $y = \sin \dfrac{5A}{2}$ (c) $y = 3 \sin 8x$

   (a) $[2, 120°]$ (b) $[1, 144°]$ (c) $[3, 45°]$

2. (a) $y = 4 \cos \dfrac{\phi}{2}$ (b) $y = \dfrac{5}{2} \cos \dfrac{3\phi}{8}$ (c) $y = 9 \cos \dfrac{7t}{3}$

   (a) $[4, 720°]$ (b) $\left[\dfrac{5}{2}, 960°\right]$ (c) $[9, 154° \, 17']$

3. (a) $y = 5 \sin^2 4A$ (b) $y = 2.4 \cos^2 \dfrac{2A}{3}$ (c) $y = \dfrac{1}{3}\left[7 \cos^2 \dfrac{5A}{6}\right]$

   (a) $[5, 45°]$ (b) $[2.4, 270°]$ (c) $\left[\dfrac{7}{3}, 216°\right]$

4. (a) $y = 3.6 \cos(4x - 30°)$ (b) $y = 5 \sin\left(\dfrac{4t}{9} + \dfrac{\pi}{4}\right)$

   (c) $y = 8 \sin^2\left(\dfrac{2\phi}{3} - \dfrac{\pi}{3}\right)$

   (a) $[3.6, 90°]$ (b) $[5, 810°]$ (c) $[8, 270°]$

5. Sketch the following graphs from $t = 0$ to $t = 2\pi$ seconds showing relevant details:

(a) $y = 5 \sin 3t$

(b) $y = 7 \cos 2t$

(c) $y = 2 \sin\left(t + \dfrac{4\pi}{9}\right)$

(d) $y = 4 \cos\left(t - \dfrac{\pi}{5}\right)$

(e) $y = 6 \sin^2 2t$

(f) $y = 1.5 \cos^2 t$

(The graphs are shown in Fig. 14.15)

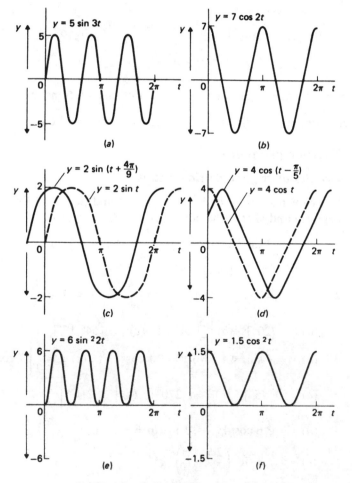

Fig. 14.15 Solution to Problem 5

Periodic functions of the form $y = A \sin(\omega t \pm \alpha)$

In Problems 6 to 9 find the amplitude, periodic time, frequency and phase angle (stating whether it is leading or lagging sin ωt) of the alternating quantities given.

6. $i = 60 \sin(50\pi t + 0.36)$
     [60, 0.04 s, 25 Hz, 20° 38′ leading 60 sin 50πt]

7. $v = 25 \sin(400\pi t - 0.231)$
     [25, 0.005 s, 200 Hz, 13° 14′ lagging 25 sin 400πt]

8. $y = 35 \sin(40 - 0.6)$
     [35, 0.157 s, 6.37 Hz, 34° 23′ lagging 35 sin 40t]

9. $x = 10 \sin(314.2t + 0.468)$
     [10, 0.02 s, 50 Hz, 26° 49′ leading 10 sin 314.2t]

10. A sinusoidal current has a maximum value of 25 A and a frequency of 60 Hz. At $t = 0$, the current is zero. Express the instantaneous current i, in the form $i = A \sin \omega t$
     [$i = 25 \sin(120\pi t)$]

11. An oscillating mechanism has a maximum displacement of 4.0 m and a frequency of 50 Hz. At $t = 0$ the displacement is 120 cm. Express the displacement in the general form $A \sin(\omega t \pm \alpha)$ [4.0 sin(100πt + 0.305)]

12. An alternating voltage v has a periodic time of 0.01 s and a maximum value of 30 volts. When $t = 0$, $v = -20$ V. Express the instantaneous voltage in the form $v = A \sin(\omega t \pm \alpha)$
     [30 sin(200πt − 0.73)]

13. The voltage in an alternating current circuit at any time t seconds is given by $e = 45.0 \sin 50t$. Find the first two times when the voltage is (a) 10.0 V and (b) 25.0 V
     (a) [0.004 48 s and 0.058 3 s]   (b) [0.011 8 s and 0.051 1 s]

14. The current in an a.c. circuit at any time t seconds is given by:
     $i = 55.0 \sin(100\pi t + 0.410)$ amperes. Find

     (a) the value of the current when $t = 0$,
     (b) the value of the current when $t = 0.005$ s,
     (c) the time when the current first reaches 32.0 A,
     (d) the time when the current is first maximum.

     Sketch one cycle of the oscillation showing important points.
     (a) [21.92 A]   (b) [50.44 A]   (c) [0.000 672 s]
     (d) [0.003 695 s]

15. The instantaneous value of voltage in an a.c. circuit at any time t seconds is given by:

     $v = 200.0 \sin(50\pi t - 0.683)$ volts. Find

    (a) the amplitude, periodic time, frequency and phase angle (with reference to 200.0 sin 50πt) of the function,

    (b) the voltage when $t = 0$,

    (c) the voltage when $t = 0.01$ s,

    (d) the times in the first cycle when the voltage is 100.0 V,

    (e) the times in the first cycle when the voltage is -58.0 V,

    (f) the first time when the voltage is a maximum.

Sketch the curve showing important points.

(a) [200 V, 0.04 s, 25 Hz, 39° 8′ lag]   (b) [-126.2 V]

(c) [155.1 V]   (d) [0.007 681 s and 0.021 01 s]

(e) [0.002 48 s and 0.026 22 s]   (f) [0.014 35 s]

Trigonometric approximations for small angles

16.   Show that if x is expressed in radians then:

$$\sin x \approx \tan x \approx x \text{ and } \cos x \approx 1 - \frac{x^2}{2}$$

when (a) $x = 0.5°$ (b) $x = 1.7°$ (c) $x = 4° 20′$

Chapter 15

Combining sinusoidal waveforms

15.1 Combination of two periodic functions

It is often necessary (especially in electrical alternating current theory and also when adding forces and other vectors in mechanics) to find the single phasor which could replace two or more separate phasors. The resulting single phasor is known as the resultant and there are a number of methods by which this may be found. Two such methods are:

(a) by plotting the periodic functions graphically, or
(b) by using phasors.

15.2 Combination of two periodic functions by sketching and plotting graphs

One method of obtaining the resultant waveform is by sketching the separate functions on the same axes using the same scales and then adding (or subtracting) ordinates at regular intervals. This method is shown in Problems 1 and 2.

Another method of obtaining the resultant waveform is by drawing up a table of values before plotting the resultant waveform. This method is shown in Problems 3 and 4. If sine waves of the same frequency (and hence period) are combined then a sine wave will result, this resultant having the same frequency as the single phasors. If, however, sine waves of different frequencies are combined then a sine wave will not result. The resultant will, however, be a periodic function although asymmetric.

Worked problems on combination of periodic functions by sketching and plotting graphs

Problem 1. Sketch the graph of $y = 2 \sin A$ from $A = 0°$ to $A = 360°$. On the same axes sketch $y = 3 \cos A$. By adding ordinates sketch $y_R = 2 \sin A + 3 \cos A$ and obtain a sinusoidal expression for this resultant waveform.

$y = 2 \sin A$, $y = 3 \cos A$ and $y_R = 2 \sin A + 3 \cos A$ are shown in Fig. 15.1

239

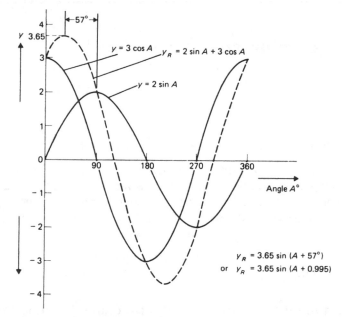

Fig. 15.1   Graphs of $y = 2 \sin A$, $y = 3 \cos A$ and
$y = 2 \sin A + 3 \cos A$

When adding ordinates at every 15° interval a pair of dividers is useful.

The resultant waveform has the same period, i.e. 360°, as the single phasors. The amplitude of the resultant (i.e. the maximum value) is 3.65. The resultant waveform leads the graph of $y = 2 \sin A$ by 57° or 0.995 rad. Hence the sinusoidal expression describing the resultant waveform is

$$y_R = \mathbf{3.65 \sin(A + 57°)} \text{ or } y_R = \mathbf{3.65 \sin(A + 0.995)}$$

Problem 2. The instantaneous values of two alternating currents are given by $i_1 = 15 \sin \omega t$ amperes and $i_2 = 12 \sin\left(\omega t + \dfrac{\pi}{3}\right)$ amperes
By sketching i_1 and i_2 on the same axes, using the same scale, over one cycle, obtain an expression for $i_1 - i_2$

$i_1 = 15 \sin \omega t$, $i_2 = 12 \sin\left(\omega t + \dfrac{\pi}{3}\right)$ and $i_1 - i_2 = i_R$ are shown in Fig. 15.2
Care must be taken when subtracting values of ordinates, especially when at least one of the ordinates is negative.

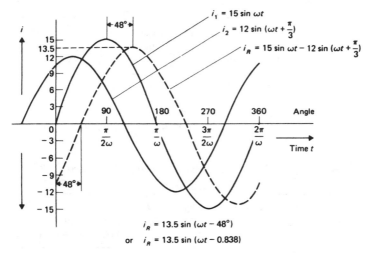

Fig. 15.2   Graphs of $i = 15 \sin \omega t$, $i = 12 \sin\left(\omega t + \dfrac{\pi}{3}\right)$ and

$$i_R = 15 \sin \omega t - 12 \sin\left(\omega t + \frac{\pi}{3}\right)$$

For example, in Fig. 15.2 at 150° the value of i_1 is 8 amperes and the value of i_2 is -5 amperes. The resultant $i_1 - i_2$ is $8 - -5$, i.e. $8 + 5$ or $+13$

The resultant waveform i_R has the same period, i.e. 360°, as the single current waveforms. The amplitude of i_R is 13.5 amperes and it is seen to lag 48° or 0.838 rad behind $i_1 = 15 \sin \omega t$. Hence

$$i_1 - i_2 = i_R = 13.5 \sin(\omega t - 48°)$$

or $$i_R = 13.5 \sin(\omega t - 0.838)$$

Problem 3. By drawing up a table plot the waveform
$y = 3 \sin 2x - \cos 3x$

x	0	15	30	45	60	75	90	105	120
$2x$	0	30	60	90	120	150	180	210	240
$\sin 2x$	0	0.500	0.866	1.00	0.866	0.500	0	-0.500	-0.866
$3 \sin 2x$	0	1.50	2.60	3.00	2.60	1.50	0	-1.50	-2.60
$3x$	0	45	90	135	180	225	270	315	360
$\cos 3x$	1.00	0.707	0	-0.707	-1.00	-0.707	0	0.707	1.00
$(3 \sin 2x - \cos 3x)$	-1.00	0.79	2.60	3.71	3.60	2.21	0	-2.21	-3.60

x	135	150	165	180	195	210	225	240
$2x$	270	300	330	360	390	420	450	480
$\sin 2x$	−1.00	−0.866	−0.500	0	0.500	0.866	1.00	0.866
$3 \sin 2x$	−3.00	−2.60	−1.50	0	1.50	2.60	3.00	2.60
$3x$	405	450	495	540	585	630	675	720
$\cos 3x$	0.707	0	−0.707	−1.00	−0.707	0	0.707	1.00
$(3 \sin 2x - \cos 3x)$	−3.71	−2.60	−0.79	1.00	2.21	2.60	2.29	1.60

x	255	270	285	300	315	330	345	360
$2x$	510	540	570	600	630	660	690	720
$\sin 2x$	0.500	0	−0.500	−0.866	−1.00	−0.866	−0.500	0
$3 \sin 2x$	1.50	0	−1.50	−2.60	−3.00	−2.60	−1.50	0
$3x$	765	810	855	900	945	990	1 035	1 080
$\cos 3x$	0.707	0	−0.707	−1.00	−0.707	0	0.707	1.00
$(3 \sin 2x - \cos 3x)$	0.79	0	−0.79	−1.60	−2.29	−2.60	−2.21	−1.00

The waveform $y = 3 \sin 2x - \cos 3x$ is shown in Fig. 15.3

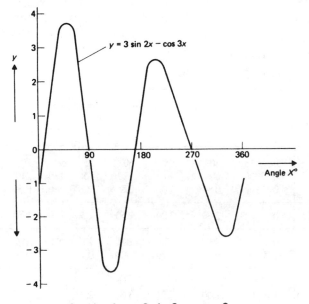

Fig. 15.3　Graph of $y = 3 \sin 2x - \cos 3x$

The waveform is not sinusoidal since the components forming it have different frequencies (and thus periods). The waveform has a period of 360° or 2π radians. When two sine waves of different periods are combined the resultant sine wave will always have a period equal to the lowest common multiple (L.C.M.) of the two component periods.

In the above example the period of $3 \sin 2x$ is 180° and the period of $\cos 3x$ is 120°. The L.C.M. of 180° and 120° is 360°. Hence the period of $y = 3 \sin 2x - \cos 3x$ is 360°.

Problem 4. Draw to scale a graph representing

$$v = 3 \sin \omega t + 2 \cos\left(\omega t - \frac{5\pi}{18}\right)$$

Express v in the general form $v = A \sin(\omega t \pm \alpha)$

A table of values must first be produced:

$-\dfrac{5\pi}{18}$ radians is the same as $-\dfrac{5\pi}{18} \times \dfrac{180°}{\pi} = -50°$

ωt (degrees)	0	15	30	45	60	75	90
$\sin \omega t$	0	0.259	0.500	0.707	0.866	0.966	1.00
$3 \sin \omega t$	0	0.78	1.50	2.12	2.60	2.90	3.00
$\omega t - 50°$	-50	-35	-20	-5	10	25	40
$\cos(\omega t - 50°)$	0.64	0.82	0.94	0.996	0.98	0.91	0.77
$2 \cos(\omega t - 50°)$	1.28	1.64	1.88	1.99	1.96	1.82	1.54
$3 \sin \omega t$ $\quad + 2 \cos(\omega t - 50°)$	1.28	2.42	3.38	4.11	4.56	4.72	4.54

ωt (degrees)	105	120	135	150	165	180
$\sin \omega t$	0.966	0.866	0.707	0.500	0.259	0
$3 \sin \omega t$	2.90	2.60	2.12	1.50	0.78	0
$\omega t - 50°$	55	70	85	100	115	130
$\cos(\omega t - 50°)$	0.57	0.34	0.087	-0.17	-0.42	-0.64
$2 \cos(\omega t - 50°)$	1.14	0.68	0.17	-0.34	-0.84	-1.28
$3 \sin \omega t$ $\quad + 2 \cos(\omega t - 50°)$	4.04	3.28	2.29	1.16	-0.06	-1.28

ωt (degrees)	195	210	225	240	255	270
$\sin \omega t$	-0.259	-0.500	-0.707	-0.866	-0.966	-1.00
$3 \sin \omega t$	-0.78	-1.50	-2.12	-2.60	-2.90	-3.00
$\omega t - 50°$	145	160	175	190	205	220
$\cos(\omega t - 50°)$	-0.82	-0.94	-0.996	-0.98	-0.91	-0.77
$2\cos(\omega t - 50°)$	-1.64	-1.88	-1.99	-1.96	-1.82	-1.54
$3 \sin \omega t$ $+ 2\cos(\omega t - 50°)$	-2.42	-3.38	-4.11	-4.56	-4.72	-4.54

ωt (degrees)	285	300	315	330	345	360
$\sin \omega t$	-0.966	-0.866	-0.707	-0.500	-0.259	0
$3 \sin \omega t$	-2.90	-2.60	-2.12	-1.50	-0.78	0
$\omega t - 50°$	235	250	265	280	295	310
$\cos(\omega t - 50°)$	-0.57	-0.34	-0.087	0.17	0.42	0.64
$2\cos(\omega t - 50°)$	-1.14	-0.68	-0.17	0.34	0.84	1.28
$3 \sin \omega t$ $+ 2\cos(\omega t - 50°)$	-4.04	-3.28	-2.29	-1.16	0.06	1.28

A graph of $v = 3 \sin \omega t + 2 \cos\left(\omega t - \dfrac{5\pi}{18}\right)$ is shown in Fig. 15.4

The maximum value of the resultant sine wave, i.e. the amplitude, is 4.73. The resultant wave is $15°$ or $\dfrac{\pi}{12}$ radians ahead of a sine wave which starts at $0°$. Hence

$$v = 3 \sin \omega t + 2 \cos\left(\omega t - \frac{5\pi}{18}\right) = 4.73 \sin\left(\omega t + \frac{\pi}{12}\right)$$

Further problems on the combination of two periodic functions by sketching and plotting graphs may be found in Section 15.4, page 251.

15.3   Combination of two periodic functions of the same frequency by using phasors

The resultant of two periodic functions of the same frequency may be found from their relative positions when the time is zero. For example, if $y_1 = 3 \sin \omega t$ and $y_2 = 4 \sin\left(\omega t - \dfrac{\pi}{4}\right)$ then each may be represented as rotating

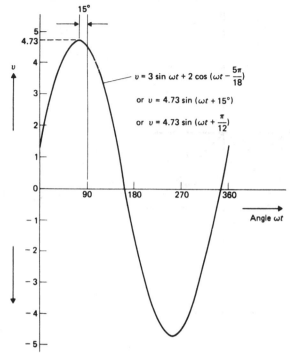

Fig. 15.4   Graph of $v = 3 \sin \omega t + 2 \cos\left(\omega t - \dfrac{5\pi}{18}\right)$

vectors (or phasors) of maximum values 3 and 4 respectively, with y_2 lagging y_1 by $\dfrac{\pi}{4}$ radians or 45° as shown in Fig. 15.5, their positions being taken when the time is zero.

When the sum or difference of two periodic functions of the same frequency is required it is usual to represent the phasors as shown in Fig. 15.6

Procedure to find the resultant phasor $y_R = y_1 + y_2$, *where* $y_1 = 3 \sin \omega t$ *and* $y_2 = 4 \sin\left(\omega t - \dfrac{\pi}{4}\right)$

(i) By drawing

1. Draw y_1 horizontal 3 units long, i.e. length 0a of Fig. 15.7.
2. Join y_2 to the end of y_1 (i.e. join y_2 to the arrow head of y_1) at an angle of 45° lagging and 4 units long.
3. The resultant is given by the length y_R and its phase angle ϕ may be measured with respect to y_1. Figure 15.7 is known as the phasor diagram.

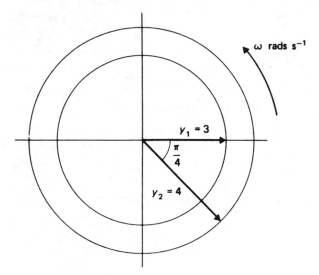

Fig. 15.5

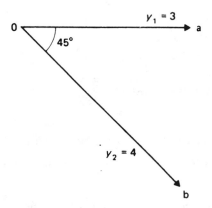

Fig. 15.6

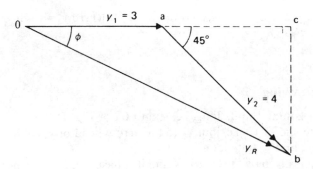

Fig. 15.7

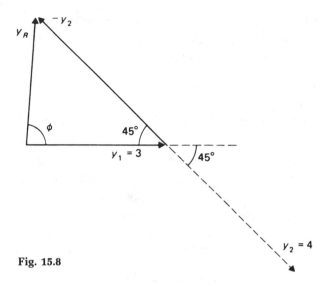

Fig. 15.8

(ii) By calculation

Sketch a phasor diagram as in Fig. 15.7, then:

Either 1. Use the cosine rule to calculate y_R and then the sine rule to calculate angle ϕ.

Or 2. Calculate the horizontal component of y_R (i.e. length 0c of Fig. 15.7) then the vertical component of y_R (i.e. length bc), and then use Pythagoras's theorem to calculate y_R and trigonometric ratios to calculate angle ϕ.

If the resultant phasor $y_R = y_1 - y_2$ is required then y_2 is drawn in the opposite direction to that shown in Fig. 15.7. This is shown in Fig. 15.8. Once the phasor diagram has been established y_R and angle ϕ may be measured or calculated.

Worked problems on finding the resultant of two periodic functions of the same frequency by using phasors

Problem 1. If $y_1 = 3 \sin \omega t$ and $y_2 = 4 \sin\left(\omega t - \dfrac{\pi}{4}\right)$ obtain an expression for the resultant $y_R = y_1 + y_2$ (a) by drawing and (b) by calculation.

(a) **By drawing**
From the phasor diagram shown in Fig. 15.7 y_R is measured

as 6.5 units and angle ϕ as 26°. Hence, by drawing,

$$y_R = 6.5 \sin(\omega t - 26°)$$

or $y_R = 6.5 \sin(\omega t - 0.454)$

(b) **By calculation**

Method 1

Using the cosine rule: $y_R{}^2 = 3^2 + 4^2 - 2(3)(4) \cos 135°$

$$= 9 + 16 - 24 \cos 135°$$

$$= 9 + 16 + 16.97 = 41.97$$

Therefore $\qquad y_R = \sqrt{41.97} = \textbf{6.48 units}$

Using the sine rule: $\dfrac{4}{\sin \phi} = \dfrac{6.48}{\sin 135°}$

Therefore $\qquad \sin \phi = \dfrac{4 \sin 135°}{6.48} = 0.436\ 5$

Therefore $\qquad$ angle $\phi = \textbf{25° 53'}$

Method 2

Referring to Fig. 15.7:

The horizontal component $0c = 0a + ac$

$$= 3 + 4 \cos 45°$$

$$= 5.828 \text{ units}$$

The vertical component $\qquad = bc$

$$= 4 \sin 45°$$

$$= 2.828 \text{ units}$$

Using Pythagoras's theorem: $0b^2 = 0c^2 + bc^2$

Therefore $y_R{}^2 = (5.828)^2 + (2.828)^2$

$$= 33.97 + 7.998$$

$$= 41.97$$

$$y_R = \sqrt{41.97} = \textbf{6.48 units}$$

$\tan \phi = \dfrac{bc}{0c} = \dfrac{2.828}{5.828} = 0.485\ 2$

Therefore **angle $\phi = $ 25° 53'**

Therefore y_R lags y_1 by 25° 53'

Hence $y_R = 6.48 \sin(\omega t - 25° 53')$ or

$$y_R = 6.48 \sin(\omega t - 0.452)$$

Problem 2. Two alternating currents are given by $i_1 = 2.0 \sin \omega t$ amperes and $i_2 = 3.0 \cos \omega t$ amperes. Obtain a sinusoidal expression for $i_1 + i_2$

To obtain the resultant $i_1 + i_2$ both components need to be of the form
$A \sin(\omega t \pm \alpha)$. Thus $3.0 \cos \omega t$ needs to be changed to this form.
$3.0 \cos \omega t = 3.0 \sin(\omega t + 90°)$ since a cosine curve leads a
sine curve by $90°$ or $\dfrac{\pi}{2}$ radians.

The relative positions of i_1 and i_2 at $t = 0$ are shown
in Fig. 15.9(a) and the phasor diagram is shown in
Fig. 15.9(b)

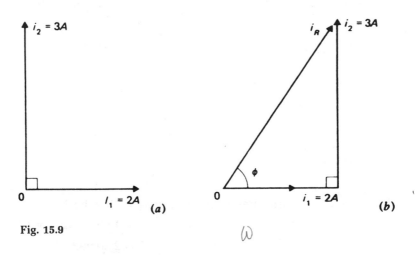

Fig. 15.9   (i)

Using Pythagoras's theorem: $i_R = \sqrt{(i_1 + i_2)^2} = \sqrt{(2.0^2 + 3.0^2)}$
$$= \sqrt{13.0} = 3.61 \text{ amperes}$$
$$\tan \phi = \frac{3.0}{2.0}$$
$$\text{angle } \phi = 56° \ 19' \text{ or } 0.983 \text{ radians}$$
Therefore i_R is leading i_1 by $56° \ 19'$ or 0.983 radians
Hence $i_R = 2.0 \sin \omega t + 3.0 \cos \omega t = 3.61 \sin(\omega t + 56° \ 19')$
or $i_R = 3.61 \sin(\omega t + 0.983)$ amperes

Problem 3. Two alternating voltages are given by $v_1 = 15.0 \sin \omega t$
volts and $v_2 = 12.0 \sin\left(\omega t + \dfrac{\pi}{3}\right)$ volts. Obtain sinusoidal
expressions for (a) $v_1 + v_2$ and (b) $v_1 - v_2$

(a)   The relative positions of v_1 and v_2 at $t = 0$ are shown
      in Fig. 15.10(a) and the phasor diagram is shown in
      Fig. 15.10(b)

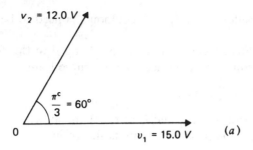

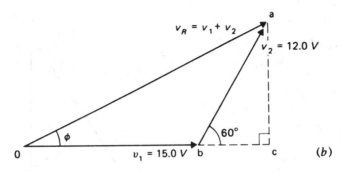

Fig. 15.10

The horizontal component of $v_R = 0b + bc$
$$= 15.0 + 12.0 \cos 60°$$
$$= 21.0 \text{ volts}$$

The vertical component of v_R $= ac = 12.0 \sin 60°$
$$= 10.39 \text{ volts}$$

$$v_R \, (= 0c) = \sqrt{[(21.0)^2 + (10.39)^2]}$$
$$= \sqrt{548.95} = \textbf{23.43 volts}$$

$$\tan \phi = \frac{ac}{0c} = \frac{10.39}{21.0} = 0.494 \, 8$$

$$\phi = \textbf{26° 19′}$$

Therefore v_R leads v_1 by 26° 19′ or 0.459 radians

Hence $v_R = 15.0 \sin \omega t + 12.0 \sin\left(\omega t + \frac{\pi}{3}\right)$

$$= \textbf{23.43} \sin(\omega t + \textbf{26° 19′})$$
$$= \textbf{23.43} \sin(\omega t + \textbf{0.459})$$

(b)   To find the resultant $v_R = v_1 - v_2$ the phasor v_2 is reversed in direction. The phasor diagram is shown in Fig. 15.11

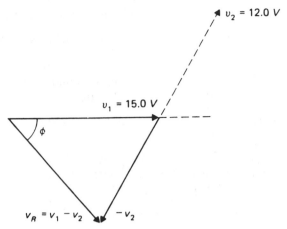

Fig. 15.11

Using the cosine rule: $v_R^2 = v_1^2 + v_2^2 - 2v_1v_2 \cos 60°$

$$= 15.0^2 + 12.0^2 - 2(15.0)(12.0)(0.5)$$

$$= 189.0$$

$$v_R = \sqrt{189.0} = \textbf{13.75 volts}$$

Using the sine rule: $\dfrac{12.0}{\sin \phi} = \dfrac{13.75}{\sin 60°}$

$$\sin \phi = \frac{12.0 \sin 60°}{13.75} = 0.755\ 8$$

from which   **angle $\phi = 49° 6'$**

Therefore v_R lags v_1 by $49° 6'$ or 0.857 radians

Hence $v_R = 15.0 \sin \omega t - 12.0 \sin\left(\omega t + \dfrac{\pi}{3}\right)$

$$= \textbf{13.75 } \sin(\omega t - 49° 6')$$

$$= \textbf{13.75 } \sin(\omega t - 0.857)$$

Further problems on finding the resultant of two periodic functions of the same frequency by using phasors may be found in the following Section 15.4.

15.4   Further problems

1.   Sketch the curve $y = 3.0 \sin A$ from $A = 0°$ to $A = 360°$. On the same axes sketch $y = \cos A$. By adding ordinates sketch $y = 3.0 \sin A + \cos A$ and obtain a sinusoidal expression for this waveform.
     [$3.16 \sin(A + 18° 26')$ or $3.16 \sin(A + 0.322)$]

2. The instantaneous values of two alternating voltages are given
    by $v_1 = 5.0 \sin \omega t$ and $v_2 = 9.0 \sin\left(\omega t - \dfrac{\pi}{6}\right)$. By sketching
    v_1 and v_2 on the same axes, using the same scale, over one
    cycle, obtain an expression for (a) $v_1 + v_2$ and (b) $v_1 - v_2$
    (a) [$13.6 \sin(\omega t - 0.338)$]   (b) [$5.30 \sin(\omega t + 2.126)$]

3. By drawing up a table of values plot the waveform
    $$y = 4.0 \sin\left(\omega t + \frac{\pi}{3}\right) + 3.0 \cos\left(\omega t - \frac{\pi}{4}\right)$$
    Express y in the general form $y = A \sin(\omega t \pm \alpha)$
    [$6.94 \sin(\omega t + 0.935)$]

4. By calculation obtain an expression for
    $$y_1 = 4.0 \sin \omega t \text{ and } y_2 = 6.0 \sin\left(\omega t - \frac{\pi}{3}\right)$$
    [$8.72 \sin(\omega t - 0.638)$]

Either by drawing or by calculation express the combination
of periodic functions in Problems 5 and 6 in the form
$A \sin(\omega t \pm \alpha)$

5. $5.0 \sin \omega t + 2.0 \sin\left(\omega t - \dfrac{\pi}{6}\right)$ [$6.81 \sin(\omega t - 0.147)$]

6. $16 \sin \omega t - 10 \sin\left(\omega t + \dfrac{\pi}{8}\right)$ [$7.77 \sin(\omega t - 0.515)$]

7. Two alternating currents are given by $i_1 = \sin\left(\omega t + \dfrac{\pi}{5}\right)$ and
    $i_2 = 2.0 \sin\left(\omega t - \dfrac{\pi}{6}\right)$. Obtain an expression for $i_1 - i_2$ in the
    form $A \sin(\omega t \pm \alpha)$ [$1.84 \sin(\omega t + 2.097)$]

8. Two voltages, $4 \cos \omega t$ and $-3 \sin \omega t$, are inputs to an
    analogue circuit. Find an expression for the output voltage in
    the form $A \sin(\omega t \pm \alpha)$ if this is given by the addition of the
    two inputs.   [$5 \sin(\omega t + 126° 52')$ or $5 \sin(\omega t + 2.214)$]

9. Two alternating voltages are given by $v_1 = 25.0 \sin \omega t$ and
    $v_2 = 16.0 \sin\left(\omega t - \dfrac{\pi}{4}\right)$. Obtain sinusoidal expressions for
    (a) $v_1 + v_2$ and (b) $v_1 - v_2$
    (a) [$38.04 \sin(\omega t - 0.302)$]   (b) [$17.76 \sin(\omega t + 0.691)$]

10. In the theory of transmission of polarised light, waveforms
     have the equations

$$x_1 = \cos \omega t \qquad\qquad y_1 = k \sin \omega t$$
$$x_2 = k^2 \cos(\omega t + \alpha) \quad y_2 = -k \sin(\omega t + \alpha)$$

Determine the resultant of $x_1 + x_2$ and $y_1 + y_2$

$$\left[\begin{array}{l} x_1 + x_2 = \sqrt{(1 + k^4 + 2k^2 \cos \alpha)} \cos(\omega t + \beta) \\ \text{where } \beta = \arcsin\left\{\dfrac{k^2 \sin \alpha}{\sqrt{(1 + k^4 + 2k^2 \cos \alpha)}}\right\} \\ y_1 + y_2 = k\sqrt{[2(1 - \cos \alpha)]} \sin(\omega t - \beta) \\ \text{where } \beta = \arcsin\left\{\dfrac{\sin \alpha}{\sqrt{[2(1 - \cos \alpha)]}}\right\} \end{array}\right]$$

Chapter 16

Compound angles

16.1 Compound angle formulae

An electrical voltage v can be expressed as $v = 200 \sin(\omega t + 0.47)$ volts. Similarly, the displacement x of a body from a fixed point can be expressed as $x = 7 \sin(2t - 0.62)$ metres. The angles $(\omega t + 0.47)$ and $(2t - 0.62)$ are called **compound angles** because they are the sum or difference of two angles.

It is often useful in engineering and science to express the trigonometrical ratios of compound angles in terms of their two component angles. These are often called the **addition and subtraction formulae** or just **compound angle formulae**. One method of deriving the formula is shown below.

Consider the triangle XYZ shown in Fig. 16.1. The line YW is constructed perpendicular to XZ. Thus two right-angled triangles, YZW and YWX, are produced. Let the sides of the two triangles be labelled a, b, c, d and e as shown, and let $\angle$ WYZ be A and $\angle$ XYW be B:

Area of triangle YZW = $\frac{1}{2} de \sin A$
Area of triangle YWX = $\frac{1}{2} ce \sin B$
Area of triangle XYZ = $\frac{1}{2} cd \sin(A + B)$

But area of triangle XYZ = area of triangle YZW + area of triangle YWX, i.e.

$$\tfrac{1}{2} cd \sin(A + B) = \tfrac{1}{2} de \sin A + \tfrac{1}{2} ce \sin B$$

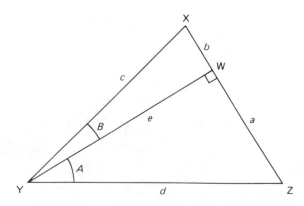

Fig. 16.1

Dividing each term by $\frac{1}{2}cd$ gives

$$\sin(A + B) = \frac{e}{c}\sin A + \frac{e}{d}\sin B$$

But $\dfrac{e}{c} = \cos B$ and $\dfrac{e}{d} = \cos A$. Hence

$$\sin (A + B) = \sin A \cos B + \cos A \sin B \qquad (1)$$

Applying the cosine rule to triangle XYZ gives

$$(a + b)^2 = c^2 + d^2 - 2cd \cos(A + B)$$

from which

$$\cos(A + B) = \frac{c^2 + d^2 - (a + b)^2}{2cd} = \frac{c^2 + d^2 - [a^2 + 2ab + b^2]}{2cd}$$

$$= \frac{(c^2 - b^2) + (d^2 - a^2) - 2ab}{2cd}$$

But $c^2 - b^2 = e^2$ and $d^2 - a^2 = e^2$ by Pythagoras's theorem. Hence

$$\cos(A + B) = \frac{e^2 + e^2 - 2ab}{2cd} = \frac{2e^2 - 2ab}{2cd} = \frac{e^2 - ab}{cd}$$

$$= \frac{e^2}{cd} - \frac{ab}{cd} = \frac{e}{d}\frac{e}{c} - \frac{a}{d}\frac{b}{c}$$

Hence $\cos(A + B) = \cos A \cos B - \sin A \sin B \qquad (2)$

Replacing B by $-B$ in equation (1) and equation (2) and remembering that $\sin(-B) = -\sin B$ and $\cos(-B) = +\cos B$:

$$\sin(A - B) = \sin A \cos B - \cos A \sin B \qquad (3)$$
and $$\cos(A - B) = \cos A \cos B + \sin A \sin B \qquad (4)$$

$$\tan(A + B) = \frac{\sin(A + B)}{\cos(A + B)} = \frac{\sin A \cos B + \cos A \sin B}{\cos A \cos B - \sin A \sin B}$$

Dividing numerator and denominator by $\cos A \cos B$ gives

$$\tan(A + B) = \frac{\dfrac{\sin A \cos B}{\cos A \cos B} + \dfrac{\cos A \sin B}{\cos A \cos B}}{\dfrac{\cos A \cos B}{\cos A \cos B} - \dfrac{\sin A \sin B}{\cos A \cos B}} = \frac{\dfrac{\sin A}{\cos A} + \dfrac{\sin B}{\cos B}}{1 - \dfrac{\sin A}{\cos A}\dfrac{\sin B}{\cos B}}$$

i.e. $$\tan(A + B) = \frac{\tan A + \tan B}{1 - \tan A \tan B} \qquad (5)$$

By similar reasoning it may be shown that

$$\tan(A - B) = \frac{\tan A - \tan B}{1 + \tan A \tan B} \tag{6}$$

The addition and subtraction formulae (i.e. equations (1) to (6)), proved above for acute angles, are, in fact, true for all values of A and B. It is very important to realise that $\sin(A + B)$ is **not** equal to $(\sin A + \sin B)$ and $\cos(A - B)$ is **not** equal to $(\cos A - \cos B)$, and so on.

Summary

$$\sin(A \pm B) = \sin A \cos B \pm \cos A \sin B$$
$$\cos(A \pm B) = \cos A \cos B \mp \sin A \sin B$$

$$\tan(A \pm B) = \frac{\tan A \pm \tan B}{1 \mp \tan A \tan B}$$

(In the cosine compound-angle formula care should be taken in particular with the signs.)

Worked problems on compound angle formulae

Problem 1. Verify (a) that the compound-angle addition formulae are true when $A = 30°$ and $B = 45°$, and (b) that the compound-angle subtraction formulae are true when $A = 203°$ and $B = 145°$.

(a) $\sin(A + B) = \sin A \cos B + \cos A \sin B$
When $A = 30°$ and $B = 45°$
$\sin(30° + 45°) = \sin 30° \cos 45° + \cos 30° \sin 45°$
$$= (0.500\ 0)(0.707\ 1) + (0.866\ 0)(0.707\ 1)$$
$$= 0.353\ 6 + 0.612\ 3$$
i.e.   **$\sin 75° = 0.965\ 9$**
This may be verified using a calculator.
$\cos(A + B) = \cos A \cos B - \sin A \sin B$
When $A = 30°$ and $B = 45°$
$\cos(30° + 45°) = \cos 30° \cos 45° - \sin 30° \sin 45°$
$$= (0.866\ 0)(0.707\ 1) - (0.500\ 0)(0.707\ 1)$$
$$= 0.612\ 3 - 0.353\ 6$$
i.e.   **$\cos 75° = 0.258\ 7$**
This may be verified to 3-significant-figure accuracy using a calculator.

$$\tan(A + B) = \frac{\tan A + \tan B}{1 - \tan A \tan B}$$

When $A = 30°$ and $B = 45°$

$$\tan(30° + 45°) = \frac{\tan 30° + \tan 45°}{1 - \tan 30° \tan 45°} = \frac{(0.577\,35) + (1)}{1 - (0.577\,35)(1)}$$

$$= \frac{1.577\,35}{0.422\,65} = 3.732\,05$$

i.e.   **$\tan 75° = 3.732\,1$**

This may be verified using a calculator.

(b) $\sin(A - B) = \sin A \cos B - \cos A \sin B$

When $A = 203°$ and $B = 145°$

$\sin(203° - 145°) = \sin 203° \cos 145° - \cos 203° \sin 145°$

$$= (-0.390\,73)(-0.819\,15)$$
$$- (-0.920\,50)(0.573\,58)$$
$$= +0.320\,07 + 0.527\,98$$

i.e.   **$\sin 58° = 0.848\,1$**

This may be verified to 3-significant-figure accuracy using a calculator.

$\cos(A - B) = \cos A \cos B + \sin A \sin B$

When $A = 203°$ and $B = 145°$

$\cos(203° - 145°) = \cos 203° \cos 145° + \sin 203° \sin 145°$

$$= (-0.920\,50)(-0.819\,15)$$
$$+ (-0.390\,73)(0.573\,58)$$
$$= +0.754\,03 - 0.224\,11$$

i.e.   **$\cos 58° = 0.529\,9$**

This may be verified using a calculator.

$$\tan(A - B) = \frac{\tan A - \tan B}{1 + \tan A \tan B}$$

When $A = 203°$ and $B = 145°$

$$\tan(203° - 145°) = \frac{\tan 203° - \tan 145°}{1 + \tan 203° \tan 145°}$$

$$= \frac{0.424\,47 - (-0.700\,21)}{1 + (0.424\,47)(-0.700\,21)} = \frac{1.124\,68}{0.702\,78}$$

i.e.   **$\tan 58° = 1.600\,3$**

This may be verified using a calculator.

For any chosen values of A and B the compound-angle addition and subtraction formulae are valid.

Problem 2. Use the compound-angle addition and subtraction formulae to simplify the following expressions:

(a) $\sin 52° \cos 29° + \cos 52° \sin 29°$

(b) $\cos 46° \cos 11° + \sin 46° \sin 11°$ (c) $\sin(180° + X)$

(d) $\cos\left(\dfrac{3\pi}{2} + Y\right)$ (e) $\cos(A - B) - \cos(A + B)$

(a) Since $\sin A \cos B + \cos A \sin B = \sin(A + B)$
then $\sin 52° \cos 29° + \cos 52° \sin 29° = \sin(52° + 29°)$
Hence **$\sin 52° \cos 29° + \cos 52° \sin 29° = \sin 81°$**

(b) Since $\cos A \cos B + \sin A \sin B = \cos(A - B)$
then $\cos 46° \cos 11° + \sin 46° \sin 11° = \cos(46° - 11°)$
Hence **$\cos 46° \cos 11° + \sin 46° \sin 11° = \cos 35°$**

(c) $\sin(180° + X) = \sin 180° \cos X + \cos 180° \sin X$
$$= (0)(\cos X) + (-1)(\sin X)$$
Hence **$\sin(180° + X) = -\sin X$**

(d) $\cos\left(\dfrac{3\pi}{2} + Y\right) = \cos\dfrac{3\pi}{2}\cos Y - \sin\dfrac{3\pi}{2}\sin Y$
$$= (0)(\cos Y) - (-1)(\sin Y)$$
Hence **$\cos\left(\dfrac{3\pi}{2} + Y\right) = \sin Y$**

(e) **$\cos(A - B) - \cos(A + B)$** $= [\cos A \cos B + \sin A \sin B]$
$$- [\cos A \cos B - \sin A \sin B]$$
$$= \mathbf{2 \sin A \sin B}$$

Problem 3. (a) Given that $\cos A = \dfrac{4}{5}$ and $\sin B = \dfrac{15}{17}$, where A and B are acute angles, find the values of $\sin(A + B)$ and $\cos(A - B)$.
(b) Given that $\sin C = 0.400\,0$ and $\cos D = 0.600\,0$, where C and D are acute angles, find the values of $\sin(C - D)$ and $\cos(C + D)$.

(a) $\cos A = \dfrac{4}{5} = \dfrac{\text{adjacent side}}{\text{hypotenuse}}$. The opposite side is found by the theorem of Pythagoras, i.e. opposite side of triangle $= \sqrt{(5^2 - 4^2)} = 3$. Hence $\sin A = \dfrac{3}{5}$. Similarly, since $\sin B = \dfrac{15}{17} = \dfrac{\text{opposite side}}{\text{hypotenuse}}$, the adjacent side of the triangle $= \sqrt{(17^2 - 15^2)} = 8$. Hence
$$\cos B = \dfrac{8}{17}$$

$\sin(A + B) = \sin A \cos B + \cos A \sin B$
$$= \left(\dfrac{3}{5}\right)\left(\dfrac{8}{17}\right) + \left(\dfrac{4}{5}\right)\left(\dfrac{15}{17}\right)$$
$$= \dfrac{24}{85} + \dfrac{60}{85} = \dfrac{84}{85}$$

$\cos(A - B) = \cos A \cos B + \sin A \sin B$

$$= \left(\frac{4}{5}\right)\left(\frac{8}{17}\right) + \left(\frac{3}{5}\right)\left(\frac{15}{17}\right)$$

$$= \frac{32}{85} + \frac{45}{85} = \frac{77}{85}$$

(b) If $\sin C = 0.400\ 0$ then, by calculator, $C = 23.58°$ or $23°\ 35'$

Then $\cos C = \cos 23°\ 35' = 0.916\ 5$

If $\cos D = 0.600\ 0$ then, by calculator, $D = 53.13°$ or $53'\ 8'$

Then $\sin D = \sin 53°\ 8' = 0.800\ 0$

$\sin(C - D) = \sin C \cos D - \cos C \sin D$

$\qquad\qquad = (0.400\ 0)(0.600\ 0) - (0.916\ 5)(0.800\ 0)$

$\qquad\qquad = 0.240\ 0 - 0.733\ 2$

$\qquad\qquad = \boldsymbol{-0.493\ 2}$

$\cos(C + D) = \cos C \cos D - \sin C \sin D$

$\qquad\qquad = (0.916\ 5)(0.600\ 0) - (0.400\ 0)(0.800\ 0)$

$\qquad\qquad = 0.549\ 9 - 0.320\ 0$

$\qquad\qquad = \boldsymbol{0.229\ 9}$

Problem 4. Find the values of (a) $\sin 15°$ and $\cos 15°$, and
(b) $\sin 75°$ and $\cos 75°$, without using a calculator. Leave answers
in surd form.

(a) $\sin 15° = \sin(60° - 45°) = \sin 60° \cos 45° - \cos 60° \sin 45°$

$$= \left(\frac{\sqrt{3}}{2}\right)\left(\frac{1}{\sqrt{2}}\right) - \left(\frac{1}{2}\right)\left(\frac{1}{\sqrt{2}}\right)$$

$$= \frac{\sqrt{3} - 1}{2\sqrt{2}}$$

$\cos 15° = \cos(60° - 45°) = \cos 60° \cos 45° + \sin 60° \sin 45°$

$$= \left(\frac{1}{2}\right)\left(\frac{1}{\sqrt{2}}\right) + \left(\frac{\sqrt{3}}{2}\right)\left(\frac{1}{\sqrt{2}}\right)$$

$$= \frac{1 + \sqrt{3}}{2\sqrt{2}}$$

(b) $\sin 75° = \sin(45° + 30°) = \sin 45° \cos 30° + \cos 45° \sin 30°$

$$= \left(\frac{1}{\sqrt{2}}\right)\left(\frac{\sqrt{3}}{2}\right) + \left(\frac{1}{\sqrt{2}}\right)\left(\frac{1}{2}\right)$$

$$= \frac{\sqrt{3} + 1}{2\sqrt{2}}$$

$$\cos 75° = \cos(45° + 30°) = \cos 45° \cos 30° - \sin 45° \sin 30°$$

$$= \left(\frac{1}{\sqrt{2}}\right)\left(\frac{\sqrt{3}}{2}\right) - \left(\frac{1}{\sqrt{2}}\right)\left(\frac{1}{2}\right)$$

$$= \frac{\sqrt{3} - 1}{2\sqrt{2}}$$

It may be proved, by calculator, that $\cos \theta = \sin(90 - \theta)$ or $\sin \theta = \cos(90 - \theta)$. Thus $\cos 15° = \sin(90° - 15°) = \sin 75°$ and $\sin 15° = \cos(90° - 15°) = \cos 75°$, which is shown in this problem.

Further problems on compound angles may be found in Section 16.7 (Problems 1 to 18), page 279.

16.2   Conversion of $a \sin \omega t + b \cos \omega t$ into $R \sin(\omega t + \alpha)$

If $R \sin(\omega t + \alpha)$ is expanded using the compound-angle addition formula then

$$R \sin(\omega t + \alpha) = R[\sin \omega t \cos \alpha + \cos \omega t \sin \alpha]$$
$$= R \sin \omega t \cos \alpha + R \cos \omega t \sin \alpha$$

Let $a \sin \omega t + b \cos \omega t = (R \cos \alpha) \sin \omega t + (R \sin \alpha) \cos \omega t$. Equating the coefficients of $\sin \omega t$ gives

$$a = R \cos \alpha, \text{ i.e. } \cos \alpha = \frac{a}{R}$$

Equating the coefficients of $\cos \omega t$ gives

$$b = R \sin \alpha, \text{ i.e. } \sin \alpha = \frac{b}{R}$$

Therefore, if the values of a and b are known, the values of R and α can be calculated.

In Fig. 16.2 the relationships between the constants a, b, R and α are shown. By the theorem of Pythagoras

$$R = \sqrt{(a^2 + b^2)}$$

and from trigonometrical ratios

$$\alpha = \arctan \frac{b}{a}$$

When the resultant of an expression such as $y = a \sin \omega t + b \cos \omega t$ is required it may be expressed readily in the form $R \sin(\omega t + \alpha)$ by using the compound-angle addition formula.

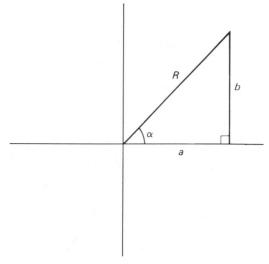

Fig. 16.2

In Chapter 15 it was shown that the resultant of an expression such as $y = a \sin \omega t + b \cos \omega t$ could be obtained either (i) by plotting graphs of $y_1 = a \sin \omega t$ and $y_2 = b \cos \omega t$ on the same axes and then adding ordinates at intervals or (ii) by drawing or calculating the resultant of two phasors from their relative positions when the time is zero. The conversion of $a \sin \omega t + b \cos \omega t$ to $R \sin(\omega t + \alpha)$ gives a third method of obtaining the resultant of two sine waves. It is quite possible to convert $a \sin \omega t \pm b \cos \omega t$ into one of four forms, i.e. $R \sin(\omega t + \alpha)$, $R \sin(\omega t - \alpha)$, $R \cos(\omega t + \alpha)$ or $R \cos(\omega t - \alpha)$. This is achieved by expanding any one of the latter expressions by using addition or subtraction formulae and then equating the coefficients of $\sin \omega t$ and $\cos \omega t$ (see Problem 3). Trigonometrical equations of the type $a \sin \omega t + b \cos \omega t = C$ may be solved by converting the left-hand side into the form $R \sin(\omega t + \alpha)$ (see Problems 5 and 6).

Worked problems on converting $a \sin \omega t \pm b \cos \omega t$ into the form $R \sin(\omega t \pm \alpha)$

Problem 1. Find an expression for $y = 4 \sin \omega t + 3 \cos \omega t$ in the general form $y = R \sin(\omega t \pm \alpha)$: (a) by plotting graphs of $y_1 = 4 \sin \omega t$ and $y_2 = 3 \cos \omega t$ on the same axes and then adding ordinates, (b) by phasor addition, and (c) by using the compound-angle addition formula.

(a)   Graphs of $y_1 = 4 \sin \omega t$ and $y_2 = 3 \cos \omega t$ are shown in Fig. 16.3, with their resultant $y = 4 \sin \omega t + 3 \cos \omega t$

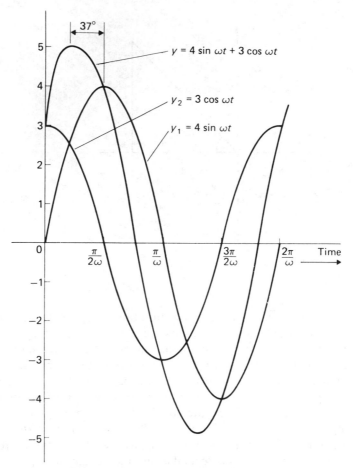

Fig. 16.3 Graphs of $y_1 = 4 \sin \omega t$, $y_2 = 3 \cos \omega t$ and
$y = 4 \sin \omega t + \cos \omega t$

obtained by adding ordinates at 15° intervals. Hence,
by drawing, **$y = 5 \sin(\omega t + 37°)$ or $y = 5 \sin(\omega t + 0.65)$**
since $37° \equiv 0.65$ radians.

(b) The relative positions of $y_1 = 4 \sin \omega t$ and $y_2 = 3 \cos \omega t$ at
$t = 0$ are shown in Fig. 16.4(a). From the phasor diagram
shown in Fig. 16.4(b) the resultant is obtained by drawing
or by calculation. By calculation
$y = \sqrt{(4^2 + 3^2)} = 5$
$\alpha = \arctan \frac{3}{4} = 36.87°$ or 36° 52′ or 0.644 radians
Hence, by phasor addition,
$y = 5 \sin(\omega t + 36° 52′)$ or $y = 5 \sin(\omega t + 0.644)$

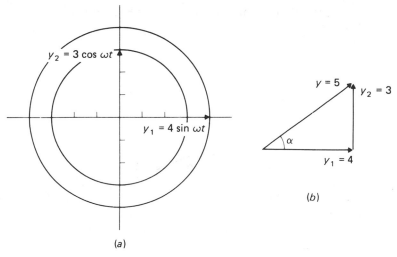

(a)

(b)

Fig. 16.4

(c)   If $4 \sin \omega t + 3 \cos \omega t = R \sin(\omega t + \alpha)$ then

$$4 \sin \omega t + 3 \cos \omega t = R[\sin \omega t \cos \alpha + \cos \omega t \sin \alpha]$$
$$= (R \cos \alpha) \sin \omega t + (R \sin \alpha) \cos \omega t$$

Equating coefficients gives

$$4 = R \cos \alpha, \text{ i.e. } \cos \alpha = \frac{4}{R} \text{ and } 3 = R \sin \alpha, \text{ i.e. } \sin \alpha = \frac{3}{R}$$

Hence R and α are calculated from Fig. 16.2 when $a = 4$ and $b = 3$:

$$R = \sqrt{(4^2 + 3^2)} = 5 \text{ and } \alpha = \arctan \tfrac{3}{4} = 36° \, 52' \text{ or } 216° \, 52'$$

From Fig. 16.2 R is in the first quadrant. Hence $\alpha = 216° \, 52'$ is neglected. If a diagram is always drawn the quadrant in which R lies is established. There will thus be only one possible value of α. Hence, by using the compound-angle addition formula,

$4 \sin \omega t + 3 \cos \omega t = 5 \sin(\omega t + 36° \, 52')$ or $5 \sin(\omega t + 0.644)$

Problem 2. Express $6.0 \sin \omega t - 2.5 \cos \omega t$ in the form $R \sin(\omega t + \alpha)$

If $6.0 \sin \omega t - 2.5 \cos \omega t = R \sin(\omega t + \alpha)$ then

$$6.0 \sin \omega t - 2.5 \cos \omega t = R[\sin \omega t \cos \alpha + \cos \omega t \sin \alpha]$$
$$= (R \cos \alpha) \sin \omega t + (R \sin \alpha) \cos \omega t$$

Equating coefficients gives

$$6.0 = R \cos \alpha, \text{ i.e. } \cos \alpha = \frac{6.0}{R} \text{ and } -2.5 = R \sin \alpha, \text{ i.e. } \sin \alpha = \frac{-2.5}{R}$$

Hence R and α are calculated from Fig. 16.5

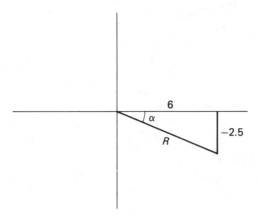

Fig. 16.5

$$R = \sqrt{[(6.0)^2 + (-2.5)^2]} = 6.50$$

and $\alpha = \arctan \dfrac{-2.5}{6} = -22.62°$ or $-22° \, 37'$

(since R is in the fourth quadrant) or -0.395 rad

Hence **6.0 sin ωt − 2.5 cos ωt = 6.50 sin(ωt − 22° 37′)**

or 6.50 sin(ωt − 0.395)

Problem 3. Express $5 \sin \omega t + 12 \cos \omega t$ in the form
(a) $R \sin(\omega t + \alpha)$ and (b) $R \cos(\omega t + \alpha)$
When do the maximum values first occur?

(a)   If $5 \sin \omega t + 12 \cos \omega t = R \sin(\omega t + \alpha)$ then

$5 \sin \omega t + 12 \cos \omega t = R \sin \omega t \cos \alpha + \cos \omega t \sin \alpha$

$= (R \cos \alpha) \sin \omega t + (R \sin \alpha) \cos \omega t$

Equating coefficients gives

$5 = R \cos \alpha$, i.e. $\mathbf{cos\ \alpha = \dfrac{5}{R}}$ and $12 = R \sin \alpha$, i.e. $\mathbf{sin\ \alpha = \dfrac{12}{R}}$

Since $\cos \alpha$ and $\sin \alpha$ are both positive, R is in the first quadrant.

$$R = \sqrt{(5^2 + 12^2)} = 13$$

and $\alpha = \arctan \dfrac{12}{5} = 67.38°$ or $67° \, 23'$ or 1.176 rad

Hence **5 sin ωt + 12 cos ωt = 13 sin(ωt + 67° 23′)**

or 13 sin(ωt + 1.176)

The maximum value of $5 \sin \omega t + 12 \cos \omega t$ is thus 13 and this occurs first when $(\omega t + 67° \, 23')$ is equal to $90°$ (since $\sin 90° = 1$).
Hence **the maximum value first occurs when $\omega t = 22° \, 37'$**

(b)   If $5 \sin \omega t + 12 \cos \omega t = R \cos(\omega t + \alpha)$ then

$$5 \sin \omega t + 12 \cos \omega t = R[\cos \omega t \cos \alpha - \sin \omega t \sin \alpha]$$
$$= (R \cos \alpha) \cos \omega t - (R \sin \alpha) \sin \omega t$$

Equating coefficients gives

$$5 = -R \sin \alpha, \text{ i.e. } \mathbf{\sin \alpha} = -\frac{5}{R} \text{ and } 12 = R \cos \alpha, \text{ i.e. } \mathbf{\cos \alpha} = \frac{12}{R}$$

Since $\sin \alpha$ is negative and $\cos \alpha$ is positive, R is in the fourth quadrant (similar to Fig. 16.5).

$$R = \sqrt{(12^2 + 5^2)} = 13$$

and $\alpha = \arctan \dfrac{-5}{12} = -22.62°$ or $-22° \, 37'$ or -0.395 rad

Hence **$5 \sin \omega t + 12 \cos \omega t = 13 \cos(\omega t - 22° \, 37')$**

or **$13 \sin(\omega t - 0.395)$**

The maximum value of 13 occurs when $\cos(\omega t - 22° \, 37')$ is 1, i.e. when $(\omega t - 22° \, 37')$ is equal to $0°$ (since $\cos 0° = 1$).
Hence **the maximum value first occurs when $\omega t = 22° \, 37'$**

Problem 4. Express $-4.50 \sin \omega t - 2.90 \cos \omega t$ in the form $R \sin(\omega t + \alpha)$

If $-4.50 \sin \omega t - 2.90 \cos \omega t = R \sin(\omega t + \alpha)$ then

$$-4.50 \sin \omega t - 2.90 \cos \omega t = R[\sin \omega t \cos \alpha + \cos \omega t \sin \alpha]$$
$$= (R \cos \alpha) \sin \omega t + (R \sin \alpha) \cos \omega t$$

Equating coefficients gives

$$-4.50 = R \cos \alpha, \text{ i.e. } \mathbf{\cos \alpha} = \frac{-4.50}{R}$$

and $-2.90 = R \sin \alpha$, i.e. $\mathbf{\sin \alpha} = \dfrac{-2.90}{R}$

From Fig. 16.6

$$R = \sqrt{[(-4.50)^2 + (-2.90)^2]} = 5.354$$

and $\alpha = \arctan \dfrac{-2.90}{-4.50} = 212.80°$ or $212° \, 48'$ (since R is in the third quadrant) or 3.714 rad

Hence **$-4.50 \sin \omega t - 2.90 \cos \omega t = 5.354 \sin(\omega t + 212° \, 48')$**

or **$5.354 \sin(\omega t + 3.714)$**

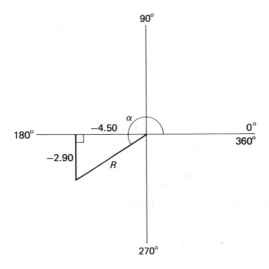

Fig. 16.6

(This answer may also be stated as $5.354 \sin(\omega t - 147° \ 12')$.
Note that when $a \sin \omega t + b \cos \omega t$ is converted into the form
$R \sin(\omega t \pm \alpha)$ the angle α is measured from $0°$. Thus when R lies
in the second or third quadrants α is **not** an acute angle.)

Problem 5. Solve the equation $2.0 \sin \theta + 4.0 \cos \theta = 3.0$ for
$0° \leqslant \theta \leqslant 360°$

If $2.0 \sin \theta + 4.0 \cos \theta = R \sin(\theta + \alpha)$ then

$$2.0 \sin \theta + 4.0 \cos \theta = R[\sin \theta \cos \alpha + \cos \theta \sin \alpha]$$
$$= (R \cos \alpha) \sin \theta + (R \sin \alpha) \cos \theta$$

Equating coefficients gives

$2.0 = R \cos \alpha$, i.e. $\boldsymbol{\cos \alpha = \dfrac{2.0}{R}}$ and $4.0 = R \sin \alpha$, i.e. $\boldsymbol{\sin \alpha = \dfrac{4.0}{R}}$

Since $\cos \alpha$ and $\sin \alpha$ are both positive, R is in the first quadrant

$R = \sqrt{[(2.0)^2 + (4.0)^2]} = 4.47$ and $\alpha = \arctan \dfrac{4.0}{2.0} = 63° \ 26'$

Hence $2.0 \sin \theta + 4.0 \cos \theta = 4.47 \sin(\theta + 63° \ 26')$. But from the
original equation $2.0 \sin \theta + 4.0 \cos \theta = 3.0$. Therefore

$4.47 \sin(\theta + 63° \ 26') = 3.0$

$$\sin(\theta + 63° \ 26') = \frac{3.0}{4.47}$$

$$\theta + 63° \ 26' = \arcsin \frac{3.0}{4.47} = 42° \ 9' \text{ or } 137° \ 51'$$

Hence $\theta = 42° \, 9' - 63° \, 26' = -21° \, 17'$ (i.e. $338° \, 43'$)

or $\theta = 137° \, 51' - 63° \, 26' = 74° \, 25'$

The solutions are checked by substitution into the original equation. Thus

$2.0 \sin 74° \, 25' + 4.0 \cos 74° \, 25' = 2.0(0.963 \, 2) + 4.0(0.268 \, 6)$

$= 1.926 \, 4 + 1.074 \, 4$

$= 3.000 \, 8$

Similarly

$2.0 \sin 338° \, 43' + 4.0 \cos 338° \, 43' = 2.0(-0.363 \, 0) + 4.0(0.931 \, 8)$

$= -0.726 \, 0 + 3.727 \, 2$

$= 3.001 \, 2$

Taking answers correct to three significant figures (and remembering that values of θ are calculated correct to the nearest minute) then $2.0 \sin \theta + 4.0 \cos \theta = 3.0$ is satisfied by $\theta = 74° \, 25'$ and $\theta = 338° \, 43'$

Hence **the solutions of the equation 2.0 sin θ + 4.0 cos θ = 3.0 between 0° and 360° are 74° 25' and 338° 43'**

Problem 6. Solve the equation $7.0 \cos x - 9.0 \sin x - 7.6 = 0$ for values of x between 0° and 360°

If $7.0 \cos x - 9.0 \sin x - 7.6 = 0$ then

$7.0 \cos x - 9.0 \sin x = 7.6$

If $7.0 \cos x - 9.0 \sin x = R \sin(x + \alpha)$ then

$7.0 \cos x - 9.0 \sin x = R[\sin x \cos \alpha + \cos x \sin \alpha]$

$= (R \cos \alpha) \sin x + (R \sin \alpha) \cos x$

Equating coefficients gives

$7.0 = R \sin \alpha$, i.e. $\sin \alpha = \dfrac{7.0}{R}$ and $-9.0 = R \cos \alpha$, i.e. $\cos \alpha = \dfrac{-9.0}{R}$

From Fig. 16.7

$R = \sqrt{[(-9.0)^2 + (7.0)^2]} = 11.4$

and $\alpha = \arctan \dfrac{7.0}{-9.0} = 142° \, 8'$ (since R is in the second quadrant)

Hence $7.0 \cos x - 9.0 \sin x = 11.4 \sin(x + 142° \, 8') = 7.6$

$\sin(x + 142° \, 8') = \dfrac{7.6}{11.4}$

$x + 142° \, 8' = \arcsin \dfrac{7.6}{11.4} = 41° \, 49'$ or $138° \, 11'$

Hence $x = 41° \, 49' - 142° \, 8' = -100° \, 19' = 259° \, 41'$

or $x = 138° \, 11' - 142° \, 8' = -3° \, 57' = 356° \, 3'$

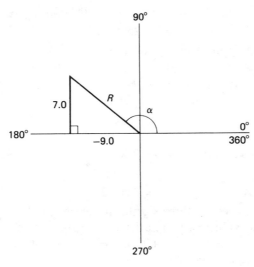

Fig. 16.7

The solutions are checked by substitution into the original equation. Thus

$7.0 \cos x - 9.0 \sin x - 7.6 = 7.0 \cos 259° 41' - 9.0 \sin 259° 41' - 7.6$

$= (7.0)(-0.179\ 1) - (9.0)(-0.983\ 8) - 7.6$

$= (-1.253\ 7) - (-8.854\ 2) - 7.6$

$= -1.253\ 7 + 8.854\ 2 - 7.6$

$= 0.000\ 5$

Similarly, $7.0 \cos 356° 3' - 9.0 \sin 356° 3' - 7.6$

$= (7.0)(0.997\ 6) - (9.0)(-0.068\ 9) - 7.6$

$= (6.983\ 2) - (-0.620\ 1) - 7.6$

$= 6.983\ 2 + 0.620\ 1 - 7.6$

$= 0.003\ 3$

Taking answers correct to three significant figures (and remembering that the values of x are calculated to the nearest minute) then $7.0 \cos x - 9.0 \sin x - 7.6 = 0$ is satisfied by $x = 259° 41'$ and $x = 356° 3'$

Hence **the solutions of the equation 7.0 cos x − 9.0 sin x − 7.6 = 0 between 0° and 360° are 259°41′ and 356° 3′**

Further problems on converting a sin ωt + *b* cos ωt *into R* sin(ωt + α) *form may be found in Section 16.7 (Problems 19 to 45), page 281.*

16.3   Double angles

(a)   If, in the compound-angle formula for $\sin(A + B)$, we let $B = A$ then

$$\sin(A + A) = \sin A \cos A + \cos A \sin A$$

i.e. $\mathbf{\sin 2A = 2 \sin A \cos A}$ (7)

This is the double-angle formula for sine. It follows from equation (7) that

$$\sin 4A = 2 \sin 2A \cos 2A$$

$$\sin 6x = 2 \sin 3x \cos 3x$$

$$\sin 9x = 2 \sin \frac{9x}{2} \cos \frac{9x}{2} \text{ and so on}$$

(b)   If, in the compound-angle formula for $\cos(A + B)$, we let $B = A$ then

$$\cos(A + A) = \cos A \cos A - \sin A \sin A$$

i.e. $\mathbf{\cos 2A = \cos^2 A - \sin^2 A}$ (8)

It was shown in Chapter 14 of *Technician Mathematics 2* that $\cos^2 A + \sin^2 A = 1$, from which $\cos^2 A = 1 - \sin^2 A$ or $\sin^2 A = 1 - \cos^2 A$. From this two further formulae for the cosine of the double angle can be produced. From equation (8) $\cos 2A = \cos^2 A - \sin^2 A$. Hence, if $\sin^2 A = 1 - \cos^2 A$ then

$$\cos 2A = \cos^2 A - (1 - \cos^2 A)$$

i.e. $\mathbf{\cos 2A = 2 \cos^2 A - 1}$ (9)

Also, if $\cos^2 A = 1 - \sin^2 A$ then

$$\cos 2A = (1 - \sin^2 A) - \sin^2 A$$

i.e. $\mathbf{\cos 2A = 1 - 2 \sin^2 A}$ (10)

Thus $\cos 2A = \cos^2 A - \sin^2 A = 2 \cos^2 A - 1 = 1 - 2 \sin^2 A$. There are therefore three double-angle formulae for cosine. It follows from equations (8) to (10) that

$$\cos 4x = \cos^2 2x - \sin^2 2x \text{ or } 2\cos^2 2x - 1 \text{ or } 1 - 2\sin^2 2x$$

$$\cos 7t = \cos^2 \frac{7}{2}t - \sin^2 \frac{7}{2}t \text{ or } 2\cos^2 \frac{7}{2}t - 1 \text{ or } 1 - 2\sin^2 \frac{7}{2}t \text{ and so on}$$

(c)   If, in the compound-angle formula for $\tan(A + B)$, we let $B = A$ then

$$\tan(A + A) = \frac{\tan A + \tan A}{1 - \tan A \tan A}$$

i.e. $\mathbf{\tan 2A = \dfrac{2 \tan A}{1 - \tan^2 A}}$ (11)

This is the double-angle formula for tangent. It follows from equation (11) that

$$\tan 6A = \frac{2\tan 3A}{1 - \tan^2 3A}, \quad \tan 3x = \frac{2\tan \frac{3}{2}x}{1 - \tan^2 \frac{3}{2}x}, \text{ and so on}$$

Worked problems on double angles

Problem 1. Given $\cos \theta = \dfrac{5}{13}$, where θ is an acute angle, evaluate (a) $\sin 2\theta$, (b) $\cos 2\theta$ and (c) $\tan 2\theta$, without evaluating angle θ, each correct to 4 decimal places

$$\cos \theta = \frac{5}{13} = \frac{\text{adjacent}}{\text{hypotenuse}}.$$

By Pythagoras's theorem the opposite side $= \sqrt{(13^2 - 5^2)} = 12$.
Hence $\sin \theta = \dfrac{12}{13}$ and $\tan \theta = \dfrac{12}{5}$

(a) $\sin 2\theta = 2\sin\theta\cos\theta = 2\left(\dfrac{12}{13}\right)\left(\dfrac{5}{13}\right) = \dfrac{120}{169} = \mathbf{0.710\,1}$

(b) $\cos 2\theta = \cos^2\theta - \sin^2\theta$

$$= \left(\frac{5}{13}\right)^2 - \left(\frac{12}{13}\right)^2 = \frac{25}{169} - \frac{144}{169} = \frac{-119}{169} = \mathbf{-0.704\,1}$$

$$\left(\text{or } \cos 2\theta = 2\cos^2\theta - 1 = 2\left(\frac{5}{13}\right)^2 - 1 \right.$$

$$= \frac{50}{169} - 1 = \frac{50}{169} - \frac{169}{169} = \frac{-119}{169} = \mathbf{-0.704\,1}$$

$$\text{or } \cos 2\theta = 1 - 2\sin^2\theta = 1 - 2\left(\frac{12}{13}\right)^2$$

$$\left. = 1 - \frac{288}{169} = \frac{169}{169} - \frac{288}{169} = \frac{-119}{169} = \mathbf{-0.704\,1} \right)$$

(c) $\tan 2\theta = \dfrac{2\tan\theta}{1 - \tan^2\theta} = \dfrac{2(12/5)}{1 - (12/5)^2}$

$$= \frac{24/5}{1 - 144/25} = \frac{24/5}{-119/25} = \left(-\frac{24}{5}\right)\left(\frac{25}{119}\right)$$

$$= \left(-\frac{24}{1}\right)\left(\frac{5}{119}\right) = -\frac{120}{119} = \mathbf{-1.008\,4}$$

Problem 2. The third harmonic of a voltage waveform is given by $V_3 \sin 3t$. Express the third harmonic in terms of the first harmonic, $\sin t$, when $V_3 = 1$

When $V_3 = 1$, $V_3 \sin 3t = \sin 3t = \sin(2t + t)$
$$= \sin 2t \cos t + \cos 2t \sin t$$
from the compound angle formula for $\sin(A + B)$

Hence $\sin 3t = \sin 2t \cos t + \cos 2t \sin t$
$$= (2 \sin t \cos t) \cos t + (1 - 2 \sin^2 t) \sin t$$
from the double-angle formula
$$= 2 \sin t \cos^2 t + \sin t - 2 \sin^3 t$$
$$= 2 \sin t(1 - \sin^2 t) + \sin t - 2 \sin^3 t$$
since $\cos^2 t = 1 - \sin^2 t$
$$= 2 \sin t - 2 \sin^3 t + \sin t - 2 \sin^3 t$$
i.e. $\quad$ **$\sin 3t = 3 \sin t - 4 \sin^3 t$**

Problem 3. Prove that $\dfrac{1 + \cos 2\theta}{\sin^2 \theta} = 2 \cot^2 \theta$

L.H.S. $= \dfrac{1 + \cos 2\theta}{\sin^2 \theta} = \dfrac{1 + (2 \cos^2 \theta - 1)}{\sin^2 \theta} = \dfrac{2 \cos^2 \theta}{\sin^2 \theta}$

$= 2 \cot^2 \theta = $ R.H.S. $\left(\text{since } \dfrac{\cos \theta}{\sin \theta} = \cot \theta \right)$

Problem 4. Show that $2 \operatorname{cosec} 2A \cos 2A = \cot A - \tan A$

L.H.S. $= 2 \operatorname{cosec} 2A \cos 2A = \dfrac{2 \cos 2A}{\sin 2A}$ $\left(\text{since } \operatorname{cosec} 2A = \dfrac{1}{\sin 2A} \right)$

$= \dfrac{2(\cos^2 A - \sin^2 A)}{2 \sin A \cos A} = \dfrac{\cos^2 A - \sin^2 A}{\sin A \cos A}$

$= \dfrac{\cos^2 A}{\sin A \cos A} - \dfrac{\sin^2 A}{\sin A \cos A} = \dfrac{\cos A}{\sin A} - \dfrac{\sin A}{\cos A}$

$= \cot A - \tan A = $ R.H.S.

Further problems on double angles may be found in Section 16.7 (Problems 46 to 54), page 283.

16.4 Resolution of vector quantities

Quantities such as force or velocity may be represented by a vector R at a particular angle α, say, to the horizontal, as shown in Fig. 16.8(a). A vector, in fact, represents the magnitude and direction of a quantity.

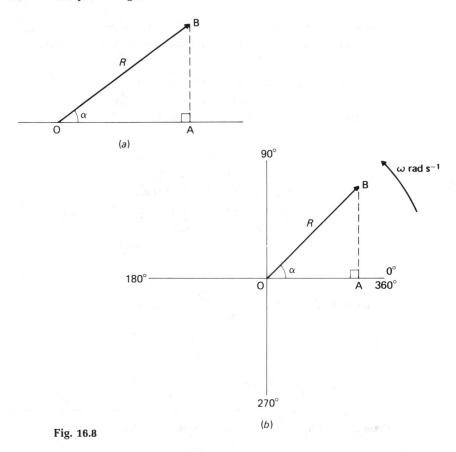

Fig. 16.8

Alternating quantities such as current or voltage which are sinusoidal may be represented by $R \sin(\omega t + \alpha)$, where R is a rotating vector (or phasor), as shown in Fig. 16.8(b). Phasors are discussed in Section 14.2 and the general form of a sine wave is discussed in Section 14.3.

Each of the vectors in Fig. 16.8 may be replaced by two components, one in the horizontal direction and the other in the vertical direction, whose vector sum is equal to the original vector, R, in magnitude and direction. Such components are often referred to as **rectangular components** of R. With reference to Fig. 16.8, the horizontal component of R is OA = $R \cos \alpha$ and the vertical component of R is AB = $R \sin \alpha$.

The resultant of a number of vectors, each having different magnitudes and directions, may be calculated by algebraically summing the horizontal components of the separate vectors and the vertical components of the separate vectors, and then using the theorem of Pythagoras.

The convention used for angular measurement is: anticlockwise is positive and clockwise is negative.

Worked problems on the resolution of vector quantities

Problem 1. A trolley on a horizontal track is being pulled by a cable at an angle of 25° with the direction of the track in the same plane as the track. The tension in the cable is 100 N. Calculate (*a*) the force tending to move the trolley forward and (*b*) the sideways thrust on the track.

(*a*) The force tending to move the trolley forward is given by OA in the plan view shown in Fig. 16.9 (i.e. the horizontal component of 100 N):
OA = 100 cos 25° = **90.63 N**

(*b*) The sideways thrust on the track is given by AB in Fig. 16.9 (i.e. the vertical component of 100 N):
AB = 100 sin 25° = **42.26 N**

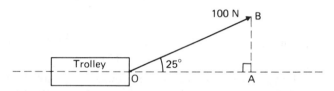

Fig. 16.9

Problem 2. Four coplanar forces act at a point O, as shown in Fig. 16.10. Calculate the value and the direction of the resultant force.

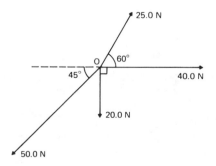

Fig. 16.10

It will be assumed that, similar to the convention used in drawing graphs, the horizontal components of the forces are positive when they are acting towards the right of point O, and that vertical components are positive when they are acting upwards from O, and vice versa.

For the 40.0 N force
Horizontal component $= 40.0 \cos 0° = 40.0$ N
  Vertical component $= 40.0 \sin 0° = 0$

For the 25.0 N force
Horizontal component $= 25.0 \cos 60° = 12.50$ N
  Vertical component $= 25.0 \sin 60° = 21.65$ N

For the 50.0 N force
Horizontal component $= -50.0 \cos 45° = -35.36$ N
  Vertical component $= -50.0 \sin 45° = -35.36$ N

For the 20.0 N force
Horizontal component $= 20.0 \cos 90° = 0$
  Vertical component $= -20.0 \sin 90° = -20.00$ N
Resultant horizontal component $= 40.0 + 12.50 - 35.36 + 0$
$$= +17.14 \text{ N}$$
  Resultant vertical component $= 0 + 21.65 - 35.36 - 20.00$
$$= -33.71 \text{ N}$$

The resultant horizontal and vertical components are shown as OA and AB respectively in Fig. 16.11, and the resultant force R is represented by OB

$$R = \sqrt{[(17.14)^2 + (-33.71)^2]} = 37.82 \text{ N}$$
$$\text{and } \alpha = \arctan \frac{-33.71}{17.14} = -63° 3'$$

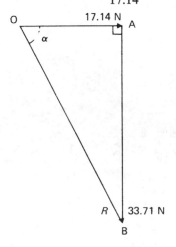

Fig. 16.11

Hence **a force of 37.82 N acting at an angle of −63° 3′ to the horizontal will have the same effect as the four coplanar forces shown in Fig. 16.10**

Problem 3. The instantaneous values of two alternating voltages are given by

$$v_1 = 30.0 \sin\left(\omega t - \frac{\pi}{4}\right) \text{ volts and } v_2 = 40.0 \cos\left(\omega t + \frac{\pi}{3}\right) \text{ volts.}$$

Calculate an expression for $v_1 + v_2$ by resolution of phasors.

The space diagram representing the alternating voltages is shown in Fig. 16.12

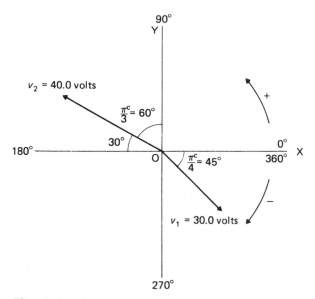

Fig. 16.12

If the axis OX is assumed to represent $\sin \omega t$ then axis OY represents $\cos \omega t \left[\text{since } \cos \omega t = \sin\left(\omega t + \frac{\pi}{2}\right) \right]$

For $v_1 = 30.0$ volts
Horizontal component = $30.0 \cos 45° = 21.21$ volts
   Vertical component = $-30.0 \sin 45° = -21.21$ volts

For $v_2 = 40.0$ volts
Horizontal component = $-40.0 \cos 30° = -34.64$ volts
   Vertical component = $40.0 \sin 30° = 20.0$ volts

Resultant horizontal component $= 21.21 - 34.64 = -13.43$ volts
Resultant vertical component $= -21.21 + 20.00 = -1.21$ volts

The resultant horizontal and vertical components are shown in Fig. 16.13

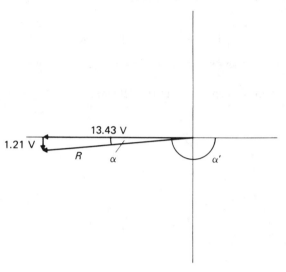

Fig. 16.13

$$R = \sqrt{[(-1.21)^2 + (-13.43)^2]} = 13.48 \text{ volts}$$
$$\text{and } \alpha = \arctan \frac{-1.21}{-13.43} = 5° \, 9'$$

Hence the obtuse angle
α' (shown in Fig. 16.13) $= 180° - 5° \, 9' = 174° \, 51' = 3.052$ radians

Hence $v_1 + v_2 = \mathbf{13.48 \sin(\omega t - 174° \, 51')}$ **volts or**
$\mathbf{13.48 \sin(\omega t - 3.052)}$ **volts**

Further problems on the resolution of vector quantities may be found in Section 16.7 (Problems 55 to 63), page 284.

16.5   Changing products of sines and cosines into sums or differences of sines and cosines

From Section 16.1

$\sin(A + B) = \sin A \cos B + \cos A \sin B$
and $\sin(A - B) = \sin A \cos B - \cos A \sin B$

Therefore
$$\sin(A + B) + \sin(A - B) = 2 \sin A \cos B$$
i.e.
$$\sin A \cos B = \tfrac{1}{2}[\sin(A + B) + \sin(A - B)] \qquad (12)$$
and
$$\sin(A + B) - \sin(A - B) = 2 \cos A \sin B$$
i.e.
$$\cos A \sin B = \tfrac{1}{2}[\sin(A + B) - \sin(A - B)] \qquad (13)$$
Similarly

$$\cos(A + B) = \cos A \cos B - \sin A \sin B$$
and
$$\cos(A - B) = \cos A \cos B + \sin A \sin B$$

Therefore
$$\cos(A + B) + \cos(A - B) = 2 \cos A \cos B$$
i.e.
$$\cos A \cos B = \tfrac{1}{2}[\cos(A + B) + \cos(A - B)] \qquad (14)$$
and
$$\cos(A + B) - \cos(A - B) = -2 \sin A \sin B$$
i.e.
$$\sin A \sin B = -\tfrac{1}{2}[\cos(A + B) - \cos(A - B)] \qquad (15)$$

Worked problems on changing products of sines and cosines into sums or differences of sines and cosines

Problem 1. Express $\sin 3\theta \cos 2\theta$ as a sum or difference of sines or cosines.

From equation (12), $\sin 3\theta \cos 2\theta = \tfrac{1}{2}[\sin(3\theta + 2\theta) + \sin(3\theta - 2\theta)]$
$$= \tfrac{1}{2}[\sin 5\theta + \sin \theta]$$

Problem 2. Express $\cos 4x \sin x$ as a sum or difference of sines or cosines.

From equation (13), $\cos 4x \sin x = \tfrac{1}{2}[\sin(4x + x) - \sin(4x - x)]$
$$= \tfrac{1}{2}[\sin 5x - \sin 3x]$$

Problem 3. Express $\cos 5t \cos 3t$ as a sum or difference of sines or cosines.

From equation (14), $\cos 5t \cos 3t = \tfrac{1}{2}[\cos(5t + 3t) + \cos(5t - 3t)]$
$$= \tfrac{1}{2}[\cos 8t + \cos 2t]$$

Problem 4. Express $\sin 45° \sin 30°$ as a sum or difference of sines or cosines.

From equation (15),
$$\sin 45° \sin 30° = -\tfrac{1}{2}[\cos(45° + 30°) - \cos(45° - 30°)]$$
$$= -\tfrac{1}{2}[\cos 75° - \cos 15°]$$
$$= \tfrac{1}{2}[\cos 15° - \cos 75°]$$

Further problems on changing products of sines and cosines into sums or differences of sines and cosines may be found in Section 16.7 (Problems 64 to 77), page 286.

16.6 Changing sums or differences of sines and cosines into products of sines and cosines

In the compound-angle formulae let $A + B = X$ and $A - B = Y$. Then

$$A = \frac{X + Y}{2} \text{ and } B = \frac{X - Y}{2}$$

Thus, instead of $\sin(A + B) + \sin(A - B) = 2 \sin A \cos B$, we have

$$\sin X + \sin Y = 2 \sin\left[\frac{X + Y}{2}\right] \cos\left[\frac{X - Y}{2}\right] \tag{16}$$

Similarly

$$\sin X - \sin Y = 2 \cos\left[\frac{X + Y}{2}\right] \sin\left[\frac{X - Y}{2}\right] \tag{17}$$

$$\cos X + \cos Y = 2 \cos\left[\frac{X + Y}{2}\right] \cos\left[\frac{X - Y}{2}\right] \tag{18}$$

and $$\cos X - \cos Y = -2 \sin\left[\frac{X + Y}{2}\right] \sin\left[\frac{X - Y}{2}\right] \tag{19}$$

Worked problems on changing sums or differences of sines and cosines into products of sines and cosines

---

Problem 1. Express $\sin 7\theta + \sin 3\theta$ as a product.

From equation (16),

$$\sin 7\theta + \sin 3\theta = 2 \sin\left[\frac{7\theta + 3\theta}{2}\right] \cos\left[\frac{7\theta - 3\theta}{2}\right]$$

$$= 2 \sin 5\theta \cos 2\theta$$

Problem 2. Express $\sin 50° - \sin 20°$ as a product.

From equation (17),

$$\sin 50° - \sin 20° = 2 \cos\left[\frac{50° + 20°}{2}\right] \sin\left[\frac{50° - 20°}{2}\right]$$

$$= 2 \cos 35° \sin 15°$$

Problem 3. Express cos 6x + cos 2x as a product.

From equation (18),

$$\cos 6x + \cos 2x = 2\cos\left[\frac{6x + 2x}{2}\right]\cos\left[\frac{6x - 2x}{2}\right]$$

$$= \mathbf{2\cos 4x \cos 2x}$$

Problem 4. Express cos 20° − cos 46° as a product.

From equation (19),

$$\cos 20° - \cos 46° = -2\sin\left[\frac{20° + 46°}{2}\right]\sin\left[\frac{20° - 46°}{2}\right]$$

$$= -2\sin 33° \sin(-13°)$$

$$= \mathbf{2\sin 33° \sin 13°}$$

Further problems on changing sums or differences of sines and cosines into products of sines and cosines may be found in the following Section 16.7 (Problems 73 to 80), page 286.

16.7   Further problems

Compound angles

1. Verify that
   $$\sin(A + B) = \sin A \cos B + \cos A \sin B$$
   $$\cos(A + B) = \cos A \cos B - \sin A \sin B$$
   $$\tan(A + B) = \frac{\tan A + \tan B}{1 - \tan A \tan B}$$
   when (a) $A = 15°$ and $B = 35°$, (b) $A = 26° 37'$ and $B = 41° 17'$, and (c) $A = 148°$ and $B = 74°$

2. Verify that
   $$\sin(A - B) = \sin A \cos B - \cos A \sin B$$
   $$\cos(A - B) = \cos A \cos B + \sin A \sin B$$
   $$\tan(A - B) = \frac{\tan A - \tan B}{1 + \tan A \tan B}$$
   when (a) $A = 11°$ and $B = 52°$, (b) $A = 36° 13'$ and $B = 45° 56'$, and (c) $A = 97°$ and $B = 114°$

3. Reduce the following to the sine of one angle:
   (a) sin 41° cos 34° + cos 41° sin 34°
   (b) sin 3x cos 4x + cos 3x sin 4x
   (c) sin 49° cos 53° + cos 49° sin 53°
   (a) [sin 75°]   (b) [sin 7x]   (c) [sin 102° (= sin 78°)]

4. Reduce the following to the sine of one angle:
   (a) $\sin 73° \cos 24° - \cos 73° \sin 24°$
   (b) $\sin 151° \cos 82° - \cos 151° \sin 82°$
   (c) $\sin 7\theta \cos 2\theta - \cos 7\theta \sin 2\theta$
   (a) [$\sin 49°$]   (b) [$\sin 69°$]   (c) [$\sin 5\theta$]

5. Reduce the following to the cosine of one angle:
   (a) $\cos 35° \cos 27° - \sin 35° \sin 27°$
   (b) $\cos 4t \cos 2t - \sin 4t \sin 2t$
   (c) $\cos 64° \cos 48° - \sin 64° \sin 48°$
   (a) [$\cos 62°$]   (b) [$\cos 6t$]   (c) [$\cos 112° (= -\cos 68°)$]

6. Reduce the following to the cosine of one angle:
   (a) $\cos 64° \cos 39° + \sin 64° \sin 39°$
   (b) $\cos \dfrac{2\pi}{5} \cos \dfrac{\pi}{4} + \sin \dfrac{2\pi}{5} \sin \dfrac{\pi}{4}$
   (c) $\cos 164° \cos 71° + \sin 164° \sin 71°$
   (a) [$\cos 25°$]   (b) $\left[\cos \dfrac{3\pi}{20} (=\cos 27°) \right]$
   (c) [$\cos 93° (= -\cos 87°)$]

In Problems 7 to 10 use the addition and subtraction formulae to simplify the expressions.

7. (a) $\sin 66° \cos 41° - \cos 66° \sin 41°$
   (b) $\cos 2\omega t \cos 3\omega t - \sin 2\omega t \sin 3\omega t$
   (a) [$\sin 25°$]   (b) [$\cos 5\omega t$]

8. (a) $\sin 4x \cos x + \cos 4x \sin x$
   (b) $\cos 78° \cos 22° + \sin 78° \sin 22°$
   (a) [$\sin 5x$]   (b) [$\cos 56°$]

9. (a) $\sin \dfrac{\pi}{4} \cos \dfrac{\pi}{6} + \cos \dfrac{\pi}{4} \sin \dfrac{\pi}{6}$ (b) $\cos 8\alpha \cos 3\alpha + \sin 8\alpha \sin 3\alpha$
   (a) $\left[\sin \dfrac{5\pi}{12} (=\sin 75°) \right]$ (b) [$\cos 5\alpha$]

10. (a) $\sin \omega t \cos(\omega t - \alpha) - \cos \omega t \sin(\omega t - \alpha)$
    (b) $\cos(A - B) \cos(A + B) - \sin(A - B) \sin(A + B)$
    (a) [$\sin \alpha$]   (b) [$\cos 2A$]

11. Prove that (a) $\sin\left[\dfrac{\pi}{2} + x \right] = \cos x$
    (b) $\sin(y + 60°) + \sin(y + 120°) = \sqrt{3} \cos y$
    (c) $\cos(90° + \alpha) = -\sin \alpha$

12. Prove that
    (a) $\cos(\theta + 45°) - \cos(\theta - 135°) = \sqrt{2}(\cos \theta - \sin \theta)$
    (b) $-\sin(\pi + x) = \sin x$ (c) $\dfrac{\cos(2\pi - \phi)}{\cos(3\pi/2 + \phi)} = \cot \phi$

13. If $\sin A = \dfrac{40}{41}$ and $\cos B = \dfrac{5}{13}$ find $\sin(A + B)$ and $\sin(A - B)$

$$\left[\dfrac{308}{533}(= 0.577\ 9),\ \dfrac{92}{533}(= 0.172\ 6)\right]$$

14. If $\cos E = 0.500\ 0$ and $\cos F = 0.300\ 0$ find $\sin(E + F)$ and $\cos(E + F)$ $[0.736\ 8,\ -0.676\ 1]$

15. If $\sin C = 0.843\ 2$ and $\cos D = 0.732\ 8$ find $\cos(C + D)$ and $\cos(C - D)$ $[-0.179\ 8,\ 0.967\ 7]$

16. Find the value of $\sin 70°$ given $\sin 36° = 0.587\ 8$ and $\cos 34° = 0.829\ 0$ $[0.939\ 7]$

17. Find the value of $\cos 56°$ given $\sin 24° = 0.406\ 7$ and $\sin 32° = 0.529\ 9$ $[0.559\ 2]$

18. Find the values of $\sin 105°$ and $\cos 105°$ without using a calculator. Assume that
$$\sin 45° = \cos 45° = \dfrac{1}{\sqrt{2}},\ \sin 60° = \dfrac{\sqrt{3}}{2}\ \text{and}\ \cos 60° = \dfrac{1}{2}$$
$$\left[\dfrac{1 + \sqrt{3}}{2\sqrt{2}}(= 0.965\ 9),\ \dfrac{1 - \sqrt{3}}{2\sqrt{2}}(= -0.258\ 8)\right]$$

Conversion of $a \sin \omega t + b \cos \omega t$ into $R \sin(\omega t \pm \alpha)$

19. (a) Plot graphs of $y_1 = 5 \sin \omega t$ and $y_2 = 12 \cos \omega t$ over one cycle on the same axes and using the same scales. By adding ordinates plot $y = 5 \sin \omega t + 12 \cos \omega t$ and express the resultant y_r in the form $y_r = R \sin(\omega t \pm \alpha)$

    (b) With reference to the relative positions of y_1 and y_2 at $t = 0$, obtain the resultant y_r by phasor addition.

    (c) Obtain the resultant y_r by converting $5 \sin \omega t + 12 \cos \omega t$ into the form $R \sin(\omega t \pm \alpha)$ using the trigonometrical addition formula.

    $[13 \sin(\omega t + 67° 23')$ or $13 \sin(\omega t + 1.176)]$

In Problems 20 to 26 change the functions into the form $R \sin(\omega t \pm \alpha)$

20. $8 \sin \omega t + 15 \cos \omega t$
    $[17 \sin(\omega t + 61° 56')$ or $17 \sin(\omega t + 1.081)]$

21. $3 \sin \omega t + 4 \cos \omega t$
    $[5 \sin(\omega t + 53° 8')$ or $5 \sin(\omega t + 0.927)]$

22. $3 \sin \omega t - 4 \cos \omega t$
    $[5 \sin(\omega t - 53° 8')$ or $5 \sin(\omega t - 0.927)]$

23. $-2.00 \sin \omega t + 3.00 \cos \omega t$
    $[3.606 \sin(\omega t + 123° 41')$ or $3.606 \sin(\omega t + 2.159)]$

24. $-5 \sin \omega t - 12 \cos \omega t$
    $[13 \sin(\omega t + 247° 23')$ or $13 \sin(\omega t + 4.318)$ or
    $13 \sin(\omega t - 112° 37')$ or $13 \sin(\omega t - 1.966)]$

25. $6.60 \sin \omega t + 11.80 \cos \omega t$
    $[13.52 \sin(\omega t + 60° 47')$ or $13.52 \sin(\omega t + 1.061)]$

26. $-12.62 \sin \omega t - 6.92 \cos \omega t$
    $[14.39 \sin(\omega t + 208° 44')$ or $14.39 \sin(\omega t + 3.643)$ or
    $14.39 \sin(\omega t - 151° 16')$ or $14.39 \sin(\omega t - 2.640)]$

In Problems 27 to 31 change the functions into the form
$R \cos(\omega t \pm \alpha)$

27. $8.00 \cos \omega t - 5.00 \sin \omega t$
    $[9.434 \cos(\omega t + 32° 1')$ or $9.434 \cos(\omega t + 0.559)]$

28. $4 \cos \omega t + 3 \sin \omega t$
    $[5 \cos(\omega t - 36° 52')$ or $5 \cos(\omega t - 0.644)]$

29. $6.00 \sin \omega t - 4.00 \cos \omega t$
    $[7.211 \cos(\omega t + 236° 19')$ or $7.211 \cos(\omega t + 4.125)$ or
    $7.211 \cos(\omega t - 123° 41')$ or $7.211 \cos(\omega t - 2.159)]$

30. $-19.6 \cos \omega t - 12.4 \sin \omega t$
    $[23.19 \cos(\omega t + 147° 41')$ or $23.19 \cos(\omega t + 2.578)]$

31. $13.00 \sin \omega t - 5.00 \cos \omega t$
    $[13.93 \cos(\omega t + 248° 58')$ or $13.93 \cos(\omega t + 4.345)$ or
    $13.93 \cos(\omega t - 111° 2')$ or $13.93 \cos(\omega t - 1.938)]$

32. Solve the equations (a) $3 \sin \theta - 5 \cos \theta = 4$ and
    (b) $15 \sin \theta + 11 \cos \theta = 7$ for $0° \leqslant \theta \leqslant 360°$
    (a) $[102° 21'$ or $195° 43']$ (b) $[121° 39'$ or $345° 51']$

33. Solve the following equations for all values of A between
    $0°$ and $360°$:
    (a) $4 \cos A + 3 \sin A = 5$ (b) $42 \cos A - 19 \sin A = 24$
    (c) $23 \sin A + 14 \cos A = 12$ (a) $[36° 52']$
    (b) $[34° 18'$ or $277° 2']$ (c) $[122° 12'$ or $355° 8']$

34. Solve the equations (a) $5 \sin \phi + 7 \cos \phi - 3 = 0$ and
    (b) $17 \cos \phi = 4 + 8 \sin \phi$ for values of ϕ between $0°$ and $360°$
    (a) $[105° 7'$ or $325° 57']$ (b) $[52° 31'$ or $257° 5']$

35. Find the roots of the equation $7.32 \cos x - 5.62 = 3.81 \sin x$
    in the range $0°$ to $360°$ $[19° 34'$ or $285° 26']$

36. Solve the equation $100 \sin y + 250 \cos y = 190.4$ for values
    of y between $0°$ and $360°$ $[66° 48'$ or $336° 48']$

37. Find the maximum value of $8 \sin \phi + 7 \cos \phi$ and find the
    smallest positive value of ϕ at which it occurs.
    $[10.63, 48° 49']$

38. Alternating currents are given by

$i_1 = 3.00 \sin\left(10\pi t - \dfrac{\pi}{6}\right)$ and $i_2 = 8.00 \sin\left(10\pi t + \dfrac{\pi}{3}\right)$.

Express $i_1 + i_2$ in the form $y = R \sin(10\pi t \pm \alpha)$
Find also the frequency of the resultant function.
[$8.544 \sin(10\pi t + 39° \ 27')$ or $8.544 \sin(10\pi t + 0.689)$, 5 Hz]

39. Express $2.00 \sin\left(\phi + \dfrac{\pi}{4}\right) + 3.00 \cos\left(\phi - \dfrac{\pi}{6}\right)$ in the form

$a \sin \phi + b \cos \phi$ and then convert this into the form
$R \cos(\phi \pm \alpha)$
[$2.914 \sin \phi + 4.012 \cos \phi$; $4.959 \cos(\phi - 36°)$ or
$4.959 \cos(\phi - 0.628)$]

40. The third harmonic of a wave motion is given by
$5.5 \cos 3\theta - 7.2 \sin 3\theta$. Express this in the form
$A \cos(3\theta + \alpha)$
[$9.06 \cos(3\theta + 52° \ 37')$ or $9.06 \cos(3\theta + 0.918)$]

41. A voltage V is given by $V = I(R \sin at - aL \cos at)$.
Express the voltage in the form $IZ \sin(at - \phi)$

$$\left[I\sqrt{[R^2 + (aL)^2]} \sin\left(at - \arctan \dfrac{aL}{R}\right) \right]$$

42. The displacement x metres of a body from a fixed point
about which it is oscillating is given by the expression
$x = 3.6 \sin 2t + 4.2 \cos 2t$, where t is the time in seconds.
Express x in the form $R \sin(2t + \alpha)$
[$5.53 \sin(2t + 49° \ 24')$ or $5.53 \sin(2t + 0.862)$]

43. Find the sum of the voltages $v_1 = 50 \sin 100\pi t$ and
$v_2 = 30 \cos 100\pi t$ in the form $R \sin(100\pi t + \alpha)$
[$58.3 \sin(100\pi t + 30° \ 58')$ or $58.3 \sin(100\pi t + 0.540)$]

44. Alternating currents are given by $i_1 = 5.0 \sin \omega t$ and
$i_2 = 15.0 \cos \omega t$. Calculate the maximum value of $i_1 + i_2$
and its phase angle relative to i_1 [15.8; $71° \ 34'$]

45. Two voltages, $4 \cos \omega t$ and $-3 \sin \omega t$, are inputs to an
analogue circuit. Find an expression for the output voltage
if this is given by the addition of the two inputs.
[$5 \sin(\omega t + 126° \ 52')$ or $5 \sin(\omega t + 2.214)$]

Double angles

46. Given $\sin \theta = \frac{4}{5}$, where θ is an acute angle, evaluate
(a) $\sin 2\theta$, (b) $\cos 2\theta$ and (c) $\tan 2\theta$, without evaluating angle θ
(a) [0.96]   (b) [-0.28]   (c) [$-3.428\ 6$]

47. Given $\sin t = 0.8$, where t is an acute angle, evaluate, correct to 4 decimal places, (a) $\sin 2t$, (b) $\cos 2t$, (c) $\tan 2t$, (d) $\sin 4t$ and (e) $\cos \dfrac{t}{2}$, without evaluating angle t

(a) [0.960 0]   (b) [−0.280 0]   (c) [−3.428 6]
(d) [−0.537 6]   (e) [0.894 4]

48. The power p in an electrical circuit is given by $p = \dfrac{v^2}{R}$. If $v = V\cos\theta$, show that $p = \dfrac{V^2}{2R}(1 + \cos 2\theta)$

In Problems 49 to 53 prove the given identities.

49. $\dfrac{1 - \cos 2t}{\sin 2t} = \tan t$

50. $1 - \dfrac{\cos 2A}{\cos^2 A} = \tan^2 A$

51. $\dfrac{\cos 2x}{\sin 4x} = \dfrac{1}{2}\operatorname{cosec} 2x$

52. $\operatorname{cosec} 2\theta + \cot 2\theta = \cot\theta$

53. $\dfrac{(\tan 2\theta)(1 - \tan\theta)}{2\tan\theta} = \dfrac{1}{(1 + \tan\theta)}$

54. The third harmonic of a current waveform is given by $I_3 \cos 3t$. Show that when $I_3 = 1$, $\cos 3t = 4\cos^3 t - 3\cos t$

Resolution of vector quantities

55. A barge is being towed along a canal by a rope inclined at 30° to the direction of the canal. The tension in the rope is 175 N. Calculate the values of the rectangular components of this tension, one component being in the direction of the canal. [151.6 N in direction of canal, 87.50 N at right angles to direction of canal]

56. Calculate, using resolution of forces, the magnitude and direction of the resultant of the three coplanar forces given below, when they are acting at a point. Force A, 10.0 N acting horizontally to the right; force B, 6.0 N inclined at an angle of 70° to force A; and force C, 13.0 N inclined at an angle of 135° to force A [15.1 N, 79° 5′ to the horizontal]

57. Find the magnitude and direction of the resultant of the coplanar forces listed below, which are acting at a point. Force A, 3.0 N acting vertically upwards; force B, 2.0 N acting at an angle of 110° to force A; force C, 5.0 N acting at an angle of 290° to force A; and force D, 7.0 N acting horizontally to the right.   [10.6 N, 22° 18′ to the horizontal]

58. Forces of 200 kN acting horizontally to the right, 300 kN inclined at 60° to the horizontal, 600 kN inclined at 120° to the horizontal, 400 kN inclined at 180° to the horizontal and 500 kN inclined at 300° to the horizontal are acting at a point. Find the magnitude and direction of the resultant of these forces by resolution of forces.
[360.6 kN, 106° 6′ to the horizontal]

59. Find the magnitude and direction of the resultant of the following coplanar forces which are acting at a point. Force A, 3.0 kN inclined at 30° to the horizontal to the right; force B, 5.0 kN inclined at 105° to force A; force C, 6.0 kN inclined at 45° to force B; and force D, 4.0 kN inclined at 120° to force C [5.18 kN, 162° 20′ to the horizontal]

60. The voltage in a circuit is the resultant of the three voltages shown in Fig. 16.14. By resolution of the voltages calculate the magnitude of the resultant.   [73.93 volts]

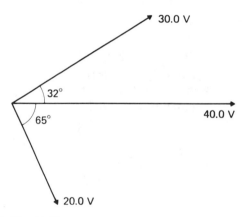

Fig. 16.14

61. The instantaneous values of two alternating currents are given by $i_1 = 15.0 \sin\left(\omega t + \dfrac{\pi}{3}\right)$ amperes and $i_2 = 9.0 \sin\left(\omega t - \dfrac{\pi}{6}\right)$ amperes. By resolving each current into its horizontal and vertical components, obtain a sinusoidal expression for $i_1 + i_2$ in the form $R \sin(\omega t + \alpha)$
[$17.49 \sin(\omega t + 29° 2′)$ amperes or $17.49 \sin(\omega t + 0.507)$ amperes]

62. The instantaneous values of two alternating voltages are given by $v_1 = 42.5 \sin\left(\omega t - \dfrac{\pi}{3}\right)$ volts and

   $v_2 = 22.5 \sin\left(\omega t - \dfrac{\pi}{4}\right)$ volts. Calculate, by resolving the voltages horizontally and vertically, an expression for $v_1 + v_2$
   $[64.50 \sin(\omega t - 0.956\ 8)$ volts$]$

63. Obtain, by resolution, a sinusoidal expression in the form $R \sin(\omega t + \alpha)$ for the resultant of the alternating-current expression

   $$i = 7.2 \sin\left(\omega t - \frac{\pi}{5}\right) + 5.5 \cos\left(\omega t + \frac{4\pi}{9}\right)$$
   $$+ 6.8 \sin\left(\omega t + \frac{4\pi}{3}\right) \text{ amperes}$$
   $[9.64 \sin(\omega t - 1.89)$ amperes$]$

Changing products into sums or differences

Express as sums or differences the following:
64. $\sin 6t \cos t$ $[\frac{1}{2}(\sin 7t + \sin 5t)]$
65. $\cos 5x \sin 3x$ $[\frac{1}{2}(\sin 8x - \sin 2x)]$
66. $4 \cos 4\theta \cos 2\theta$ $[2(\cos 6\theta + \cos 2\theta)]$
67. $\sin 7\alpha \sin 3\alpha$ $[-\frac{1}{2}(\cos 10\alpha - \cos 4\alpha)]$
68. $2 \cos 60° \sin 30°$ $[\sin 90° - \sin 30°]$
69. $\sin \dfrac{\pi}{3} \cos \dfrac{\pi}{6}$ $\left[\dfrac{1}{2}\left(\sin \dfrac{\pi}{2} + \sin \dfrac{\pi}{6}\right)\right]$
70. $\sin 72° \sin 22°$ $[-\frac{1}{2}(\cos 94° - \cos 50°)]$
71. $3 \cos 78° \cos 48°$ $[\frac{3}{2}(\cos 126° + \cos 30°)]$
72. In an alternating-current circuit a voltage $v = 6 \sin \omega t$ and a current $i = 5 \sin\left(\omega t - \dfrac{\pi}{4}\right)$. Find an expression for the instantaneous power p in the circuit in time t, given that $p = vi$, expressing the answer as a sum or difference of sines or cosines. $\left[p = 15\left\{\cos \dfrac{\pi}{4} - \cos\left(2\omega t - \dfrac{\pi}{4}\right)\right\}\right]$

Changing sums or differences into products

Express as products the following:
73. $\sin 4x + \sin x$ $\left[2 \sin \dfrac{5x}{2} \cos \dfrac{3x}{2}\right]$

74. $\sin 11\theta - \sin 7\theta$ $[2 \cos 9\theta \sin 2\theta]$

75. $\cos 6t + \cos 4t$ $[2 \cos 5t \cos t]$

76. $\frac{1}{4}(\cos 5\alpha - \cos 2\alpha)$ $\left[-\frac{1}{2} \sin \frac{7\alpha}{2} \sin \frac{3\alpha}{2} \right]$

77. $\sin 40° - \sin 22°$ $[2 \cos 31° \sin 9°]$

78. $\cos \frac{\pi}{4} + \cos \frac{\pi}{6}$ $\left[2 \cos \frac{5\pi}{24} \cos \frac{\pi}{24} \right]$

79. $\frac{1}{2}(\sin 75° + \sin 15°)$ $[\sin 45° \cos 30°]$

80. $\cos 82° - \cos 38°$ $[-2 \sin 60° \sin 22°]$

Chapter 17

Vectors

17.1 Introduction

Many physical quantities such as time, volume, density, mass, energy and temperature are defined entirely by a numerical value together with the appropriate units. Such quantities are called **scalar quantities** or just **scalars**, and obey the fundamental laws of algebra. Other physical quantities such as displacement, velocity, acceleration, force, moment and momentum are defined in terms of both a numerical value and a direction in space. These quantities are called **vector quantities** or just **vectors**.

A vector quantity representing, say, a force, has two properties:

(i) the magnitude or length of the vector is directly proportional to the magnitude of the force, and
(ii) the direction of the vector is in the same direction as the line of action of the force.

Vector quantities are indicated by using smaller letters and **bold** type and line 0a shown in Fig. 17.1 is **0a**, that is vector 0a, where the magnitude of line 0a is directly proportional to 9 newtons and its direction is from 0 to a, i.e. **0a** is 9 newtons at an angle of 45° to the horizontal.

17.2 Vector addition and subtraction

Vectors can be added or subtracted by means of triangle or parallelogram laws. For example, Fig. 17.2(a) shows a force of F_2 newtons inclined at an angle of $\theta°$ to a force of F_1 newtons. These forces can be added vectorially

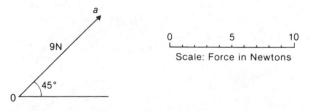

Fig. 17.1   Vector **0a** = 9N at 45° to the horizontal

by the 'noise-to-tail' method. The force F_1 is represented by line $0a$ in Fig. 17.2(b) and is drawn F_1 units long in the direction of the F_1 newton force. From its nose (arrow-head), ar is drawn inclined at an angle of $\theta°$ to $0a$ and F_2 units long. The resultant vector is $0r$.

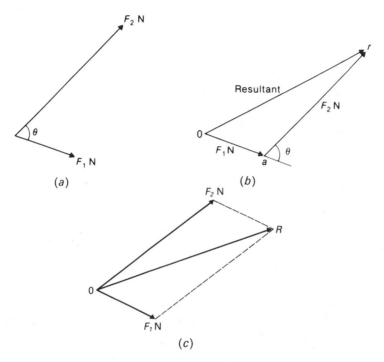

Fig. 17.2   Nose-to-tail and parallelogram methods of adding vectors
$0a + ar = 0r$

The vector equation is $0a + ar = 0r$.

Alternatively, by drawing lines parallel to F_1 and F_2 from the noses of F_2 and F_1 respectively, and letting the point of intersection of these parallel lines be R, gives $0R$ as the magnitude and direction of the resultant of adding F_1 and F_2 as shown in Fig. 17.2(c). This is called the 'parallelogram' method of vector addition.

The vector $(-ar)$ is the vector represented by line ra, that is, a minus sign in front of a vector has the effect of rotating it through $180°$ with respect to its positive direction, i.e. reversing its direction. Thus, $0a + (-ar) = 0a + ra$.

A force of F_2 newtons is shown inclined at an angle of $\theta°$ to a force of F_1 newtons in Fig. 17.3(a). The vector representing the force of F_2 newtons can be subtracted from the vector representing the force of F_1 newtons by using the nose-to-tail method. $0a$ is drawn F_1 units long in the direction of the F_1 newton force in Fig. 17.3(b). ra is drawn in the direction of the F_2 newton force with its nose coinciding with *a*. The resultant is vector **0r**. The vector equation is:

$$0a - ar = 0a + ra = 0r$$

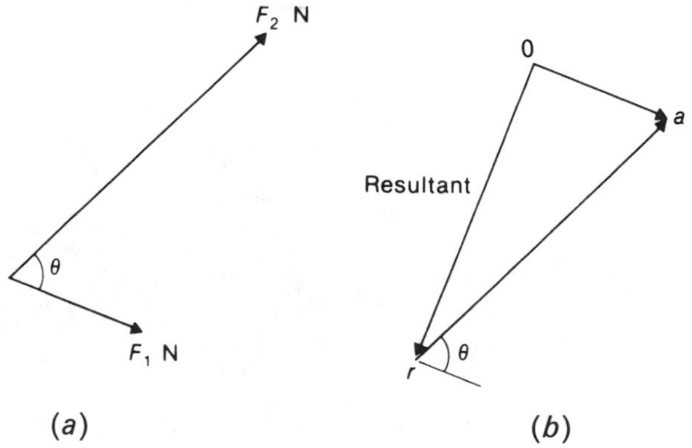

(*a*)   (*b*)

Fig. 17.3   Vector subtraction **0a** − **ar** = **0a** + **ra** = **0r**

The magnitude or numerical value of a vector is shown by using modulus lines. Thus the magnitude of **0a** is indicated by |**0a**| and |**0a**| = 0a. When vector **0a** is multiplied by a scalar quantity, say *k*, the resultant is the vector *k* **0a** which has a magnitude of *k* |**0a**| and acts in the direction of **0a**.

A vectorial approach is used quite extensively in solving problems in mechanics.

Worked problems on vector addition and subtraction

Problem 1. A force of 5 N is inclined at an angle of 45° to a second force of 8 N, both forces acting at a point. Determine the magnitude of the resultant of these two forces and the direction of the resultant with respect to the 8 N force by both the 'triangle' and the 'parallelogram' methods.

The forces are shown in Fig. 17.4(*a*). Although the 8 N force is shown as a horizontal line, it could actually have been drawn in any direction. Using the 'nose-to-tail' method, a line 8 units long is drawn horizontally to give vector **0a** in Fig. 17.4(*b*). To the nose of this vector, **ar** is drawn 5 units long at an angle of 45° to **0a**. The resultant of vector addition is **0r** and, by measurement, is 12 units long and at an angle of 17° to the 8 N force. This is the 'triangle' method.

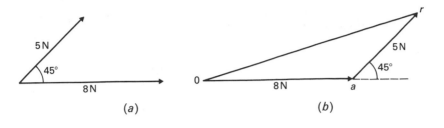

(a)   (b)

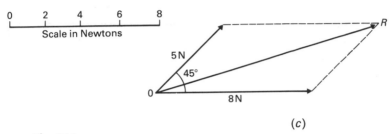

(c)

Fig. 17.4

Figure 17.4(c) uses the 'parallelogram' method in which lines
are drawn parallel to the 8 N and 5 N forces from the noses of the
5 N and 8 N forces respectively. These intersect at R. Vector **0R**
gives the magnitude and direction of the resultant vector addition
and as obtained by the 'triangle' method is 12 units long at an
angle of 17° to the 8 N force.
  Thus by both methods, the resultant of vector addition is a
force of 12 N at an angle of 17° to the 8 N force.

Problem 2. Calculate the resultant of the two forces in Problem 1.

From the resolution of vector quantities explained in Section 16.4,
and with reference to Fig. 17.4:

Total horizontal component of the two forces,

$H = 8 \cos 0° + 5 \cos 45°$

$= 8 + 3.536 = 11.536$ N

Total vertical component of the two forces,

$V = 8 \sin 0° + 5 \sin 45°$

$= 0 + 3.536 = 3.536$ N

The magnitude of the resultant force $= \sqrt{(H^2 + V^2)}$

$$= \sqrt{(11.536^2 + 3.536^2)}$$

$$= 12.07 \text{ N}$$

The direction of the resultant force $= \arctan\left(\dfrac{V}{H}\right)$

$$= \arctan\left(\dfrac{3.536}{11.536}\right)$$

$$= 17.04° \text{ or } 17° \, 2'$$

Thus the resultant of the two forces is a single vector of 12.07 N at 17° 2′ to the 8 N vector.

Problem 3. Use a graphical method to determine the magnitude and direction of the resultant of the three velocities shown in Fig. 17.5

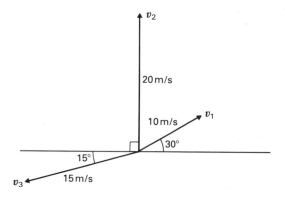

Fig. 17.5

It is usually easier to use the 'nose-to-tail' method when more than two vectors are being added. The order in which the vectors are added does not matter. In this case the order taken is v_1, then v_2, then v_3 but the same result would have been obtained if the order had been, say, v_3, v_1 and then v_2.

v_1 is drawn 10 units long at at angle of 30° to the horizontal, shown as **0a** in Fig. 17.6. v_2 is added to v_1 by drawing a line 20 units long vertically upwards from a, shown as **ab**. Finally, v_3 is added to $v_1 + v_2$ by drawing a line 15 units long at an angle of 195° from b, shown as **br**. The resultant of vector addition is **0r** and by measurement is 21.9 units long at an angle of 105° to the horizontal.

Thus, $v_1 + v_2 + v_3$ is **21.9 m/s at 105° to the horizontal**.

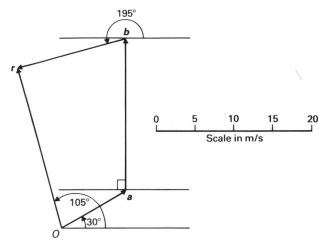

Fig. 17.6

Problem 4. Calculate the resultant of the three velocities in Problem 3.

With reference to Fig. 17.5:

Total horizontal component of the three velocities,

$H = 10 \cos 30° + 20 \cos 90° + 15 \cos 195°$

$= 8.6603 + 0 + (-14.4889) = -5.828\,6$

Total vertical component of the three velocities,

$V = 10 \sin 30° + 20 \sin 90° + 15 \sin 195°$

$= 5 + 20 + (-3.882\,3) = 21.117\,7$

The magnitude of the resultant velocity

$$= \sqrt{(H^2 + V^2)} = \sqrt{[(-5.828\,6)^2 + (21.117\,7)^2]}$$
$$= 21.91 \text{ m/s}$$

Direction of the resultant velocity $= \arctan\left(\dfrac{V}{H}\right)$

$$= \arctan\left(\dfrac{21.117\,7}{-5.828\,6}\right)$$

$$= 105.41° \text{ or } 105° \ 25'$$

(Note that the resultant velocity is in the second quadrant.)

Thus the resultant of the three velocities is a single vector of 21.91 m/s at 105° 25′ to the horizontal.

Problem 5. Accelerations of $a_1 = 2$ m/s² at 90° and $a_2 = 3.5$ m/s² at 150° act at a point. Determine (*a*) $\boldsymbol{a_1} + \boldsymbol{a_2}$ and (*b*) $\boldsymbol{a_1} - \boldsymbol{a_2}$, each by (i) drawing a scale vector diagram and (ii) by calculation.

(*a*) (i) A scale vector diagram is shown in Fig. 17.7, showing the resultant $\boldsymbol{a_1} + \boldsymbol{a_2}$

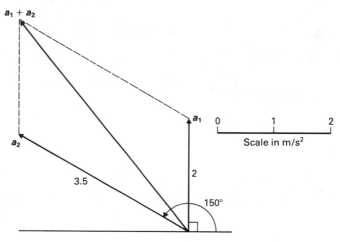

Fig. 17.7

By measurement, $\boldsymbol{a_1} + \boldsymbol{a_2} = $ **4.8 mš² at 129°**

(ii) By calculation:

Total horizontal component of the two accelerations,

$H = 2 \cos 90° + 3.5 \cos 150°$

$= 0 + (-3.031\,1) = -3.031\,1$

Total vertical component of the two accelerations,

$V = 2 \sin 90° + 3.5 \sin 150° = 2 + 1.75$

$= 3.75$

Magnitude of the resultant acceleration

$= \sqrt{(H^2 + V^2)} = \sqrt{[(-3.031\,1)^2 + (3.75)^2]}$

$= 4.82$ m/s²

Direction of the resultant acceleration

$= \arctan\left(\dfrac{V}{H}\right) = \arctan\left(\dfrac{3.75}{-3.031\,1}\right)$

$= 128.95°$ or $128°\ 57'$

(i.e. in the second quadrant)

Thus $\boldsymbol{a_1} + \boldsymbol{a_2} = $ **4.82 m/s² at 128° 57′**

(b)   (i) The scale vector diagram for $a_1 - a_2$ is shown in Fig. 17.8

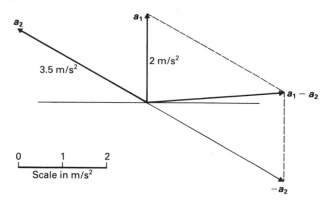

Fig. 17.8

By measurement, $a_1 - a_2 = 3$ ms^2 at 5°

(ii) By calculation:

Total horizontal component of $a_1 - a_2$, i.e. $a_1 + (-a_2)$,

$$H = 2 \cos 90° + 3.5 \cos(150° - 180°)$$

$$= 0 + 3.031\ 1 = 3.031\ 1$$

Total vertical component of $a_1 - a_2$,

$$V = 2 \sin 90° + 3.5 \sin(150° - 180°)$$

$$= 2 + (-1.75) = 0.25$$

Magnitude of
$$a_1 - a_2 = \sqrt{(H^2 + V^2)} = \sqrt{(3.031\ 1^2 + 0.25^2)}$$

$$= 3.04 \text{ m/s}^2$$

Direction of $a_1 - a_2 = \arctan\left(\dfrac{V}{H}\right)$

$$= \arctan\left(\frac{0.25}{3.031\ 1}\right) = 4.71° \text{ or } 4° \ 43'$$

Thus $a_1 - a_2 = 3.04$ m/s^2 at 4° 43′

Problem 6. If $v_1 = 20$ units at 70°, $v_2 = 30$ units at 160° and $v_3 = 25$ units at 210°, calculate the resultant of (a) $v_1 - v_2 + v_3$ and (b) $v_2 - v_1 - v_3$

(a)   The vectors v_1, v_2 and v_3 are shown in Fig. 17.9

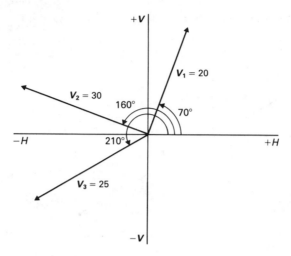

Fig. 17.9

Horizontal component H of $v_1 - v_2 + v_3 = v_1 + (-v_2) + v_3$

$\quad = 20 \cos 70° + 30 \cos(160° - 180°) + 25 \cos 210°$

$\quad = 6.840\,4 + 28.190\,8 + (-21.650\,6) = 13.380\,6$

Vertical component V of $v_1 - v_2 + v_3$

$\quad = 20 \sin 70° + 30 \sin(160° - 180°) + 25 \sin 210°$

$\quad = 18.793\,9 + (-10.260\,6) + (-12.5) = -3.966\,7$

Magnitude of resultant $= \sqrt{(H^2 + V^2)}$

$$= \sqrt{[(13.380\,6)^2 + (-3.966\,7)^2]} = 13.96$$

Direction of resultant $= \arctan\left(\dfrac{V}{H}\right)$

$$= \arctan\left(\dfrac{-3.966\,7}{13.380\,6}\right) = -16.51° \text{ or}$$
$$-16° \, 31'$$

(i.e. in the fourth quadrant)

Thus $v_1 - v_2 + v_3 =$ **13.96 units at $-16° \, 31'$**

or   **13.96 units at $343° \, 29'$**

(b) Horizontal component H of $v_2 - v_1 - v_3$

$\quad = 30 \cos 160° + 20 \cos(70° - 180°) + 25 \cos(210° - 180°)$

$\quad = -28.190\,8 + (-6.840\,4) + 21.650\,6$

$\quad = -13.380\,6$

Vertical component V of $v_2 - v_1 - v_3$

$\quad = 30 \sin 160° + 20 \sin(70° - 180°) + 25 \sin(210° - 180°)$

$\quad = 10.260\,6 + (-18.793\,9) + 12.5$

$\quad = 3.966\,7$

Magnitude of resultant $= \sqrt{(H^2 + V^2)}$

$\quad = \sqrt{[(-13.380\,6)^2 + (3.966\,7)^2]} = 13.96$ units

Direction of resultant $= \arctan\left(\dfrac{V}{H}\right)$

$\quad = \arctan\left(\dfrac{3.966\,7}{-13.380\,6}\right) = 163.49°$ or $163°\ 29'$

(i.e. in the second quadrant)

Thus $v_2 - v_1 - v_3 =$ **13.96 units at 163° 29′**

(The result is as expected, since $v_2 - v_1 - v_3 = -(v_1 - v_2 + v_3)$ and the vector 13.96 units at 163° 29′ is minus times the vector 13.96 units at $-16°\ 31'$)

Further problems on vector addition and subtraction may be found in Section 17.6 (Problems 1 to 8), page 313.

17.3   Scalar products and unit vectors

Scalar or dot products

Definition

The **scalar** or **dot** product of two vectors is found by multiplying the product of their magnitudes by the cosine of the angle between them. Thus, with reference to Fig. 17.10, if **0a** represents a force F_1 and **0b** represents a force F_2, the scalar product is written as **0a·0b**. From the definition given above:

$\mathbf{0a \cdot 0b} = |\mathbf{0a}|\,|\mathbf{0b}| \cos\theta$

$\qquad = (0a)(0b) \cos\theta \qquad\qquad\qquad (1)$

$\qquad = F_1 F_2 \cos\theta$

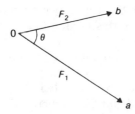

Fig. 17.10   Scalar product: $\mathbf{0a \cdot 0b} = (0a)(0b) \cos\theta = F_1 F_2 \cos\theta$

In algebra $a \times b = b \times a$, and this law is also true for vectors. This can be shown as follows. It is seen from Fig. 17.11(a) that $F_2 \cos \theta$ is the magnitude of the projection of F_2 on F_1, thus

$$\mathbf{0a} \cdot \mathbf{0b} = (\mathbf{0a})(\text{projection of } \mathbf{0b} \text{ on } \mathbf{0a})$$
$$= (F_1)(F_2 \cos \theta) = F_1 F_2 \cos \theta$$

Similarly, it is seen from Fig. 17.11(b) that $F_1 \cos \theta$ is the magnitude of the projection of F_1 on F_2, thus

$$\mathbf{0b} \cdot \mathbf{0a} = (\mathbf{0b})(\text{projection of } \mathbf{0a} \text{ on } \mathbf{0b})$$
$$= (F_2)(F_1 \cos \theta) = F_1 F_2 \cos \theta$$

It follows that $\mathbf{0a} \cdot \mathbf{0b} = \mathbf{0b} \cdot \mathbf{0a}$

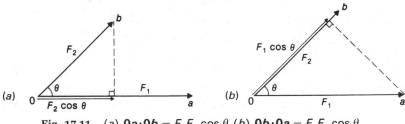

Fig. 17.11   (a) $\mathbf{0a} \cdot \mathbf{0b} = F_1 F_2 \cos \theta$ (b) $\mathbf{0b} \cdot \mathbf{0a} = F_1 F_2 \cos \theta$

Unit vectors

A unit vector is a vector whose magnitude is unity. For any vector $\mathbf{0a}$, the unit vector is $\dfrac{\mathbf{0a}}{\mathbf{0a}}$ in the direction of $\mathbf{0a}$. Unit vectors are frequently used to specify directions in space. To completely specify the direction of a vector with reference to some point, three unit vectors, mutually at right angles to each other and associated with the x-, y- and z-axes of a cartesian reference system, are used. In this context, the three unit vectors are called the **unit triad**. In Fig. 17.12, Ox, Oy and Oz represent the three axes mutually at right angles. $\mathbf{0i}$, $\mathbf{0j}$ and $\mathbf{0k}$ represent the three unit vectors. The system usually adopted is a 'right-handed system' and the vectors are labelled in such a way that the head of a right-threaded screw being screwed along Oz in the direction of $\mathbf{0k}$ turns from Ox to Oy.

An alternative notation to the geometrical '$\mathbf{0a}$' is usually used in analytical work on vectors, using a single letter only. In this notation $\mathbf{0a}$ is written as a and $a\mathbf{0}$ as $-a$. The unit triad vectors shown in Fig. 17.12 are i, j and k in this notation. This notation will be largely adopted for the remainder of this text. Any point P can be specified by its distance from planes Oyz, Oxz and Oxy, these distances being shown as a, b and c respectively in Fig. 17.12. By vector addition, using the nose-to-tail method and the geometry of Fig. 17.12:

$$\mathbf{OP} = \mathbf{OA} + \mathbf{AB} + \mathbf{BP}$$

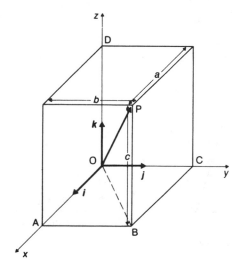

Fig. 17.12

From Fig. 17.12, $|\mathbf{OP}| = OP = \sqrt{(OB^2 + BP^2)}$

$$= \sqrt{(OA^2 + AB^2 + BP^2)}$$

$$= \sqrt{(a^2 + b^2 + c^2)}$$

That is, the magnitude of **OP** is $\sqrt{(a^2 + b^2 + c^2)}$
OP is called the **position vector** of P with respect to O. Using the unit triad notation and single letters for vectors, if **OP** $= \mathbf{r}$, then, position vector

$$\mathbf{r} = a\mathbf{i} + b\mathbf{j} + c\mathbf{k}$$

and $|\mathbf{r}| = \sqrt{(a^2 + b^2 + c^2)}$ (2)

For example, the vector $\mathbf{r} = 3\mathbf{i} + 4\mathbf{j} - 4\mathbf{k}$ is a position vector and is shown in Fig. 17.13. From equation (2) above, the magnitude of $\mathbf{r}$, i.e. $|\mathbf{r}|$, is given by

$$|\mathbf{r}| = r = \sqrt{(3^2 + 4^2 + (-4)^2)}$$

$$= \sqrt{(9 + 16 + 16)} = \sqrt{41}$$

To determine the angle between two vectors

One of the ways of expressing the scalar product of two vectors, say $\mathbf{a}$ and $\mathbf{b}$, is given in equation (1), and is

$$\mathbf{a} \cdot \mathbf{b} = ab \cos \theta$$

where a and b are the magnitudes of $\mathbf{a}$ and $\mathbf{b}$ respectively and θ is the acute

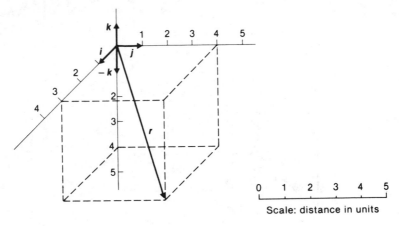

Fig. 17.13 $r = 3i + 4j - 4k$

angle between the positive directions of **a** and **b**. Thus

$$\cos \theta = \frac{\boldsymbol{a} \cdot \boldsymbol{b}}{ab}$$

Let the position vectors **a** and **b** be $a_1 i + a_2 j + a_3 k$ and $b_1 i + b_2 j + b_3 k$ respectively. Then

$$\boldsymbol{a} \cdot \boldsymbol{b} = (a_1 i + a_2 j + a_3 k) \cdot (b_1 i + b_2 j + b_3 k)$$

$$= a_1 i \cdot b_1 i + a_1 i \cdot b_2 j + a_1 i \cdot b_3 k + a_2 j \cdot b_1 i + a_2 j \cdot b_2 j + \cdots$$

It is shown in Worked Problem 1(b) following that these terms can be written as:

$$\boldsymbol{a} \cdot \boldsymbol{b} = a_1 b_1 i \cdot i + a_1 b_2 i \cdot j + a_1 b_3 i \cdot k + a_2 b_1 i \cdot j + a_2 b_2 j \cdot j + \cdots$$

By the definition of a scalar product ($\boldsymbol{a} \cdot \boldsymbol{b} = ab \cos \theta$), for the unit vectors i, j and k having a magnitude of unity and being mutually at right angles:

$$i \cdot i = (1)(1) \cos 0 = 1, \text{ since } \theta = 0°$$

$$i \cdot j = (1)(1) \cos 90° = 0, \text{ since } \theta = 90°$$

and $i \cdot k = (1)(1) \cos 90° = 0$

Thus $\boldsymbol{a} \cdot \boldsymbol{b} = (a_1 b_1)(1) + (a_1 b_2)(0) + (a_1 b_3)(0) + (a_2 b_1)(0)$

$$+ (a_2 b_2)(1) + \cdots$$

i.e. $\boldsymbol{a} \cdot \boldsymbol{b} = a_1 b_1 + a_2 b_2 + a_3 b_3$ (3)

Substituting for $\boldsymbol{a} \cdot \boldsymbol{b}$ in the equation for $\cos \theta$, gives:

$$\cos \theta = \frac{a_1 b_1 + a_2 b_2 + a_3 b_3}{ab}$$

But from equation (2), $a = |a| = \sqrt{(a_1^2 + a_2^2 + a_3^2)}$

and $$b = |b| = \sqrt{(b_1^2 + b_2^2 + b_3^2)}$$

Thus, $\cos \theta = \dfrac{a \cdot b}{|a|\,|b|} = \dfrac{a_1 b_1 + a_2 b_2 + a_3 b_3}{\sqrt{(a_1^2 + a_2^2 + a_3^2)}\sqrt{(b_1^2 + b_2^2 + b_3^2)}}$ (4)

To determine the angle between vectors a and b, where, for example, $a = 2i - j + 3k$ and $b = -4i + 6j + 5k$, the constants in equation (4) are $a_1 = 2$, $a_2 = -1$, $a_3 = 3$, $b_1 = -4$, $b_2 = 6$ and $b_3 = 5$, giving

$$\cos \theta \frac{(2)(-4) + (-1)(6) + (3)(5)}{\sqrt{(2^2 + (-1)^2 + 3^2)}\sqrt{((-4)^2 + 6^2 + 5^2)}}$$

$$= \frac{1}{\sqrt{14}\sqrt{77}} = \frac{1}{32.83} = 0.030\,46$$

Hence $\theta = 88.25°$ or $88° \; 15'$

Work done

A typical application of scalar products is that of determining the work done by a force when moving a body. The amount of work done is the product of the applied force and the distance moved in the direction of the applied force. With reference to Fig. 17.14, if force F applied at A moves the point of application to A′, through vector displacement d, then the work done by F is given by $F \cdot d$. Thus, if a constant force of $F = i + 2j - 5k$ newtons acts on a body at a point and the point is displaced by $d = 2i + 3j - k$ metres, then from equation (3):

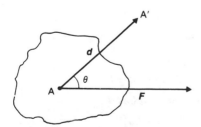

Fig. 17.14   Work done $= F \cdot d$

work done, $F \cdot d = (1)(2) + (2)(3) + (-5)(-1)$

$$= 13 \text{ N m}$$

Worked problems on scalar products and unit vectors

Problem 1. Show that (a) $p \cdot (q + r) = p \cdot q + p \cdot r$
(b) $k(p \cdot q) = (kp) \cdot q = p \cdot (kq)$

(a) By the definition of a scalar product:

$p \cdot q = p(q \cos \theta)$, where θ is the angle between p and q

$= p$ (the magnitude of the projection of q on p)

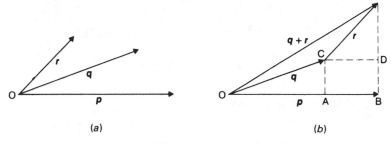

(a)   (b)

Fig. 17.15 (a) p, q and r; (b) $p \cdot q + p \cdot r = p \cdot (q + r)$

Let p, q and r be any vectors as shown in Fig. 17.15(a). Using the nose-to-tail method, $(q + r)$ is as shown in Fig. 17.15(b). Then

$$p \cdot q + p \cdot r = p(OA) + p(CD)$$
$$= p(OA) + p(AB) = p(OA + AB)$$
$$= p(OB)$$
$$= p \text{ (projection of } (q + r) \text{ on } p)$$

i.e. $p \cdot q + p \cdot r = p \cdot (q + r)$

(b) By the definition of a scalar product:

$p \cdot q = pq \cos \theta$, where θ is the angle between p and q

Thus $k(p \cdot q) = kpq \cos \theta$

Similarly, $(kp) \cdot q = (kp)q \cos \theta = kpq \cos \theta$

and $p \cdot (kq) = p(kq) \cos \theta = kpq \cos \theta$

Hence $k(p \cdot q) = (kp) \cdot q = p \cdot (kq)$

Problem 2. Find vector r joining points A and B when point A has coordinates $(3, -4, 5)$ and point B has coordinates $(2, 1, 0)$. Also find $|r|$, the magnitude of r

Let O be the origin, i.e. its coordinates are $(0, 0, 0)$

The position vectors of A and B are given by:

$$\mathbf{OA} = 3\mathbf{i} - 4\mathbf{j} + 5\mathbf{k}$$

and $\mathbf{OB} = 2\mathbf{i} + \mathbf{j}$

By the addition laws of vectors $\mathbf{OA} + \mathbf{AB} = \mathbf{OB}$, hence

$$\mathbf{r} = \mathbf{AB} = \mathbf{OB} - \mathbf{OA}$$

i.e. $\mathbf{r} = \mathbf{AB} = (2\mathbf{i} + \mathbf{j}) - (3\mathbf{i} - 4\mathbf{j} + 5\mathbf{k})$

$$= -\mathbf{i} + 5\mathbf{j} - 5\mathbf{k}$$

From equation (2), the magnitude of $\mathbf{r}$, $|\mathbf{r}| = \sqrt{(a^2 + b^2 + c^2)}$

$$= \sqrt{((-1)^2 + 5^2 + (-5)^2)}$$

$$= \sqrt{51}$$

Problem 3. If $\mathbf{p} = \mathbf{i} + 3\mathbf{j} - \mathbf{k}$ and $\mathbf{q} = 2\mathbf{i} + \mathbf{j} + 4\mathbf{k}$, determine:

(a) $\mathbf{p} \cdot \mathbf{q}$
(b) $\mathbf{p} + \mathbf{q}$
(c) $|\mathbf{p} + \mathbf{q}|$
(d) $|\mathbf{p}| + |\mathbf{q}|$

and (e) the angle between $\mathbf{p}$ and $\mathbf{q}$

(a) $\mathbf{p} \cdot \mathbf{q}$

From equation (3), if $\mathbf{p} = a_1\mathbf{i} + a_2\mathbf{j} + a_3\mathbf{k}$

and $\mathbf{q} = b_1\mathbf{i} + b_2\mathbf{j} + b_3\mathbf{k}$

then $\mathbf{p} \cdot \mathbf{q} = a_1 b_1 + a_2 b_2 + a_3 b_3$

When $\mathbf{p} = \mathbf{i} + 3\mathbf{j} - \mathbf{k}, a_1 = 1, a_2 = 3, a_3 = -1$

and when $\mathbf{q} = 2\mathbf{i} + \mathbf{j} + 4\mathbf{k}, b_1 = 2, b_2 = 1, b_3 = 4$

Hence $\mathbf{p} \cdot \mathbf{q} = (1)(2) + (3)(1) + (-1)(4)$

i.e. $\mathbf{p} \cdot \mathbf{q} = 1$

(b) $\mathbf{p} + \mathbf{q}$

If the unit vector $\mathbf{i}$ is considered, $\mathbf{p}$ contributes one unit of length in this direction and $\mathbf{q}$ two units of length in this direction. Hence, the resulting distance in the $\mathbf{i}$ direction is $1 + 2$, i.e. $3\mathbf{i}$. Similarly, the resulting distance in the $\mathbf{j}$ direction is $3 + 1$, i.e. $4\mathbf{j}$, and that in the $\mathbf{k}$ direction is $(-1) + 4$, i.e. $3\mathbf{k}$

Hence, when $\mathbf{p} = \mathbf{i} + 3\mathbf{j} - \mathbf{k}$ and $\mathbf{q} = 2\mathbf{i} + \mathbf{j} + 4\mathbf{k}$

$$\mathbf{p} + \mathbf{q} = (1 + 2)\mathbf{i} + (3 + 1)\mathbf{j} + (-1 + 4)\mathbf{k}$$

$$= 3\mathbf{i} + 4\mathbf{j} + 3\mathbf{k}$$

(c) $|\mathbf{p} + \mathbf{q}|$

The magnitude or modulus of $\mathbf{p} + \mathbf{q}$, i.e.

$|\mathbf{p} + \mathbf{q}| = |3\mathbf{i} + 4\mathbf{j} + 3\mathbf{k}|$ from part (b). From equation (2)

$|\mathbf{p} + \mathbf{q}| = \sqrt{(3^2 + 4^2 + 3^2)} = \sqrt{34}$

(d) $|p| + |q|$

From equation (2), if $a = a_1 i + a_2 j + a_3 k$,

$|a| = \sqrt{(a_1^2 + a_2^2 + a_3^2)}$. Thus, when $p = i + 3j - k$

$|p| = \sqrt{(1^2 + 3^2 + (-1)^2)}$

$= \sqrt{11}$

Similarly, $|q| = |2i + j + 4k| = \sqrt{(2^2 + 1^2 + 4^2)} = \sqrt{21}$

Hence, $|p| + |q| = \sqrt{11} + \sqrt{21} = 7.899$, correct to three decimal places.

(e) The angle between p and q

If θ is the angle between p and q, then from equation (4):

$$\cos \theta = \frac{p \cdot q}{|p||q|}$$

From part (a), the value of the numerator is 1 and from part (d) the value of the denominator is $\sqrt{11}\sqrt{21}$

Thus $\cos \theta = \dfrac{1}{\sqrt{11}\sqrt{21}}$

$= 0.065\,80$

Hence $\theta = 86.23° = 86° \ 14'$

Problem 4. Find the work done by a force of F newtons acting at point A on a body, when A is displaced to point B, the coordinates of A and B being $(3, 1, -2)$ and $(4, -1, 0)$ metres respectively, and when $F = -i - 2j - k$ newtons.

If a vector displacement from A to B is d, then the work done is $F \cdot d$ newton metres or joules. The position vector **OA** is $3i + j - 2k$ (see Worked Problem 2) and **OB** is $4i - j$

AB $= d = $ **OB** $- $ **OA** (see Worked Problem 2). Thus

$d = (4i - j) - (3i + j - 2k)$

$= i - 2j + 2k$

Work done $= F \cdot d = (-1)(1) + (-2)(-2) + (-1)(2)$

from equation (3)

$= 1 \text{ N m or joule}$

Further problems on scalar products and unit vectors may be found in Section 17.6 (Problems 9 to 18), page 315.

17.4   Vector products

Definition

The **vector** or **cross product** of a and b is c, where the magnitude of c is $ab \sin \theta$, θ being the acute angle between the position direction of a and b. The direction of c is perpendicular to both a and b, such that a, b and c form a right-handed system of vectors. Thus if a right-handed screw is screwed along c with its head at the origin of the vectors, the head rotates from a to b, as shown in Figure 17.16. Typical applications of vector products as defined above to problems in mechanics are dealt with at the end of this section.

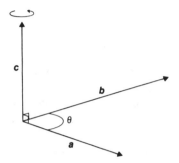

Fig. 17.16   Vector product: $c = a \cdot b$

A vector product is shown as

$$c = a \times b, \text{ called } a \text{ cross } b$$

From the definition of a vector product,

$$|c| = |a \times b| = |a| \, |b| \sin \theta = ab \sin \theta \tag{5}$$

An alternative equation for $|a \times b|$ may be derived as follows:

Squaring equation (5) gives $|c|^2 = |a|^2 \, |b|^2 \sin^2 \theta$

$$= |a|^2 \, |b|^2 (1 - \cos^2 \theta)$$

$$= |a|^2 \, |b|^2 - |a|^2 \, |b|^2 \cos^2 \theta$$

But from Section 17.2, equation (3)

$$a \cdot b = ab \cos \theta$$

hence $a \cdot a = aa \cos \theta = a^2 \cos \theta = |a|^2 \cos \theta$

But for the vectors (a and a), $\theta = 0$, $\cos \theta = 1$, hence

$$a \cdot a = |a|^2$$

i.e. $|a| = \sqrt{(a \cdot a)}$

Also, from Section 17.2, equation (4)

$$\cos \theta = \frac{a \cdot b}{ab}$$

Hence $(a^2 b^2 \cos^2 \theta) = \dfrac{a^2 b^2 (a \cdot b)^2}{a^2 b^2} = (a \cdot b)^2$

Thus, substituting for $|a|^2$, $|b|^2$ and $a^2 b^2 \cos^2 \theta$ in:

$$|c|^2 = |a|^2 |b|^2 - |a|^2 |b|^2 \cos^2 \theta$$

gives $|c|^2 = (a \cdot a)(b \cdot b) - (a \cdot b)^2$

i.e. $|a \times b| = \sqrt{[(a \cdot a)(b \cdot b) - (a \cdot b)^2]}$ (6)

Basic relationships

In Section 17.2, it is shown that the scalar product of two vectors a and b is such that $a \cdot b = b \cdot a$, that is, scalar products are said to be commutative. However, the vector product of a and b, $a \times b$, is **not** commutative, i.e. $a \times b \neq b \times a$. The magnitudes of both $a \times b$ and $b \times a$ are the same, that is, $ab \sin \theta$, but from the definition of the direction of c, where $a \times b = c$, it follows that $a \times b = -b \times a$, and the vectors are as shown in Fig. 17.17

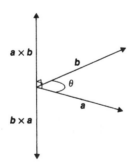

Fig. 17.17 $a \cdot b = -b \cdot a$

The other basic relationships of vector products are similar to those of scalar products, i.e.

$$a \times (b + c) = a \times b + a \times c$$

and $k(a \times b) = (ka) \times b = a \times (kb)$

Unit vector relationships

The unit vector relationships for vector products are as follows:

(a) $i \times i$ (and $j \times j$, $k \times k$)
$i \times i$ corresponds to $a \times b$ where $a = 1i + 0j + 0k$
and $b = 1i + 0j + 0k$
Hence, $i \times i = ab \sin \theta = (1)(1)(\sin 0) = 0$, since the angle θ is 0
Thus $i \times i = j \times j = k \times k = 0$

(b) $i \times j$ (and $j \times k$, $k \times i$, $j \times i$, ...)
$i \times j$ corresponds to $a \times b$ where $a = 1i + 0j + 0k$
and $b = 0i + 1j + 0k$
Hence $i \times j = ab \sin \theta = (1)(1)(\sin 90) = 1$, since angle θ is
90°. The direction of $i \times j$ is perpendicular to both i and j, that is, along
k and of magnitude 1. It follows that $i \times k = k$. Similarly it may be
shown that $j \times k = i$, and that $k \times i = j$, i.e.

$$i \times j = k, j \times k = i, k \times i = j$$

This cyclic order must be adhered to, since vector products are not
commutative, i.e. if $i \times j = k$, then $j \times i = -k$, and so on.

The vector product in terms of the unit vectors

For vectors $a = a_1i + a_2j + a_3k$ and
$b = b_1i + b_2j + b_3k$, the vector product
$a \times b = (a_1i + a_2j + a_3k) \times (b_1i + b_2j + b_3k)$

i.e. $a \times b = a_1b_1 i \times i + a_1b_2 i \times j + a_1b_3 i \times k + a_2b_1 j \times i + a_2b_2 j \times j$
$+ a_2b_3 j \times k + a_3b_1 k \times i + a_3b_2 k \times j + a_3b_3 k \times k$

Since $i \times i = 0$ and $i \times j = k$ and so on,

then $a \times b = a_1b_2k - a_1b_3j - a_2b_1k + a_2b_3i + a_3b_1j - a_3b_2i$
i.e. $a \times b = (a_2b_3 - a_3b_2)i + (a_3b_1 - a_1b_3)j + (a_1b_2 - a_2b_1)k$

This relationship is best remembered in determinant form, which is introduced in Chapter 9 of *Technician mathematics 4/5*. In determinant form:

$$a \times b = \begin{vmatrix} i & j & k \\ a_1 & a_2 & a_3 \\ b_1 & b_2 & b_3 \end{vmatrix} \tag{7}$$

Thus for $a = 2i - j + 3k$ and $b = i + 4j - 2k$

$$a \times b = \begin{vmatrix} i & j & k \\ 2 & -1 & 3 \\ 1 & 4 & -2 \end{vmatrix}$$

A 3 by 3 determinant is evaluated as follows:

$$\begin{vmatrix} a_1 & b_1 & c_1 \\ a_2 & b_2 & c_2 \\ a_3 & b_3 & c_3 \end{vmatrix} = a_1 \begin{vmatrix} b_2 & c_2 \\ b_3 & c_3 \end{vmatrix} - b_1 \begin{vmatrix} a_2 & c_2 \\ a_3 & c_3 \end{vmatrix} + c_1 \begin{vmatrix} a_2 & b_2 \\ a_3 & b_3 \end{vmatrix}$$

Using this rule gives:

$$\boldsymbol{a} \times \boldsymbol{b} = \boldsymbol{i} \begin{vmatrix} -1 & 3 \\ 4 & -2 \end{vmatrix} - \boldsymbol{j} \begin{vmatrix} 2 & 3 \\ 1 & -2 \end{vmatrix} + \boldsymbol{k} \begin{vmatrix} 2 & -1 \\ 1 & 4 \end{vmatrix}$$

i.e. $\boldsymbol{a} \times \boldsymbol{b} = [(-1)(-2) - (3)(4)]\boldsymbol{i} - [(2)(-2) - (3)(1)]\boldsymbol{j}$
$$+ [(2)(4) - (-1)(1)]\boldsymbol{k}$$

That is, $\boldsymbol{a} \times \boldsymbol{b} = -10\boldsymbol{i} + 7\boldsymbol{j} + 9\boldsymbol{k}$

Typical applications of vector products

Typical applications of vector products are to moments and to angular velocity.

(*a*)   **Moments.** If M is the moment vector about a point 0 of a force vector $\boldsymbol{F}$ which has position vector $\boldsymbol{r}$ from 0, as shown in Fig. 17.18, then

$$M = \boldsymbol{r} \times \boldsymbol{F}$$

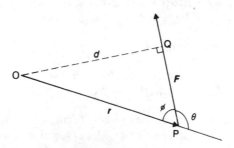

Fig. 17.18 $M = \boldsymbol{r} \cdot \boldsymbol{F}$; $M = Fd$

By the definition of the modulus of a vector given in equation (5)
$|M| = |r||F| \sin \theta$, where θ is the angle between the positive directions of $\boldsymbol{r}$ and $\boldsymbol{F}$

But with reference to Fig. 17.18, $|r| = OP$ and $\sin \theta = \sin(180 - \theta)$
$$= \sin \phi$$

Hence $|M| = |F|(OP \sin \theta)$

From triangle OPQ, $\sin \phi = \dfrac{OQ}{OP} = \dfrac{d}{OP}$, i.e. $OP \sin \phi = d$

Hence $|M| = |F| d = Fd$, which is the basic concept of the moment of a force in mechanics.

(b) **Angular velocity**. If v is the velocity vector of a point P on a body rotating about a fixed axis, ω is its angular velocity vector and r is the position vector of P, then

$$v = \omega \times r$$

The angular velocity vector, ω, has a magnitude of ω and its direction is along the axis. With reference to Fig. 17.19, if $QP = d$ and angle QOP, between ω and r, is θ, then

$$|v| = |\omega| |r| \sin \theta$$

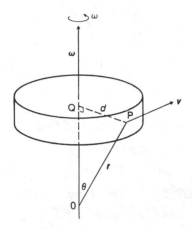

Fig. 17.19 $v = \omega \cdot r$

But $|r| = OP$ and $\sin \theta = \dfrac{QP}{OP} = \dfrac{d}{OP}$, hence $|r| \sin \theta = d$

Hence $|v| = \omega d$

The relationship between linear and angular velocity, i.e. $v = \omega d$, is a basic concept in dynamics.

Worked problems on vector products

Problem 1. If $p = 3i + 2j - k$, $q = i - j + \dfrac{k}{2}$ and $r = i - k$, find

(a) $p \times q$, (b) $|p \times \subset r|$, (c) $(2p - q) \times r$ and (d) $p \times (2r \times q)$

(a) $p \times q$

From equation (7), when $a = a_1 i + a_2 j + a_3 k$ and
$$b = b_1 i + b_2 j + b_3 k, \text{ then}$$

$$a \times b = \begin{vmatrix} i & j & k \\ a_1 & a_2 & a_3 \\ b_1 & b_2 & b_3 \end{vmatrix}$$

Thus, when $p = 3i + 2j - k$ and $q = i - j + \frac{1}{2}k$

$$p \times q = \begin{vmatrix} i & j & k \\ 3 & 2 & -1 \\ 1 & -1 & \frac{1}{2} \end{vmatrix}$$

i.e. $p \times q = i \begin{vmatrix} 2 & -1 \\ -1 & \frac{1}{2} \end{vmatrix} - j \begin{vmatrix} 3 & -1 \\ 1 & \frac{1}{2} \end{vmatrix} + k \begin{vmatrix} 3 & 2 \\ 1 & -1 \end{vmatrix}$

$= i(1 - 1) - j(\frac{3}{2} - (-1)) + k(-3 - 2)$

$= -\frac{5}{2}j - 5k$

(b) $|p \times r|$

From equation (6), $|a \times b| = \sqrt{[(a \cdot a)(b \cdot b) - (a \cdot b)^2]}$

Hence $\qquad |p \times r| = \sqrt{[(p \cdot p)(r \cdot r) - (p \cdot r)^2]}$

where $\qquad p = 3i + 2j - k$ and $r = i - k$

Then $\qquad p \cdot p = (3)(3) + (2)(2) + (-1)(-1)$ from equation (3)

$= 9 + 4 + 1 = 14$

$r \cdot r = (1)(1) + (0)(0) + (-1)(-1) = 2$

and $\qquad p \cdot r = (3)(1) + (2)(0) + (-1)(-1) = 4$

Hence $|p \times r| = \sqrt{[(14)(2) - (4)^2]}$

$= \sqrt{12}$

(c) $(2p - q) \times r = [2(3i + 2j - k) - (i - j + \frac{1}{2}k)] \times (i - k)$

$= (5i + 5j - 2\frac{1}{2}k) \times (i - k)$

i.e. $(2p - q) \times r = \begin{vmatrix} i & j & k \\ 5 & 5 & -2\frac{1}{2} \\ 1 & 0 & -1 \end{vmatrix}$

$= i(-5 - 0) - j(-5 + 2\frac{1}{2}) + k(0 - 5)$

$= -5i + 2\frac{1}{2}j - 5k$

(d) $p \times (2r \times q)$

$2r \times q = (2i - 2k) \times (i - j + \tfrac{1}{2}k)$

$$= \begin{vmatrix} i & j & k \\ 2 & 0 & -2 \\ 1 & -1 & \tfrac{1}{2} \end{vmatrix}$$

$= i(0 - 2) - j(1 + 2) + k(-2 - 0)$

$= -2i - 3j - 2k$

Hence $p \times (2r \times q) = (3i + 2j - k) \times (-2i - 3j - 2k)$

$$= \begin{vmatrix} i & j & k \\ 3 & 2 & -1 \\ -2 & -3 & -2 \end{vmatrix}$$

$= i(-4 - 3) - j(-6 - 2) + k(-9 + 4)$

$= -7i + 8j - 5k$

Problem 2. A force of $F = 2i + j - k$ newtons acts on a line passing through a point A. Find moment M and its magnitude M, of the force F about a point B when A has coordinates $(1, 2, -3)$ metres and B has coordinates $(0, 1, -1)$ metres.

Let r be the vector **BA**. From Worked Problem 2 on page 302,
OA $= i + 2j - 3k$ and **OB** $= j - k$

Then **BA** = **OA** − **OB**

i.e. $r = (i + 2j - 3k) - (j - k) = i + j - 2k$

Moment, $M = r \times F$

$$= (i + j - 2k) \times (2i + j - k)$$

$$= \begin{vmatrix} i & j & k \\ 1 & 1 & -2 \\ 2 & 1 & -1 \end{vmatrix}$$

$= i(-1 + 2) - j(-1 + 4) + k(1 - 2)$

$= i - 3j - k \text{ N m}$

From equation (6) $M = |M| = |r \times F|$

$$= \sqrt{[(r \cdot r)(F \cdot F) - (r \cdot F)^2]}$$

$r \cdot r = (1)(1) + (1)(1) + (-2)(-2) = 6$

$F \cdot F = (2)(2) + (1)(1) + (-1)(-1) = 6$

$r \cdot F = (1)(2) + (1)(1) + (-2)(-1) = 5$

Hence, $M = \sqrt{[(6)(6) - 5^2]} = \sqrt{11} \text{ N m}$

Problem 3. A sphere of radius 10 centimetres with its centre at the origin, $(0, 0, 0)$, rotates about the z-axis with an angular velocity vector $\omega = i - 5j + 4k$ radians per second. Determine the velocity vector v and its magnitude for point **A** in the sphere when the position vector of **A** is $3i - j + 2k$ centimetres.

The velocity vector $v = \omega \times r$ where ω is the angular velocity vector and r is the position vector. Hence

$$v = \omega \times r = (i - 5j + 4k) \times (3i - j + 2k)$$

i.e. $\qquad v = \begin{vmatrix} i & j & k \\ 1 & -5 & 4 \\ 3 & -1 & 2 \end{vmatrix}$

$$= i(-10 + 4) - j(2 - 12) + k(-1 + 15)$$

$$= -6i + 10j + 14k$$

From equation (6):

$$v = |\omega \times r| = \sqrt{[(\omega \cdot \omega)(r \cdot r) - (\omega \cdot r)^2]}$$

$$\omega \cdot \omega = 1^2 + (-5)^2 + 4^2 = 42$$

$$r \cdot r = 3^2 + (-1)^2 + 2^2 = 14$$

$$\omega \cdot r = (1)(3) + (-5)(-1) + (4)(2) = 16$$

Hence, $v = \sqrt{[(42)(14) - 16^2]}$

$$= 18.22 \text{ cm/s}$$

Further problems on vector products may be found in Section 17.6 (Problems 19 to 28), page 315.

17.5   Summary of scalar and vector product formulae

The formulae given below refer to two vectors:

$a = a_1 i + a_2 j + a_3 k$ and $b = b_1 i + b_2 j + b_3 k$, having an angle of θ between the positive directions of the vectors.

Scalar or dot products

By definition, the scalar or dot product is given by:

$$a \cdot b = |a| \, |b| \cos \theta = ab \cos \theta \qquad (1)$$

The modulus of a, $|a| = a = \sqrt{(a_1^2 + a_2^2 + a_3^2)} \qquad (2)$

The value of the scalar product is:

$$a \cdot b = a_1 b_1 + a_2 b_2 + a_3 b_3 \qquad (3)$$

The angle between two vectors is given by

$$\cos \theta = \frac{a \cdot b}{|a|\,|b|} = \frac{a \cdot b}{ab} = \frac{a_1 b_1 + a_2 b_2 + a_3 b_3}{\sqrt{(a_1^2 + a_2^2 + a_3^2)}\sqrt{(b_1^2 + b_2^2 + b_3^2)}} \qquad (4)$$

Vector or cross products

The modulus of a vector or cross product is given by:

$$|a \times b| = |a|\,|b| \sin \theta = ab \sin \theta \qquad (5)$$

Alternatively, in terms of scalar products.

$$|a \times b| = \sqrt{[(a \cdot a)(b \cdot b) - (a \cdot b)^2]} \qquad (6)$$

The vector resulting from a vector product is given by:

$$a \times b = \begin{vmatrix} i & j & k \\ a_1 & a_2 & a_3 \\ b_1 & b_2 & b_3 \end{vmatrix} \qquad (7)$$

The applications of scalar and vector products to mechanics at this stage are very limited and the one or two obvious applications have been included in the worked problems in the various sections. It will not be until differentiation and integration of vector quantities have been mastered at a higher level that their real potential will become apparent.

17.6   Further problems

Vector addition and subtraction

1. Forces of 15 N and 32 N act at a point and are inclined at 90° to each other. Determine, by drawing, the resultant force and its direction relative to the 32 N force.   [35.3 N at 25°]
2. Calculate the magnitude and direction of velocities of 4 m/s at 20° and 9 m/s at 120° when acting simultaneously at a point.   [9.19 m/s at 94° 38′]
3. Forces P, Q and R are coplanar and act at a point. Force P is 10 kN at 75°, Q is 6 kN at 150° and R is 8 kN at 195°. Determine graphically the resultant force. [14.8 kN at 134°]
4. Three forces of 5 N, 8 N and 13 N act as shown in Fig. 17.20 Calculate the magnitude of the resulting force and its direction relative to the 5 N force. [14.72 N at −14° 43′ (i.e. 345° 17′)]

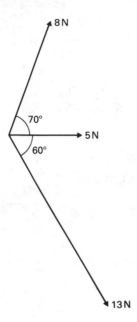

Fig. 17.20

5. A load of 4.32 N is lifted by two strings, making angles of 18° and 30° with the vertical. If for this system the vectors representing the forces form a closed triangle when in equilibrium, calculate the tensions in the strings.
   [1.80 N, 2.91 N]

6. Accelerations of $a_1 = 4 \text{ m/s}^2$ at 45° and $a_2 = 7.5 \text{ m/s}^2$ at 175° act at a point. Determine (i) $a_1 + a_2$ and (ii) $a_1 - a_2$
   (i) [5.80 m/s² at 143° 8']   (ii) [10.53 m/s² at 11° 55']

7. If $f_1 = 15$ units at 55°, $f_2 = 20$ units at 145° and $f_3 = 30$ units at 250°, calculate the resultant of $f_2 - f_1 - f_3$
   [31.08 units at 118° 17']

8. The acceleration of a body is due to four coplanar accelerations. These are 5 m/s² due west, 4 m/s² due north, 7 m/s² to the south-east and 8 m/s² to the south-west. Calculate the resultant acceleration and its direction.
   [8.73 m/s² at 229° 11']

Scalar products and unit vectors

In Problems 9 to 14: $p = i + 3k$
$$q = 2i + j - k$$
$$r = -i + 4j + 5k$$
Determine the quantities stated.

9.   (a) $p \cdot q$ $[-1]$
     (b) $p \cdot r$ $[14]$
     (c) $q \cdot r$ $[-3]$
10. (a) $|p|$ $[\sqrt{10}]$
     (b) $|q|$ $[\sqrt{6}]$
     (c) $|r|$ $[\sqrt{42}]$
11. (a) $q \cdot (p + r)$ $[-4]$
     (b) $r \cdot (p + q)$ $[11]$
12. $3p \cdot (q + 2r)$ $[81]$
13. (a) $|p + q|$ $[\sqrt{14}]$
     (b) $|p| + |q|$ $[\sqrt{10} + \sqrt{6} \approx 5.612]$
14. The angle between
     (a) p and q $[97° \ 25']$
     (b) p and r $[46° \ 55']$
     (c) p and $-q$ $[82° \ 35']$
     (d) q and $(p - r)$ $[80° \ 24']$

In Problems 15 to 18, find the work done for the data given by a
constant force of F newtons acting on a body at point A, when A
is displaced to point B, in metres.

15. $F = 2i - 3k$, A: $(0, 1, 2)$, B: $(3, 0, -4)$ [24 N m]
16. $F = -i + 2j - k$, A: $(2, 0, 0)$, B: $(-1, -2, -3)$ [2 N m]
17. $F = 4i - 5j$, A: $(2, 1, -3)$, B: $(4, 4, 1)$ $[-7$ N m]
18. $F = 3i - j + k$, A: $(0, 1, 0)$, B: $(4, 2, 1)$ [12 N m]

Vector products

In Problems 19 to 24: $p = 2i + 3k$
$$q = i - 2j + k$$
$$r = -2i + 4j - 3k$$
Determine the quantities stated.

19. (a) $p \times q$ $[6i + j - 4j]$
     (b) $q \times p$ $[-6i - j + 4k]$
20. (a) $|p \times r|$ $[14.42]$
     (b) $|r \times q|$ $[2.236]$
21. (a) $2p \times 4r$ $[32(-3i + 2k)]$
     (b) $(p + q) \times r$ $[-10i + j + 8k]$
22. (a) $p \times (q \times r)$ $[-3i + 6j + 2k]$
     (b) $(p + 2q) \times r$ $[2(-4i + j + 4k)]$

23. (a) $(r - 2p) \times \dfrac{q}{2}$ $[-7i - 1\frac{1}{2}j + 4k]$

     (b) $(2p \times 3r) \times q$ $[8(12i + 15j + 18k)]$

24.   Prove that $p \times (q \times r) + q \times (r \times p) + r \times (p \times q) = 0$

In Problems 25 and 26, a force F acts on a line through point P. Find the moment vector M and its modulus M of F about point Q, where:

25. $F = i + j$, P has coordinates (2, 1, 1) and Q has coordinates
(3, −2, 4) $[M = 3i − 3j − 4j;\ M = 5.831]$

26. $F = 2i − j + k$, P has coordinates (0, 3, 1) and Q has
coordinates (4, 0, −1) $[M = 5i + 8j − 2k;\ M = 9.644]$

In Problems 27 and 28, a circular cylinder with its centre at the origin rotates about the z-axis with vector angular velocity ω. Determine the vector velocity v and its magnitude v of point P on the cylinder for the data given and when $\omega = −5i + 2j − 7k$

27.   The position vector of P is $i + 2j$
$[v = 14i − 7j − 12k;\ v = 19.72]$

28.   The position vector of P is $i − j + 2k$
$[v = 3(−i + j + k);\ v = 5.196]$

Chapter 18

Introduction to differentiation

18.1 Introduction

Calculus is a branch of mathematics involving or leading to calculations dealing with continuously varying functions. The subject falls into two parts, namely **differential calculus** (usually abbreviated to **differentiation**) and **integral calculus** (usually abbreviated to **integration**).

The central problem of the differential calculus is the investigation of the rate of change of a function with respect to changes in the variables on which it depends.

The two main uses of integral calculus are firstly, finding such quantities as the length of a curve, the area enclosed by a curve, or the volume enclosed by a surface, and secondly, the problem of determining a variable quantity given its rate of change.

There is a close relationship between the processes of differentiation and integration, the latter being considered as the inverse of the former.

Calculus is a comparatively young branch of mathematics; its systematic development started in the middle of the seventeenth century. Since then there has been an enormous expansion in the scope of calculus and it is now used in every field of applied science as an instrument for the solution of problems of the most varied nature.

Before such uses can be investigated it is essential to grasp the basic concepts and to understand the notations used. The following text deals with this necessary preparatory work.

18.2 Functional notation

An expression such as $y = 4x^2 - 4x - 3$ contains two variables. For every value of x there is a corresponding value of y. The variable x is called the **independent variable** and y is called the **dependent variable**. y is said to be a function of x and is written as $y = f(x)$. Hence from above $f(x) = 4x^2 - 4x - 3$.

The value of the function $f(x)$ when $x = 0$ is denoted by $f(0)$. Similarly when $x = 1$ the value of the function is denoted by $f(1)$ and so on.

If $f(x) = 4x^2 - 4x - 3$

then $f(0) = 4(0)^2 - 4(0) - 3$ $= -3$

$f(1) = 4(1)^2 - 4(1) - 3$ $= -3$

$f(2) = 4(2)^2 - 4(2) - 3$ $= 5$

$f(3) = 4(3)^2 - 4(3) - 3$ $= 21$

$f(-1) = 4(-1)^2 - 4(-1) - 3 = 5$

and $f(-2) = 4(-2)^2 - 4(-2) - 3 = 21$

Figure 18.1 shows the curve $f(x) = 4x^2 - 4x - 3$ for values of x between $x = -2$ and $x = 3$. The lengths represented by $f(0)$, $f(1)$, $f(2)$, etc. are also shown.

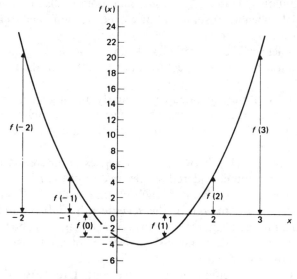

Fig. 18.1   Graph of $f(x) = 4x^2 - 4x - 3$

Worked problems on functional notation

Problem 1. If $f(x) = 5x^2 - 3x + 1$ find $f(0)$, $f(3)$, $f(-1)$, $f(-2)$ and $f(3) - f(-2)$

$f(x) = 5x^2 - 3x + 1$

$f(0) = 5(0)^2 - 3(0) + 1$ $= 1$

$f(3) = 5(3)^2 - 3(3) + 1$ $= 37$

$f(-1) = 5(-1)^2 - 3(-1) + 1 = 9$

$f(-2) = 5(-2)^2 - 3(-2) + 1 = 27$

$f(3) - f(-2) = 37 - 27$ $= 10$

Problem 2. For the curve $f(x) = 3x^2 + 2x - 9$ evaluate $f(2) \div f(1)$, $f(2 + a)$, $f(2 + a) - f(2)$ and $\dfrac{f(2 + a) - f(2)}{a}$

$f(x) = 3x^2 + 2x - 9$

$f(1) = 3(1)^2 + 2(1) - 9 = -4$

$f(2) = 3(2)^2 + 2(2) - 9 = 7$

$\begin{aligned} f(2 + a) &= 3(2 + a)^2 + 2(2 + a) - 9 \\ &= 3(4 + 4a + a^2) + 4 + 2a - 9 \\ &= 12 + 12a + 3a^2 + 4 + 2a - 9 \\ &= 7 + 14a + 3a^2 \end{aligned}$

$f(2) \div f(1) = \dfrac{f(2)}{f(1)} = \dfrac{7}{-4} = -1\tfrac{3}{4}$

$\begin{aligned} f(2 + a) - f(2) &= 7 + 14a + 3a^2 - 7 \\ &= 14a + 3a^2 \end{aligned}$

$\dfrac{f(2 + a) - f(2)}{a} = \dfrac{14a + 3a^2}{a} = 14 + 3a$

Further problems on functional notation may be found in Section 18.5 (Problems 1 to 5), page 331.

18.3   The gradient of a curve

If a tangent is drawn at a point A on a curve then the gradient of this tangent is said to be the gradient of the curve at A.

In Fig. 18.2 the gradient of the curve at A is equal to the gradient of the tangent AB.

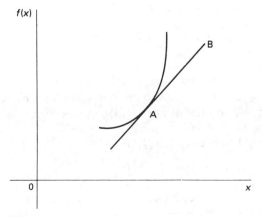

Fig. 18.2

Consider the graph of $f(x) = 2x^2$, part of which is shown in Fig. 18.3
The gradient of the chord PQ is given by:

$$\frac{QR}{PR} = \frac{QS - RS}{PR} = \frac{QS - PT}{PR}$$

At point P, $x = 1$ and at point Q, $x = 3$

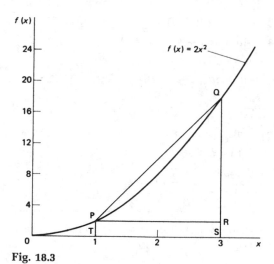

Fig. 18.3

Hence the gradient of the chord PQ $= \dfrac{f(3) - f(1)}{3 - 1}$

$$= \frac{18 - 2}{2}$$

$$= \frac{16}{2} = 8$$

More generally, for any curve (as shown in Fig. 18.4):

Gradient of PQ $= \dfrac{f(x_2) - f(x_1)}{x_2 - x_1}$

For the part of the curve $f(x) = 2x^2$ shown in Fig. 18.5 let us consider what happens as the point Q, at present at $(3, f(3))$, moves closer and closer to point P, which is fixed at $(1, f(1))$

Let Q_1 be the point on the curve $(2.5, f(2.5))$

Gradient of chord $PQ_1 = \dfrac{f(2.5) - f(1)}{2.5 - 1}$

$$= \frac{12.5 - 2}{1.5} = 7$$

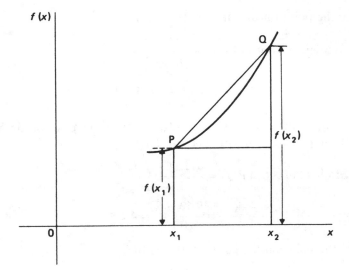

Fig. 18.4

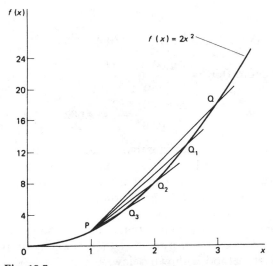

Fig. 18.5

Let Q_2 be the point on the curve $(2, f(2))$

$$\text{Gradient of chord } PQ_2 = \frac{f(2) - f(1)}{2 - 1}$$

$$= \frac{8 - 2}{1} = 6$$

Let Q_3 be the point on the curve $(1.5, f(1.5))$

$$\text{Gradient of chord } PQ_3 = \frac{f(1.5) - f(1)}{1.5 - 1}$$

$$= \frac{4.5 - 2}{0.5} = 5$$

The following points, i.e. Q_4, Q_5 and Q_6, are not shown on Fig. 18.5
Let Q_4 be the point on the curve $(1.1, f(1.1))$

$$\text{Gradient of chord } PQ_4 = \frac{f(1.1) - f(1)}{1.1 - 1}$$

$$= \frac{2.42 - 2}{0.1} = 4.2$$

Let Q_5 be the point on the curve $(1.01, f(1.01))$

$$\text{Gradient of chord } PQ_5 = \frac{f(1.01) - f(1)}{1.01 - 1}$$

$$= \frac{2.040\,2 - 2}{0.01} = 4.02$$

Let Q_6 be the point on the curve $(1.001, f(1.001))$

$$\text{Gradient of chord } PQ_6 = \frac{f(1.001) - f(1)}{1.001 - 1}$$

$$= \frac{2.004\,002 - 2}{0.001} = 4.002$$

Thus as the point Q approaches closer and closer to the point P the gradients of the chords approach nearer and nearer to the value 4. This is called the **limiting value** of the gradient of the chord and **at P the chord becomes the tangent to the curve**. Thus the limiting value of 4 is the gradient of the tangent at P.

It can be seen from the above example that deducing the gradient of the tangent to a curve at a given point by this method is a lengthy process. A much more convenient method is shown below.

18.4   Differentiation from first principles

Let P and Q be two points very close together on a curve as shown in Fig. 18.6

Let the length PR be δx (pronounced delta x), representing a small increment (or increase) in x, and the length QR, the corresponding increase in y, be δy (pronounced delta y). It is important to realise that δ and x are

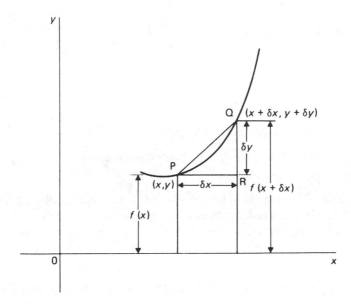

Fig. 18.6

inseparable, i.e. δx does not mean δ times x. Let P be any point on the curve with coordinates (x, y). Then Q will have the coordinates $(x + \delta x, y + \delta y)$.

The slope of the chord $PQ = \dfrac{\delta y}{\delta x}$

But from Fig. 18.6, $\delta y = (y + \delta y) - y = f(x + \delta x) - f(x)$

Hence $\dfrac{\delta y}{\delta x} = \dfrac{f(x + \delta x) - f(x)}{\delta x}$

The smaller δx becomes, the closer the gradient of the chord PQ approaches the gradient of the tangent at P. That is, as $\delta x \to 0$, the gradient of the chord $\to$ the gradient of the tangent. (Note '$\to$' means 'approaches'.) As δx approaches zero, the value of $\dfrac{\delta y}{\delta x}$ approaches what is called a **limiting value**.

There are two notations commonly used when finding the gradient of a tangent drawn to a curve.

1. The gradient of the curve at P is represented as $\lim\limits_{\delta x \to 0} \dfrac{\delta y}{\delta x}$

This is written as $\dfrac{dy}{dx}$ (pronounced dee y by dee x), i.e.

$$\frac{dy}{dx} = \lim_{\delta x \to 0} \frac{\delta y}{\delta x}$$

This way of stating the gradient of a curve is called **Leibniz notation**.

2. The gradient of the curve at P $= \lim\limits_{\delta x \to 0} \left\{ \dfrac{f(x + \delta x) - f(x)}{\delta x} \right\}$

This is written as $f'(x)$ (pronounced f dash x)

i.e. $f'(x) = \lim\limits_{\delta x \to 0} \left\{ \dfrac{f(x + \delta x) - f(x)}{\delta x} \right\}$

This way of stating the gradient of a curve is called **functional notation**.

$\dfrac{dy}{dx}$ equals $f'(x)$ and is called the **differential coefficient**, or simply the **derivative**.

The process of finding the differential coefficient is called **differentiation**.

In the following worked problems the expression for $f'(x)$, which is a definition of the differential coefficient, will be used as a starting point.

Worked problems on differentiation from first principles

Problem 1. Differentiate from first principles $f(x) = x^2$ and find the value of the gradient of the curve at $x = 3$

To 'differentiate from first principles' means 'to find $f'(x)$' by using the expression:

$f'(x) = \lim\limits_{\delta x \to 0} \left\{ \dfrac{f(x + \delta x) - f(x)}{\delta x} \right\}$

$f(x) = x^2$

$f(x + \delta x) = (x + \delta x)^2 = x^2 + 2x\delta x + \delta x^2$

$f(x + \delta x) - f(x) = x^2 + 2x\delta x + \delta x^2 - x^2$

$\qquad\qquad\qquad\quad = 2x\delta x + \delta x^2$

$\dfrac{f(x + \delta x) - f(x)}{\delta x} = \dfrac{2x\delta x + \delta x^2}{\delta x}$

$\qquad\qquad\qquad\quad = 2x + \delta x$

As $\delta x \to 0$, $\dfrac{f(x + \delta x) - f(x)}{\delta x} \to 2x + 0$

Therefore $f'(x) = \lim\limits_{\delta x \to 0} \left\{ \dfrac{f(x + \delta x) - f(x)}{\delta x} \right\} = 2x$

At $x = 3$, the gradient of the curve, i.e. $f'(x) = 2(3) = 6$

Hence if $f(x) = x^2$, $f'(x) = 2x$. The gradient at $x = 3$ is 6

Problem 2. Find the differential coefficient of $f(x) = 3x^3$, from first principles.

To 'find the differential coefficient' means to 'to find $f'(x)$' by using the expression:

$$f'(x) = \lim_{\delta x \to 0} \left\{ \frac{f(x + \delta x) - f(x)}{\delta x} \right\}$$

$f(x) = 3x^3$

$f(x + \delta x) = 3(x + \delta x)^3$

$\qquad = 3(x + \delta x)(x^2 + 2x\delta x + \delta x^2)$

$\qquad = 3(x^3 + 3x^2\delta x + 3x\delta x^2 + \delta x^3)$

$\qquad = 3x^3 + 9x^2\delta x + 9x\delta x^2 + 3\delta x^3$

$f(x + \delta x) - f(x) = 3x^3 + 9x^2\delta x + 9x\delta x^2 + 3\delta x^3 - 3x^3$

$\qquad = 9x^2\delta x + 9x\delta x^2 + 3\delta x^3$

$$\frac{f(x + \delta x) - f(x)}{\delta x} = \frac{9x^2\delta x + 9x\delta x^2 + 3\delta x^3}{\delta x}$$

$$= 9x^2 + 9x\delta x + 3\delta x^2$$

As $\delta x \to 0$, $\dfrac{f(x + \delta x) - f(x)}{\delta x} \to 9x^2 + 9x(0) + 3(0)^2$

i.e. $f'(x) = \lim_{\delta x \to 0} \left\{ \dfrac{f(x + \delta x) - f(x)}{\delta x} \right\} = 9x^2$

Problem 3. By differentiation from first principles determine $\dfrac{dy}{dx}$ for $y = 3x$

The object is to find $\dfrac{dy}{dx}$

$$\frac{dy}{dx} = f'(x) = \lim_{\delta x \to 0} \left\{ \frac{f(x + \delta x) - f(x)}{\delta x} \right\}$$

$y = f(x) = 3x$

$f(x + \delta x) = 3(x + \delta x) = 3x + 3\delta x$

$f(x + \delta x) - f(x) = 3x + 3\delta x - 3x$

$\qquad = 3\delta x$

$$\frac{f(x + \delta x) - f(x)}{\delta x} = \frac{3\delta x}{\delta x} = 3$$

Hence $\dfrac{dy}{dx} = \lim\limits_{\delta x \to 0} \left\{ \dfrac{f(x + \delta x) - f(x)}{\delta x} \right\} = 3$

Another way of writing $\dfrac{dy}{dx} = 3$ is $f'(x) = 3$

or $\dfrac{d}{dx}(3x) = 3$ since $y = 3x$

Problem 4. Find the derivative of $y = \sqrt{x}$

Let $y = f(x) = \sqrt{x}$
To 'find the derivative' means 'to find $f'(x)$'

$f'(x) = \lim\limits_{\delta x \to 0} \left\{ \dfrac{f(x + \delta x) - f(x)}{\delta x} \right\}$

$f(x) = \sqrt{x} = x^{\frac{1}{2}}$

$\quad f(x + \delta x) = (x + \delta x)^{\frac{1}{2}}$

$\quad f(x + \delta x) - f(x) = (x + \delta x)^{\frac{1}{2}} - x^{\frac{1}{2}}$

$\quad \dfrac{f(x + \delta x) - f(x)}{\delta x} = \dfrac{(x + \delta x)^{\frac{1}{2}} - x^{\frac{1}{2}}}{\delta x}$

Now from algebra, $(a - b)(a + b) = a^2 - b^2$, i.e. the difference of two squares. Therefore, in this case, multiplying both the numerator and the denominator by $[(x + \delta x)^{\frac{1}{2}} + x^{\frac{1}{2}}]$, to make the numerator of the fraction of $(a + b)(a - b)$ form, gives:

$\dfrac{f(x + \delta x) - f(x)}{\delta x} = \dfrac{[(x + \delta x)^{\frac{1}{2}} - x^{\frac{1}{2}}][(x + \delta x)^{\frac{1}{2}} + x^{\frac{1}{2}}]}{\delta x[(x + \delta x)^{\frac{1}{2}} + x^{\frac{1}{2}}]}$

$\qquad = \dfrac{[(x + \delta x)^{\frac{1}{2}}]^2 - [x^{\frac{1}{2}}]^2}{\delta x[(x + \delta x)^{\frac{1}{2}} + x^{\frac{1}{2}}]}$

$\qquad = \dfrac{(x + \delta x) - (x)}{\delta x[(x + \delta x)^{\frac{1}{2}} + x^{\frac{1}{2}}]}$

$\qquad = \dfrac{\delta x}{\delta x[(x + \delta x)^{\frac{1}{2}} + x^{\frac{1}{2}}]}$

$\qquad = \dfrac{1}{(x + \delta x)^{\frac{1}{2}} + x^{\frac{1}{2}}}$

As $\delta x \to 0$, $\dfrac{f(x + \delta x) - f(x)}{\delta x} \to \dfrac{1}{(x + 0)^{\frac{1}{2}} + x^{\frac{1}{2}}}$

Therefore $f'(x) = \lim\limits_{\delta x \to 0} \left\{ \dfrac{f(x + \delta x) - f(x)}{\delta x} \right\} = \dfrac{1}{x^{\frac{1}{2}} + x^{\frac{1}{2}}} = \dfrac{1}{2x^{\frac{1}{2}}}$

Hence if $f(x) = \sqrt{x}$, $f'(x) = \dfrac{1}{2\sqrt{x}}$ or $\tfrac{1}{2}x^{-\frac{1}{2}}$

Another way of writing this is:

If $y = \sqrt{x}$, $\dfrac{dy}{dx} = \dfrac{1}{2\sqrt{x}}$

or $\dfrac{d}{dx}(\sqrt{x}) = \dfrac{1}{2\sqrt{x}}$

Problem 5. Differentiate from first principles $f(x) = \dfrac{1}{2x}$

$$f'(x) = \lim_{\delta x \to 0} \left\{ \frac{f(x + \delta x) - f(x)}{\delta x} \right\}$$

$$f(x) = \frac{1}{2x}$$

$$f(x + \delta x) = \frac{1}{2(x + \delta x)}$$

$$f(x + \delta x) - f(x) = \frac{1}{2(x + \delta x)} - \frac{1}{2x}$$

$$= \frac{x - (x + \delta x)}{2x(x + \delta x)}$$

$$= \frac{-\delta x}{2x(x + \delta x)}$$

$$\frac{f(x + \delta x) - f(x)}{\delta x} = \frac{-\delta x}{2x(x + \delta x)\delta x}$$

$$= \frac{-1}{2x(x + \delta x)}$$

As $\delta x \to 0$, $\dfrac{f(x + \delta x) - f(x)}{\delta x} \to \dfrac{-1}{2x(x + 0)}$

Therefore $f'(x) = \lim\limits_{\delta x \to 0} \left\{ \dfrac{f(x + \delta x) - f(x)}{\delta x} \right\} = \dfrac{-1}{2x(x)} = -\dfrac{1}{2x^2}$

Problem 6. Find the differential coefficient of $y = 5$

The differential coefficient of $y = 5$ may be deduced as follows:
If a graph is drawn of $y = 5$ a straight horizontal line results and
the gradient or slope of a horizontal line is zero. Finding the
differential coefficient is, in fact, finding the slope of a curve, or, as
in this case, of a horizontal straight line.

Hence $\dfrac{dy}{dx} = 0$

This may also be shown by the conventional method since:

$$\frac{dy}{dx} = f'(x) = \lim_{\delta x \to 0} \left\{ \frac{f(x + \delta x) - f(x)}{\delta x} \right\}$$

$y = f(x) = 5$

and $f(x + \delta x) = 5$

$$\frac{dy}{dx} = f'(x) = \lim_{\delta x \to 0} \left\{ \frac{5 - 5}{\delta x} \right\}$$

$$= \frac{0}{\delta x} = 0$$

More generally, if C is any constant, then

if $f(x) = C$, $f'(x) = 0$

i.e. **if $y = C$ then $\dfrac{dy}{dx} = 0$**

Problem 7. Differentiate from first principles $f(x) = 3x^2 + 6x - 3$ and find the gradient of the curve at $x = -2$

$$f'(x) = \lim_{\delta x \to 0} \left\{ \frac{f(x + \delta x) - f(x)}{\delta x} \right\}$$

$f(x) = 3x^2 + 6x - 3$

and $\quad f(x + \delta x) = 3(x + \delta x)^2 + 6(x + \delta x) - 3$

$$= 3(x^2 + 2x\delta x + \delta x^2) + 6x + 6\delta x - 3$$

$$= 3x^2 + 6x\delta x + 3\delta x^2 + 6x + 6\delta x - 3$$

$f(x + \delta x) - f(x) = (3x^2 + 6x\delta x + 3\delta x^2 + 6x + 6\delta x - 3)$

$$- (3x^2 + 6x - 3)$$

$$= 6x\delta x + 3\delta x^2 + 6\delta x$$

and $\quad \dfrac{f(x + \delta x) - f(x)}{\delta x} = \dfrac{6x\delta x + 3\delta x^2 + 6\delta x}{\delta x}$

$$= 6x + 3\delta x + 6$$

As $\delta x \to 0$, $\dfrac{f(x + \delta x) - f(x)}{\delta x} \to 6x + 3(0) + 6$

Therefore $f'(x) = \lim_{\delta x \to 0} \left\{ \dfrac{f(x + \delta x) - f(x)}{\delta x} \right\} = 6x + 6$

At $x = -2$ the gradient of the curve, i.e. $f'(x)$, is $6(-2) + 6$, i.e. -6
Hence if $f(x) = 3x^2 + 6x - 3$, $f'(x) = 6x + 6$ and the gradient of the curve at $x = -2$ is -6

A summary of the results obtained in the above problems is tabulated below:

y or $f(x)$	$\dfrac{dy}{dx}$ or $f'(x)$
x^2	$2x$
$3x^3$	$9x^2$
$3x$	3
$x^{\frac{1}{2}}$	$\frac{1}{2}x^{-\frac{1}{2}}$
$\dfrac{1}{2x}$	$-\dfrac{1}{2x^2}$
5	0
$3x^2 + 6x - 3$	$6x + 6$

Three basic rules of differentiation emerge from these results:

Rule 1. The differential coefficient of a constant is zero.

Rule 2. $\dfrac{d}{dx}(x^n) = nx^{n-1}$

For example $\dfrac{d}{dx}(x^3) = 3x^{3-1} = 3x^2$ (as in the table)

Rule 3. Constants associated with variables are carried forward.

For example $\dfrac{d}{dx}(3x^2) = 3\dfrac{d}{dx}(x^2)$

Problem 8. Differentiate from first principles $f(x) = \dfrac{1}{5x + 3}$

$f'(x) = \lim_{\delta x \to 0} \left\{ \dfrac{f(x + \delta x) - f(x)}{\delta x} \right\}$

$f(x) = \dfrac{1}{5x + 3}$

$f(x + \delta x) = \dfrac{1}{5(x + \delta x) + 3}$

$$f(x + \delta x) - f(x) = \frac{1}{5(x + \delta x) + 3} - \frac{1}{5x + 3}$$

$$= \frac{[5x + 3] - [5(x + \delta x) + 3]}{[5(x + \delta x) + 3][5x + 3]}$$

$$= \frac{5x + 3 - 5x - 5\delta x - 3}{[5(x + \delta x) + 3][5x + 3]}$$

$$= \frac{-5\delta x}{[5(x + \delta x) + 3][5x + 3]}$$

$$\frac{f(x + \delta x) - f(x)}{\delta x} = \frac{-5\delta x}{[5(x + \delta x) + 3][5x + 3]\delta x}$$

$$= \frac{-5}{[5(x + \delta x) + 3][5x + 3]}$$

As $\delta x \to 0$, $\dfrac{f(x + \delta x) - f(x)}{\delta x} \to \dfrac{-5}{[5(x + 0) + 3][5x + 3]}$

Therefore $f'(x) = \lim\limits_{\delta x \to 0} \left\{ \dfrac{f(x + \delta x) - f(x)}{\delta x} \right\} = \dfrac{-5}{(5x + 3)(5x + 3)}$

Hence if $f(x) = \dfrac{1}{5x + 3}$, $f'(x) = \dfrac{-5}{(5x + 3)^2}$

In the above worked problems the questions have been worded in a variety of ways. The important thing to realise is that they all mean the same thing. For example, in Worked Problem 8, on differentiating from first principles $f(x) = \dfrac{1}{5x + 3}$ gives $\dfrac{-5}{(5x + 3)^2}$
This result can be expressed in a number of ways.

1. If $f(x) = \dfrac{1}{5x + 3}$ then $f'(x) = \dfrac{-5}{(5x + 3)^2}$

2. If $y = \dfrac{1}{5x + 3}$ then $\dfrac{dy}{dx} = \dfrac{-5}{(5x + 3)^2}$

3. The differential coefficient of $\dfrac{1}{5x + 3}$ is $\dfrac{-5}{(5x + 3)^2}$

4. The derivative of $\dfrac{1}{5x + 3}$ is $\dfrac{-5}{(5x + 3)^2}$

5. $\dfrac{d}{dx}\left(\dfrac{1}{5x + 3}\right) = \dfrac{-5}{(5x + 3)^2}$

Further problems on differentiating from first principles may be found in the following Section 18.5 (Problems 6 to 34).

18.5 Further problems

Functional notation

1. If $f(x) = 2x^2 - x + 3$ find $f(0)$, $f(1)$, $f(2)$, $f(-1)$ and $f(-2)$
   [3, 4, 9, 6, 13]
2. If $f(x) = 6x^2 - 4x + 7$ find $f(1)$, $f(2)$, $f(-2)$ and
   $f(1) - f(-2)$ [9, 23, 39, -30]
3. If a curve is represented by $f(x) = 2x^3 + x^2 - x + 6$ prove that

   $$f(1) = \frac{1}{3} f(2)$$

4. If $f(x) = 3x^2 + 2x - 9$ find $f(3)$, $f(3+a)$ and $\dfrac{f(3+a) - f(3)}{a}$

   [24, $3a^2 + 20a + 24$, $3a + 20$]
5. If $f(x) = 4x^3 - 2x^2 - 3x + 1$ find $f(2)$, $f(-3)$ and
   $\dfrac{f(1+b) - f(1)}{b}$ [19, -116, $4b^2 + 10b + 5$]

Differentiation from first principles

6. Sketch the curve $f(x) = 5x^2 - 6$ for values of x from
   $x = -2$ to $x = +4$. Label the coordinate $(3.5, f(3.5))$ as A.
   Label the coordinate $(1.5, f(1.5))$ as B. Join points A and B
   to form the chord AB. Find the gradient of the chord AB.
   By moving A nearer and nearer to B find the gradient of the
   tangent of the curve at B.   [25, 15]

In Problems 7 to 27 differentiate from first principles:

7. $y = x$ [1]
8. $y = 5x$ [5]
9. $y = x^2$ [$2x$]
10. $y = 7x^2$ [$14x$]
11. $y = 4x^3$ [$12x^2$]
12. $y = 2x^2 - 3x + 2$ [$4x - 3$]

13. $y = 2\sqrt{x}$ $\left[\dfrac{1}{\sqrt{x}} \text{ or } x^{-\frac{1}{2}} \right]$

14. $y = \dfrac{1}{x}$ $\left[-\dfrac{1}{x^2} \right]$

15. $y = \dfrac{5}{6x^2}$ $\left[-\dfrac{5}{3x^3} \right]$

16. $y = 19$ $[0]$

17. $f(x) = 3x$ $[3]$

18. $f(x) = \dfrac{x}{4}$ $\left[\dfrac{1}{4} \right]$

19. $f(x) = 3x^2$ $[6x]$

20. $f(x) = 14x^3$ $[42x^2]$

21. $f(x) = x^2 + 16x - 4$ $[2x + 16]$

22. $f(x) = 4x^{\frac{1}{2}}$ $\left[2x^{-\frac{1}{2}} \text{ or } \dfrac{2}{\sqrt{x}} \right]$

23. $f(x) = \dfrac{16}{17x}$ $\left[-\dfrac{16}{17x^2} \right]$

24. $f(x) = \dfrac{1}{x^3}$ $\left[-\dfrac{3}{x^4} \right]$

25. $f(x) = 8$ $[0]$

26. $f(x) = \dfrac{1}{\sqrt{x}}$ $\left[-\dfrac{1}{2\sqrt{x^3}} \text{ or } -\dfrac{1}{2}x^{-3/2} \right]$

27. $f(x) = \dfrac{1}{3x - 2}$ $\left[\dfrac{-3}{(3x - 2)^2} \right]$

28. Find $\dfrac{d}{dx}(6x^3)$ $[18x^2]$

29. Find $\dfrac{d}{dx}(3\sqrt{x} + 6)$ $\left[\dfrac{3}{2\sqrt{x}} \right]$

30. Find $\dfrac{d}{dx}(2x^{-2} + 7x^2)$ $[-4x^{-3} + 14x]$

31. Find $\dfrac{d}{dx}\left(13 - \dfrac{3}{2x} \right)$ $\left[\dfrac{3}{2x^2} \right]$

32. If E, F and G are the points (1, 2), (2, 16) and (3, 54) respectively on the graph of $y = 2x^3$, find the gradients of the tangents at the points E, F and G and the gradient of the chord EG. $[6, 24, 54, 26]$

33. Differentiate from first principles $f(x) = 5x^2 - 6x + 2$ and find the gradient of the curve at $x = 2$ $[10x - 6, 14]$

34. If $y = \dfrac{7}{2}\sqrt{x} + \dfrac{3}{x^2} - 9$ find the differential coefficient of y with

respect to x $\left[\dfrac{7}{4\sqrt{x}} - \dfrac{6}{x^3} \right]$

Chapter 19

Methods of differentiation

19.1 Differential coefficients of some mathematical functions

(i) Differential coefficient of ax^n

In the worked problems of Chapter 18 the differential coefficients of certain algebraic functions of the form $y = ax^n$ are derived from first principles and the results are summarised on page 329. The rules stated on page 329 are best remembered by the single statement that

$$\text{when } y = ax^n, \frac{dy}{dx} = anx^{n-1}$$

(ii) Differential coefficient of sin x

A graph of $y = \sin x$ is shown in Fig. 19.1(a). The slope or gradient of the curve at any point is given by $\dfrac{dy}{dx}$ and is continually changing as values of x vary from O to S. By drawing tangents to the curve at many points on the curve and measuring the gradient of the tangents, values of $\dfrac{dy}{dx}$ may be obtained for corresponding values of x and these values are shown graphically in Fig. 19.1(b). The graph of $\dfrac{dy}{dx}$ against x so produced (called the derived curve) is a graph of $y = \cos x$. It follows that

$$\text{when } y = \sin x, \frac{dy}{dx} = \cos x$$

By applying the principles of differentiation by substitution (see Section 19.3), it may also be proved that

$$\text{when } y = \sin ax, \frac{dy}{dx} = a \cos ax$$

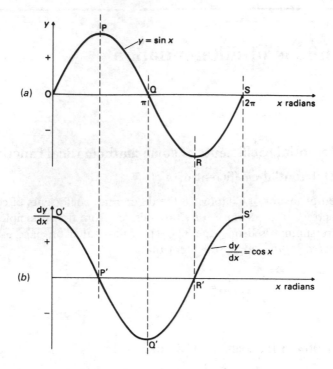

Fig. 19.1

[An alternative method of reasoning the shape of the derived curve of $y = \sin x$ is as follows. By examining the curve of $y = \sin x$ in Fig. 19.1(a), the following observations can be made:

(i)   at point O, the gradient is positive and at its steepest, giving a maximum positive value, shown by O' in Fig. 19.1(b),
(ii) between O and P, values of the gradient are positive and decreasing in value, as values of x approach P,
(iii) at point P, the tangent is a horizontal line, hence the gradient is zero, shown as P' in Fig. 19.1(b),
(iv) between P and Q, the gradient is negative and increasing in numerical value as x approaches point Q,
(v)   at point Q the gradient is negative and at its steepest, giving a maximum negative value, shown by Q' in Fig. 19.1(b).

Similarly, points R' and S' may be reasoned out for the negative half cycle of the curve $y = \sin x$.]

(iii) Differential coefficient of cos x

When graphs of $y = \cos x$ and its derived curve $\left(\dfrac{dy}{dx} \text{ against } x\right)$ are drawn in a similar way to those for $y = \sin x$ shown in (ii) above, the derived curve is a graph of $(-\sin x)$. Thus

$$\text{when } y = \cos x, \frac{dy}{dx} = -\sin x$$

By applying the principles of differentiation by substitution (see Section 19.3), it may be proved that

$$\text{when } y = \cos ax, \frac{dy}{dx} = -a \sin ax$$

(iv) Differential coefficient of e^x

A graph of $y = e^x$ is shown in Fig. 19.2(a). The slope or gradient of the curve at any point is given by $\dfrac{dy}{dx}$ and is continually changing. By drawing tangents

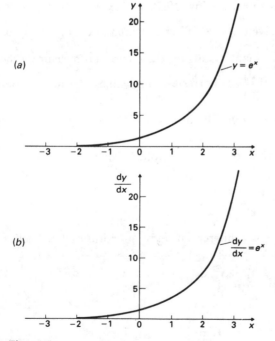

Fig. 19.2

to the curve at many points on the curve and measuring the gradient of the tangents, values of $\dfrac{dy}{dx}$ for corresponding values of x may be obtained. These values are shown graphically in Fig. 19.2(b). The graph of $\dfrac{dy}{dx}$ against x so produced is identical to the original graph of $y = e^x$. It follows that

$$\text{when } y = e^x, \frac{dy}{dx} = e^x$$

By applying the principles of differentiation by substitution (see Section 19.3), it may be proved that

$$\text{when } y = e^{ax}, \frac{dy}{dx} = a\,e^{ax}$$

This is as expected since by definition, the exponential function e^x, is a function whose rate of change is proportional to the original function. In the case of $y = e^x$, $a = 1$.

(v) Differential coefficient of ln x

A graph of $y = \ln x$ is shown in Fig. 19.3(a). The slope or gradient of the curve at any point is given by $\dfrac{dy}{dx}$ and is continually changing. By drawing tangents to the curve at many points on the curve and measuring the slope of the tangents, values of $\dfrac{dy}{dx}$ for corresponding values of x may be obtained. These values are shown graphically in Fig. 19.3(b). The graph of $\dfrac{dy}{dx}$ against x so produced is the graph of $\dfrac{dy}{dx} = \dfrac{1}{x}$. It follows that

$$\text{when } y = \ln x, \frac{dy}{dx} = \frac{1}{x}$$

By applying the principles of differentiation by substitution (see Section 19.3), it may be proved that

$$\text{when } y = \ln ax, \frac{dy}{dx} = \frac{1}{x}$$

$$\left(\text{note that when } y = \ln ax, \frac{dy}{dx} \neq \frac{1}{ax} \right)$$

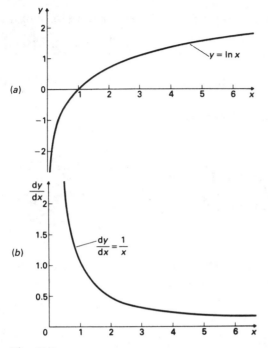

Fig. 19.3

Summary

y or $f(x)$	$\dfrac{dy}{dx}$ or $f'(x)$
(i) ax^n	anx^{n-1}
(ii) $\sin ax$	$a \cos ax$
(iii) $\cos ax$	$-a \sin ax$
(iv) e^{ax}	$a\,e^{ax}$
(v) $\ln ax$	$\dfrac{1}{x}$

For functions containing two or more terms added together or subtracted, the rules of algebra also apply to calculus. Thus, the differential coefficient of a sum or a difference is the differential coefficient of the terms added or subtracted (see Problem 2). Thus, when

$$f(x) = g(x) + h(x) - j(x)$$

then $f'(x) = g'(x) + h'(x) - j'(x)$

The differential coefficients obtained in the following worked problems are deduced using only the above summary.

Worked problems on differential coefficients of common functions

---

Problem 1. Find the differential coefficient of: (a) $5x^4$ (b) $\dfrac{3}{x^2}$ (c) $4\sqrt{x}$

When $f(x) = ax^n$, $f'(x) = anx^{n-1}$
(a) $f(x) = 5x^4$
$\qquad f'(x) = (5)(4)x^{4-1} = 20x^3$

(b) $f(x) = \dfrac{3}{x^2} = 3x^{-2}$
$\qquad f'(x) = (3)(-2)x^{-2-1} = -6x^{-3}$

i.e. $f'(x) = \dfrac{-6}{x^3}$

(c) $f(x) = 4\sqrt{x} = 4x^{\frac{1}{2}}$
$\qquad f'(x) = (4)(\frac{1}{2})x^{\frac{1}{2}-1} = 2x^{-\frac{1}{2}}$

i.e. $f'(x) = \dfrac{2}{\sqrt{x}}$

Problem 2. Differentiate $2x^3 + 7x + \dfrac{1}{3x^2} - \dfrac{4}{x^3} + \dfrac{4}{3}\sqrt{x^3} - 8$ with respect to x

$f(x) = 2x^3 + 7x + \dfrac{1}{3x^2} - \dfrac{4}{x^3} + \dfrac{4}{3}\sqrt{x^3} - 8$

$\qquad = 2x^3 + 7x + \dfrac{x^{-2}}{3} - 4x^{-3} + \dfrac{4}{3}x^{\frac{3}{2}} - 8$

$f'(x) = (2)(3)x^{3-1} + (7)x^{1-1} + \dfrac{(-2)}{3}x^{-2-1} - (4)(-3)x^{-3-1}$

$\qquad + \left(\dfrac{4}{3}\right)\left(\dfrac{3}{2}\right)x^{\frac{3}{2}-1} - 0$

$\qquad = 6x^2 + 7 - \dfrac{2}{3}x^{-3} + 12x^{-4} + 2x^{\frac{1}{2}}$

i.e. $f'(x) = 6x^2 + 7 - \dfrac{2}{3x^3} + \dfrac{12}{x^4} + 2\sqrt{x}$

Problem 3. (a) If $f(x) = 2\sin x$ find $f'(x)$

(b) If $y = 5\cos x$ find $\dfrac{dy}{dx}$

(a)   When $f(x) = \sin x$, $f'(x) = \cos x$
When $f(x) = 2 \sin x$
then $f'(x) = 2 \cos x$

(b)   When $y = \cos x$, $\dfrac{dy}{dx} = -\sin x$

When $y = 5 \cos x$

then $\dfrac{dy}{dx} = -5 \sin x$

Problem 4. Differentiate: (a) e^{6t} (b) $5e^{-3t}$ with respect to t

When $f(t) = e^{at}$, $f'(t) = a\, e^{at}$
(a) $f(t) = e^{6t}$
      $f'(t) = 6\, e^{6t}$
(b) $f(t) = 5e^{-3t}$
      $f'(t) = (5)(-3)e^{-3t} = -15\, e^{-3t}$

Problem 5. Find the differential coefficient of: (a) $\ln 4x$ (b) $3 \ln 2x$

When $f(x) = \ln ax$, $f'(x) = \dfrac{1}{x}$

(a) $f(x) = \ln 4x, f'(x) = \dfrac{1}{x}$

(b) $f(x) = 3 \ln 2x, f'(x) = \dfrac{3}{x}$

Problem 6. If $g = 3.2t^5 - 3 \sin t - 5e^{7t} + \sqrt[3]{t^4} + 6$, find $\dfrac{dg}{dt}$

$g = 3.2t^5 - 3 \sin t - 5e^{7t} + t^{\frac{4}{3}} + 6$

$\dfrac{dg}{dt} = (3.2)(5)t^4 - 3 \cos t - (5)(7)e^{7t} + \left(\dfrac{4}{3}\right)t^{\frac{1}{3}} + 0$

$= 16t^4 - 3 \cos t - 35e^{7t} + \dfrac{4}{3}t^{\frac{1}{3}}$

i.e. $\dfrac{dg}{dt} = 16t^4 - 3 \cos t - 35e^{7t} + \dfrac{4}{3}\sqrt[3]{t}$

Problem 7. $f(x) = 4 \ln(2.6x) - \dfrac{3}{\sqrt[3]{x^2}} + \dfrac{4}{e^{3x}} + \dfrac{1}{5} - 2 \cos x$. Find $f'(x)$

$$f(x) = 4 \ln(2.6x) - 3x^{-\frac{2}{3}} + 4e^{-3x} + \frac{1}{5} - 2 \cos x$$

$$f'(x) = \frac{4}{x} - (3)\left(-\frac{2}{3}\right)x^{-\frac{5}{3}} + (4)(-3)e^{-3x} + 0 - (-2 \sin x)$$

i.e. $f'(x) = \dfrac{4}{x} + \dfrac{2}{\sqrt[3]{x^5}} - \dfrac{12}{e^{3x}} + 2 \sin x$

Problem 8. Find the gradient of the curve $y = \dfrac{3}{2\sqrt{x}}$ at the point

$\left(4, \dfrac{3}{4}\right)$

$y = \dfrac{3}{2\sqrt{x}} = \dfrac{3}{2}x^{-\frac{1}{2}}$

Gradient $= \dfrac{dy}{dx} = \left(\dfrac{3}{2}\right)\left(-\dfrac{1}{2}\right)x^{-\frac{3}{2}} = -\dfrac{3}{4\sqrt{x^3}}$

When $x = 4$, gradient $= -\dfrac{3}{4\sqrt{4^3}} = -\dfrac{3}{4(8)}$

i.e. **Gradient** $= -\dfrac{3}{32}$

Problem 9. Find the coordinates of the point on the curve
$y = 3\sqrt[3]{x^2}$ where the gradient is 1

$y = 3\sqrt[3]{x^2} = 3x^{\frac{2}{3}}$

Gradient $= \dfrac{dy}{dx} = (3)(\frac{2}{3})x^{-\frac{1}{3}} = \dfrac{2}{\sqrt[3]{x}}$

If the gradient is equal to 1, then $1 = \dfrac{2}{\sqrt[3]{x}}$

i.e. $\sqrt[3]{x} = 2$

$x = 2^3 = 8$

When $x = 8$, $y = 3\sqrt[3]{8^2} = 3(4) = 12$

Hence the gradient is 1 at the point (8, 12)

Problem 10. If $f(x) = \dfrac{4x^3 - 8x^2 + 6x}{2x}$ find the coordinates of the
point at which the gradient is: (a) zero and (b) 4

$f(x) = \dfrac{4x^3 - 8x^2 + 6x}{2x} = \dfrac{4x^3}{2x} - \dfrac{8x^2}{2x} + \dfrac{6x}{2x}$

$= 2x^2 - 4x + 3$

The derivative, $f'(x)$, gives the gradient of the curve.
Hence $f'(x) = (2)(2)x^1 - 4$

$$= 4x - 4$$

(a) When $f'(x)$ is zero

$4x - 4 = 0$

i.e. $x = 1$

When $x = 1$, $y = \dfrac{4x^3 - 8x^2 + 6x}{2x} = \dfrac{4(1)^3 - 8(1)^2 + 6(1)}{2(1)}$

i.e. $y = 1$

Hence the coordinates of the point where the gradient is zero are **(1, 1)**

(b) When $f'(x)$ is 4

$4x - 4 = 4$

i.e. $x = 2$

When $x = 2$, $y = 2(2)^2 - 4(2) + 3 = 3$

Hence the coordinates of the point where the gradient is 4 are **(2, 3)**

Further problems on differential coefficients of common functions may be found in Section 19.5 (Problems 1 to 30), page 356.

19.2   Differentiation of products and quotients

(ii) Differentiation of a product

The function $y = 3x^2 \sin x$ is a product of two terms in x, i.e. $3x^2$ and $\sin x$.
Let $u = 3x^2$ and $v = \sin x$.

Let x increase by a small increment δx, causing incremental changes in u, v and y of δu, δv and δy respectively.

Then $y = (3x^2)(\sin x)$

$$= (u)(v)$$

$y + \delta y = (u + \delta u)(v + \delta v)$

$$= uv + v\delta u + u\delta v + \delta u\delta v$$

$(y + \delta y) - (y) = uv + v\delta u + u\delta v + \delta u\delta v - uv$

$$\delta y = v\delta u + u\delta v + \delta u\delta v$$

Dividing both sides by δx gives:

$$\frac{\delta y}{\delta x} = v\frac{\delta u}{\delta x} + u\frac{\delta v}{\delta x} + \frac{\delta u}{\delta x}\delta v$$

As $\delta x \to 0$ then $\delta u \to 0$, $\delta v \to 0$ and $\delta y \to 0$

However, the fact that δu and δx, for example, both approach zero does not mean that $\dfrac{\delta u}{\delta x}$ will approach zero.

Ratios of small quantities, such as $\dfrac{\delta u}{\delta x}, \dfrac{\delta v}{\delta x}$ or $\dfrac{\delta y}{\delta x}$ can be significant.

Consider two lines AB and AC meeting at A and whose intersecting angle (i.e. $\angle$ BAC) is any value.

If $AB = \delta y = 1$ unit, say, and $AC = \delta x = 2$ units, then the ratio

$$\frac{\delta y}{\delta x} = \frac{1}{2}$$

This ratio of $\frac{1}{2}$ is still the same whether the unit of δy and δx is in, say, kilometres or millimetres. No matter how small δy or δx is made, the ratio is still $\frac{1}{2}$. Thus when $\delta y \to 0$ and when $\delta x \to 0$, the ratio $\dfrac{\delta y}{\delta x}$ is still a significant value.

As $\delta x \to 0$, $\dfrac{\delta u}{\delta x} \to \dfrac{du}{dx}, \dfrac{\delta v}{\delta x} \to \dfrac{dv}{dx}, \dfrac{\delta y}{\delta x} \to \dfrac{dy}{dx}$ and $\delta v \to 0$

Hence $\dfrac{dy}{dx} = v\dfrac{du}{dx} + u\dfrac{dv}{dx}$

This is known as the **product rule**.

Summary

When $y = uv$ and u and v are functions of x, then
$$\frac{dy}{dx} = v\frac{du}{dx} + u\frac{dv}{dx}$$

Using functional notation: When $F(x) = f(x)g(x)$ then:

$$F'(x) = f(x)g'(x) + g(x)f'(x)$$

Applying the product rule to $y = 3x^2 \sin x$:

let $u = 3x^2$ and $v = \sin x$

Then $\dfrac{dy}{dx} = (\sin x)\dfrac{d}{dx}(3x^2) + (3x^2)\dfrac{d}{dx}(\sin x)$

$= (\sin x)(6x) + (3x^2)(\cos x)$

$= 6x \sin x + 3x^2 \cos x$

i.e. $\dfrac{dy}{dx} = 3x(2 \sin x + x \cos x)$

From the above it should be noted that the differential coefficient of a product **cannot** be obtained merely by differentiating each term and multiplying the two answers together. The above formula must be used whenever differentiating products.

(ii) Differentiation of a quotient

The function $y = \dfrac{3 \cos x}{5x^3}$ is a quotient of two terms in x, i.e. 3 cos x and $5x^3$.

Let $u = 3 \cos x$ and $v = 5x^3$.

Let x increase by a small increment δx causing incremental changes in u, v and y of δu, δv and δy respectively.

Then
$$y = \frac{3 \cos x}{5x^3}$$

$$= \frac{u}{v}$$

$$y + \delta y = \frac{u + \delta u}{v + \delta v}$$

$$(y + \delta y) - (y) = \frac{u + \delta u}{v + \delta v} - \frac{u}{v}$$

$$= \frac{uv + v\delta u - uv - u\delta v}{v(v + \delta v)}$$

i.e.
$$\delta y = \frac{v\delta u - u\delta v}{v^2 + v\delta v}$$

Dividing both sides by δx gives:

$$\frac{\delta y}{\delta x} = \frac{v \dfrac{\delta u}{\delta x} - u \dfrac{\delta v}{\delta x}}{v^2 + v\delta v}$$

As $\delta x \to 0$, $\dfrac{\delta u}{\delta x} \to \dfrac{du}{dx}$, $\dfrac{\delta v}{\delta x} \to \dfrac{dv}{dx}$, $\dfrac{\delta y}{\delta x} \to \dfrac{dy}{dx}$ and $\delta v \to 0$

Hence
$$\frac{dy}{dx} = \frac{v \dfrac{du}{dx} - u \dfrac{dv}{dx}}{v^2}$$

This is known as the **quotient rule**.

Summary

When $y = \dfrac{u}{v}$ and u and v are functions of x, then

$$\frac{dy}{dx} = \frac{v\dfrac{du}{dx} - u\dfrac{dv}{dx}}{v^2}$$

Using functional notation:

When $F(x) = \dfrac{f(x)}{g(x)}$, then $F'(x) = \dfrac{g(x)f'(x) - f(x)g'(x)}{[g(x)]^2}$

Applying the quotient rule to $y = \dfrac{3\cos x}{5x^3}$:

Let $u = 3\cos x$ and $v = 5x^3$

Then $\dfrac{dy}{dx} = \dfrac{(5x^3)\dfrac{d}{dx}(3\cos x) - (3\cos x)\dfrac{d}{dx}(5x^3)}{(5x^3)^2}$

$$= \frac{(5x^3)(-3\sin x) - (3\cos x)(15x^2)}{25x^6}$$

$$= \frac{-15x^2(x\sin x + 3\cos x)}{25x^6}$$

i.e. $\dfrac{dy}{dx} = \dfrac{-3(x\sin x + 3\cos x)}{5x^4}$

From above it should be noted that the differential coefficient of a quotient **cannot** be obtained by merely differentiating each term and dividing the numerator by the denominator. The above formula **must** be used when differentiating quotients.

The first step when differentiating a product such as $y = uv$ or a quotient such as $y = \dfrac{u}{v}$ is to decide clearly which is the u part and which is the v part. When this has been decided differentiation involves substitution into the appropriate formula.

Worked problems on differentiating products and quotients

Problem 1. Find the differential coefficient of $5x^2 \cos x$

Let $y = 5x^2 \cos x$

Also, let $u = 5x^2$ and $v = \cos x$

Then $\dfrac{du}{dx} = 10x$ and $\dfrac{dv}{dx} = -\sin x$

Then $\dfrac{dy}{dx} = v\dfrac{du}{dx} + u\dfrac{dv}{dx}$

$\qquad = (\cos x)(10x) + (5x^2)(-\sin x)$

$\qquad = 10x \cos x - 5x^2 \sin x$

i.e. $\dfrac{dy}{dx} = 5x(2 \cos x - x \sin x)$

Problem 2. Differentiate $3e^{2b} \sin b$ with respect to b

Let $F(b) = 3e^{2b} \sin b$

Let $f(b) = 3e^{2b}$ and $g(b) = \sin b$

then $f'(b) = 6e^{2b}$ and $g'(b) = \cos b$

Then $F'(b) = g(b)f'(b) + f(b)g'(b)$

$\qquad = (\sin b)(6e^{2b}) + (3e^{2b})(\cos b)$

$\qquad = 6e^{2b} \sin b + 3e^{2b} \cos b$

i.e. $F'(b) = 3e^{2b}[2 \sin b + \cos b]$

Problem 3. If $y = 7\sqrt{x} \ln 4x$ find $\dfrac{dy}{dx}$

$y = 7x^{\frac{1}{2}} \ln 4x$

Let $u = 7x^{\frac{1}{2}}$ and $v = \ln 4x$

then $\dfrac{du}{dx} = \dfrac{7}{2}x^{-\frac{1}{2}}$ and $\dfrac{dv}{dx} = \dfrac{1}{x}$

Then $\dfrac{dy}{dx} = v\dfrac{du}{dx} + u\dfrac{dv}{dx}$

$\qquad = (\ln 4x)\left[\dfrac{7}{2}x^{-\frac{1}{2}}\right] + [7x^{\frac{1}{2}}]\left[\dfrac{1}{x}\right]$

$\qquad = \dfrac{7}{2\sqrt{x}} \ln 4x + \dfrac{7}{\sqrt{x}}$

i.e. $\dfrac{dy}{dx} = \dfrac{7}{2\sqrt{x}}(\ln 4x + 2)$

Problem 4. Find the differential coefficients of: (a) tan x (b) cot x (c) sec x (d) cosec x

(a) Let $y = \tan x = \dfrac{\sin x}{\cos x}$

Differentiation of tan x is treated as a quotient with $u = \sin x$ and $v = \cos x$.

Then $\dfrac{du}{dx} = \cos x$ and $\dfrac{dv}{dx} = -\sin x$

$$\frac{dy}{dx} = \frac{v\dfrac{du}{dx} - u\dfrac{dv}{dx}}{v^2}$$

$$= \frac{(\cos x)(\cos x) - (\sin x)(-\sin x)}{(\cos x)^2}$$

$$= \frac{(\cos^2 x + \sin^2 x)}{(\cos x)^2}$$

$$= \frac{1}{\cos^2 x} \quad (\text{since } \cos^2 x + \sin^2 x = 1)$$

i.e. $\dfrac{dy}{dx} = \sec^2 x$

Hence, when $y = \tan x$, $\dfrac{dy}{dx} = \sec^2 x$

or, when $f(x) = \tan x$, $f'(x) = \sec^2 x$

(b) Let $y = \cot x = \dfrac{\cos x}{\sin x}$

Differentiation of cot x is treated as a quotient with $u = \cos x$ and $v = \sin x$.

Then $\dfrac{du}{dx} = -\sin x$ and $\dfrac{dv}{dx} = \cos x$

$$\frac{dy}{dx} = \frac{v\dfrac{du}{dx} - u\dfrac{dv}{dx}}{v^2}$$

$$= \frac{(\sin x)(-\sin x) - (\cos x)(\cos x)}{(\sin x)^2}$$

$$= \frac{-(\sin^2 x + \cos^2 x)}{\sin^2 x}$$

$$= \frac{-1}{\sin^2 x}$$

i.e. $\dfrac{dy}{dx} = -\operatorname{cosec}^2 x$

Hence, when $y = \cot x$, $\dfrac{dy}{dx} = -\operatorname{cosec}^2 x$

or, when $f(x) = \cot x$, $f'(x) = -\operatorname{cosec}^2 x$

(c) Let $y = \sec x = \dfrac{1}{\cos x}$

Differentiation of $\sec x$ is treated as a quotient with $u = 1$ and $v = \cos x$.

Then $\dfrac{du}{dx} = 0$ and $\dfrac{dv}{dx} = -\sin x$

$$\dfrac{dy}{dx} = \dfrac{v\dfrac{du}{dx} - u\dfrac{dv}{dx}}{v^2}$$

$$= \dfrac{(\cos x)(0) - (1)(-\sin x)}{(\cos x)^2}$$

$$= \dfrac{\sin x}{\cos^2 x}$$

$$= \left[\dfrac{1}{\cos x}\right]\left[\dfrac{\sin x}{\cos x}\right]$$

i.e. $\dfrac{dy}{dx} = \sec x \tan x$

Hence, when $y = \sec x$, $\dfrac{dy}{dx} = \sec x \tan x$

or, when $f(x) = \sec x$, $f'(x) = \sec x \tan x$

(d) Let $y = \operatorname{cosec} x = \dfrac{1}{\sin x}$

Differentiation of $\operatorname{cosec} x$ is treated as a quotient with $u = 1$ and $v = \sin x$.

Then $\dfrac{du}{dx} = 0$ and $\dfrac{dv}{dx} = \cos x$

$$\dfrac{dy}{dx} = \dfrac{v\dfrac{du}{dx} - u\dfrac{dv}{dx}}{v^2}$$

$$= \dfrac{(\sin x)(0) - (1)(\cos x)}{(\sin x)^2}$$

$$= \frac{-\cos x}{\sin^2 x}$$

$$= -\left[\frac{1}{\sin x}\right]\left[\frac{\cos x}{\sin x}\right]$$

i.e. $\dfrac{dy}{dx} = -\operatorname{cosec} x \cot x$

Hence, when $y = \operatorname{cosec} x,\ \dfrac{dy}{dx} = -\operatorname{cosec} x \cot x$

or, when $f(x) = \operatorname{cosec} x,\ f'(x) = -\operatorname{cosec} x \cot x$

The differential coefficients of the six trigonometrical ratios may thus be summarised as below:

y or $f(x)$	$\dfrac{dy}{dx}$ or $f'(x)$
1. $\sin x$	$\cos x$
2. $\cos x$	$-\sin x$
3. $\tan x$	$\sec^2 x$
4. $\sec x$	$\sec x \tan x$
5. $\operatorname{cosec} x$	$-\operatorname{cosec} x \cot x$
6. $\cot x$	$-\operatorname{cosec}^2 x$

Problem 5. If $f(t) = \dfrac{4e^{7t}}{\sqrt[3]{t^2}}$ find $f'(t)$

$$f(t) = \frac{4e^{7t}}{t^{\frac{2}{3}}}$$

Let $g(t) = 4e^{7t}$ and $h(t) = t^{\frac{2}{3}}$

then $g'(t) = 28e^{7t}$ and $h'(t) = \frac{2}{3}t^{-\frac{1}{3}}$

$$f'(t) = \frac{h(t)g'(t) - g(t)h'(t)}{[h(t)]^2} = \frac{(t^{\frac{2}{3}})(28e^{7t}) - (4e^{7t})(\frac{2}{3}t^{-\frac{1}{3}})}{(t^{\frac{2}{3}})^2}$$

$$= \frac{28t^{\frac{2}{3}}e^{7t} - \frac{8}{3}t^{-\frac{1}{3}}e^{7t}}{t^{\frac{4}{3}}} = \frac{28t^{\frac{2}{3}}e^{7t}}{t^{\frac{4}{3}}} - \frac{8t^{-\frac{1}{3}}e^{7t}}{3t^{\frac{4}{3}}}$$

$$= 28t^{-\frac{2}{3}}e^{7t} - \frac{8}{3}t^{-\frac{5}{3}}e^{7t}$$

$$= \frac{4}{3}e^{7t}t^{-\frac{5}{3}}(21t - 2)$$

i.e. $f'(t) = \dfrac{4e^{7t}}{3\sqrt[3]{t^5}}(21t - 2)$

(Note that initially, $f(t) = \dfrac{4e^{7t}}{t^{\frac{2}{3}}}$ could have been treated as a

product $f(t) = 4e^{7t}t^{-\frac{2}{3}}$)

Problem 6. Find the coordinates of the points on the curve

$y = \dfrac{\frac{1}{3}(5 - 6x)}{3x^2 + 2}$ where the gradient is zero.

$y = \dfrac{\frac{1}{3}(5 - 6x)}{3x^2 + 2}$

Let $u = \frac{1}{3}(5 - 6x)$ and $v = 3x^2 + 2$

then $\dfrac{du}{dx} = -2$ and $\dfrac{dv}{dx} = 6x$

$$\dfrac{dy}{dx} = \dfrac{v\dfrac{du}{dx} - u\dfrac{dv}{dx}}{v^2} = \dfrac{(3x^2 + 2)(-2) - \frac{1}{3}(5 - 6x)(6x)}{(3x^2 + 2)^2}$$

$$= \dfrac{-6x^2 - 4 - 10x + 12x^2}{(3x^2 + 2)^2} = \dfrac{6x^2 - 10x - 4}{(3x^2 + 2)^2}$$

When the gradient is zero, $\dfrac{dy}{dx} = 0$

Hence $6x^2 - 10x - 4 = 0$

$\qquad\qquad 2(3x^2 - 5x - 2) = 0$

$\qquad\qquad 2(3x + 1)(x - 2) = 0$

i.e. $x = -\frac{1}{3}$ or $x = 2$

Substituting in the original equation for y:

When $x = -\frac{1}{3}$, $y = \dfrac{\frac{1}{3}[5 - 6(-\frac{1}{3})]}{3(-\frac{1}{3})^2 + 2} = \dfrac{\frac{7}{3}}{\frac{7}{3}} = 1$

When $x = 2$, $y = \dfrac{\frac{1}{3}[5 - 6(2)]}{3(2)^2 + 2} = \dfrac{-\frac{7}{3}}{14} = -\frac{1}{6}$

Hence the coordinates of the points on the curve $y = \dfrac{\frac{1}{3}(5 - 6x)}{3x^2 + 2}$

where the gradient is zero are $(-\frac{1}{3}, 1)$ and $(2, -\frac{1}{6})$

Problem 7. Differentiate $\dfrac{\sqrt{x} \sin x}{2e^{4x}}$ with respect to x

The function $\dfrac{\sqrt{x} \sin x}{2e^{4x}}$ is a quotient, although the numerator (i.e.

$\sqrt{x} \sin x$) is a product.

Let $\qquad y = \dfrac{x^{\frac{1}{2}} \sin x}{2e^{4x}}$

Let $\qquad u = x^{\frac{1}{2}} \sin x$ and $v = 2e^{4x}$

then $\qquad \dfrac{du}{dx} = (x^{\frac{1}{2}})(\cos x) + (\sin x)(\tfrac{1}{2}x^{-\frac{1}{2}})$

and $\qquad \dfrac{dv}{dx} = 8e^{4x}$

$$\dfrac{dy}{dx} = \dfrac{v\dfrac{du}{dx} - u\dfrac{dv}{dx}}{v^2}$$

$$= \dfrac{(2e^{4x})(x^{\frac{1}{2}}\cos x + \tfrac{1}{2}x^{-\frac{1}{2}}\sin x) - (x^{\frac{1}{2}}\sin x)(8e^{4x})}{(2e^{4x})^2}$$

Dividing throughout by $2e^{4x}$ gives:

$$\dfrac{dy}{dx} = \dfrac{x^{\frac{1}{2}}\cos x + \tfrac{1}{2}x^{-\frac{1}{2}}\sin x - 4x^{\frac{1}{2}}\sin x}{2e^{4x}}$$

Hence $\qquad \dfrac{dy}{dx} = \dfrac{\sqrt{x}\cos x + \sin x\left(\dfrac{1}{2\sqrt{x}} - 4\sqrt{x}\right)}{2e^{4x}}$

or $\qquad \dfrac{dy}{dx} = \dfrac{\sqrt{x}\cos x + \left(\dfrac{1 - 8x}{2\sqrt{x}}\right)\sin x}{2e^{4x}}$

Further problems on differentiating products and quotients may be found in Section 19.5 (Problems 31 to 64), page 358.

19.3   Differentiation by substitution

The function $y = (4x - 3)^7$ can be differentiated by first multiplying $(4x - 3)$ by itself seven times, and then differentiating each term produced in turn. This would be a long process. In this type of function a substitution is made.

Let $u = 4x - 3$, then instead of $y = (4x - 3)^7$ we have $y = u^7$

An important rule that is used when differentiating by substitution is:

$$\boxed{\dfrac{dy}{dx} = \dfrac{dy}{du} \cdot \dfrac{du}{dx}}$$

This is often known as the **chain rule**.

From above, $y = (4x - 3)^7$

If $u = 4x - 3$ then $y = u^7$

Thus $\dfrac{dy}{dx} = 7u^6$ and $\dfrac{du}{dx} = 4$

Hence since $\dfrac{dy}{dx} = \dfrac{dy}{du} \cdot \dfrac{du}{dx}$

$$\dfrac{dy}{dx} = (7u^6)(4) = 28u^6$$

Rewriting $u = 4x - 3$, $\dfrac{dy}{dx} = \mathbf{28(4x - 3)^6}$

Since y is a function of u, and u is a function of x, then y is a 'function of a function' of x. The method of obtaining differential coefficients by making substitutions is often called the **'function of a function process'**.

Worked problems on differentiation by substitution

Problem 1. Differentiate $\sin(6x + 1)$

Let $\quad y = \sin(6x + 1)$
and $\quad u = 6x + 1$

Then $\quad y = \sin u$, giving $\dfrac{dy}{du} = \cos u$

and $\quad \dfrac{du}{dx} = 6$

Using the 'differentiation by substitution' formula: $\dfrac{dy}{dx} = \dfrac{dy}{du} \cdot \dfrac{du}{dx}$

gives $\dfrac{dy}{dx} = (\cos u)(6) = 6 \cos u$

Rewriting $u = 6x + 1$ gives:

$$\dfrac{dy}{dx} = \mathbf{6 \cos(6x + 1)}$$

Note that this result could have been obtained by first differentiating the trigonometric function (i.e. differentiating $\sin f(x)$) giving $\cos f(x)$ and then multiplying by the differential coefficient of $f(x)$, i.e. 6

Problem 2. Find the differential coefficient of $(3t^4 - 2t)^5$

Let $\quad y = (3t^4 - 2t)^5$
and $\quad u = 3t^4 - 2t$

Then $\quad y = u^5$, giving $\dfrac{dy}{du} = 5u^4$

and $\dfrac{du}{dt} = 12t^3 - 2$

Using the 'chain rule': $\dfrac{dy}{dt} = \dfrac{dy}{du} \cdot \dfrac{du}{dt}$ gives $\dfrac{dy}{dt} = (5u^4)(12t^3 - 2)$

Rewriting $u = 3t^4 - 2t$ gives:

$\dfrac{dy}{dt} = 5(3t^4 - 2t)^4(12t^3 - 2)$

Note that this result could have been obtained by first differentiating the bracket, giving $5[f(x)]^4$ and then multiplying this result by the differential coefficient of $f(x)$ (i.e. $(12t^3 - 2)$)

Problem 3. If $y = 5 \operatorname{cosec}(3\sqrt{x} + 2x)$ find $\dfrac{dy}{dx}$

$y = 5 \operatorname{cosec}(3\sqrt{x} + 2x)$

Let $\quad u = (3\sqrt{x} + 2x)$ then $\dfrac{du}{dx} = \dfrac{3}{2\sqrt{x}} + 2$

Thus $\quad y = 5 \operatorname{cosec} u$ and $\dfrac{dy}{du} = -5 \operatorname{cosec} u \cot u$

Now $\dfrac{dy}{dx} = \dfrac{dy}{du} \cdot \dfrac{du}{dx} = (-5 \operatorname{cosec} u \cot u)\left(\dfrac{3}{2\sqrt{x}} + 2\right)$

Rewriting $u = 3\sqrt{x} + 2x$ gives:

$\dfrac{dy}{dx} = -5\left(\dfrac{3}{2\sqrt{x}} + 2\right) \operatorname{cosec}(3\sqrt{x} + 2x) \cot(3\sqrt{x} + 2x)$

In a similar way to Problem 1, this result could have been obtained by first differentiating $5 \operatorname{cosec} f(x)$ giving $-5 \operatorname{cosec} f(x) \cot f(x)$ and then multiplying this result by the differential coefficient of $f(x)$

Problem 4. If $p = 2 \tan^5 v$ find $\dfrac{dp}{dv}$

$p = 2 \tan^5 v$

Let $\quad u = \tan v$ then $\dfrac{du}{dv} = \sec^2 v$

Then $\quad p = 2u^5$ and $\dfrac{dp}{du} = 10u^4$

Now $\dfrac{dp}{dv} = \dfrac{dp}{du} \cdot \dfrac{du}{dv} = (10u^4)(\sec^2 v)$

Rewriting $u = \tan v$ gives:

$$\frac{\mathrm{d}p}{\mathrm{d}v} = 10(\tan v)^4 \sec^2 v$$

$$\frac{\mathrm{d}p}{\mathrm{d}v} = \mathbf{10\ tan^4\ v\ sec^2\ v}$$

In a similar way to Problem 2, this result could have been obtained by firstly differentiating the bracket (i.e. differentiating $2[f(v)]^5$) giving $10[f(v)]^4$ and then multiplying this result by the differential coefficient of $f(v)$

Problem 5. Write down the differential coefficients of the following:
(a) $\sqrt{(4x^2 + x - 3)}$ (b) $2\sec^3 t$ (c) $4\cot(5g^2 + 2)$
(d) $\sqrt{(4x^3 + 2)^3}\cos(3x^2 + 2)$

(a)　$f(x) = \sqrt{(4x^2 + x - 3)} = (4x^2 + x - 3)^{\frac{1}{2}}$

　　$f'(x) = \frac{1}{2}(4x^2 + x - 3)^{-\frac{1}{2}}(8x + 1)$

　　$= \dfrac{8x + 1}{2\sqrt{(4x^2 + x - 3)}}$

(b)　$f(t) = 2\sec^3 t = 2(\sec t)^3$

　　$f'(t) = 6(\sec t)^2(\sec t \tan t)$

　　$= \mathbf{6\ sec^3\ t\ tan\ t}$

(c)　$f(g) = 4\cot(5g^2 + 2)$
　　$f'(g) = 4[-\operatorname{cosec}^2(5g^2 + 2)](10g)$

　　$= \mathbf{-40g\ cosec^2(5g^2 + 2)}$

(d)　$f(x) = \sqrt{(4x^3 + 2)^3}\cos(3x^2 + 2)$

　　$= (4x^3 + 2)^{\frac{3}{2}}\cos(3x^2 + 2)$　(i.e. a product)

　　$f'(x) = [\cos(3x^2 + 2)][\frac{3}{2}(4x^3 + 2)^{\frac{1}{2}}(12x^2)]$

　　$\quad + [(4x^3 + 2)^{\frac{3}{2}}][(-\sin(3x^2 + 2))(6x)]$

　　$= \mathbf{6x\sqrt{(4x^3 + 2)}[3x\ cos(3x^2 + 2) - (4x^3 + 2)\ sin(3x^2 + 2)]}$

Further problems on differentiation by substitution may be found in Section 19.5 (Problems 65 to 120), page 360.

19.4 Successive differentiation

When a function, say, $y = f(x)$, is differentiated, the differential coefficient is written as $f'(x)$ or $\dfrac{\mathrm{d}y}{\mathrm{d}x}$

If the expression is differentiated again, the second differential coefficient or the second derivative is obtained. This is written as $f''(x)$ (pronounced 'f double-dash x') or $\dfrac{d^2y}{dx^2}$ (pronounced 'dee two y by dee x squared'). Similarly, if differentiated again the third differential coefficient or third derivative is obtained, and is written as $f'''(x)$ or $\dfrac{d^3y}{dx^3}$, and so on.

Worked problems on successive differentiation

Problem 1. If $f(x) = 3x^4 + 2x^3 + x - 1$, find $f'(x)$ and $f''(x)$

$$f(x) = 3x^4 + 2x^3 + x - 1$$
$$f'(x) = (3)(4)x^3 + (2)(3)x^2 + 1 - 0$$
$$= 12x^3 + 6x^2 + 1$$
$$f''(x) = (12)(3)x^2 + (6)(2)x + 0$$
$$= 36x^2 + 12x$$

Problem 2. If $y = \dfrac{4}{3}x^3 - \dfrac{2}{x^2} + \dfrac{1}{3x} - \sqrt{x}$ find $\dfrac{d^2y}{dx^2}$ and $\dfrac{d^3y}{dx^3}$

$$y = \frac{4}{3}x^3 - \frac{2}{x^2} + \frac{1}{3x} - \sqrt{x}$$

$$= \frac{4}{3}x^3 - 2x^{-2} + \frac{1}{3}x^{-1} - x^{\frac{1}{2}}$$

$$\frac{dy}{dx} = [\tfrac{4}{3}](3)x^2 - (2)(-2)x^{-3} + \tfrac{1}{3}(-1)x^{-2} - [\tfrac{1}{2}]x^{-\frac{1}{2}}$$

$$= 4x^2 + 4x^{-3} - \tfrac{1}{3}x^{-2} - \tfrac{1}{2}x^{-\frac{1}{2}}$$

$$\frac{d^2y}{dx^2} = (4)(2)x + (4)(-3)x^{-4} - [\tfrac{1}{3}](-2)x^{-3} - [\tfrac{1}{2}][-\tfrac{1}{2}]x^{-\frac{3}{2}}$$

$$= 8x - 12x^{-4} + \tfrac{2}{3}x^{-3} + \tfrac{1}{4}x^{-\frac{3}{2}}$$

i.e. $\dfrac{d^2y}{dx^2} = 8x - \dfrac{12}{x^4} + \dfrac{2}{3x^3} + \dfrac{1}{4\sqrt{x^3}}$

$$\frac{d^3y}{dx^3} = 8 - (12)(-4)x^{-5} + [\tfrac{2}{3}](-3)x^{-4} + [\tfrac{1}{4}][-\tfrac{3}{2}]x^{-\frac{5}{2}}$$

$$= 8 + 48x^{-5} - 2x^{-4} - \tfrac{3}{8}x^{-\frac{5}{2}}$$

i.e. $\dfrac{d^3y}{dx^3} = 8 + \dfrac{48}{x^5} - \dfrac{2}{x^4} - \dfrac{3}{8\sqrt{x^5}}$

Problem 3. Evaluate $f'(t)$ and $f''(t)$, correct to 3 decimal places when $t = \dfrac{1}{2}$ given $f(t) = 3 \ln \cos 2t$

$$f(t) = 3 \ln \cos 2t$$

$$f'(t) = 3\left(\frac{1}{\cos 2t}\right)(-2 \sin 2t)$$

$$= -6 \tan 2t$$

When $\quad t = \dfrac{1}{2}, \; f'(t) = -6 \tan 1 = -6(1.557\,4) = \mathbf{-9.344}$

$$f''(t) = -6(\sec^2 2t)2$$

$$= -12 \sec^2 2t$$

When $\quad t = \dfrac{1}{2}, \; f''(t) = -12(3.425\,5) = \mathbf{-41.106}$

Problem 4. If $y = A\,e^{2x} + B\,e^{-3x}$ prove that $\dfrac{d^2 y}{dx^2} + \dfrac{dy}{dx} - 6y = 0$

$$y = A\,e^{2x} + B\,e^{-3x}$$

$$\frac{dy}{dx} = 2A\,e^{2x} - 3B\,e^{-3x}$$

$$\frac{d^2 y}{dx^2} = 4A\,e^{2x} + 9B\,e^{-3x}$$

$$6y = 6(A\,e^{2x} + B\,e^{-3x}) = 6A\,e^{2x} + 6B\,e^{-3x}$$

Substituting into $\dfrac{d^2 y}{dx^2} + \dfrac{dy}{dx} - 6y$ gives:

$$(4A\,e^{2x} + 9B\,e^{-3x}) + (2A\,e^{2x} - 3B\,e^{-3x}) - (6A\,e^{2x} + 6B\,e^{-3x})$$

$$= 4A\,e^{2x} + 9B\,e^{-3x} + 2A\,e^{2x} - 3B\,e^{-3x} - 6A\,e^{2x} - 6B\,e^{-3x} = 0$$

Thus $\dfrac{d^2 y}{dx^2} + \dfrac{dy}{dx} - 6y = 0$

(Note that an equation of the form $\dfrac{d^2 y}{dx^2} + \dfrac{dy}{dx} - 6y = 0$ is known as a 'differential equation' and such equations are discussed in Chapters 33 and 34.)

Further problems on successive differentiation may be found in the following Section 19.5 (Problems 121 to 143), page 362.

19.5   Further problems

Differentiation of common functions

Find the differential coefficients with respect to x of the functions in Problems 1 to 6.

1. (a) x^4 (b) x^6 (c) x^9 (d) $x^{3.2}$ (e) $x^{4.7}$
    (a) $[4x^3]$ (b) $[6x^5]$ (c) $[9x^8]$ (d) $[3.2x^{2.2}]$ (e) $[4.7x^{3.7}]$

2. (a) $3x^3$ (b) $4x^7$ (c) $2x^{10}$ (d) $4.6x^{1.5}$ (e) $6x^{5.4}$
    (a) $[9x^2]$ (b) $[28x^6]$ (c) $[20x^9]$ (d) $[6.9x^{0.5}]$
    (e) $[32.4x^{4.4}]$

3. (a) x^{-2} (b) x^{-3} (c) x^{-5} (d) $\dfrac{1}{x}$ (e) $-\dfrac{1}{x^3}$ (f) $\dfrac{1}{x^{10}}$

    (a) $[-2x^{-3}]$ (b) $[-3x^{-4}]$ (c) $[-5x^{-6}]$ (d) $\left[-\dfrac{1}{x^2}\right]$

    (e) $\left[\dfrac{3}{x^4}\right]$ (f) $\left[-\dfrac{10}{x^{11}}\right]$

4. (a) $4x^{-1}$ (b) $-5x^{-4}$ (c) $3x^{-7}$ (d) $-\dfrac{6}{x^2}$ (e) $\dfrac{4}{3x^5}$ (f) $\dfrac{2}{5x^{1.4}}$

    (a) $[-4x^{-2}]$ (b) $[20x^{-5}]$ (c) $[-21x^{-8}]$ (d) $\left[\dfrac{12}{x^3}\right]$

    (e) $\left[-\dfrac{20}{3x^6}\right]$ (f) $\left[\dfrac{-2.8}{5x^{2.4}}\right]$

5. (a) $x^{\frac{7}{2}}$ (b) $x^{\frac{3}{4}}$ (c) $x^{-\frac{3}{2}}$ (d) $\dfrac{1}{x^{\frac{1}{2}}}$ (e) $-\dfrac{1}{x^{\frac{4}{3}}}$ (f) $\dfrac{2}{3x^{\frac{7}{4}}}$

    (a) $[\frac{7}{2}x^{\frac{5}{2}}]$ (b) $[\frac{3}{4}x^{-\frac{1}{4}}]$ (c) $[-\frac{3}{2}x^{-\frac{5}{2}}]$ (d) $\left[\dfrac{-1}{2x^{\frac{3}{2}}}\right]$ (e) $\left[\dfrac{4}{3x^{\frac{7}{3}}}\right]$

    (f) $\left[\dfrac{-7}{6x^{\frac{11}{4}}}\right]$

6. (a) $\dfrac{\sqrt{x}}{2}$ (b) $\sqrt{x^3}$ (c) $\sqrt[3]{x^2}$ (d) $4\sqrt{x^5}$ (e) $\dfrac{3}{5\sqrt{x^7}}$ (f) $\dfrac{-1}{2\sqrt[4]{x^9}}$

    (a) $\left[\dfrac{1}{4\sqrt{x}}\right]$ (b) $\left[\dfrac{3}{2}\sqrt{x}\right]$ (c) $\left[\dfrac{2}{3\sqrt[3]{x}}\right]$ (d) $[10\sqrt{x^3}]$

    (e) $\left[\dfrac{-21}{10\sqrt{x^9}}\right]$ (f) $\left[\dfrac{9}{8\sqrt[4]{x^{13}}}\right]$

Differentiate the functions in Problems 7 to 26 with respect to the variable:

7. (a) $4u^3$ (b) $\frac{3}{2}t^4$ (a) $[12u^2]$ (b) $[6t^3]$
8. (a) $5v^2$ (b) $1.4z^5$ (a) $[10v]$ (b) $[7z^4]$

9. (a) $\dfrac{4}{a}$ (b) $\dfrac{3}{2S^2}$ (a) $\left[-\dfrac{4}{a^2}\right]$ (b) $\left[-\dfrac{3}{S^3}\right]$

10. (a) $\dfrac{7}{4y^3}$ (b) $3m^{-4}$ (a) $\left[-\dfrac{21}{4y^4}\right]$ (b) $[-12m^{-5}]$

11. (a) $\sqrt{b}$ (b) $5\sqrt{c^3}$ (a) $\left[\dfrac{1}{2\sqrt{b}}\right]$ (b) $\left[\dfrac{15}{2}\sqrt{c}\right]$

12. (a) $\dfrac{1}{\sqrt{e}}$ (b) $g^{\frac{5}{3}}$ (a) $\left[-\dfrac{1}{2\sqrt{e^3}}\right]$ (b) $\left[\frac{5}{3}g^{\frac{2}{3}}\right]$

13. (a) $4\sqrt[3]{k^2}$ (b) $\dfrac{3}{5\sqrt[4]{x^5}}$ (a) $\left[\dfrac{8}{3\sqrt[3]{k}}\right]$ (b) $\left[\dfrac{-3}{4\sqrt[4]{x^9}}\right]$

14. $5x^2 - \dfrac{1}{\sqrt{x^7}}$ $\left[10x + \dfrac{7}{2\sqrt{x^9}}\right]$

15. $3\left(2u - u^{-\frac{1}{2}} + \dfrac{4}{5u}\right)$ $\left[3\left(2 + \dfrac{u^{-\frac{3}{2}}}{2} - \dfrac{4}{5u^2}\right)\right]$

16. $\dfrac{1}{x}\left(3x^3 - \dfrac{2}{x} + \dfrac{\sqrt{x}}{5} + 1\right)$ $\left[6x + \dfrac{4}{x^3} - \dfrac{1}{10\sqrt{x^3}} - \dfrac{1}{x^2}\right]$

17. $\dfrac{3x^2 - 2\sqrt{x} - 5\sqrt[4]{x^3}}{x^2}$ $\left[\dfrac{3}{\sqrt{x^5}} + \dfrac{25}{4\sqrt[4]{x^9}}\right]$

18. $(t + 1)^2$ $[2(t + 1)]$
19. $(3\theta - 1)^2$ $[6(3\theta - 1)]$
20. $(f - 1)^4$ $[4(f^3 - 3f^2 + 3f - 1)$ or $4(f - 1)^3]$
21. (a) $5\sin\theta$ (b) $4\cos x$ (a) $[5\cos\theta]$ (b) $[-4\sin x]$
22. (a) $3(\sin t + 2\cos t)$ (b) $7\sin x - 2\cos x$
   (a) $[3(\cos t - 2\sin t)]$ (b) $[7\cos x + 2\sin x]$
23. (a) e^{3x} (b) e^{-4y} (a) $[3e^{3x}]$ (b) $[-4e^{-4y}]$
24. (a) $6e^{2x}$ (b) $\dfrac{4}{e^{7t}}$ (a) $[12e^{2x}]$ (b) $\left[\dfrac{-28}{e^{7t}}\right]$
25. (a) $3(e^{8y} - e^{3y})$ (b) $-2(3e^{9x} - 4e^{-2x})$
   (a) $[3(8e^{8y} - 3e^{3y})]$ (b) $[-2(27e^{9x} + 8e^{-2x})]$
26. (a) $\ln 5b$ (b) $4\ln 3g$ (a) $\left[\dfrac{1}{b}\right]$ (b) $\left[\dfrac{4}{g}\right]$
27. Find the gradient of the curve $y = 4x^3 - 3x^2 + 2x - 4$ at the
   points $(0, -4)$ and $(1, -1)$ $[2, 8]$
28. What are the coordinates of the point on the graph of
   $y = 5x^2 - 2x + 1$ where the gradient is zero. $[(\frac{1}{5}, \frac{4}{5})]$
29. Find the point on the curve $f(\theta) = 4\sqrt[3]{\theta^4} + 2$ where the
   gradient is $10\frac{2}{3}$ $[(8, 66)]$

30. If $f(x) = \dfrac{5x^2}{2} - 6x + 3$ find the coordinates at the point at

which the gradient is: (*a*) 4 and (*b*) -6
(*a*) [(2, 1)]   (*b*) [(0, 3)]

Differentiation of products and quotients

Differentiate the products in Problems 31 to 45 with respect to the variable and express your answers in their simplest form:

31. $3x^3 \sin x$ $[3x^2(x \cos x + 3 \sin x)]$

32. $\sqrt{t^3} \cos t$ $[\sqrt{t}(\tfrac{3}{2} \cos t - t \sin t)]$

33. $(3x^2 - 4x + 2)(2x^3 + x - 1)$
$[(30x^4 - 32x^3 + 21x^2 - 14x + 6)]$

34. $2 \sin \theta \cos \theta$ $[2(\cos^2 \theta - \sin^2 \theta)]$

35. $5e^{2a} \sin a$ $[5e^{2a}(\cos a + 2 \sin a)]$

36. $e^{7y} \cos y$ $[e^{7y}(7 \cos y - \sin y)]$

37. $b^3 \ln 2b$ $[b^2(1 + 3 \ln 2b)]$

38. $3\sqrt{x}\, e^{4x}$ $\left[3e^{4x}\left(\dfrac{8x + 1}{2\sqrt{x}}\right)\right]$

39. $e^t \ln t$ $\left[e^t\left(\dfrac{1}{t} + \ln t\right)\right]$

40. $e^{2d}(4d^2 - 3d + 1)$ $[e^{2d}(8d^2 + 2d - 1)]$

41. $3\sqrt{f^5} \ln 5f$ $[3\sqrt{f^3}(1 + \tfrac{5}{2} \ln 5f)]$

42. $2 \sin g \ln g$ $\left[2\left(\dfrac{1}{g} \sin g + \ln g \cos g\right)\right]$

43. $6e^{5m} \sin m$ $[6e^{5m}(\cos m + 5 \sin m)]$

44. $\sqrt{x}(1 + \sin x)$ $\left[\dfrac{2x \cos x + \sin x + 1}{2\sqrt{x}}\right]$

45. $e^v \ln v \sin v$ $\left[e^v\left\{(\sin v + \cos v) \ln v + \dfrac{\sin v}{v}\right\}\right]$

Differentiate the quotients in Problems 46 to 62 with respect to the variable and express your answers in their simplest form:

46. $\dfrac{4x}{x^2 - 1}$ $\left[\dfrac{-4(x^2 + 1)}{(x^2 - 1)^2}\right]$

47. $\dfrac{2t - 1}{3t^2 + 5t}$ $\left[\dfrac{5 + 6t - 6t^2}{(3t^2 + 5t)^2}\right]$

48. $\dfrac{2x^2 - 6x + 2}{3x^2 + 2x - 1}$ $\left[\dfrac{2(11x^2 - 8x + 1)}{(3x^2 + 2x - 1)^2}\right]$

49. $\dfrac{3e^{2\theta}}{4\theta^2 - 3}$ $\left[\dfrac{6e^{2\theta}(4\theta^2 - 4\theta - 3)}{(4\theta^2 - 3)^2}\right]$

50. $\dfrac{3u^4 + 2u^2 - 1}{4e^{5u}}$ $\left[\dfrac{-15u^4 + 12u^3 - 10u^2 + 4u + 5}{4e^{5u}} \right]$

51. $\dfrac{4 \sin c}{5c^2 + 2c}$ $\left[\dfrac{4(5c^2 + 2c) \cos c - 4(10c + 2) \sin c}{(5c^2 + 2c)^2} \right]$

52. $\dfrac{4\sqrt[3]{f^7}}{3 \sin f}$ $\left[\dfrac{4(\sqrt[3]{f^4})(7 \sin f - 3f \cos f)}{9 \sin^2 f} \right]$

53. $\dfrac{6 \cos h}{h^3 + 4}$ $\left[\dfrac{-6\{(h^3 + 4) \sin h + 3h^2 \cos h\}}{(h^3 + 4)^2} \right]$

54. $\dfrac{\sqrt{k^3}}{\cos k}$ $\left[\dfrac{\sqrt{k}(\frac{3}{2} \cos k + k \sin k)}{\cos^2 k} \right]$

55. $\dfrac{4e^{6x}}{\sin x}$ $\left[\dfrac{4e^{6x}(6 \sin x - \cos x)}{\sin^2 x} \right]$

56. $\dfrac{3 \ln \frac{5}{2}n}{n^2 + 2n}$ $\left[\dfrac{3(n + 2) - 6(n + 1) \ln \dfrac{5n}{2}}{(n^2 + 2n)^2} \right]$

57. $\dfrac{3\sqrt{x} + x}{\frac{7}{2} \ln 4x}$ $\left[\dfrac{\left(\dfrac{3}{2\sqrt{x}} + 1 \right) \ln 4x - \left(\dfrac{3}{\sqrt{x}} + 1 \right)}{\frac{7}{2}(\ln 4x)^2} \right]$

58. $\dfrac{\ln 6y}{6 \sin y}$ $\left[\dfrac{\dfrac{1}{y} \sin y - \ln 6y \cos y}{6 \sin^2 y} \right]$

59. $\dfrac{x^2 \ln 4x}{3 \sin x}$ $\left[\dfrac{x \ln 4x(2 \sin x - x \cos x) + x \sin x}{3 \sin^2 x} \right]$

60. $\dfrac{2\sqrt{t}}{\ln 3t \cos t}$ $\left[\dfrac{(\ln 3t \cos t + 2t \ln 3t \sin t - 2 \cos t)}{\sqrt{t}(\ln 3t \cos t)^2} \right]$

61. $\dfrac{x^2 \sec x}{e^{2x}}$ $\left[\dfrac{x \sec x}{e^{2x}} (x \tan x + 2 - 2x) \right]$

62. $\dfrac{k}{e^k \cosec k}$ $\left[\dfrac{1 + k(\cot k - 1)}{e^k \cosec k} \right]$

63. Find the slope of the curve $y = x e^{-2x}$ at the point $\left(\dfrac{1}{2}, \dfrac{1}{2e} \right)$ [0]

64. Calculate the gradient of the curve $f(x) = \dfrac{3x^4 - 2\sqrt{x^3} + 2}{5x^2 + 1}$

   at the points $(0, 2)$ and $(1, \frac{1}{2})$ $[0, \frac{2}{3}]$

Differentiation by substitution

Find the differential coefficients of the functions in Problems 65 to 120 with respect to the variable and express your answers in their simplest form.

65. $\sin 4x$ $[4 \cos 4x]$
66. $3 \tan 4x$ $[12 \sec^2 4x]$
67. $\cos 3t$ $[-3 \sin 3t]$
68. $5 \sec 2\theta$ $[10 \sec 2\theta \tan 2\theta]$
69. $4 \operatorname{cosec} 5\mu$ $[-20 \operatorname{cosec} 5\mu \cot 5\mu]$
70. $6 \cot 3\alpha$ $[-18 \operatorname{cosec}^2 3\alpha]$
71. $4 \cos(2x - 5)$ $[-8 \sin(2x - 5)]$
72. $\operatorname{cosec}(5t - 1)$ $[-5 \operatorname{cosec}(5t - 1) \cot(5t - 1)]$
73. $(t^3 - 2t + 3)^4$ $[4(t^3 - 2t + 3)^3(3t^2 - 2)]$

74. $\sqrt{(2v^3 - v)}$ $\left[\dfrac{6v^2 - 1}{2\sqrt{(2v^3 - v)}} \right]$

75. $\sin(3x - 2)$ $[3 \cos(3x - 2)]$
76. $3 \tan(5y - 1)$ $[15 \sec^2(5y - 1)]$
77. $4 \cos(6x + 5)$ $[-24 \sin(6x + 5)]$
78. $(1 - 2u^2)^7$ $[-28u(1 - 2u^2)^6]$

79. $\dfrac{1}{2n^2 - 3n + 1}$ $\left[\dfrac{3 - 4n}{(2n^2 - 3n + 1)^2} \right]$

80. $\sin^2 t$ $[2 \sin t \cos t]$
81. $3 \cos^2 x$ $[-6 \cos x \sin x]$

82. $\dfrac{1}{(2g - 1)^6}$ $\left[\dfrac{-12}{(2g - 1)^7} \right]$

83. $3 \operatorname{cosec}^2 x$ $[-6 \operatorname{cosec}^2 x \cot x]$
84. $6 \cos^3 t$ $[-18 \cos^2 t \sin t]$
85. $\frac{3}{2} \cot(6x - 2)$ $[-9 \operatorname{cosec}^2(6x - 2)]$

86. $\sqrt{(4x^3 + 2x^2 - 5x)}$ $\left[\dfrac{12x^2 + 4x - 5}{2\sqrt{(4x^3 + 2x^2 - 5x)}} \right]$

87. $2 \sin^4 h$ $[8 \sin^3 h \cos h]$

88. $\dfrac{3}{(x^2 + 6x - 1)^5}$ $\left[\dfrac{-30(x + 3)}{(x^2 + 6x - 1)^6} \right]$

89. $(x^2 - x + 1)^{12}$ $[12(x^2 - x + 1)^{11}(2x - 1)]$

90. $5e^{x-5}$ $[5e^{x-5}]$

91. $\ln(3p - 1)$ $\left[\dfrac{3}{3p - 1}\right]$

92. $15 \ln\left(\dfrac{x}{3} + 5\right)$ $\left[\dfrac{15}{x + 15}\right]$

93. $3 \sec 5g$ $[15 \sec 5g \tan 5g]$

94. $4 \operatorname{cosec}(2k - 1)$ $[-8 \operatorname{cosec}(2k - 1) \cot(2k - 1)]$

95. $7\beta \tan 4\beta$ $[7(4\beta \sec^2 4\beta + \tan 4\beta)]$

96. $\sqrt{x} \sec \dfrac{x}{3}$ $\left[\dfrac{\sqrt{x}}{3} \sec \dfrac{x}{3} \tan \dfrac{x}{3} + \dfrac{1}{2\sqrt{x}} \sec \dfrac{x}{3}\right]$

97. $2e^{5l} \operatorname{cosec} 3l$ $[2e^{5l} \operatorname{cosec} 3l(5 - 3 \cot 3l)]$

98. $\ln 5v \cot v$ $\left[\dfrac{\cot v}{v} - \operatorname{cosec}^2 v \ln 5v\right]$

99. $\dfrac{\sec 2t}{(t - 1)}$ $\left[\dfrac{\sec 2t}{(t - 1)^2} \{2(t - 1) \tan 2t - 1\}\right]$

100. $(x^2 + 1) \sin(2x^2 - 3)$
$[2x\{2(x^2 + 1) \cos(2x^2 - 3) + \sin(2x^2 - 3)\}]$

101. $(2x - 1)^9 \cos 4x$
$[-2(2x - 1)^8\{2(2x - 1) \sin 4x - 9 \cos 4x\}]$

102. $\dfrac{2t}{\tan 3t}$ $\left[\dfrac{2(\tan 3t - 3t \sec^2 3t)}{\tan^2 3t}\right]$

103. $\dfrac{1}{\operatorname{cosec}(4v + 1)}$ $[4 \cos(4v + 1)]$

104. $\dfrac{\sin(6x - 5)}{\sqrt{(x^2 - 1)}}$ $\left[\dfrac{6(x^2 - 1) \cos(6x - 5) - x \sin(6x - 5)}{\sqrt{(x^2 - 1)^3}}\right]$

105. $\dfrac{3 \cot x}{\ln 2x}$ $\left[\dfrac{-(3 \operatorname{cosec}^2 x \ln 2x + \dfrac{3}{x} \cot x)}{(\ln 2x)^2}\right]$

106. $\dfrac{5 \tan 3b}{\sqrt{b}}$ $\left[\dfrac{5}{2\sqrt{b^3}} (6b \sec^2 3b - \tan 3b)\right]$

107. $\dfrac{(3\theta^2 - 2)}{4 \sec 2\theta}$ $\left[\dfrac{3\theta - (3\theta^2 - 2) \tan 2\theta}{2 \sec 2\theta}\right]$

108. $\dfrac{3e^{7x-1}}{(x - 1)^9}$ $\left[\dfrac{3e^{7x-1}(7x - 16)}{(x - 1)^{10}}\right]$

109. $\dfrac{\sqrt{(3c^2 + 4c - 1)}}{2 \ln 5c}$ $\left[\dfrac{2c(3c + 2) \ln 5c - 2(3c^2 + 4c - 1)}{4c\sqrt{(3c^2 + 4c - 1)}(\ln 5c)^2} \right]$

110. $3 \cos(2x^2 + 1)$ $[-12x \sin(2x^2 + 1)]$

111. $5 \sec(5x^3 - 2x^2 + 2)$
    $[5x(15x - 4) \sec(5x^3 - 2x^2 + 2) \tan(5x^3 - 2x^2 + 2)]$

112. $\sin^2(2d - 1)$ $[4 \sin(2d - 1) \cos(2d - 1)]$

113. $3 \tan\sqrt{(4x - 2)}$ $\left[\dfrac{6 \sec^2\sqrt{(4x - 2)}}{\sqrt{(4x - 2)}} \right]$

114. $e^{\sec g}$ $[\sec g \tan g \, e^{\sec g}]$

115. $3e^{\cosec(2x-1)}$ $[-6 \cosec(2x - 1) \cot(2x - 1) \, e^{\cosec(2x-1)}]$

116. $(3x^3 + 2x)^2 \sin\sqrt{(x^2 - 1)}$

$\left[(3x^3 + 2x) \left\{ \dfrac{x(3x^3 + 2x)}{\sqrt{(x^2 - 1)}} \cos\sqrt{(x^2 - 1)} \right. \right.$

$\left. \left. + 2(9x^2 + 2) \sin\sqrt{(x^2 - 1)} \right\} \right]$

117. $(x + 3)^9 \cos^4(x^2 + 2)$
    $[(x + 3)^8 \cos^3(x^2 + 2)\{9 \cos(x^2 + 2) - 8x(x + 3) \sin(x^2 + 2)\}]$

118. $\dfrac{\sqrt{(3x^2 - 2)}}{\cosec^2(3x^2 - 2)}$ $\left[\dfrac{3x\{1 + 4(3x^2 - 2) \cot(3x^2 - 2)\}}{\sqrt{(3x^2 - 2)} \cosec^2(3x^2 - 2)} \right]$

119. $\ln\sqrt{(\cosec t)}$ $[-\frac{1}{2} \cot t]$

120. $\dfrac{3e^{3x^2 + 2x + 1}}{\ln(\cos x)}$ $\left[\dfrac{3e^{3x^2 + 2x + 1}}{[\ln(\cos x)]^2} \{(6x + 2) \ln(\cos x) + \tan x\} \right]$

Successive differentiation

121. If $y = 5x^3 - 6x^2 + 2x - 6$ find $\dfrac{d^2y}{dx^2}$ $[30x - 12]$

122. Find $f''(x)$ given $f(x) = \dfrac{5}{x} + \sqrt{x} - \dfrac{5}{\sqrt{x^5}} + 8$

$\left[\dfrac{10}{x^3} - \dfrac{1}{4\sqrt{x^3}} - \dfrac{175}{4\sqrt{x^9}} \right]$

123. Given $f(\theta) = 3 \sin 4\theta - 2 \cos 3\theta$ find $f'(\theta)$, $f''(\theta)$ and $f'''(\theta)$
    $[f'(\theta) = 6(2 \cos 4\theta + \sin 3\theta); \ f''(\theta) = 6(3 \cos 3\theta - 8 \sin 4\theta)$
    $f'''(\theta) = -6(32 \cos 4\theta + 9 \sin 3\theta)]$

124. If $m = (6p + 1)\left(\dfrac{1}{p} - 3 \right)$ find $\dfrac{d^2m}{dp^2}$ and $\dfrac{d^3m}{dp^3}$ $\left[\dfrac{2}{p^3}; \dfrac{-6}{p^4} \right]$

In Problems 125 to 135 find the second differ   al coefficient
with respect to the variable.

125. $3 \ln 5g$ $\left[-\dfrac{3}{g^2} \right]$

126. $(x - 2)^5$ $[20(x - 2)^3]$

127. $3 \sin t - \cos 2t$ $[4 \cos 2$

128. $3 \tan 2y + 4 \cot 3y$
$[24(\sec^2 2y \tan 2y + 3 \operatorname{cosec}^2 3$

129. $(3m^2 - 2)^6$ $[36(3m^2 - 2)^4(3$

130. $\dfrac{1}{(2r - 1)^7}$ $\left[\dfrac{224}{(2r - 1)^9} \right]$

131. $3 \cos^2 \theta$ $[6(\sin^2 \theta - \cos^2 \theta)]$

132. $\frac{1}{2} \cot(3x - 1)$ $[9 \operatorname{cosec}^2(3x - 1) \cot(3x - 1)]$

133. $4 \sin^5 n$ $[20 \sin^3 n(4 \cos^2 n - \sin^2 n)]$

134. $3x^2 \sin 2x$ $[6(1 - 2x^2) \sin 2x + 24x \cos 2x]$

135. $\dfrac{\sin t}{2t^2}$ $\left[\dfrac{1}{2t^4} \{(6 - t^2) \sin t - 4t \cos t\} \right]$

136. $x = 3t^2 - 2\sqrt{t} + \dfrac{1}{t} - 6$. Evaluate $\dfrac{d^2x}{dt^2}$ when $t = 1$ $[8\frac{1}{2}]$

137.   Evaluate $f''(\theta)$ when $\theta = 0$ given $f(\theta) = 5 \sec 2\theta$ $[20]$

138.   If $y = \cos \alpha - \sin \alpha$ evaluate α when $\dfrac{d^2y}{d\alpha^2}$ is zero. $\left[\dfrac{\pi}{4} \right]$

139.   If $y = A e^x - B e^{-x}$ prove that: $\dfrac{e^x}{2} \left\{ \dfrac{d^2y}{dx^2} + \dfrac{dy}{dx} \right\} - e^x y = B$

140.   Show that $\dfrac{d^2b}{dS^2} + 6 \dfrac{db}{dS} + 25b = 0$ when $b = e^{-3S} \sin 4S$

141.   Show that $x = 2t e^{-2t}$ satisfies the equation:
$\dfrac{d^2x}{dt^2} + 4 \dfrac{dx}{dt} + 4x = 0$

142.   If $y = 3x^3 + 2x - 4$ prove that:
$\dfrac{d^3y}{dx^3} + \dfrac{2}{9} \dfrac{d^2y}{dx^2} + x \dfrac{dy}{dx} - 3y = 30$

143.   Show that the differential equation $\dfrac{d^2y}{dx^2} - 8 \dfrac{dy}{dx} + 41y = 0$ is
satisfied when $y = 2e^{4x} \cos 5x$

Chapter 20

Applications of differentiation

20.1 Velocity and acceleration

Let a car move a distance x metres in a time t seconds along a straight road. If the velocity v of the car is constant then

$$v = \frac{x}{t}\,\text{m/s}$$

i.e. the gradient of the distance/time graph shown in Fig. 20.1(a) is constant. If, however, the velocity of the car is not constant then the distance/time graph will not be a straight line. It may be as shown in Fig. 20.1(b).

The average velocity over a small time δt and distance δx is given by the gradient of the chord CD, i.e. the average velocity over time $\delta t = \dfrac{\delta x}{\delta t}$.

As $\delta t \to 0$, the chord CD becomes a tangent, such that at point C the velocity v is given by:

$$v = \frac{dx}{dt}$$

Hence the velocity of the car at any instant t is given by the gradient of the distance/time graph. If an expression for the distance x is known in terms of time t then the velocity is obtained by differentiating the expression.

The acceleration a of the car is defined as the rate of change of velocity.

With reference to the velocity/time graph shown in Fig. 20.1(c), let δv be the change in v and δt the corresponding time interval, then:

$$\text{average acceleration } a = \frac{\delta v}{\delta t}$$

As $\delta t \to 0$, the chord EF becomes a tangent such that at point E the acceleration is given by:

$$a = \frac{dv}{dt}$$

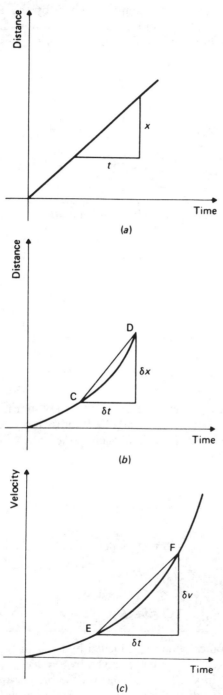

(a)

(b)

(c)

Fig. 20.1

Hence the acceleration of the car at any instant t is given by the gradient of the velocity/time graph. If an expression for velocity v is known in terms of time t then the acceleration is obtained by differentiating the expression.

$\quad$ Acceleration, $a = \dfrac{dv}{dt}$; but $v = \dfrac{dx}{dt}$. Hence $a = \dfrac{d}{dt}\left(\dfrac{dx}{dt}\right)$

$\quad$ which is written as: $\mathbf{a = \dfrac{d^2x}{dt^2}}$

Thus acceleration is given by the second differential coefficient of x with respect to t.

Summary

> If a body moves a distance x metres in a time t seconds then:
>
> $$\text{distance } x = f(t)$$
>
> $$\text{velocity } v = f'(t) \text{ or } \frac{dx}{dt}$$
>
> $$\text{and acceleration } a = f''(t) \text{ or } \frac{d^2x}{dt^2}$$

Worked problems on velocity and acceleration

Problem 1. The distance x metres moved by a body in a time t seconds is given by $x = 2t^3 + 3t^2 - 6t + 2$. Express the velocity and acceleration in terms of t and find their values when $t = 4$ seconds.

Distance $x = 2t^3 + 3t^2 - 6t + 2$ metres

Velocity $v = \dfrac{dx}{dt} = 6t^2 + 6t - 6$ metres per second

Acceleration $a = \dfrac{d^2x}{dt^2} = 12t + 6$ metres per second squared

After 4 seconds, $v = 6(4)^2 + 6(4) - 6$

$\qquad\qquad\qquad = 96 + 24 - 6 = \mathbf{114 \text{ m/s}}$

$\qquad\qquad a = 12(4) + 6 = \mathbf{54 \text{ m/s}^2}$

Problem 2. If the distance s metres travelled by a car in time t seconds after the brakes are applied is given by $s = 15t - \frac{5}{3}t^2$: (a) what is the speed (in km/h) at the instant the brakes are applied, and (b) how far does the car travel before it stops?

(a) Distance $s = 15t - \frac{5}{3}t^2$

Velocity $v = \dfrac{ds}{dt} = 15 - \frac{10}{3}t$

At the instant the brakes are applied, $t = 0$
 Hence velocity = 15 m/s

$15 \text{ m/s} = \dfrac{15}{1\,000}(60 \times 60) \text{ km/h} = \textbf{54 km/h}$

(b) When the car finally stops, the velocity is zero,
 i.e. $v = 15 - \frac{10}{3}t = 0$
 i.e. $15 = \frac{10}{3}t$ or $t = 4.5$ seconds
 Hence the distance travelled before the car stops is given by:
 $s = 15t - \frac{5}{3}t^2$
 $\quad = 15(4.5) - \frac{5}{3}(4.5)^2$
 $\quad = \textbf{33.75 m}$

Problem 3. The distance x metres moved by a body in t seconds is given by:
$x = 3t^3 - \frac{11}{2}t^2 + 2t + 5$
Find:
(a) its velocity after t seconds,
(b) its velocity at the start and after 4 seconds,
(c) the value of t when the body comes to rest,
(d) its acceleration after t seconds,
(e) its acceleration after 2 seconds,
(f) the value of t when the acceleration is 16 m/s² and
(g) the average velocity over the third second.

(a) Distance $x = 3t^3 - \frac{11}{2}t^2 + 2t + 5$

Velocity $v = \dfrac{dx}{dt} = 9t^2 - 11t + 2$

(b) Velocity at the start means the velocity when $t = 0$,
 i.e. $v_0 = 9(0)^2 - 11(0) + 2 = \textbf{2 m/s}$
 Velocity after 4 seconds, $v_4 = 9(4)^2 - 11(4) + 2 = \textbf{102 m/s}$

(c) When the body comes to rest, $v = 0$
 i.e. $9t^2 -- 11t + 2 = 0$
 $\quad (9t - 2)(t - 1) = 0$
 $\quad\quad t = \frac{2}{9}\text{ s or } t = \textbf{1 s}$

(d) Acceleration $a = \dfrac{d^2x}{dt^2} = (18t - 11)$

(e) Acceleration after 2 seconds, $a_2 = 18(2) - 11 = \mathbf{25\ m/s^2}$

(f) When the acceleration is 16 m/s^2 then

$$18t - 11 = 16$$
$$18t = 16 + 11 = 27$$
$$t = \tfrac{27}{18} = \mathbf{1\tfrac{1}{2}\ seconds}$$

(g) Distance travelled in the third second = (distance travelled after 3 s) − (distance travelled after 2 s)

$$= [3(3)^3 - \tfrac{11}{2}(3)^2 + 2(3) + 5] - [3(2)^3 - \tfrac{11}{2}(2)^2 + 2(2) + 5]$$
$$= 42\tfrac{1}{2} - 11$$
$$= 31\tfrac{1}{2}\ \text{m}$$

Average velocity over the third second $= \dfrac{\text{distance travelled}}{\text{time interval}}$

$$= \frac{31\tfrac{1}{2}\ \text{m}}{1\ \text{s}}$$
$$= \mathbf{31\tfrac{1}{2}\ m/s}$$

(Note that should a negative value occur for velocity it merely means that the body is moving in the direction opposite to that with which it started. Also if a negative value occurs for acceleration it indicates a deceleration (or a retardation).)

Further problems on velocity and acceleration may be found in Section 20.4 (Problems 1 to 15), page 385.

20.2   Maximum and minimum values

Consider the curve shown in Fig. 20.2

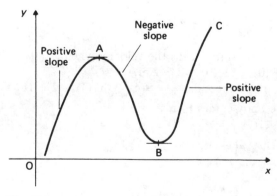

Fig. 20.2

The slope of the curve (i.e. $\dfrac{dy}{dx}$) between points O and A is positive. The slope of the curve between points A and B is negative and the slope between points B and C is again positive.

At point A the slope is zero and as x increases, the slope of the curve changes from positive just before A to negative just after. Such a point is called a **maximum value**.

At point B the slope is also zero and, as x increases, the slope of the curve changes from negative just before B to positive just after. Such a point is called a **minimum value**.

Points such as A and B are given the general name of **turning-points**.

Maximum and minimum values can be confusing inasmuch as they suggest that they are the largest and smallest values of a curve. However, by their definition this is not so. A maximum value occurs at the 'crest of a wave' and the minimum value at the 'bottom of a valley'. In Fig. 20.2 the point C has a larger y-ordinate value than A and point O has a smaller y ordinate than B. Points A and B are turning-points and are given the special names of maximum and minimum values respectively.

Summary

1. At a maximum point the slope $\dfrac{dy}{dx} = 0$ and changes from positive just before the maximum point to negative just after.

2. At a minimum point the slope $\dfrac{dy}{dx} = 0$ and changes from negative just before the minimum point to positive just after.

Consider the function $y = x^3 - x^2 - 5x + 6$
The turning-points (i.e. the maximum and minimum values) may be determined without going through the tedious process of drawing up a table of values and plotting the graph.

If $y = x^3 - x^2 - 5x + 6$

then $\dfrac{dy}{dx} = 3x^2 - 2x - 5$

Now at a maximum or minimum value $\dfrac{dy}{dx} = 0$

Hence $3x^2 - 2x - 5 = 0$ for a maximum or minimum value
$(3x - 5)(x + 1) = 0$

i.e. *Either* $3x - 5 = 0$ giving $x = \frac{5}{3}$

   or $x + 1 = 0$ giving $x = -1$

For each value of the independent variable x there is a corresponding value of the dependent variable y.

When $x = \frac{5}{3}$, $y = [\frac{5}{3}]^3 - [\frac{5}{3}]^2 - 5[\frac{5}{3}] + 6 = -\frac{13}{27}$

When $x = -1$, $y = (-1)^3 - (-1)^2 - 5(-1) + 6 = 9$

Hence turning-points occur at $(\frac{5}{3}, -\frac{13}{27})$ and $(-1, 9)$

The next step is to determine which of the points is a maximum and which is a minimum. There are two methods whereby this may be achieved.

Method 1

Consider firstly the point $(\frac{5}{3}, -\frac{13}{27})$

$$\frac{dy}{dx} = 3x^2 - 2x - 5 = (3x - 5)(x + 1)$$

If x is slightly less than $\frac{5}{3}$ then $(3x - 5)$ becomes negative, $(x + 1)$ remains positive, making $\frac{dy}{dx} = (-) \times (+) = $ negative

If x is slightly greater than $\frac{5}{3}$ then $(3x - 5)$ becomes positive, $(x + 1)$ remains positive, making $\frac{dy}{dx} = (+) \times (+) = $ positive

Hence the slope is negative just before $(\frac{5}{3}, -\frac{13}{27})$ and positive just after. This is thus a **minimum** value.

Consider now the point $(-1, 9)$

$$\frac{dy}{dx} = (3x - 5)(x + 1)$$

If x is slightly less than -1 (for example -1.1) then $(3x - 5)$ remains negative, $(x + 1)$ becomes negative, making $\frac{dy}{dx} = (-) \times (-) = $ positive

If x is slightly greater than -1 (for example -0.9) then $(3x - 5)$ remains negative, $(x + 1)$ becomes positive, making $\frac{dy}{dx} = (-) \times (+) = $ negative

Hence the slope is positive just before $(-1, 9)$ and negative just after. This is thus a **maximum** value.

Figure 20.3 shows a graph of $y = x^3 - x^2 - 5x + 6$ with the maximum value at $(-1, 9)$ and the minimum at $(\frac{5}{3}, -\frac{13}{27})$

Method 2

When passing through a maximum value, $\frac{dy}{dx}$ changes from positive, through zero, to negative. By convention, moving from a positive value

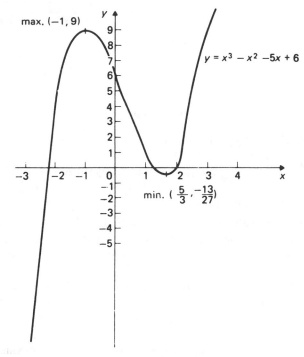

Fig. 20.3   Graph of $y = x^3 - x^2 - 5x + 6$

to a negative value is moving in a negative direction. Hence the rate of change of $\dfrac{dy}{dx}$ is negative.

i.e. $\dfrac{d}{dx}\left(\dfrac{dy}{dx}\right) = \dfrac{d^2y}{dx^2}$ **is negative at a maximum value**

Similarly, when passing through a minimum value, $\dfrac{dy}{dx}$ changes from negative, through zero, to positive. By convention, moving from a negative value to a positive value is moving in a positive direction. Hence the rate of change of $\dfrac{dy}{dx}$ is positive.

i.e. $\dfrac{d^2y}{dx^2}$ **is positive at a minimum value**

Thus, in the above example, to distinguish between the points $(\frac{5}{3}, -\frac{13}{27})$ and $(-1, 9)$ the second differential is required.

Since $\dfrac{\mathrm{d}y}{\mathrm{d}x} = 3x^2 - 2x - 5$

then $\dfrac{\mathrm{d}^2 y}{\mathrm{d}x^2} = 6x - 2$

When $x = \frac{5}{3}, \dfrac{\mathrm{d}^2 y}{\mathrm{d}x^2} = 6[\frac{5}{3}] - 2 = +8$ which is **positive**

Hence $(\frac{5}{3}, -\frac{13}{27})$ **is a minimum point**

When $x = -1, \dfrac{\mathrm{d}^2 y}{\mathrm{d}x^2} = 6(-1) - 2 = -8$ which is **negative**

Hence $(-1, 9)$ is a maximum point

The actual numerical value of the second differential is insignificant for maximum and minimum values – the sign is the important factor. There are thus two methods of distinguishing between maximum and minimum values. Normally, the second method, that of determining the sign of the second differential, is preferred but sometimes the first method, that of examining the sign of the slope just before and just after the turning-point, is necessary because the second differential coefficient is too difficult to obtain.

It is possible to have a turning-point, the slope on either side of which is the same. This point is given the special name of a **point of inflexion**. At a point of inflexion $\dfrac{\mathrm{d}^2 y}{\mathrm{d}x^2}$ is zero.

Maximum and minimum points and points of inflexion are given the general term of **stationary points**. Examples of each are shown in Fig. 20.4

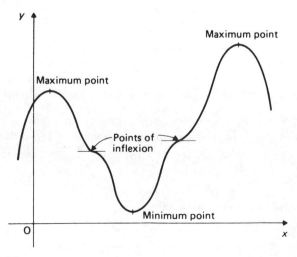

Fig. 20.4

Procedure for finding and distinguishing between stationary points

(i)   If $y = f(x)$, find $\dfrac{dy}{dx}$

(ii) Let $\dfrac{dy}{dx} = 0$ and solve for the value(s) of x

(iii) Substitute the value(s) of x into the original equation, $y = f(x)$, to obtain the y-ordinate value(s). Hence the coordinates of the stationary points are established.

(iv) Find $\dfrac{d^2y}{dx^2}$

or

Determine the sign of the gradient of the curve just before and just after the stationary point(s).

(v)   Substitute values of x into $\dfrac{d^2y}{dx^2}$. If the result is:

(*a*) positive – the point is a minimum value
(*b*) negative – the point is a maximum value
(*c*) zero – the point is a point of inflexion

or

If the sign change for the gradient of the curve is:
(*a*) positive to negative – the point is a maximum value
(*b*) negative to positive – the point is a minimum value
(*c*) positive to positive, or
(*d*) negative to negative – the point is a point of inflexion

Worked problems on maximum and minimum values

Problem 1. Find the coordinates of the maximum and minimum values of the graph of $y = \dfrac{2x^3}{3} - 5x^2 + 12x - 7$ and distinguish between them.

From the above procedure:

(i) $y = \dfrac{2x^3}{3} - 5x^2 + 12x - 7$

$\dfrac{dy}{dx} = 2x^2 - 10x + 12$

(ii) $\dfrac{dy}{dx} = 0$ at a turning-point.

Therefore $2x^2 - 10x + 12 = 0$
$$2(x^2 - 5x + 6) = 0$$
$$2(x - 2)(x - 3) = 0$$

Hence $x = 2$ or $x = 3$

(iii)   When $x = 2$, $y = \frac{2}{3}(2)^3 - 5(2)^2 + 12(2) - 7 = 2\frac{1}{3}$
When $x = 3$, $y = \frac{2}{3}(3)^3 - 5(3)^2 + 12(3) - 7 = 2$
The coordinates of the turning-points are thus $(2, 2\frac{1}{3})$ and $(3, 2)$

(iv) $\dfrac{dy}{dx} = 2x^2 - 10x + 12$

$\dfrac{d^2y}{dx^2} = 4x - 10$

(v)   When $x = 2$, $\dfrac{d^2y}{dx^2} = -2$, which is negative, giving a
maximum value

When $x = 3$, $\dfrac{d^2y}{dx^2} = +2$, which is positive, giving a
minimum value

Hence **the point $(2, 2\frac{1}{3})$ is a maximum value and the point $(3, 2)$ a
minimum value.**

Note that with a quadratic equation there will be one
turning-point. With a cubic equation (i.e. one containing a highest
term of power 3) there may be two turning-points (i.e. one less
than the highest power), and so on.

Problem 2. Locate the turning points on the following curves and
distinguish between maximum and minimum values: (a) $x(5 - x)$
(b) $2t - e^t$ (c) $2(\theta - \ln \theta)$

(a)   Let $y = x(5 - x) = 5x - x^2$

$\dfrac{dy}{dx} = 5 - 2x = 0$ for a maximum or minimum value.

i.e. $x = 2\frac{1}{2}$
When $x = 2\frac{1}{2}$, $y = 2\frac{1}{2}(5 - 2\frac{1}{2}) = 6\frac{1}{4}$
Hence a turning-point occurs at $(2\frac{1}{2}, 6\frac{1}{4})$

$\dfrac{d^2y}{dx^2} = -2$, which is negative, giving a maximum value.

Hence $(2\frac{1}{2}, 6\frac{1}{4})$ is a maximum point.

(b) Let $y = 2t - e^t$

$$\frac{dy}{dt} = 2 - e^t = 0 \text{ for a maximum or minimum value.}$$

i.e. $2 = e^t$

$\ln 2 = t$

$t = 0.693\,1$

When $t = 0.693\,1$, $y = 2(0.693\,1) - 2 = -0.613\,8$
Hence a turning-point occurs at $(0.693\,1, -0.613\,8)$

$$\frac{d^2y}{dt^2} = -e^t$$

When $t = 0.693\,1$, $\dfrac{d^2y}{dt^2} = -2$, which is negative, giving a maximum value.
Hence **(0.693 1, −0.613 8) is a maximum point.**

(c) Let $y = 2(\theta - \ln \theta) = 2\theta - 2 \ln \theta$

$$\frac{dy}{d\theta} = 2 - \frac{2}{\theta} = 0 \text{ for a maximum or minimum value.}$$

i.e. $\theta = 1$
When $\theta = 1$, $y = 2 - 2 \ln 1 = 2$
Hence a turning-point occurs at $(1, 2)$

$$\frac{d^2y}{d\theta^2} = +\frac{2}{\theta^2}$$

When $\theta = 1$, $\dfrac{d^2y}{d\theta^2} = +2$, which is positive, giving a minimum value.
Hence **(1, 2) is a minimum point.**

Problem 3. Find the maximum and minimum values of the function

$$f(p) = \frac{(p-1)(p-6)}{(p-10)}$$

$$f(p) = \frac{(p-1)((p-6)}{(p-10)} = \frac{p^2 - 7p + 6}{(p-10)} \text{ (i.e. a quotient)}$$

$$f'(p) = \frac{(p-10)(2p-7) - (p^2 - 7p + 6)(1)}{(p-10)^2}$$

$$= \frac{(2p^2 - 27p + 70) - (p^2 - 7p + 6)}{(p-10)^2}$$

$$= \frac{p^2 - 20p + 64}{(p - 10)^2}$$

$$= \frac{(p - 4)(p - 16)}{(p - 10)^2} = 0 \text{ for a maximum or minimum value.}$$

Therefore $(p - 4)(p - 16) = 0$

i.e. $p = 4$ or $p = 16$

When $p = 4$, $f(p) = \dfrac{(3)(-2)}{(-6)} = 1$

When $p = 16$, $f(p) = \dfrac{(15)(10)}{(6)} = 25$

Hence there are turning-points at (4, 1) and (16, 25)

To use the second-derivative approach in this case would result in a complicated and long expression. Thus the slope is investigated just before and just after the turning-point.

It will be easier to use the factorised version of $f'(p)$

i.e. $f'(p) = \dfrac{(p - 4)(p - 16)}{(p - 10)^2}$

Consider the point (4, 1):

When p is just less than 4, $f'(p) = \dfrac{(-)(-)}{(+)}$, i.e. positive.

When p is just greater than 4, $f'(p) = \dfrac{(+)(-)}{(+)}$, i.e. negative.

Since the gradient changes from positive to negative the point (4, 1) is a maximum.

Consider the point (16, 25):

When p is just less than 16, $f'(p) = \dfrac{(+)(-)}{(+)}$, i.e. negative.

When p is just greater than 16, $f'(p) = \dfrac{(+)(+)}{(+)}$, i.e. positive.

Since the gradient changes from negative to positive the point (16, 25) is a minimum.

Since, in the question, the maximum and minimum values are asked for (and not the coordinates of the turning-points) the answers are: **maximum value = 1, minimum value = 25**

Problem 4. Find the maximum and minimum values of

$y = 1.25 \cos 2\theta + \sin \theta$ for values of θ between 0 and $\dfrac{\pi}{2}$ inclusive,

given $\sin 2\theta = 2 \sin \theta \cos \theta$.

$y = 1.25 \cos 2\theta + \sin \theta$

$\dfrac{dy}{d\theta} = -2.50 \sin 2\theta + \cos \theta = 0$ for a maximum or minimum value.

But $\sin 2\theta = 2 \sin \theta \cos \theta$

Therefore $-2.50(2 \sin \theta \cos \theta) + \cos \theta = 0$

$$-5.0 \sin \theta \cos \theta + \cos \theta = 0$$

$$\cos \theta(-5.0 \sin \theta + 1) = 0$$

Hence $\cos \theta = 0$, i.e. $\theta = 90°$ or $270°$

or $-5.0 \sin \theta + 1 = 0$, i.e. $\sin \theta = \frac{1}{5}$

$\theta = 11° \, 32'$ or $168° \, 28'$

Thus within the range $\theta = 0$ to $\theta = \dfrac{\pi}{2}$ inclusive, turning-points

occur at $11° \, 32'$ and $90°$

$\dfrac{d^2y}{d\theta^2} = -5.0 \cos 2\theta - \sin \theta$

When $\theta = 11° \, 32'$, $\dfrac{d^2y}{d\theta^2} = -4.80$, i.e. it is negative, giving a

maximum value.

When $\theta = 90°$, $\dfrac{d^2y}{d\theta^2} = 4$, i.e. it is positive, giving a minimum value.

$y_{max} = 1.25 \cos 2(11° \, 32') + \sin(11° \, 32') = \mathbf{1.35}$

$y_{min} = 1.25 \cos 2(90°) + \sin(90°) = \mathbf{-0.25}$

Further problems on maximum and minimum values may be found in Section 20.4 (Problems 16 to 41), page 387.

20.3   Practical problems involving maximum and minimum values

There are many practical problems on maximum and minimum values in engineering and science which can be solved using the method(s) shown in Section 20.2. Often the quantity whose maximum or minimum value is required appears at first to be a function of more than one variable. It is thus necessary to eliminate all but one of the variables, and this is often the only difficult part of its solution. Once the quantity has been expressed in terms of a single variable, the procedure is identical to that used in Section 20.2.

Worked problems on practical problems involving maximum and minimum values

---

Problem 1. A rectangular area is formed using a piece of wire 36 cm long. Find the length and breadth of the rectangle if it is to enclose the maximum possible area.

Let the dimension of the rectangle be x and y.
Perimeter of rectangle $= 2x + 2y = 36$

i.e. $x + y = 18$ (1)

Since it is the maximum area that is required a formula for the area A must be obtained in terms of one variable only.

Area $A = xy$

From equation (1) $y = 18 - x$

Hence $A = x(18 - x) = 18x - x^2$

Now that an expression for the area has been obtained in terms of one variable it can be differentiated with respect to that variable.

$$\frac{dA}{dx} = 18 - 2x = 0 \text{ for a maximum or minimum value.}$$

i.e. $x = 9$

$$\frac{d^2A}{dx^2} = -2, \text{ which is negative, giving a maximum value.}$$

$y = 18 - x = 18 - 9 = 9$

Hence **the length and breadth of the rectangle of maximum area are both 9 cm**, i.e. a square gives the maximum possible area for a given perimeter length. When the perimeter of a rectangle is 36 cm the maximum area possible is 81 cm^2

Problem 2. Find the area of the largest piece of rectangular ground that can be enclosed by 1 km of fencing if part of an existing straight wall is used as one side.

There are a large number of possible rectangular areas which can be produced from 1 000 m of fencing. Three such possibilities are shown in Fig. 20.5(a) where AB represents the existing wall. All three rectangles have different areas. There must be one particular condition which gives a maximum area.

Let the dimensions of any rectangle be x and y as shown in Fig. 20.5(b)

Then $2x + y = 1\,000$ (2)

Area of rectangle, $A = xy$ (3)

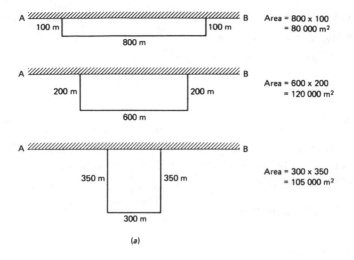

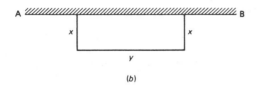

Fig. 20.5

Since it is the **maximum area** that is required, a formula for the area A must be obtained in terms of one variable only.

From equation (2) $y = 1\,000 - 2x$

Hence
$$A = x(1\,000 - 2x) = 1\,000x - 2x^2$$

$$\frac{\mathrm{d}A}{\mathrm{d}x} = 1\,000 - 4x = 0 \text{ for a maximum or minimum value}$$

i.e.
$$x = 250$$

$$\frac{\mathrm{d}^2 A}{\mathrm{d}x^2} = -4, \text{ which is negative, giving a maximum value}$$

When $x = 250$, $y = 1\,000 - 2(250) = 500$
Hence the maximum possible area $= xy \text{ m}^2 = (250)(500) \text{ m}^2$
$$= \mathbf{125\,000 \text{ m}^2}$$

Problem 3. A lidless, rectangular box with square ends is to be made from a thin sheet of metal. What is the least area of the metal for which the volume is $4\frac{1}{2} \text{ m}^3$?

Let the dimensions of the box be x metres by x metres by y metres.

Volume of box $= x^2 y = 4\frac{1}{2}$ (4)

Surface area A of box consists of: two ends $= 2x^2$

$$\text{two sides} = 2xy$$

$$\text{base} = xy$$

$A = 2x^2 + 2xy + xy = 2x^2 + 3xy$ (5)

Since it is the **least (i.e. minimum in this case) area** that is required, a formula for the area A must be obtained in terms of one variable only.

From equation (4), $y = \dfrac{4\frac{1}{2}}{x^2} = \dfrac{9}{2x^2}$

Substituting $y = \dfrac{9}{2x^2}$ in equation (5) gives:

$$A = 2x^2 + 3x\left(\frac{9}{2x^2}\right) = 2x^2 + \frac{27}{2x}$$

$$\frac{\mathrm{d}A}{\mathrm{d}x} = 4x - \frac{27}{2x^2} = 0 \text{ for a maximum or minimum value}$$

$$4x = \frac{27}{2x^2}$$

$x^3 = \frac{27}{8}$, i.e. $x = \frac{3}{2}$

$$\frac{\mathrm{d}^2 A}{\mathrm{d}x^2} = 4 + \frac{27}{x^3}$$

When $x = \frac{3}{2}$, $\dfrac{\mathrm{d}^2 A}{\mathrm{d}x^2} = 4 + \dfrac{27}{(\frac{3}{2})^3} = +12$, which is positive, giving a minimum (or least) value

When $x = \frac{3}{2}$, $y = \dfrac{9}{2x^2} = \dfrac{9}{2(\frac{3}{2})^2} = 2$

Therefore area $A = 2x^2 + 3xy = 2[\frac{3}{2}]^2 + 3[\frac{3}{2}](2) = 13\frac{1}{2}$

Hence **the least possible area of metal required to form a rectangular box with square ends of volume $4\frac{1}{2}$ m^3 is $13\frac{1}{2}$ m^2**

Problem 4. Find the base radius and height of the cylinder of maximum volume which can be cut from a sphere of radius 10.0 cm

A cylinder of radius r and height h is shown in Fig. 20.6 enclosed in a sphere of radius $R = 10.0$ cm

Volume of cylinder, $V = \pi r^2 h$ (6)

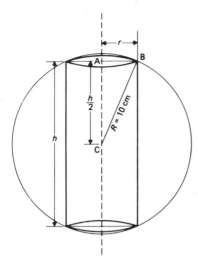

Fig. 20.6

Using the theorem of Pythagoras on the triangle ABC of Fig. 20.6 gives:

$$r^2 + \left[\frac{h}{2}\right]^2 = R^2$$

i.e. $r^2 + \dfrac{h^2}{4} = 100$ (7)

Since it is the **maximum volume** that is required, a formula for the volume V must be obtained in terms of one variable only.

From equation (7), $r^2 = 100 - \dfrac{h^2}{4}$

Substituting $r^2 = 100 - \dfrac{h^2}{4}$ in equation (6) gives:

$$V = \pi\left[100 - \frac{h^2}{4}\right]h = 100\pi h - \frac{\pi h^3}{4}$$

$\dfrac{dV}{dh} = 100\pi - \dfrac{3\pi}{4}h^2 = 0$ for a maximum or minimum value

$100\pi = \dfrac{3\pi}{4}h^2$

$h^2 = \dfrac{400}{3}$

$h = 11.55$ cm ($h = -11.55$ cm is neglected for obvious reasons)

$$\frac{d^2V}{dh^2} = -\tfrac{3}{2}\pi h$$

When $h = 11.55$, $\dfrac{d^2V}{dh^2} = -\tfrac{3}{2}\pi(11.55) = -54.43$ which is negative, giving a maximum value.

From equation (2), $r^2 = 100 - \dfrac{h^2}{4}$

$$r = \sqrt{\left(100 - \frac{h^2}{4}\right)} = 8.164 \text{ cm}$$

Hence the cylinder having the largest volume that can be cut from a sphere of radius 10.0 cm is one in which the base radius is 8.164 cm and the height is 11.55 cm

---

Problem 5. A piece of wire 4.0 m long is cut into two parts one of which is bent into a square and the other bent into a circle. Find the radius of the circle if the sum of their areas is a minimum.

Let the square be of side x m and the circle of radius r m.
The sum of the perimeters of the square and circle is given by:

$$4x + 2\pi r = 4 \qquad (8)$$

or $2x + \pi r = 2$

Total area A of the two shapes, $A = x^2 + \pi r^2$ $\qquad (9)$

Since it is the **minimum area** that is required a formula for the area A must be obtained in terms of one variable only.

From equation (8), $x = \dfrac{2 - \pi r}{2}$

Substituting $x = \dfrac{2 - \pi r}{2}$ in equation (9) gives:

$$A = \left(\frac{2 - \pi r}{2}\right)^2 + \pi r^2 = \frac{4 - 4\pi r + \pi^2 r^2}{4} + \pi r^2$$

i.e. $A = 1 - \pi r + \dfrac{\pi^2 r^2}{4} + \pi r^2$

$$\frac{dA}{dr} = -\pi + \frac{\pi^2 r}{2} + 2\pi r = 0 \text{ for a maximum or minimum value}$$

i.e. $\pi = r\left[\dfrac{\pi^2}{2} + 2\pi\right]$

$$r = \frac{\pi}{\left(\dfrac{\pi^2}{2} + 2\pi\right)} = \frac{1}{\left(\dfrac{\pi}{2} + 2\right)} = 0.280 \text{ m}$$

$\dfrac{d^2A}{dr^2} = \dfrac{\pi^2}{2} + 2\pi = 11.22$, which is positive, giving a minimum value

Hence **for the sum of the areas of the square and circle to be a minimum the radius of the circle must be 28.0 cm**

Problem 6. Find the base radius of a cylinder of maximum volume which can be cut from a cone of height 12 cm and base radius 9 cm

A cylinder of base radius r cm and height h cm is shown enclosed in a cone of height 12 cm and base radius 9 cm in Fig. 20.7

Volume of the cylinder, $V = \pi r^2 h$ (10)

By similar triangles: $\dfrac{12 - h}{r} = \dfrac{12}{9}$ (11)

Since it is the **maximum volume** that is required a formula for the volume V must be obtained in terms of one variable only.

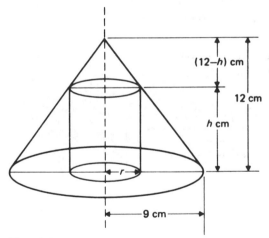

Fig. 20.7

From equation (11), $9(12 - h) = 12r$

$$108 - 9h = 12r$$

$$h = \frac{108 - 12r}{9}$$

Substituting for h in equation (10) gives:

$$V = \pi r^2 \left[\frac{108 - 12r}{9} \right] = 12\pi r^2 - \frac{4\pi r^3}{3}$$

$\dfrac{dV}{dr} = 24\pi r - 4\pi r^2 = 0$ for a maximum or minimum value

i.e. $4\pi r(6 - r) = 0$

Therefore $r = 0$ or $r = 6$

$\dfrac{d^2 V}{dr^2} = 24\pi - 8\pi r$

When $r = 0, \dfrac{d^2 V}{dr^2}$ is positive, giving a minimum value (which we
would expect).

When $r = 6, \dfrac{d^2 V}{dr^2} = 24\pi - 48\pi = -24\pi$, which is negative, giving
a maximum value

Hence **a cylinder of maximum volume having a base radius of 6 cm
can be cut from a cone of height 12 cm and base radius 9 cm**

Problem 7. A rectangular sheet of metal which measures 24.0 cm
by 16.0 cm has squares removed from each of the four corners so
that an open box may be formed. Find the maximum possible
volume for the box.

The squares which are to be removed are shown shaded and
having side x cm in Fig. 20.8

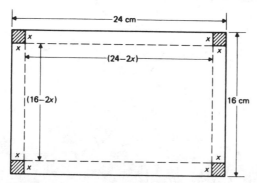

Fig. 20.8

To form a box the metal has to be bent upwards along the
broken lines.

The dimensions of the box will be: length $= (24.0 - 2x)$ cm
breadth $= (16.0 - 2x)$ cm
height $= x$ cm

If volume of box is V cm³, $V = (24.0 - 2x)(16.0 - 2x)(x)$

$$= 384x - 80x^2 + 4x^3$$

$\dfrac{dV}{dx} = 384 - 160x + 12x^2 = 0$ for a maximum or minimum value

i.e. $4(3x^2 - 40x + 96) = 0$

$$x = \frac{40 \pm \sqrt{[(-40)^2 - 4(3)(96)]}}{6}$$

$x = $ **10.194 cm** or $x = $ **3.139 cm**

Since the breadth $= (16.0 - 2x)$ cm, $x = 10.194$ cm is an impossible solution to this problem and is thus neglected.

Hence $x = 3.139$ cm

$$\frac{d^2V}{dx^2} = -160 + 24x$$

When $x = 3.139$ cm, $\dfrac{d^2V}{dx^2} = -160 + 24(3.139) = -84.66$ which is

negative, giving a maximum value.

The dimensions of the box are: length $= 24.0 - 2(3.139) = 17.72$ cm

breadth $= 16.0 - 2(3.139) = 9.722$ cm

height $= 3.139$ cm

Maximum volume $= (17.72)(9.722)(3.139) = $ **540.8 cm³**

Further problems on practical maximum and minimum problems may be found in the following Section 20.4 (Problems 42 to 69), page 388. There are also some further typical practical differentiation examples (Problems 70 to 81), page 391.

20.4 Further problems

Velocity and acceleration

1. The distance x metres moved by a body in a time t seconds is given by $x = 4t^3 - 3t^2 + 5t + 2$. Express the velocity and acceleration in terms of t and find their values when $t = 3$ s
   $[v = 12t^2 - 6t + 5; v_3 = 95 \text{ m/s}; a = 24t - 6; a_3 = 66 \text{ m/s}^2]$
2. A body obeys the equation $x = 3t - 20t^2$ where x is in metres and t is in seconds. Find expressions for velocity and acceleration. Find also its velocity and acceleration when $t = 1$ s
   $[v = 3 - 40t; v_1 = -37 \text{ m/s}; a = -40; a_1 = -40 \text{ m/s}^2]$
3. If the distance x metres travelled by a vehicle in t seconds after the brakes are applied is given by: $x = 22.5t - \frac{5}{6}t^2$, then

what is the speed in km/h when the brakes are applied? How far does the vehicle travel before it stops?

[81 km/h; 151.9 m]

4. An object moves in a straight line so that after t seconds its distance x metres from a fixed point on the line is given by $x = \frac{2}{3}t^3 - 5t^2 + 8t - 6$. Obtain an expression for the velocity and acceleration of the object after t seconds and hence calculate the values of t when the object is at rest.

[$v = 2t^2 - 10t + 8$; $a = 4t - 10$; $t = 1$ s or 4 s]

In Problems 5 to 9, x denotes the distance in metres of a body moving in a straight line, from a fixed point on the line and t denotes the time in seconds measured from a certain instant. Find the velocity and acceleration of the body when t has the given values. Find also the values of t when the body is momentarily at rest.

5. $x = \frac{4}{3}t^3 - 4t^2 + 3t - 2$; $t = 2$ [3 m/s; 8 m/s²; $t = \frac{1}{2}$ or $1\frac{1}{2}$ s]

6. $x = 3\cos 2t$; $t = \dfrac{\pi}{4}$ [-6 m/s; 0; $t = 0, \dfrac{\pi}{2}, \pi, \dfrac{3\pi}{2}$, etc.]

7. $x = t^4 - \frac{1}{2}t^2 + 1$; $t = 1$ [3 m/s; 11 m/s²; $t = 0$ or $\pm\frac{1}{2}$ s]

8. $x = 4t + 2\cos 2t$; $t = 0$

[4 m/s; -8 m/s²; $t = \dfrac{\pi}{4}, \dfrac{5\pi}{4}, \dfrac{9\pi}{4}, \dfrac{13\pi}{4}$, etc.]

9. $x = \dfrac{t^4}{4} - \frac{5}{3}t^3 + 3t^2 + 5$; $t = 2$ [0; -2 m/s²; $t = 0, 2$ or 3 s]

10. The distance s metres moved by a point in t seconds is given by $s = 5t^3 + 4t^2 - 3t + 2$. Find:
    (a) expressions for velocity and acceleration in terms of t,
    (b) the velocity and acceleration after 3 seconds, and
    (c) the average velocity over the fourth second.

    (a) [$v = (15t^2 + 8t - 3)$ m/s, $a = (30t + 8)$ m/s²]
    (b) [156 m/s; 98 m/s²]   (c) [210 m/s]

11. The distance x metres moved by a body in t seconds is given by $x = \frac{16}{3}t^3 - 32t^2 + 39t - 16$. Find:
    (a) the velocity and acceleration at the start,
    (b) the velocity and acceleration at $t = 3$ seconds,
    (c) the values of t when the body is at rest,
    (d) the value of t when the acceleration is 16 m/s², and
    (e) the distance travelled in the second second.

    (a) [39 m/s; -64 m/s²]   (b) [-9 m/s; 32 m/s²]
    (c) [$t = \frac{3}{4}$ or $3\frac{1}{4}$ s]   (d) [$2\frac{1}{2}$ s]
    (e) [$-19\frac{2}{3}$ m (i.e. in the opposite direction to that in which the body initially moved)]

12. The displacement y centimetres of the slide valve of an engine is given by the expression $y = 2.6 \cos 5\pi t + 3.8 \sin 5\pi t$. Find an expression for the velocity v of the valve and evaluate the velocity (in metres per second) when $t = 20$ ms
    $[v = 5\pi(3.8 \cos 5\pi t - 2.6 \sin 5\pi t); 0.442 \text{ m/s}]$

13. At any time t seconds the distance x metres of a particle moving in a straight line from a fixed point is given by:
    $x = 5t + \ln(1 - 2t)$. Find:
    (a) expressions for the velocity and acceleration in terms of t,
    (b) the initial velocity and acceleration,
    (c) the velocity and acceleration after 2 s, and
    (d) the time when the velocity is zero.

    (a) $\left[\left(5 - \dfrac{2}{(1-2t)}\right) \text{m/s}; \left(\dfrac{-4}{(1-2t)^2}\right) \text{m/s}^2\right]$

    (b) $[3 \text{ m/s}; -4 \text{ m/s}^2]$ (c) $[5\frac{2}{3} \text{ m/s}; -\frac{4}{9} \text{ m/s}^2]$ (d) $[\frac{3}{10} \text{ s}]$

14. If the equation $\theta = 12\pi + 27t - 3t^2$ gives the angle in radians through which a wheel turns in t seconds, find how many seconds the wheel takes to come to rest. Calculate the angle turned through in the last second of movement.
    $[4\frac{1}{2} \text{ s}; 3 \text{ radians}]$

15. A missile fired from ground level rises s metres in t seconds, and $s = 75t - 12.5t^2$. Determine:
    (a) the initial velocity of the missile,
    (b) the time when the height of the missile is a maximum,
    (c) the maximum height reached, and
    (d) the velocity with which the missile strikes the ground.
    (a) $[75 \text{ m/s}]$ (b) $[3 \text{ s}]$ (c) $[112.5 \text{ m}]$ (d) $[75 \text{ m/s}]$

Maximum and minimum values

In Problems 16 to 20 find the turning-points and distinguish between them by examining the sign of the slope on either side.

16. $y = 2x^2 - 4x$ 　　　[min. $(1, -2)$]
17. $y = 3t^2 - 2t + 6$ 　　　[min. $(\frac{1}{3}, 5\frac{2}{3})$]
18. $x = \theta^3 - 3\theta + 3$ 　　　[max. $(-1, 5)$; min. $(1, 1)$]
19. $y = 3x^3 + 6x^2 + 3x - 2$ 　　　[max. $(-1, -2)$; min. $(-\frac{1}{3}, -2\frac{4}{9})$]
20. $y = 7t^3 - 4t^2 - 5t + 6$ 　　　[max. $(-\frac{1}{3}, 6\frac{26}{27})$; min. $(\frac{5}{7}, 2\frac{46}{49})$]

Locate the turning-points on the curves in Problems 21 to 39 and determine whether they are maximum or minimum points.

21. $y = x(7 - x)$ 　　　[max. $(3\frac{1}{2}, 12\frac{1}{4})$]
22. $y = 4x^2 - 2x + 3$ 　　　[min. $(\frac{1}{4}, 2\frac{3}{4})$]
23. $y = 2x^3 + 7x^2 + 4x - 3$ 　　　[max. $(-2, 1)$; min. $(-\frac{1}{3}, -3\frac{17}{27})$]
24. $2pq = 18p^2 + 8$ 　　　[max. $(-\frac{2}{3}, -12)$; min. $(\frac{2}{3}, 12)$]

25. $y = 3t - e^{-t}$ [min. $(-1.098\,6, -0.295\,8)$]
26. $x = 3 \ln \theta - 4\theta$ [max. $(0.75, -3.863)$]
27. $S = 5t^3 - \frac{3}{2}t^2 - 12t + 6$ [max. $(-\frac{4}{5}, 12\frac{2}{25})$; min. $(1, -2\frac{1}{2})$]
28. $y = 4x - 2 \ln x$ [min. $(0.5, 3.386)$]
29. $y = 3x - e^x$ [max. $(1.098\,6, 0.295\,8)$]

30. $p = \dfrac{(q-1)(q-3)}{q}$

[max. $(-1.732, -7.464)$; min. $(1.732, -0.535\,9)$]

31. $y = \dfrac{(x-2)(x-5)}{(x-6)}$ [max. $(4, 1)$; min. $(8, 9)$]

32. $y = 3 \sin \theta - 4 \cos \theta$ in the range θ to 2π
[max. 5 at $143° \, 8'$; min. -5 at $323° \, 8'$]

33. $y = 4 \cos 2\theta + 3 \sin \theta$ in the range 0 to $\dfrac{\pi}{2}$ inclusive, given

that $\sin 2\theta = 2 \sin \theta \cos \theta$
[max. 4.281 2 at $10° \, 48'$; min. -1 at $90°$]
34. $V = l^2(l-1)$ [max. $(0, 0)$; min. $(\frac{2}{3}, \frac{-4}{27})$]

35. $y = 8x + \dfrac{1}{2x^2}$ [min. $(\frac{1}{2}, 6)$]

36. $x = t^3 + \dfrac{t^2}{2} - 2t + 4$ [max. $(-1, 5\frac{1}{2})$; min. $(\frac{2}{3}, 3\frac{5}{27})$]

37. $y = \dfrac{3x}{(x-1)(x-4)}$ [max. $(2, -3)$; min. $(-2, -\frac{1}{3})$]

38. $y = (x-1)^3 + 3x(x-2)$ [max. $(-1, 1)$; min. $(1, -3)$]

39. $y = \frac{1}{2} \ln(\sin x) - \sin x$ in the range 0 to $\dfrac{\pi}{4}$

[max. $-0.846\,6$ at $30°$]
40. (a) If $p + q = 7$, find the maximum value of $3pq + q^2$
(b) If $3a - 2b = 5$, find the least value of $2a^2b$
(a) $[55\frac{1}{8}]$ (b) $[-2\frac{14}{243}$ or $-2.057\,6]$
41. The sum of a number and its reciprocal is to be a minimum. Find the number.   [1]

Practical maximum and minimum problems

42. Find the maximum area of a rectangular piece of ground that can be enclosed by 200 m of fencing.   [$2\,500$ m²]
43. A rectangular area is formed using a piece of wire of length 26 cm. Find the dimensions of the rectangle if it is to enclose the maximum possible area.   [$6\frac{1}{2}$ cm by $6\frac{1}{2}$ cm]

44. A shell is projected upwards with a speed of 12 m/s and the distance vertically s metres is given by $s = 12t - 3t^2$, where t is the time in seconds. Find the maximum height reached.   [12 m]

45. A length of 42 cm of thin wire i   t into a rectangular shape with one side repeated. Finu the largest area that can be enclosed.   [73.5 cm²]

46. Find the area of the largest piece of rectangular ground that can be enclosed by 800 m of fencing if part of an existing wall is used on one side.   [80 000 m²]

47. The bending moment M of a beam of length l at a distance a from one end is given by $M = \dfrac{Wa}{2}(l - a)$, where W is the load per unit length. Find the maximum bending moment.
$$\left[\frac{Wl^2}{8}\right]$$

48. A lidless box with square ends is to be made from a thin sheet of metal. What is the least area of the metal for which the volume of the box is 6.64 m³?   [17.50 m²]

49. Find the height and the radius of a cylinder of volume 150 cm³ which has the least surface area.
[5.759 cm; 2.879 cm]

50. Find the height of a right circular cylinder of greatest volume which can be cut from a sphere of radius R $\left[\dfrac{2R}{\sqrt{3}}\right]$

51. The power P developed in a resistor R by a battery of e.m.f. E and internal resistance r is given by $P = \dfrac{E^2 R}{(R + r)^2}$

Differentiate P with respect to R and show that the power is a maximum when $R = r$

52. A piece of wire 5.0 m long is cut into two parts, one of which is bent into a square and the other into a circle. Find the diameter of the circle if the sum of their areas is a minimum.
[0.700 m]

53. Find the height of a cylinder of maximum volume which can be cut from a cone of height 15 cm and base radius 7.5 cm
[5 cm]

54. An alternating current is given by $i = 100 \sin(50\pi t + 0.32)$ amperes, where t is the time in seconds. Determine the maximum value of the current and the time when this maximum first occurs.   [100 amperes when $t = 7.96$ ms]

55. A frame for a box kite with a square cross-section is made

of 16 pieces of wood as shown in Fig. 20.9. Find the maximum volume of the frame if a total length of 12 m of wood is used.   [$\frac{4}{9}$ m³]

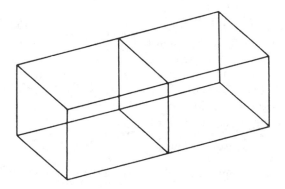

Fig. 20.9

56. A rectangular box with a lid which covers the top and front has a volume of 150 cm³ and the length of the base is to be $1\frac{1}{2}$ times the height. Find the dimensions of the box so that the surface area shall be a minimum.
     [3.816 cm by 5.724 cm by 6.867 cm]

57. The force F required to move a body along a rough horizontal plane is given by $F = \dfrac{\mu W}{\cos\theta + \mu\sin\theta}$ where μ is the coefficient of friction and θ the angle to the direction of F. If F varies with θ show that F is a minimum when $\tan\theta = \mu$

58. A closed cylindrical container has a surface area of 300 cm². Find its dimensions for maximum volume.
     [radius = 3.989 cm; height = 7.981 cm]

59. A rectangular block of metal, with a square cross-section, has a total surface area of 240 cm². Find the maximum volume of the block of metal.   [253.0 cm³]

60. The displacement s metres in a damped harmonic oscillation is given by $s = 4e^{-2t}\sin 2t$, where t is the time in milliseconds. Find the values of t to give maximum displacements. $\left[\dfrac{\pi}{8}, \dfrac{5\pi}{8}, \dfrac{9\pi}{8}, \text{ and so on}\right]$

61. A square sheet of metal of side 25.0 cm has squares cut from each corner, so that an open box may be formed. Find the surface area and the volume of the box if the volume is to be a maximum.   [555.6 cm²; 1 157 cm³]

62. A right circular cylinder of maximum volume is to be cut

from a sphere of radius 14.0 cm. Determine the base
diameter and the height of the cylinder.
[22.86 cm; 16.17 cm]

63. The speed v of a signal transmitted through a cable is given
by $v = kx^2 \ln \dfrac{1}{x}$, where x is the ratio of the inner to the outer
diameters of the core and k is a constant. Find the value of x
for maximum speed of the transmitted signal.
$[x = e^{-\frac{1}{2}} = 0.606\,5]$

64. An open rectangular box with square ends is fitted with an
overlapping lid which covers the whole of the square ends,
the open top and the front face. Find the maximum volume
of the box if 8.0 m² of metal are used altogether.
[0.871 m³]

65. An electrical voltage E is given by:
$E = 12.0 \sin 50\pi t + 36.0 \cos 50\pi t$ volts, where t is the time in
seconds. Determine the maximum value of E
[37.95 volts]

66. The velocity v of a piston of a reciprocating engine can be
expressed by $v = 2\pi nr \left(\dfrac{\sin 2\theta}{16} + \sin \theta \right)$, where n and r are
constants. Find the value of θ between $0°$ and $360°$ that
makes the velocity a maximum. (Note: $\cos 2\theta = 2\cos^2 \theta - 1$)
[83° 2′]

67. The periodic time T of a compound pendulum of variable
height h is given by $T = 2\pi \sqrt{\left[\dfrac{h^2 + k^2}{gh}\right]}$, where k and g are
constants. Find the minimum value of T
$\left[T_{\min} = 2\pi \sqrt{\left(\dfrac{2k}{g}\right)} \text{ when } h = k \right]$

68. The heat capacity (C) of carbon monoxide varies with
absolute temperature (T) as shown:
$C = 26.53 + 7.70 \times 10^{-3}T - 1.17 \times 10^{-6}T^2$. Determine the
maximum value of C and the temperature at which it occurs.
$[C = 39.20, T = 3.291 \times 10^3]$

69. The electromotive force (E) of the Clark cell is given by
$E = 1.4 - 0.001\,2(T - 288) - 0.000\,007(T - 288)^2$ volts.
Determine the maximum value of E [1.451 4 volts]

Practical differentiation

70. The length l metres of a certain rod at temperature $t\,°C$ is
given by $l = 1 + 0.000\,02t + 0.000\,000\,2t^2$. Find the rate at

which l increases with respect to t $\left(\text{i.e. } \dfrac{dl}{dt}\right)$ when the temperature is: (a) 100 °C, and (b) 300 °C

(a) $[0.00006 \text{ m °C}^{-1}]$ (b) $[0.00014 \text{ m °C}^{-1}]$

71. An alternating voltage v volts is given by $v = 125 \sin 80t$, where t is the time in seconds. Calculate the rate of change of voltage $\left(\text{i.e. } \dfrac{dv}{dt}\right)$ when $t = 20$ ms

[−292 volts per second]

72. In a first-order reaction the concentration c after time t is governed by the relation $c = ae^{-kt}$, where a is the initial concentration and k is the rate constant. Show that the value of $\dfrac{dc}{dt}$ is given by $-kc$

73. A displacement s metres is given by $s = \frac{1}{3}t^3 - 4t^2 + 15t$, where t is the time in seconds. At what times is the velocity v $\left(\text{i.e. } \dfrac{ds}{dt}\right)$ zero? Determine the acceleration $\left(\text{i.e. } \dfrac{dv}{dt}\right)$ when $t = 6$ seconds. $[t = 3 \text{ s or } 5 \text{ s; } 4 \text{ m/s}^2]$

74. The luminous intensity, I candelas, of a lamp at different voltages V is given by $I = 4 \times 10^{-4}V^2$. Find the voltage at which the light is increasing at a rate of 0.2 candelas per volt $\left(\text{i.e. when } \dfrac{dI}{dV} = 0.2\right)$ [250 V]

75. The relationship between pressure p and volume v is given by $pv^n = k$, where n and k are constants. Prove that
$$v^{n+1}\frac{dp}{dv} + kn = 0$$

76. An alternating current i amperes is given by $i = 70 \sin 2\pi ft$, where f is the frequency in hertz and t the time in seconds. Find the rate of change of current $\left(\text{i.e. } \dfrac{di}{dt}\right)$ when $t = 20$ ms, given that the frequency is 50 hertz.
[7000π A/s]

77. Newton's law of cooling is given by $\theta = \theta_0 e^{-kt}$, where the excess of temperature at zero time is θ_0 °C and at time t seconds is θ °C. Determine the rate of change of temperature $\left(\text{i.e. } \dfrac{d\theta}{dt}\right)$, given that $\theta_0 = 15$ °C, $k = -0.017$ and $t = 60$ seconds. $[0.707 \text{ °C s}^{-1}]$

78. A coil has a self inductance L of 2 henrys and a resistance R of 100 ohms. A d.c. supply voltage E of 100 volts is applied

to the coil. The instantaneous current i amperes is given by

$i = \dfrac{E}{R}(1 - e^{-Rt/L})$. Find: (a) the rate at which the current

increases at the moment of switching on $\left(\text{i.e. } \dfrac{di}{dt} \text{ when } t = 0\right)$,

and (b) the rate after 10 ms (a) [50 A/s] (b) [30.3 A/s]

79. A fully charged capacitor C of 0.1 microfarads has a
potential difference V of 150 volts between its plates. The
capacitor is then discharged through a resistor R of 1
megaohm. If the potential difference v across the plate at any
time t seconds after closing the circuit is given by
$v = Ve^{-t/RC}$ calculate:

(a) the initial rate of loss of voltage $\left(\text{i.e. } \dfrac{dv}{dt} \text{ at } t = 0\right)$,

(b) the rate after 0.1 seconds.
(a) [1 500 V/s] (b) [552 V/s]

80. The pressure p of the atmosphere at height h above ground
level is given by $p = p_0 e^{-h/c}$, where p_0 is the pressure at
ground level and c is a constant.
Determine the rate of change of pressure with height

$\left(\text{i.e. } \dfrac{dp}{dh}\right)$ when p_0 is 1.013×10^5 pascals and c is 6.062×10^4 at

1 500 metres. [−1.630 Pa/m]

81. The displacement s cm of the end of a stiff spring at time t
seconds is given by $s = ae^{-kt} \sin 2\pi ft$. Find the velocity v

$\left(\text{i.e. } \dfrac{ds}{dt}\right)$ and acceleration a $\left(\text{i.e. } \dfrac{d^2s}{dt^2}\right)$ of the end of the

spring after 3 seconds if $a = 4$, $k = 0.8$ and $f = 2$
[4.56 cm/s, −7.30 cm/s²]

Chapter 21

Introduction to integration

21.1 The process of integration

The process of integration reverses the process of differentiation. In differentiation, if $f(x) = x^2$ then $f'(x) = 2x$. Since integration reverses the process of moving from $f(x)$ to $f'(x)$, it follows that the integral of $2x$ is x^2, i.e. it is the process of moving from $f'(x)$ to $f(x)$. Similarly, if $y = x^3$ then $\dfrac{dy}{dx} = 3x^2$. Reversing this process shows that the integral of $3x^2$ is x^3.

Integration is also a process of summation or adding parts together and an elongated 'S', shown as $\int$, is used to replace the words 'the integral of'. Thus

$$\int 2x = x^2 \quad \text{and} \quad \int 3x^2 = x^3$$

In differentiation, the differential coefficient $\dfrac{dy}{dx}$ or $\dfrac{d}{dx}[f(x)]$ indicates that a function of x is being differentiated with respect to x, the dx indicating this. In integration, the variable of integration is shown by adding d (the variable) after the function to be integrated. Thus $\int 2x \, dx$ means 'the integral of $2x$ with respect to x' and $\int 3u^2 \, du$ means 'the integral of $3u^2$ with respect to u'. It follows that $\int y \, dx$ means 'the integral of y with respect to x' and since only functions of x can be integrated with respect to x, y must be expressed as a function of x before the process of integration can be performed.

The arbitrary constant of integration

The differential coefficient of x^2 is $2x$, hence $\int 2x \, dx = x^2$. Also the differential coefficient of $x^2 + 3$ is $2x$, hence $\int 2x \, dx = x^2 + 3$. Since the differential coefficient of any constant is zero, it follows that the differential coefficient of $x^2 + c$, where c is any constant, is $2x$. To allow for the possible presence of this constant, whenever the process of integration is performed the constant should be added to the result. Hence $\int 2x \, dx = x^2 + c$. c is called the arbitrary **constant of integration** and it is important to include it in all work involving the process of determining integrals. Its omission will result in obtaining incorrect solutions in later work, such as in the solution of differential equations (see Chapters 33 and 34).

21.2   The general solution of integrals of the form x^n

$$\int x^n \, dx = \frac{x^{n+1}}{n+1} + c$$

In order to integrate x^n it is necessary to

(a)   increase the power of x by 1, i.e. the power of x^n is raised by 1 to x^{n+1},
(b)   divide by the new power of x, i.e. x^{n+1} is divided by $n+1$, and
(c)   add the arbitrary constant of integration, c.

Thus to integrate x^4, the power of x is increased by 1 to x^{n+1} or x^{4+1}, i.e. x^5, and the term is divided by $(n+1)$ or $(4+1)$, i.e. 5. So the integral of x^4 is $\dfrac{x^5}{5} + c$. In the general solution of $\int x^n \, dx$, given above, n may be a positive or negative integer or fraction, or zero, with just one exception, that being $n = -1$.

It was shown in differentiation that $\dfrac{d}{dx}(\ln x) = \dfrac{1}{x}$. Thus

$$\int \frac{1}{x} \, dx \left(= \int x^{-1} \, dx \right) = \ln x + c$$

More generally

$$\frac{d}{dx}(\ln ax) = \frac{d}{dx}(\ln x + \ln a) = \frac{1}{x} + 0$$

Therefore $\displaystyle\int \frac{1}{x} \, dx = \ln ax + c$

Rules of integration

Three of the basic rules of integration are:

(i)   The integral of a constant k is $kx + c$. For example,

$$\int 5 \, dx = \int 5x^0 \, dx \text{ since } x^0 = 1$$

Applying the standard integral $\displaystyle\int x^n \, dx = \frac{x^{n+1}}{n+1} + c$ gives

$$\int 5 \, dx = \frac{5x^{0+1}}{0+1} + c = 5x + c$$

(ii)   As in differentiation, constants associated with variables are carried
forward, i.e. they are not involved in the integration. For example,

$$\int 3x^4 \, dx = 3 \int x^4 \, dx = 3\left(\frac{x^5}{5}\right) + c = \frac{3}{5}x^5 + c$$

(iii) As in differentiation, the rules of algebra apply where functions of a
variable are added or subtracted. For example,

$$\int (x^2 + x^5) \, dx = \int x^2 \, dx + \int x^5 \, dx = \frac{x^3}{3} + \frac{x^6}{6} + c$$

and
$$\int (2x^3 + 4) \, dx = 2 \int x^3 \, dx + \int 4 \, dx = 2\left(\frac{x^4}{4}\right) + 4x + c$$

$$= \frac{x^4}{2} + 4x + c$$

It should be noted that only one constant c is included since any sum of
arbitrary constants gives another arbitrary constant. Combining rule (ii) with
the standard integral for x^n gives

$$\int ax^n \, dx = \frac{ax^{n+1}}{n+1} + c$$

where a and n are constants and n is **not** equal to -1.

Integrals written in this form are called 'indefinite integrals', since their
precise value cannot be found (i.e. c cannot be calculated) unless additional
information is provided. (In differentiation there are special rules for
multiplication and division of functions. However, there are no such special
rules for multiplication and division in integration.)

21.3   Definite integrals

Limits can be applied to integrals and such integrals are then called 'definite
integrals'. The increase in the value of the integral $(x^2 - 3)$ as x increases
from 1 to 2 can be written as

$$\left[\int (x^2 - 3) \, dx\right]_1^2$$

However, this is invariably abbreviated by showing the value of the upper
limit at the top of the integral sign and the value of the lower limit at the
bottom, i.e.

$$\left[\int (x^2 - 3) \, dx\right]_1^2 = \int_1^2 (x^2 - 3) \, dx$$

The integral is evaluated as for an indefinite integral and then placed in the square brackets of the limit operator. Thus

$$\int_1^2 (x^2 - 3)\, dx = \left[\frac{x^3}{3} - 3x + c \right]_1^2$$

$$= \left[\frac{(2)^3}{3} - 3(2) + c \right] - \left[\frac{(1)^3}{3} - 3(1) + c \right]$$

$$= (\tfrac{8}{3} - 6 + c) - (\tfrac{1}{3} - 3 + c)$$

$$= 2\tfrac{2}{3} - 6 - \tfrac{1}{3} + 3 = -\tfrac{2}{3}$$

The arbitrary constant of integration, c, always cancels out when limits are applied to an integral and it is not usually shown when evaluating a definite integral.

21.4 Integrals of sin *ax*, cos *ax*, sec² *ax* and *e^{ax}*

Since integration is the reverse process to that of differentiation the following standard integrals may be deduced:

(*a*) $\dfrac{d}{dx} (\sin x) = \cos x$

Hence $\displaystyle\int \cos x\, dx = \sin x + c$

More generally $\dfrac{d}{dx} (\sin ax) = a \cos ax$

Hence $\displaystyle\int a \cos ax\, dx = \sin ax + c$

$$\int \cos ax\, dx = \frac{1}{a} \sin ax + c$$

(*b*) $\dfrac{d}{dx} (\cos x) = -\sin x$

Hence $\displaystyle\int -\sin x\, dx = \cos x + c$

$$\int \sin x\, dx = -\cos x + c$$

More generally $\dfrac{d}{dx} (\cos ax) = -a \sin ax$

Hence $\displaystyle\int -a \sin ax\, dx = \cos ax + c$

$$\int \sin ax\, dx = -\frac{1}{a} \cos ax + c$$

(c) $\dfrac{d}{dx}(\tan x) = \sec^2 x$

Hence $\displaystyle\int \sec^2 x \, dx = \tan x + c$

More generally $\dfrac{d}{dx}(\tan ax) = a \sec^2 ax$

Hence $\displaystyle\int a \sec^2 ax \, dx = \tan ax + c$

$$\int \sec^2 ax \, dx = \frac{1}{a}\tan ax + c$$

(d) $\dfrac{d}{dx}(e^x) = e^x$

Hence $\displaystyle\int e^x \, dx = e^x + c$

More generally $\dfrac{d}{dx}(e^{ax}) = a e^{ax}$

Hence $\displaystyle\int a e^{ax} \, dx = e^{ax} + c$

$$\int e^{ax} \, dx = \frac{1}{a} e^{ax} + c$$

Summary of standard integrals

1.	$\displaystyle\int ax^n \, dx = \dfrac{ax^{n+1}}{n+1} + c$ (except where $n = -1$)
2.	$\displaystyle\int \cos ax \, dx = \dfrac{1}{a}\sin ax + c$
3.	$\displaystyle\int \sin ax \, dx = -\dfrac{1}{a}\cos ax + c$
4.	$\displaystyle\int \sec^2 ax \, dx = \dfrac{1}{a}\tan ax + c$
5.	$\displaystyle\int e^{ax} \, dx = \dfrac{1}{a} e^{ax} + c$
6.	$\displaystyle\int \dfrac{1}{x} \, dx = \ln x + c$

Worked problems on standard integrals

Problem 1. Integrate the following with respect to the variable:
(a) x^7, (b) $5.2y^{1.6}$ and (c) $\dfrac{2}{p^3}$

(a) $\displaystyle\int x^7\,dx = \frac{x^{7+1}}{7+1} + c = \frac{x^8}{8} + c$

(b) $\displaystyle\int 5.2y^{1.6}\,dy = \frac{5.2y^{1.6+1}}{1.6+1} + c = \frac{5.2y^{2.6}}{2.6} + c = 2.0y^{2.6} + c$

(c) $\displaystyle\int \frac{2}{p^3}\,dp = \int 2p^{-3}\,dp = \frac{2p^{-3+1}}{-3+1} + c = \frac{2p^{-2}}{-2} + c$

$\qquad = \dfrac{-1}{p^2} + c$

If the final answer of an integration is differentiated then
the original must result (otherwise an error has occurred).
For example, in (a) above

$\dfrac{d}{dx}\left(\dfrac{x^8}{8} + c\right) = \dfrac{8x^7}{8} = x^7$ (i.e. the original integral)

It will be assumed that in all future integral problems such a
check will be made.

Problem 2. Integrate with respect to the variable
(a) $\left(2x^5 - 4\sqrt{x} + \dfrac{5}{x^4} - \dfrac{2}{\sqrt{x^3}} + 6\right)$ and (b) $\left(\dfrac{4p^5 - 3 + p}{p^3}\right)$

(a) $\displaystyle\int \left(2x^5 - 4\sqrt{x} + \frac{5}{x^4} - \frac{2}{\sqrt{x^3}} + 6\right) dx$

$\qquad = \displaystyle\int (2x^5 - 4x^{\frac{1}{2}} + 5x^{-4} - 2x^{-\frac{3}{2}} + 6)\,dx$

$\qquad = \dfrac{2x^{5+1}}{(5+1)} - \dfrac{4x^{\frac{1}{2}+1}}{(\frac{1}{2}+1)} + \dfrac{5x^{-4+1}}{(-4+1)} - \dfrac{2x^{-\frac{3}{2}+1}}{(-\frac{3}{2}+1)} + 6x + c$

$\qquad = \dfrac{2x^6}{6} - \dfrac{4x^{\frac{3}{2}}}{\frac{3}{2}} + \dfrac{5x^{-3}}{-3} - \dfrac{2x^{-\frac{1}{2}}}{-\frac{1}{2}} + 6x + c$

$\qquad = \dfrac{x^6}{3} - \dfrac{8\sqrt{x^3}}{3} - \dfrac{5}{3x^3} + \dfrac{4}{\sqrt{x}} + 6x + c$

(b) $\int\left(\dfrac{4p^5 - 3 + p}{p^3}\right)dp = \int\left(\dfrac{4p^5}{p^3} - \dfrac{3}{p^3} + \dfrac{p}{p^3}\right)dp$

$$= \int(4p^2 - 3p^{-3} + p^{-2})\,dp$$

$$= \dfrac{4p^3}{3} - \dfrac{3p^{-2}}{-2} + \dfrac{p^{-1}}{-1} + c$$

$$= \dfrac{4}{3}p^3 + \dfrac{3}{2p^2} - \dfrac{1}{p} + c$$

Problem 3. Given $y = \int\left(r + \dfrac{1}{r}\right)^2 dr$, find the value of the arbitrary constant of integration if $y = \frac{1}{3}$ when $r = 1$

$$y = \int\left(r + \dfrac{1}{r}\right)^2 dr = \int\left(r^2 + 2 + \dfrac{1}{r^2}\right)dr$$

$$= \dfrac{r^3}{3} + 2r - \dfrac{1}{r} + c$$

$y = \frac{1}{3}$ when $r = 1$

Hence $\qquad\qquad \frac{1}{3} = \dfrac{(1)^3}{3} + 2(1) - \dfrac{1}{(1)} + c$

$\qquad\qquad\quad \frac{1}{3} = \frac{1}{3} + 2 - 1 + c$

$\qquad\qquad\quad c = -1$

Hence **the arbitrary constant of integration is** -1

Problem 4. Integrate with respect to the variable
(a) $4\cos 3\theta$, (b) $7\sin 2x$ and (c) $3\sec^2 5t$

(a) $\int 4\cos 3\theta\,d\theta = 4\left(\dfrac{1}{3}\sin 3\theta\right) + c = \dfrac{4}{3}\sin 3\theta + c$

(b) $\int 7\sin 2x\,dx = 7\left(-\dfrac{1}{2}\cos 2x\right) + c = -\dfrac{7}{2}\cos 2x + c$

(c) $\int 3\sec^2 5t\,dt = 3\left(\dfrac{1}{5}\tan 5t\right) + c = \dfrac{3}{5}\tan 5t + c$

Problem 5. Find (a) $\int 6\,e^{4x}\,dx$, (b) $\int\dfrac{3}{e^{2t}}\,dt$ and (c) $\int\dfrac{3}{2u}\,du$

(a) $\int 6\,e^{4x}\,dx = 6\left(\dfrac{e^{4x}}{4}\right) + c = \dfrac{3}{2}e^{4x} + c$

(b) $\displaystyle\int \frac{3}{e^{2t}}\,dt = \int 3\,e^{-2t}\,dt = 3\left(\frac{e^{-2t}}{-2}\right) + c = \frac{-3}{2}\,e^{-2t} + c = \frac{-3}{2\,e^{2t}} + c$

(c) $\displaystyle\int \frac{3}{2u}\,du = \frac{3}{2}\int \frac{1}{u}\,du = \frac{3}{2}\ln u + c$

Problem 6. Evaluate (a) $\displaystyle\int_1^3 (4x-3)^2\,dx$ and (b) $\displaystyle\int_0^4 \left(5\sqrt{b}-\frac{1}{\sqrt{b}}\right)\,db$

(a) $\displaystyle\int_1^3 (4x-3)^2\,dx = \int_1^3 (16x^2 - 24x + 9)\,dx$

$$= \left[\frac{16}{3}x^3 - 24\frac{x^2}{2} + 9x + c\right]_1^3$$

$$= \left[\frac{16}{3}(3)^3 - 12(3)^2 + 9(3) + c\right]$$

$$- \left[\frac{16}{3}(1)^3 - 12(1)^2 + 9(1) + c\right]$$

$$= (144 - 108 + 27 + c) - (5\tfrac{1}{3} - 12 + 9 + c)$$

$$= (63 + c) - (2\tfrac{1}{3} + c) = \mathbf{60\tfrac{2}{3}}$$

The arbitrary constant of integration, c, cancels out, thus showing it to be an unnecessary inclusion when evaluating definite integrals.

(b) $\displaystyle\int_0^4 \left(5\sqrt{b}-\frac{1}{\sqrt{b}}\right)\,db = \int_0^4 (5b^{\frac{1}{2}} - b^{-\frac{1}{2}})\,db$

$$= \left[\frac{5b^{\frac{3}{2}}}{\frac{3}{2}} - \frac{b^{\frac{1}{2}}}{\frac{1}{2}}\right]_0^4 = \left[\frac{10}{3}\sqrt{b^3} - 2\sqrt{b}\right]_0^4$$

$$= \left(\frac{10}{3}\sqrt{4^3} - 2\sqrt{4}\right) - \left(\frac{10}{3}\sqrt{0^3} - 2\sqrt{0}\right)$$

$$= \frac{10}{3}(8) - 2(2) - 0 = \frac{80}{3} - 4 = \mathbf{22\tfrac{2}{3}}$$

(taking positive values of square roots only)

Problem 7. Evaluate (a) $\displaystyle\int_0^{\pi/2} 4\sin 2x\,dx$, (b) $\displaystyle\int_0^1 3\cos 3t\,dt$ and

(c) $\displaystyle\int_{\pi/6}^{\pi/3} (2\sin\theta - 3\cos 2\theta + 4\sec^2\theta)\,d\theta$

(a) $\displaystyle\int_0^{\pi/2} 4 \sin 2x \, dx = \left[-\frac{4}{2} \cos 2x \right]_0^{\pi/2}$

$$= \left(-2 \cos 2\left(\frac{\pi}{2}\right) \right) - \left(-2 \cos 2(0) \right)$$

$$= (-2 \cos \pi) - (-2 \cos 0)$$

$$= (-2(-1)) - (-2(1))$$

$$= 2 + 2 = 4$$

(b) $\displaystyle\int_0^1 3 \cos 3t \, dt = \left[\frac{3}{3} \sin 3t \right]_0^1 = \left[\sin 3t \right]_0^1 = (\sin 3 - \sin 0)$

The limits in trigonometric functions are expressed in radians. Thus 'sin 3' means 'the sine of 3 radians' or $3\left(\dfrac{180}{\pi}\right)^\circ$, i.e. 171.89°. Hence

$\sin 3 - \sin 0 = \sin 171.89° - \sin 0° = 0.141\,1 - 0 = 0.141\,1$

Thus $\displaystyle\int_0^1 3 \cos 3t \, dt = 0.141\,1$

(c) $\displaystyle\int_{\pi/6}^{\pi/3} (2 \sin \theta - 3 \cos 2\theta + 4 \sec^2 \theta) \, d\theta$

$$= \left[-2 \cos \theta - \frac{3}{2} \sin 2\theta + 4 \tan \theta \right]_{\pi/6}^{\pi/3}$$

$$= \left(-2 \cos \frac{\pi}{3} - \frac{3}{2} \sin \frac{2\pi}{3} + 4 \tan \frac{\pi}{3} \right)$$

$$- \left(-2 \cos \frac{\pi}{6} - \frac{3}{2} \sin \frac{2\pi}{6} + 4 \tan \frac{\pi}{6} \right)$$

or $(-2 \cos 60° - \frac{3}{2} \sin 120° + 4 \tan 60°)$

$- (-2 \cos 30° - \frac{3}{2} \sin 60° + 4 \tan 30°)$

$= (-1 - 1.299\,0 + 6.928\,2)$

$- (-1.732\,1 - 1.299\,0 + 2.309\,4)$

$= \mathbf{5.350\,9}$

Problem 8. Evaluate (a) $\displaystyle\int_1^2 3 \, e^{4x} \, dx$ and (b) $\displaystyle\int_3^4 \frac{5}{x} \, dx$.

(a) $\displaystyle\int_1^2 3\,e^{4x}\,dx = [\tfrac{3}{4}\,e^{4x}]_1^2 = \tfrac{3}{4}\,e^8 - \tfrac{3}{4}\,e^4 = \tfrac{3}{4}\,e^4(e^4 - 1) = \mathbf{2\,195}$

(b) $\displaystyle\int_3^4 \frac{5}{x}\,dx = 5[\ln x]_3^4 = 5[\ln 4 - \ln 3] = 5\,\ln\tfrac{4}{3} = \mathbf{1.438\,4}$

Further problems on standard integrals may be found in the following Section 21.5 (Problems 1 to 65).

21.5   Further problems

In Problems 1 to 35 integrate with respect to the variable.

1. x^5 $\left[\dfrac{x^6}{6} + c\right]$

2. $2p^3$ $\left[\dfrac{p^4}{2} + c\right]$

3. $3k^6$ $[\tfrac{3}{7}k^7 + c]$

4. $4u^{2.3}$ $\left[\dfrac{4}{3.3}u^{3.3} + c\right]$

5. $x^{-2.1}$ $\left[\dfrac{-x^{-1.1}}{1.1} + c\right]$

6. $\dfrac{2}{x^2}$ $\left[\dfrac{-2}{x} + c\right]$

7. $\dfrac{3}{p}$ $[3 \ln p + c]$

8. $\sqrt{y}$ $[\tfrac{2}{3}\sqrt{y^3} + c]$

9. $2\sqrt{S^3}$ $[\tfrac{4}{5}\sqrt{S^5} + c]$

10. $\dfrac{1}{3\sqrt{t}}$ $[\tfrac{2}{3}\sqrt{t} + c]$

11. $\dfrac{4}{\sqrt[3]{k^2}}$ $[12\sqrt[3]{k} + c]$

12. $3a^3 - \tfrac{2}{3}\sqrt{a}$ $\left[\dfrac{3a^4}{4} - \tfrac{4}{9}\sqrt{a^3} + c\right]$

13. $\dfrac{-4}{v^{1.4}}$ $\left[\dfrac{10}{v^{0.4}} + c\right]$

14. $\dfrac{x}{3}(2x + \sqrt{x})$ $\left[\dfrac{2x^3}{9} + \dfrac{2}{15}\sqrt{x^5} + c\right]$

15. $\dfrac{r^3 + 2r - 1}{r^2}$ $\left[\dfrac{r^2}{2} + 2 \ln r + \dfrac{1}{r} + c\right]$

16. $(x + 2)^2$ $\left[\dfrac{x^3}{3} + 2x^2 + 4x + c\right]$

17. $(1 + \sqrt{w})^2$ $\left[w + \tfrac{4}{3}\sqrt{w^3} + \dfrac{w^2}{2} + c\right]$

18. $\sin 2\theta$ $[-\frac{1}{2}\cos 2\theta + c]$

19. $\cos 4\alpha$ $[\frac{1}{4}\sin 4\alpha + c]$

20. $2\sin 3t$ $[-\frac{2}{3}\cos 3t + c]$

21. $-4\cos 5x$ $[-\frac{4}{5}\sin 5x + c]$

22. $\sec^2 6\beta$ $[\frac{1}{6}\tan 6\beta + c]$

23. $-3\sec^2 t$ $[-3\tan t + c]$

24. $4(\cos 2\theta - 3\sin \theta)$ $[2(\sin 2\theta + 6\cos \theta) + c]$

25. e^{3x} $\left[\dfrac{e^{3x}}{3} + c\right]$

26. $2e^{-4t}$ $[-\frac{1}{2}e^{-4t} + c]$

27. $\dfrac{6}{e^t}$ $\left[-\dfrac{6}{e^t} + c\right]$

28. $3(e^x - e^{-x})$ $[3(e^x + e^{-x}) + c]$

29. $3(e^t - 1)^2$ $\left[3\left(\dfrac{e^{2t}}{2} - 2e^t + t\right) + c\right]$

30. $\dfrac{4}{e^{2x}} + e^x$ $\left[\dfrac{-2}{e^{2x}} + e^x + c\right]$

31. $\dfrac{1}{4t}$ $[\frac{1}{4}\ln t + c]$

32. $\dfrac{3}{5t} + \sqrt{t^5}$ $[\frac{3}{5}\ln t + \frac{2}{7}\sqrt{t^7} + c]$

33. $\left(\dfrac{1}{x} + x\right)^2$ $\left[-\dfrac{1}{x} + 2x + \dfrac{x^3}{3} + c\right]$

34. $3\sin 50\pi t + 4\cos 50\pi t$ $\left[\dfrac{1}{50\pi}(4\sin 50\pi t - 3\cos 50\pi t) + c\right]$

35. $(e^{2x} - 1)(e^{-2x} + 1)$ $[\frac{1}{2}(e^{2x} + e^{-2x}) + c]$

In Problems 36 to 65 evaluate the definite integrals.
(Where roots are involved in the solution, take positive values
only when evaluating.)

36. $\displaystyle\int_1^3 2\,dt$ $[4]$ 37. $\displaystyle\int_3^5 4x\,dx$ $[32]$

38. $\displaystyle\int_{-4}^2 -3u^2\,du$ $[-72]$ 39. $\displaystyle\int_{-1}^1 \frac{3}{4}f^2\,df$ $[\frac{1}{2}]$

40. $\displaystyle\int_1^4 x^{-1.5}\,dx$ $[1]$ 41. $\displaystyle\int_1^9 \frac{dx}{\sqrt{x}}$ $[4]$

42. $\displaystyle\int_2^5 \frac{4}{x}\,dx$ [3.665]

43. $\displaystyle\int_0^2 (x^2 + 2x - 1)\,dx$ [$4\frac{2}{3}$]

44. $\displaystyle\int_1^4 \left(\sqrt{r} - \frac{1}{\sqrt{r}}\right)dr$ [$2\frac{2}{3}$]

45. $\displaystyle\int_1^4 (3x^3 - 4x^2 + x - 2)\,dx$ [$108\frac{3}{4}$]

46. $\displaystyle\int_1^3 (m - 2)(m - 1)\,dm$ [$\frac{2}{3}$]

47. $\displaystyle\int_1^2 \left(\frac{1}{x^2} + \frac{1}{x} + \frac{1}{2}\right)dx$ [1.693]

48. $\displaystyle\int_{-2}^2 (3x - 1)\,dx$ [-4]

49. $\displaystyle\int_1^3 \left(\frac{2}{t^2} - 3t^2 + 4\right)dt$ [$-16\frac{2}{3}$]

50. $\displaystyle\int_0^{\pi/2} \sin\theta\,d\theta$ [1]

51. $\displaystyle\int_0^{\pi/3} 3\sin 2x\,dx$ [$2\frac{1}{4}$]

52. $\displaystyle\int_0^{\pi/6} 4\sin 3\theta\,d\theta$ [$1\frac{1}{3}$]

53. $\displaystyle\int_{\pi/6}^{\pi/3} 2\cos t\,dt$ [0.732 1]

54. $\displaystyle\int_0^1 5\sin 2\theta\,d\theta$ [3.540 4]

55. $\displaystyle\frac{1}{2}\int_1^2 \cos 3\alpha\,d\alpha$ [-0.070 1]

56. $\displaystyle\int_{0.1}^{0.6} (\tfrac{1}{4}\sin 3\beta + \tfrac{1}{2}\cos 2\beta)\,d\beta$ [0.281 9]

57. $\displaystyle\int_{-\pi/2}^{\pi/2} 3\cos\theta\,d\theta$ [6]

58. $\displaystyle\int_0^{\pi/4} 3\sec^2\theta\,d\theta$ [3]

59. $\displaystyle\int_{-1}^{1} 3 \sec^2 2t \, dt$ $\qquad$ $[-6.555]$

60. $\displaystyle\int_{1}^{2} \frac{e^{3x}}{5} \, dx$ $\qquad$ $[25.56]$

61. $\displaystyle\int_{0.4}^{0.7} 3 \, e^{2t} \, dt$ $\qquad$ $[2.744]$

62. $\displaystyle\int_{0}^{1} \frac{2}{e^{3t}} \, dt$ $\qquad$ $[0.633\ 5]$

63. $\displaystyle\int_{1}^{4} \left(\frac{t+2}{\sqrt{t}} \right) dt$ $\qquad$ $[8\frac{2}{3}]$

64. $\displaystyle\int_{1}^{3} \frac{(3x+2)(x-4)}{x} \, dx$ $\qquad$ $[-16.789]$

65. $\displaystyle\int_{0}^{1} 2\sqrt{x}(x+2)^2 \, dx$ $\qquad$ $[9.105]$

Chapter 22

Integration by algebraic substitutions

22.1   Introduction

Chapter 21 introduced the process of integration for standard integrals. However, functions which require integrating are not always in the standard integral form. In such cases it may be possible by one of five methods to change the function into a form which can be readily integrated.

The methods available are:

(*a*)   by using an algebraic substitution (see Section 22.2 of this chapter),

(*b*)   by using trigonometric identities and substitutions (see Chapter 23),

(*c*)   by using partial fractions (see Chapter 24),

(*d*)   by using the $t = \tan \dfrac{\theta}{2}$ substitution (see Chapter 25), and

(*e*)   by using integration by parts (see Chapter 26).

It should be realised that many mathematical functions cannot be integrated by any of the above methods and approximate methods have then to be used. This is explained in Chapter 27 on Numerical Integration.

22.2   Algebraic substitutions

With algebraic substitutions, the substitution usually made is to let u be equal to $f(x)$, such that $f(u)\,\mathrm{d}u$ is a standard integral.

A most important point in the use of substitution is that once a substitution has been made the original variable must be removed completely, because a variable can only be integrated with respect to itself, i.e. we cannot integrate, for example, a function of t with respect to x.

A concept that $\dfrac{\mathrm{d}u}{\mathrm{d}x}$ is a single entity (measuring the differential coefficient of u with respect to x) has been established in the work done on differentiation. Frequently in work on integration and differential equations, $\dfrac{\mathrm{d}u}{\mathrm{d}x}$ is split. Provided that when this is done, the original differential coefficient can be re-formed by applying the rules of algebra, then it is in order to do it. For example, if $\dfrac{dy}{dx} = x$ then it is in order to write $dy = x \, dx$ since dividing both sides by dx re-forms the original differential coefficient. This principle is shown in the following worked problems where it is found that integrals of the forms

$$k \int [f(x)]^n f'(x) \, dx \text{ and } k \int \dfrac{f'x}{[f(x)]^n} \, dx$$

(where k and n are constants) can both be integrated by substituting u for $f(x)$

Worked problems on integration by algebraic substitution

Problem 1. Find $\displaystyle\int \cos(5x + 2) \, dx$

Let $u = 5x + 2$

then $\dfrac{du}{dx} = 5$, i.e. $dx = \dfrac{du}{5}$

$$\int \cos(5x + 2) \, dx = \int \cos u \, \dfrac{du}{5} = \dfrac{1}{5} \int \cos u \, du$$

$$= \tfrac{1}{5}(\sin u) + c$$

Since the original integral is given in terms of x, the result should be stated in terms of x.

$u = 5x + 2$

Hence $\displaystyle\int \cos(5x + 2) \, dx = \tfrac{1}{5} \sin(5x + 2) + c$

Problem 2. Find $\displaystyle\int (4t - 3)^7 \, dt$

Let $u = 4t - 3$

then $\dfrac{du}{dt} = 4$, i.e. $dt = \dfrac{du}{4}$

$$\int (4t - 3)^7 \, dt = \int u^7 \, \dfrac{du}{4} = \dfrac{1}{4} \int u^7 \, du$$

$$= \dfrac{1}{4}\left(\dfrac{u^8}{8}\right) + c$$

$$= \dfrac{u^8}{32} + c$$

Since $u = (4t - 3)$,

$$\int (4t - 3)^7 \, dt = \tfrac{1}{32} (4t - 3)^8 + c$$

Problem 3. Integrate $\dfrac{1}{7x + 2}$ with respect to x

Let $u = 7x + 2$

then $\dfrac{du}{dx} = 7$, i.e. $dx = \dfrac{du}{7}$

$$\int \frac{1}{7x + 2} \, dx = \int \frac{1}{u} \frac{du}{7}$$

$$= \tfrac{1}{7} \ln u + c$$

Since $u = (7x + 2)$,

$$\int \frac{1}{7x + 2} \, dx = \tfrac{1}{7} \ln(7x + 2) + c$$

From Problems 1 to 3 above it may be seen that:
If 'x' in a standard integral is replaced by $(ax + b)$ where a and b are constants, then $(ax + b)$ is written for x in the result and the result is multiplied by $\dfrac{1}{a}$

For example, $\displaystyle\int (ax + b) \, dx = \dfrac{1}{2a} (ax + b)^2 + c$ and, more generally,

$$\int (ax + b)^n \, dx = \frac{1}{a(n + 1)} (ax + b)^{n+1} + c \text{ (except when } n = -1)$$

Problem 4. Integrate the following with respect to x, using the general rule (i.e. without making a substitution): (a) $3 \sin(2x - 1)$, (b) $2e^{8x + 3}$ and (c) $\dfrac{5}{9x - 2}$

(a) $\displaystyle\int 3 \sin(2x - 1) \, dx = 3(\tfrac{1}{2})[-\cos(2x - 1)] + c$

$$= -\tfrac{3}{2} \cos(2x - 1) + c$$

(b) $\displaystyle\int 2e^{8x + 3} \, dx = 2(e^{8x + 3})(\tfrac{1}{8}) + c = \tfrac{1}{4}e^{8x + 3} + c$

(c) $\displaystyle\int \frac{5}{9x - 2} \, dx = 5[\ln(9x - 2)]\tfrac{1}{9} + c = \tfrac{5}{9} \ln(9x - 2) + c$

Problem 5. Find $\dfrac{3}{2}\displaystyle\int (x^2 + 2)^6 2x \, dx$

Let $u = x^2 + 2$

then $\dfrac{du}{dx} = 2x$, i.e. $dx = \dfrac{du}{2x}$

Hence $\dfrac{3}{2}\displaystyle\int (x^2 + 2)^6 2x \, dx = \dfrac{3}{2}\int u^6 2x \, \dfrac{du}{2x} = \dfrac{3}{2}\int u^6 \, du$

The original variable, x, has been removed completely and the integral is now only in terms of u.

$\dfrac{3}{2}\displaystyle\int u^6 \, du = \dfrac{3}{2}\left(\dfrac{u^7}{7}\right) + c$

Since $u = x^2 + 2$,

$\displaystyle\int 3x(x^2 + 2)^6 \, dx = \tfrac{3}{14}(x^2 + 2)^7 + c$

Problem 6. Find $\displaystyle\int \sin \theta \cos \theta \, d\theta$

Let $u = \sin \theta$

then $\dfrac{du}{d\theta} = \cos \theta$, i.e. $d\theta = \dfrac{du}{\cos \theta}$

Hence $\displaystyle\int \sin \theta \cos \theta \, d\theta = \int u \cos \theta \, \dfrac{du}{\cos \theta} = \int u \, du = \dfrac{u^2}{2} + c$

Since $u = \sin \theta$,

$\displaystyle\int \sin \theta \cos \theta \, d\theta = \tfrac{1}{2} \sin^2 \theta + c$

Another solution to this integral is possible.
  Let $u = \cos \theta$

then $\dfrac{du}{d\theta} = -\sin \theta$, i.e. $d\theta = \dfrac{-du}{\sin \theta}$

Hence $\displaystyle\int \sin \theta \cos \theta \, d\theta = \int \sin \theta (u)\left(\dfrac{-du}{\sin \theta}\right) = -\int u \, du = -\dfrac{u^2}{2} + c$

Since $u = \cos \theta$,

$\displaystyle\int \sin \theta \cos \theta \, d\theta = -\tfrac{1}{2} \cos^2 \theta + c$

From Problems 5 and 6 above it may be seen that:

Integrals of the form $k \displaystyle\int [f(x)]^n f'(x)\,dx$ (where k is a constant) can be integrated by substituting u for $f(x)$

Problem 7. Find $\dfrac{1}{2} \displaystyle\int \dfrac{(4x+6)}{\sqrt{(2x^2+6x-1)}}\,dx$

Let $u = 2x^2 + 6x - 1$

then $\dfrac{du}{dx} = 4x + 6$, i.e. $dx = \dfrac{du}{4x+6}$

Hence $\dfrac{1}{2} \displaystyle\int \dfrac{(4x+6)}{\sqrt{(2x^2+6x-1)}}\,dx = \dfrac{1}{2} \int \dfrac{(4x+6)}{\sqrt{u}}\,\dfrac{du}{(4x+6)} = \dfrac{1}{2} \int \dfrac{du}{\sqrt{u}}$

$$= \dfrac{1}{2} \int u^{-\frac{1}{2}}\,du = \dfrac{1}{2}\left(\dfrac{u^{\frac{1}{2}}}{\frac{1}{2}}\right) + c = u^{\frac{1}{2}} + c$$

Since $u = 2x^2 + 6x - 1$,

$$\dfrac{1}{2}\int \dfrac{(4x+6)}{\sqrt{(2x^2+6x-1)}}\,dx = \sqrt{(2x^2+6x-1)} + c$$

Problem 8. Find $\displaystyle\int \tan\theta\,d\theta$

$$\int \tan\theta\,d\theta = \int \dfrac{\sin\theta}{\cos\theta}\,d\theta$$

Let $u = \cos\theta$

then $\dfrac{du}{d\theta} = -\sin\theta$, i.e. $d\theta = \dfrac{-du}{\sin\theta}$

Hence

$$\int \dfrac{\sin\theta}{\cos\theta}\,d\theta = \int \dfrac{\sin\theta}{u}\left(\dfrac{-du}{\sin\theta}\right) = -\int \dfrac{1}{u}\,du = -\ln u + c = \ln u^{-1} + c$$

Since $u = \cos\theta$,

$$\int \tan\theta\,d\theta = \ln(\cos\theta)^{-1} + c$$

$$= \ln(\sec\theta) + c$$

From Problems 7 and 8 above it may be seen that:

Integrals of the form $k \displaystyle\int \dfrac{f'(x)}{[f(x)]^n}\,dx$ (where k and n are constants) can be integrated by substituting u for $f(x)$

Problem 9. Evaluate the following:

(a) $\displaystyle\int_0^1 3\sec^2(4\theta - 1)\,d\theta$

(b) $\displaystyle\int_0^4 5x\sqrt{(2x^2 + 4)}\,dx$, taking positive values of roots only

(c) $\displaystyle\int_1^3 \frac{e^t}{3 + e^t}\,dt$

(a) $\displaystyle\int_0^1 3\sec^2(4\theta - 1)\,d\theta = [\tfrac{3}{4}\tan(4\theta - 1)]_0^1 = \tfrac{3}{4}[\tan 3 - \tan(-1)]$

$$= \tfrac{3}{4}(1.414\,86) = \mathbf{1.061\,1}$$

Note that with trigonometric functions the units of integration are **always** in radians. Hence 'tan 3' above means 'the tangent of 3 radians'.

(b) $\displaystyle\int_0^4 5x\sqrt{(2x^2 + 4)}\,dx = \int_0^4 5x(2x^2 + 4)^{\frac{1}{2}}\,dx$

Let $u = 2x^2 + 4$

then $\dfrac{du}{dx} = 4x$, i.e. $dx = \dfrac{du}{4x}$

$\displaystyle\int 5x(2x^2 + 4)^{\frac{1}{2}}\,dx = \int 5x(u^{\frac{1}{2}})\frac{du}{4x} = \frac{5}{4}\int u^{\frac{1}{2}}\,du$

$$= \frac{5}{4}\left(\frac{u^{\frac{3}{2}}}{\frac{3}{2}}\right) + c = \tfrac{5}{6}(\sqrt{u^3}) + c$$

Since $u = 2x^2 + 4$,

$\displaystyle\int_0^4 5x\sqrt{(2x^2 + 4)}\,dx = [\tfrac{5}{6}\sqrt{(2x^2 + 4)^3}]_0^4$

$$= \tfrac{5}{6}\{\sqrt{[(2(4)^2 + 4)]^3} - \sqrt{(4)^3}\}$$
$$= \tfrac{5}{6}(216 - 8),\ \text{taking positive values of}$$
$$\text{roots only}$$
$$= \mathbf{173\tfrac{1}{3}}$$

(c) $\displaystyle\int_1^3 \frac{e^t}{3 + e^t}\,dt$

Let $u = 3 + e^t$

then $\dfrac{du}{dt} = e^t$, i.e. $dt = \dfrac{du}{e^t}$

Hence $\displaystyle\int \frac{e^t}{3 + e^t}\,dt = \int \frac{e^t\,du}{u\,e^t} = \int \frac{du}{u} = \ln u + c$

Since $u = 3 + e^t$,

$$\int_1^3 \frac{e^t}{3 + e^t}\, dt = [\ln(3 + e^t)]_1^3$$

$$= [\ln(3 + e^3) - \ln(3 + e^1)]$$

$$= \ln\left[\frac{3 + e^3}{3 + e^1}\right] = \ln\left[\frac{23.086}{5.718\,3}\right]$$

$$= 1.395\,6$$

Further problems on integration by substitution may be found in the following Section 22.3 (Problems 1 to 60).

22.3   Further problems

In Problems 1 to 40 integrate with respect to the appropriate variable.

1. $\sin(3x + 2)$ $[-\frac{1}{3}\cos(3x + 2) + c]$
2. $2\cos(4t + 1)$ $[\frac{1}{2}\sin(4t + 1) + c]$
3. $3\sec^2(t + 5)$ $[3\tan(t + 5) + c]$
4. $4\sin(6\theta - 3)$ $[-\frac{2}{3}\cos(6\theta - 3) + c]$
5. $(2x + 1)^5$ $[\frac{1}{12}(2x + 1)^6 + c]$
6. $3(4S - 7)^4$ $[\frac{3}{20}(4S - 7)^5 + c]$
7. $\frac{1}{12}(9x + 5)^8$ $[\frac{1}{972}(9x + 5)^9 + c]$

8. $\dfrac{1}{3a + 1}$ $[\frac{1}{3}\ln(3a + 1) + c]$

9. $\dfrac{5}{5f - 2}$ $[\ln(5f - 2) + c]$

10. $\dfrac{7}{2x + 1}$ $[\frac{7}{2}\ln(2x + 1) + c]$

11. $\dfrac{-1}{6x + 5}$ $[-\frac{1}{6}\ln(6x + 5) + c]$

12. $\dfrac{3}{15y - 2}$ $[\frac{1}{5}\ln(15y - 2) + c]$

13. e^{3x+2} $[\frac{1}{3}e^{3x+2} + c]$
14. $4e^{7t-1}$ $[\frac{4}{7}e^{7t-1} + c]$
15. $2e^{2-3x}$ $[-\frac{2}{3}e^{2-3x} + c]$
16. $4x(2x^2 + 3)^5$ $[\frac{1}{6}(2x^2 + 3)^6 + c]$
17. $5t(t^2 - 1)^7$ $[\frac{5}{16}(t^2 - 1)^8 + c]$
18. $(3x^2 + 4)^8 x$ $[\frac{1}{54}(3x^2 + 4)^9 + c]$
19. $\sin^2\theta\cos\theta$ $[\frac{1}{3}\sin^3\theta + c]$

20. $\sin^3 t \cos t$ $\quad [\frac{1}{4}\sin^4 t + c]$
21. $2\cos^2 \beta \sin \beta$ $\quad [-\frac{2}{3}\cos^3 \beta + c]$
22. $\sec^2 \theta \tan \theta$ $\quad [\frac{1}{2}\tan^2 \theta + c]$
23. $3\tan 2x \sec^2 2x$ $\quad [\frac{3}{4}\tan^2 2x + c]$
24. $\frac{6}{5}\sin^5 \theta \cos \theta$ $\quad [\frac{1}{5}\sin^6 \theta + c]$
25. $6x\sqrt{(3x^2 + 2)}$ $\quad [\frac{2}{3}\sqrt{(3x^2 + 2)^3} + c]$
26. $(4x^2 - 1)\sqrt{(4x^3 - 3x + 1)}$ $\quad [\frac{2}{9}\sqrt{(4x^3 - 3x + 1)^3} + c]$

27. $\dfrac{3\ln t}{t}$ $\quad [\frac{3}{2}(\ln t)^2 + c]$

28. $\dfrac{6x + 2}{(3x^2 + 2x - 1)^5}$ $\quad \left[\dfrac{-1}{4(3x^2 + 2x - 1)^4} + c\right]$

29. $\dfrac{4y - 1}{(4y^2 - 2y + 5)^7}$ $\quad \left[\dfrac{-1}{12(4y^2 - 2y + 5)^6} + c\right]$

30. $\dfrac{2x}{\sqrt{(x^2 + 1)}}$ $\quad [2\sqrt{(x^2 + 1)} + c]$

31. $\dfrac{3a}{\sqrt{(3a^2 + 5)}}$ $\quad [\sqrt{(3a^2 + 5)} + c]$

32. $\dfrac{12x^2 + 1}{\sqrt{(4x^3 + x - 1)}}$ $\quad [2\sqrt{(4x^3 + x - 1)} + c]$

33. $\dfrac{r^2 - 1}{\sqrt{(r^3 - 3r + 2)}}$ $\quad [\frac{2}{3}\sqrt{(r^3 - 3r + 2)} + c]$

34. $\dfrac{3e^t}{\sqrt{(1 + e^t)}}$ $\quad [6\sqrt{(1 + e^t)} + c]$

35. $2x\sin(x^2 + 1)$ $\quad [-\cos(x^2 + 1) + c]$
36. $(4\theta + 1)\sec^2(4\theta^2 + 2\theta)$ $\quad [\frac{1}{2}\tan(4\theta^2 + 2\theta) + c]$
37. $\frac{1}{3}(4x + 1)\cos(2x^2 + x - 1)$ $\quad [\frac{1}{3}\sin(2x^2 + x - 1) + c]$
38. $4te^{2t^2 - 3}$ $\quad [e^{2t^2 - 3} + c]$
39. $3\tan \beta$ $\quad [3\ln(\sec \beta) + c]$
40. $(5x - 2)e^{5x^2 - 4x + 1}$ $\quad [\frac{1}{2}e^{5x^2 - 4x + 1} + c]$

In Problems 41 to 60 evaluate the definite integrals.

41. $\displaystyle\int_0^1 (3x - 1)^4\,dx$ $\quad [2\frac{1}{5}]$

42. $\displaystyle\int_0^2 (8x - 3)(4x^2 - 3x)^3\,dx$ $\quad [2\,500]$

43. $\displaystyle\int_1^3 x\sqrt{(x^2 + 1)}\,dx$ $\quad [9.598]$

44. $\displaystyle\int_0^{\pi/4} \sin\left(4\theta + \frac{\pi}{3}\right) d\theta \qquad [\frac{1}{4}]$

45. $\displaystyle\int_{\frac{1}{3}}^1 \sec^2(3x - 1)\, dx \qquad [-0.728\,3]$

46. $\displaystyle\int_1^2 3\cos(5t - 2)\, dt \qquad [0.508\,9]$

47. $\displaystyle\int_{\frac{1}{2}}^2 \frac{1}{(4s - 1)}\, ds \qquad [0.486\,5]$

48. $\displaystyle\int_0^2 (9x^2 - 4)\sqrt{(3x^3 - 4x)}\, dx \qquad [42\frac{2}{3}]$

49. $\displaystyle\int_1^3 \frac{4\ln x}{x}\, dx \qquad [2.413\,9]$

50. $\displaystyle\int_0^2 \frac{t}{\sqrt{(2t^2 + 1)}}\, dt \qquad [1]$

51. $\displaystyle\int_1^2 \frac{4x - 3}{(2x^2 - 3x - 1)^4}\, dx \qquad [-\frac{3}{8}]$

52. $\displaystyle\int_1^2 3\theta \sin(2\theta^2 + 1)\, d\theta \qquad [-0.059\,1]$

53. $\displaystyle\int_0^1 2te^{3t^2 - 1}\, dt \qquad [2.340\,4]$

54. $\displaystyle\int_0^{\pi/2} 3\sin^4\theta \cos\theta\, d\theta \qquad [\frac{3}{5}]$

55. $\displaystyle\int_1^2 \frac{dx}{(2x - 1)^3} \qquad [\frac{2}{9}]$

56. $\displaystyle\int_1^2 \frac{2e^{3\theta}}{e^{3\theta} - 5}\, d\theta \qquad [2.182\,5]$

57. $\displaystyle\int_0^1 2t \sec^2(3t^2)\, dt \qquad [-0.047\,5]$

58. $\displaystyle\int_1^2 x \sin(2x^2 - 1)\, dx \qquad [-0.053\,4]$

59. $\displaystyle\int_{\pi/6}^{\pi/3} \frac{2}{3}\sin t \cos^3 t\, dt \qquad [0.083\,3]$

60. $\displaystyle\int_1^2 \frac{e^{3\theta} - e^{-3\theta}}{2}\, d\theta \qquad [63.88]$

Chapter 23

Integration using trigonometric identities and substitutions

23.1 Integration of $\cos^2 x$, $\sin^2 x$, $\tan^2 x$ and $\cot^2 x$

It is shown in Chapter 16 that the compound angle addition formula $\cos(A + B)$ is given by:

$$\cos(A + B) = \cos A \cos B - \sin A \sin B.$$

If $B = A$, then $\cos 2A = \cos^2 A - \sin^2 A$ (1)

Since $\cos^2 A + \sin^2 A = 1$ then $\sin^2 A = 1 - \cos^2 A$

Hence $\cos 2A = \cos^2 A - (1 - \cos^2 A)$

i.e. $\cos 2A = 2 \cos^2 A - 1$ (2)

Also $\cos^2 A = 1 - \sin^2 A$

Hence $\cos 2A = (1 - \sin^2 A) - \sin^2 A$

i.e. $\cos 2A = 1 - 2 \sin^2 A$ (3)

From equation (2), $\cos^2 A = \frac{1}{2}(1 + \cos 2A)$ (4)

From equation (3), $\sin^2 A = \frac{1}{2}(1 - \cos 2A)$ (5)

$$\text{Thus} \int \cos^2 x \, dx = \int \frac{1}{2}(1 + \cos 2x) \, dx = \frac{1}{2}\left(x + \frac{\sin 2x}{2}\right) + c$$

$$\text{Similarly,} \int \sin^2 x \, dx = \int \frac{1}{2}(1 - \cos 2x) \, dx = \frac{1}{2}\left(x - \frac{\sin 2x}{2}\right) + c$$

From $1 + \tan^2 x = \sec^2 x$, $\tan^2 x = \sec^2 x - 1$

$$\text{Thus} \int \tan^2 x \, dx = \int (\sec^2 x - 1) \, dx = \tan x - x + c$$

From $\cot^2 x + 1 = \operatorname{cosec}^2 x$, $\cot^2 x = \operatorname{cosec}^2 x - 1$

$$\text{Thus} \int \cot^2 x \, dx = \int (\operatorname{cosec}^2 x - 1) \, dx = -\cot x - x + c$$

Problem 1. Find $(a) \int \cos^2 5t \, dt$ $(b) \int \tan^2 3\theta \, d\theta$

(a) $\displaystyle \int \cos^2 5t \, dt = \int \frac{1}{2}(1 + \cos 10t) \, dt = \frac{1}{2}\left(t + \frac{\sin 10t}{10}\right) + c$

(b) $\displaystyle \int \tan^2 3\theta \, d\theta = \int (\sec^2 3\theta - 1) \, d\theta = \tfrac{1}{3}\tan 3\theta - \theta + c$

Problem 2. Determine (a) $\displaystyle \int 3 \sin^2 2x \, dx$ (b) $\displaystyle \int 2 \cot^2 4\theta \, d\theta$

(a) $\displaystyle \int 3 \sin^2 2x \, dx = 3 \int \frac{1}{2}(1 - \cos 4x) \, dx$

$$= \frac{3}{2}\left(x - \frac{\sin 4x}{4}\right) + c$$

(b) $\displaystyle \int 2 \cot^2 4\theta \, d\theta = 2 \int (\operatorname{cosec}^2 4\theta - 1) \, dx$

$$= 2\left(-\frac{\cot 4\theta}{4} - \theta\right) + c$$

$$= -\frac{1}{2}\cot 4\theta - 2\theta + c$$

A further problem on integration of $\cos^2 x$, $\sin^2 x$, $\tan^2 x$ *and* $\cot^2 x$ *may be found in Section 23.8 (Problem 1), page 426.*

23.2 Powers of sines and cosines of the form $\int \cos^m x \sin^n x \, dx$

(i) To evaluate $\int \cos^m x \sin^n x \, dx$ when either m or n is odd (but not both), use is made of the trigonometric identity $\cos^2 x + \sin^2 x = 1$ as shown in Problems 1 and 2.

Problem 1. Find $\displaystyle \int \cos^5 x \, dx$

$$\int \cos^5 x \, dx = \int \cos x (\cos^2 x)^2 \, dx$$

$$= \int \cos x (1 - \sin^2 x)^2 \, dx$$

$$= \int \cos x (1 - 2 \sin^2 x + \sin^4 x) \, dx$$

$$= \int \cos x - 2 \sin^2 x \cos x + \sin^4 x \cos x \, dx$$

Hence the integral has been reduced to one which may be determined by inspection (see Chapter 22).

Hence $\displaystyle\int \cos^5 x \, dx = \sin x - \frac{2}{3}\sin^3 x + \frac{1}{5}\sin^5 x + c$

Problem 2. Find $\displaystyle\int \sin^3 x \cos^2 x \, dx$

$$\int \sin^3 x \cos^2 x \, dx = \int \sin x(\sin^2 x)(\cos^2 x) \, dx$$

$$= \int \sin x(1 - \cos^2 x)(\cos^2 x) \, dx$$

$$= \int (\sin x \cos^2 x - \sin x \cos^4 x) \, dx$$

$$= -\tfrac{1}{3}\cos^3 x + \tfrac{1}{5}\cos^5 x + c$$

(ii) *To evaluate* $\int \cos^m x \sin^n x \, dx$ *when m and n are both even use is made of equations (4) and (5) of Section 23.1, i.e.*

$\cos^2 x = \frac{1}{2}(1 + \cos 2x)$ *and* $\sin^2 x = \frac{1}{2}(1 - \cos 2x)$ *as shown in Problems 3 and 4.*

(It follows from these equations that $\cos^2 2x = \frac{1}{2}(1 + \cos 4x)$, $\sin^2 3x = \frac{1}{2}(1 - \cos 6x)$, and so on)

Problem 3. Find $\displaystyle\int \cos^4 x \, dx$

$$\int \cos^4 x \, dx = \int (\cos^2 x)^2 \, dx = \int [\tfrac{1}{2}(1 + \cos 2x)]^2 \, dx$$

$$= \frac{1}{4}\int (1 + 2\cos 2x + \cos^2 2x) \, dx$$

$$= \frac{1}{4}\int 1 + 2\cos 2x + \tfrac{1}{2}(1 + \cos 4x) \, dx$$

$$(\text{since } \cos 4x = 2\cos^2 2x - 1)$$

$$= \frac{1}{4}\int [\tfrac{3}{2} + 2\cos 2x + \tfrac{1}{2}\cos 4x] \, dx$$

Hence $\displaystyle\int \cos^2 x \, dx = \tfrac{1}{4}[\tfrac{3}{2}x + \sin 2x + \tfrac{1}{8}\sin 4x] + c$

Problem 4. Find $\int \sin^4 x \cos^2 x \, dx$

$$\int \sin^4 x \cos^2 x \, dx = \int (\sin^2 x)^2 \cos^2 x \, dx$$

$$= \int \left(\frac{1 - \cos 2x}{2}\right)^2 \left(\frac{1 + \cos 2x}{2}\right) dx$$

$$= \frac{1}{8} \int (1 - 2\cos 2x + \cos^2 2x)(1 + \cos 2x) \, dx$$

$$= \frac{1}{8} \int (1 - 2\cos 2x + \cos^2 2x + \cos 2x - 2\cos^2 2x$$

$$+ \cos^3 2x) \, dx$$

$$= \frac{1}{8} \int (1 - \cos 2x - \cos^2 2x + \cos^3 2x) \, dx$$

$$= \frac{1}{8} \int 1 - \cos 2x - \left(\frac{1 + \cos 4x}{2}\right)$$

$$+ \cos 2x(1 - \sin^2 2x) \, dx$$

$$= \frac{1}{8} \int [\tfrac{1}{2} - \cos 2x - \tfrac{1}{2}\cos 4x + \cos 2x$$

$$- \cos 2x \sin^2 2x] \, dx$$

$$= \frac{1}{8} \int [\tfrac{1}{2} - \tfrac{1}{2}\cos 4x - \cos 2x \sin^2 2x] \, dx$$

$$= \frac{1}{8} \left[\frac{x}{2} - \frac{\sin 4x}{8} - \frac{\sin^3 2x}{6}\right] + c$$

Hence $\int \sin^4 x \cos^2 x \, dx = \tfrac{1}{16}(x - \tfrac{1}{4}\sin 4x - \tfrac{1}{3}\sin^3 2x) + c$

Further problems on powers of sines and cosines may be found in Section 23.8 (Problems 2 to 5), page 427.

23.3   Products of sines and cosines

It is shown in Section 16.5, that:

$$\sin A \cos B = \tfrac{1}{2}[\sin(A + B) + \sin(A - B)] \tag{6}$$

$$\cos A \sin B = \tfrac{1}{2}[\sin(A + B) - \sin(A - B)] \tag{7}$$

$$\cos A \cos B = \tfrac{1}{2}[\cos(A + B) + \cos(A - B)] \tag{8}$$

$$\sin A \sin B = -\tfrac{1}{2}[\cos(A + B) - \cos(A - B)] \tag{9}$$

These formulae are used when integrating products of sines and cosines.

Problem 1. Find $\int \sin 4\theta \cos 3\theta \, d\theta$

$$\int \sin 4\theta \cos 3\theta \, d\theta = \int \tfrac{1}{2}(\sin 7\theta + \sin \theta) \, d\theta \quad \text{from equation (6)}$$
$$= \frac{1}{2} \left(-\frac{\cos 7\theta}{7} - \cos \theta \right) + c$$
$$= -\tfrac{1}{14}(\cos 7\theta + 7 \cos \theta) + c$$

Problem 2. Find $\int \cos 6x \sin 2x \, dx$

$$\int \cos 6x \sin 2x \, dx = \int \tfrac{1}{2}(\sin 8x - \sin 4x) \, dx, \quad \text{from equation (7)}$$
$$= \frac{1}{2} \left(\frac{-\cos 8x}{8} + \frac{\cos 4x}{4} \right) + c$$
$$= \tfrac{1}{16} (2 \cos 4x - \cos 8x) + c$$

Problem 3. Find $\int \cos 3t \cos t \, dt$

$$\int \cos 3t \cos t \, dt = \int \tfrac{1}{2}(\cos 4t + \cos 2t) \, dt, \quad \text{from equation (8)}$$
$$= \frac{1}{2} \left(\frac{\sin 4t}{4} + \frac{\sin 2t}{2} \right) + c$$
$$= \tfrac{1}{8}(\sin 4t + 2 \sin 2t) + c$$

Problem 4. Find $\int \sin 6x \sin 3x \, dx$

$$\int \sin 6x \sin 3x \, dx = \int -\tfrac{1}{2}(\cos 9x - \cos 3x) \, dx, \quad \text{from equation (9)}$$
$$= -\frac{1}{2} \left(\frac{\sin 9x}{9} - \frac{\sin 3x}{3} \right) + c$$
$$= -\tfrac{1}{18}(\sin 9x - 3 \sin 3x) + c$$

Further problems on products of sines and cosines may be found in Section 23.8 (Problems 6 to 9), page 427.

23.4 Integrals containing $\sqrt{(a^2 - x^2)}$ – the 'sine θ' substitution

When an integral contains a term $\sqrt{(a^2 - x^2)}$, the substitution $x = a \sin \theta$ is used. The reasons for this are made obvious by the following worked problems.

Problem 1. Find $\displaystyle\int \frac{1}{\sqrt{(a^2 - x^2)}}\,dx$

Let $x = a \sin\theta$, then $\dfrac{dx}{d\theta} = a \cos\theta$, i.e. $dx = a \cos\theta\,d\theta$

Hence $\displaystyle\int \frac{1}{\sqrt{(a^2 - x^2)}}\,dx = \int \frac{1}{\sqrt{(a^2 - a^2 \sin^2\theta)}}\,a\cos\theta\,d\theta$

$$= \int \frac{a\cos\theta\,d\theta}{\sqrt{[a^2(1 - \sin^2\theta)]}}$$

$$= \int \frac{a\cos\theta}{a\cos\theta}\,d\theta, \text{ since } 1 - \sin^2\theta = \cos^2\theta$$

$$= \int d\theta = \theta + c$$

Since $x = a\sin\theta$ then $\sin\theta = \dfrac{x}{a}$ and $\theta = \arcsin\dfrac{x}{a}$

Hence $\displaystyle\int \frac{1}{\sqrt{(a^2 - x^2)}}\,dx = \arcsin\frac{x}{a} + c$

Problem 2. Evaluate $\displaystyle\int_0^2 \frac{1}{\sqrt{(16 - x^2)}}\,dx$

Using the result of Problem 11,

$$\int_0^2 \frac{1}{\sqrt{(16 - x^2)}}\,dx = \left[\arcsin\frac{x}{4}\right]_0^2$$

$$= (\arcsin\tfrac{1}{2} - \arcsin 0)$$

$$= \frac{\pi}{6} \text{ or } 0.523\,6$$

Problem 3. Find $\displaystyle\int \sqrt{(a^2 - x^2)}\,dx$

Let $x = a\sin\theta$, then $\dfrac{dx}{d\theta} = a\cos\theta$, i.e. $dx = a\cos\theta\,d\theta$

Thus $\displaystyle\int \sqrt{(a^2 - x^2)}\,dx = \int \sqrt{(a^2 - a^2 \sin^2\theta)}(a\cos\theta\,d\theta)$

$$= \int \sqrt{[a^2(1 - \sin^2\theta)]}(a\cos\theta\,d\theta)$$

$$= \int (a\cos\theta)(a\cos\theta)\,d\theta$$

$$= \int a^2 \cos^2 \theta \, d\theta$$

$$= a^2 \int \tfrac{1}{2}(1 + \cos 2\theta) \, d\theta$$

$$= \frac{a^2}{2} \left(\theta + \frac{\sin 2\theta}{2} \right) + c$$

In the compound angle addition formula
$\sin(A + B) = \sin A \cos B + \cos A \sin B$. If $B = A$ then
$\sin 2A = 2 \sin A \cos A$

Hence $\displaystyle \int \sqrt{(a^2 - x^2)} \, dx = \frac{a^2}{2} \left(\theta + \frac{2 \sin \theta \cos \theta}{2} \right) + c$

$$= \frac{a^2}{2} (\theta + \sin \theta \cos \theta) + c$$

Since $x = a \sin \theta$ then $\sin \theta = \dfrac{x}{a}$ and $\theta = \arcsin \dfrac{x}{a}$

Also $\cos^2 \theta + \sin^2 \theta = 1$ from which $\cos \theta = \sqrt{(1 - \sin^2 \theta)}$

i.e. $\cos \theta = \sqrt{\left[1 - \left(\dfrac{x}{a} \right)^2 \right]} = \sqrt{\left(\dfrac{a^2 - x^2}{a^2} \right)} = \dfrac{\sqrt{(a^2 - x^2)}}{a}$

Hence $\displaystyle \int \sqrt{(a^2 - x^2)} \, dx = \frac{a^2}{2} \left[\arcsin \frac{x}{a} + \left(\frac{x}{a} \right) \frac{\sqrt{(a^2 - x^2)}}{a} \right] + c$

$$= \frac{a^2}{2} \arcsin \frac{x}{a} + \frac{x}{2} \sqrt{(a^2 - x^2)} + c$$

Problem 4. Evaluate $\displaystyle \int_0^3 \sqrt{(9 - x^2)} \, dx$

Using the result of Problem 13:

$$\int_0^3 \sqrt{(9 - x^2)} \, dx = \left[\frac{9}{2} \arcsin \frac{x}{3} + \frac{x}{2} \sqrt{(9 - x^2)} \right]_0^3$$

$$= \left[\frac{9}{2} \arcsin 1 + \frac{3}{2} \sqrt{(9 - 9)} \right] - \left[\frac{9}{2} \arcsin 0 + 0 \right]$$

$$= \frac{9}{2} \arcsin 1$$

'arcsin 1' means 'the angle whose sine is equal to 1', i.e. $\dfrac{\pi}{2}$ radians.

Hence $\displaystyle \int_0^3 \sqrt{(9 - x^2)} \, dx = \frac{9}{2} \times \frac{\pi}{2} = \frac{9\pi}{4}$ **or 7.068 6**

Problem 5. Find $\displaystyle\int \frac{1}{\sqrt{(5 + 4x - x^2)}}\, dx$

$$\int \frac{1}{\sqrt{(5 + 4x - x^2)}}\, dx = \int \frac{1}{\sqrt{[-(x^2 - 4x - 5)]}}\, dx$$

$$= \int \frac{1}{\sqrt{[-\{(x - 2)^2 - 9\}]}}\, dx$$

$$= \int \frac{1}{\sqrt{[(3)^2 - (x - 2)^2]}}\, dx$$

$$= \arcsin\left(\frac{x - 2}{3}\right) + c$$

Further problems on the $\sin \theta$ *substitution may be found in Section 23.8 (Problems 10 to 12), page 427.*

23.5 $\quad\displaystyle\int \frac{1}{a^2 + x^2}\, dx$ – the 'tan θ' substitution

When an integral is of the form $\dfrac{1}{a^2 + x^2}$ the substitution $x = a \tan \theta$ is used. The reason for this is made obvious by the following worked problem.

Problem 1. Find $\displaystyle\int \frac{1}{a^2 + x^2}\, dx$

Let $x = a \tan \theta$, then $\dfrac{dx}{d\theta} = a \sec^2 \theta$, i.e. $dx = a \sec^2 \theta\, d\theta$

$$\int \frac{1}{a^2 + x^2}\, dx = \int \frac{1}{a^2 + a^2 \tan^2 \theta}\, a \sec^2 \theta\, d\theta = \int \frac{a \sec^2 \theta\, d\theta}{a^2(1 + \tan^2 \theta)}$$

$$= \int \frac{a \sec^2 \theta}{a^2 \sec^2 \theta}\, d\theta, \text{ since } 1 + \tan^2 \theta = \sec^2 \theta$$

$$= \int \frac{1}{a}\, d\theta = \frac{1}{a}(\theta) + c$$

Hence $\displaystyle\int \frac{1}{a^2 + x^2}\, dx = \frac{1}{a} \arctan \frac{x}{a} + c$, since $x = a \tan \theta$

Problem 2. Evaluate $\displaystyle\int_0^1 \frac{1}{9 + 4x^2}\, dx$, correct to 4 significant figures.

$$\int_0^1 \frac{1}{9 + 4x^2}\,dx = \int_0^1 \frac{1}{4(\frac{9}{4} + x^2)}\,dx = \frac{1}{4}\int_0^1 \frac{1}{(\frac{3}{2})^2 + x^2}\,dx$$

$$= \frac{1}{4}\left[\frac{1}{(\frac{3}{2})}\arctan\frac{x}{(\frac{3}{2})}\right]_0^1$$

from the result of Problem 16

$$= \frac{1}{6}\left[\arctan\frac{2x}{3}\right]_0^1$$

$$= \frac{1}{6}\left[\arctan\frac{2}{3} - \arctan 0\right]$$

$$= \frac{1}{6}(0.588\,0) = \mathbf{0.098\,0}$$

Further problems on the tan θ *substitution may be found in Section 23.8 (Problems 13 to 14), page 428.*

23.6   Summary of trigonometric substitutions

$f(x)$	$\int f(x)\,dx$	Method
$\cos^2 x$	$\frac{1}{2}\left(x + \frac{\sin 2x}{2}\right) + c$	Use $\cos 2x = 2\cos^2 x - 1$
$\sin^2 x$	$\frac{1}{2}\left(x - \frac{\sin 2x}{2}\right) + c$	Use $\cos 2x = 1 - 2\sin^2 x$
$\tan^2 x$	$\tan x - x + c$	Use $1 + \tan^2 x = \sec^2 x$
$\cot^2 x$	$-\cot x - x + c$	Use $\cot^2 x + 1 = \mathrm{cosec}^2 x$
$\cos^m x \sin^n x$	If m or n is odd, use $\cos^2 x + \sin^2 x = 1$ If both m and n are even, use either $\cos 2x = 2\cos^2 x - 1$ or $\cos 2x = 1 - 2\sin^2 x$	
$\sin A \cos B$ $\cos A \sin B$ $\cos A \cos B$ $\sin A \sin B$	Use $\frac{1}{2}[\sin(A + B) + \sin(A - B)]$ Use $\frac{1}{2}[\sin(A + B) + \sin(A - B)]$ Use $\frac{1}{2}[\cos(A + B) + \cos(A - B)]$ Use $-\frac{1}{2}[\cos(A + B) - \cos(A - B)]$	

$f(x)$	$\int f(x)\,dx$	Method
$\dfrac{1}{\sqrt{(a^2-x^2)}}$	$\arcsin\dfrac{x}{a}+c$	Use $x=a\sin\theta$ substitution
$\sqrt{(a^2-x^2)}$	$\dfrac{a^2}{2}\arcsin\dfrac{x}{a}+\dfrac{x}{2}\sqrt{(a^2-x^2)}+c$	Use $x=a\sin\theta$ substitution
$\dfrac{1}{a^2+x^2}$	$\dfrac{1}{a}\arctan\dfrac{x}{a}+c$	Use $x=\tan\theta$ substitution

23.7 Change of limits of integration by a substitution

When evaluating definite integrals involving substitutions it is often easier to change the limits of the integral as shown in the following worked problems.

Problem 1. Evaluate $\displaystyle\int_0^5 \frac{1}{\sqrt{(25-x^2)}}\,dx$

Let $x=5\sin\theta$, then $dx=5\cos\theta\,d\theta$

When $x=5$, $\sin\theta=1$ and $\theta=\dfrac{\pi}{2}$

When $x=0$, $\sin\theta=0$ and $\theta=0$

Hence $\displaystyle\int_{x=0}^{x=5}\frac{1}{\sqrt{(25-x^2)}}\,dx=\int_{\theta=0}^{\theta=\pi/2}\frac{5\cos\theta\,d\theta}{\sqrt{[25(1-\sin^2\theta)]}}$

$$=[\theta]_0^{\pi/2}=\frac{\pi}{2}\text{ or }1.570\,8$$

Problem 2. Evaluate $\displaystyle\int_0^\infty \frac{1}{(x^2+4)}\,dx$

Let $x=2\tan\theta$, then $dx=2\sec^2\theta\,d\theta$

When $x=+\infty$, $\theta=\dfrac{\pi}{2}$ and when $x=0$, $\theta=0$

Hence $\displaystyle\int_{x=0}^{x=\infty} \frac{1}{(x^2+4)}\,dx = \int_{\theta=0}^{\theta=\pi/2} \frac{2\sec^2\theta\,d\theta}{4(\tan^2\theta+1)} = \int_0^{\pi/2} \frac{1}{2}\,d\theta$

$$= \left[\frac{\theta}{2}\right]_0^{\pi/2} = \frac{1}{2}\left(\frac{\pi}{2}-0\right) = \frac{\pi}{4} \text{ or } 0.785\,4$$

Problem 3. Evaluate $\displaystyle\int_0^1 \sqrt{(4-x^2)}\,dx$

Let $x = 2\sin\theta$, then $dx = 2\cos\theta\,d\theta$

When $x = 1$, $\sin\theta = \dfrac{1}{2}$ and $\theta = \dfrac{\pi}{6}$

When $x = 0$, $\sin\theta = 0$ and $\theta = 0$

Hence $\displaystyle\int_{x=0}^{x=1} \sqrt{(4-x^2)}\,dx = \int_{\theta=0}^{\theta=\pi/6} \sqrt{[4(1-\sin^2\theta)]}\,2\cos\theta\,d\theta$

$$= \int_0^{\pi/6} 4\cos^2\theta\,d\theta$$

$$= \int_0^{\pi/6} 4\left(\frac{1+\cos 2\theta}{2}\right)d\theta$$

$$= 2\left[\theta + \frac{\sin 2\theta}{2}\right]_0^{\pi/6}$$

$$= 2\left[\left(\frac{\pi}{6} + \frac{\sin \pi/3}{2}\right) - (0)\right]$$

$$= \frac{\pi}{3} + \frac{\sqrt{3}}{2} \text{ or } 1.913\,2$$

Further problems on evaluating definite integrals may be found in the following Section 23.8 (Problems 15 to 22), page 428.

23.8   Further problems

In Problems 1 to 14 integrate with respect to the variable.

1. (a) $\cos^2 2x$ (b) $\sin^2 3x$ (c) $\tan^2 4x$ (d) $\cot^2 5x$

(a) $\left[\dfrac{1}{2}\left(x + \dfrac{\sin 4x}{4}\right) + c\right]$ (b) $\left[\dfrac{1}{2}\left(x - \dfrac{\sin 6x}{6}\right) + c\right]$

(c) $\left[\dfrac{1}{4}\tan 4x - x + c\right]$ (d) $\left[-\dfrac{1}{5}\cot 5x - x + c\right]$

Powers of sines and cosines

2. (a) $\cos^3 \theta$ (b) $\sin^3 2\theta$

(a) $\left[\sin \theta - \dfrac{\sin^3 \theta}{3} + c \right]$ (b) $\left[\dfrac{1}{6} \cos^3 2\theta - \dfrac{1}{2} \cos 2\theta + c \right]$

3. (a) $\cos^3 t \sin^2 t$ (b) $\sin^4 x \cos^3 x$
    (a) $[\frac{1}{3} \sin^3 t - \frac{1}{5} \sin^5 t + c]$ (b) $[\frac{1}{5} \sin^5 x - \frac{1}{7} \sin^7 x + c]$

4. (a) $\sin^4 2x$ (b) $\cos^2 x \sin^2 x$

(a) $\left[\dfrac{3x}{8} - \dfrac{1}{8} \sin 4x + \dfrac{1}{64} \sin 8x + c \right]$ (b) $\left[\dfrac{x}{8} - \dfrac{1}{32} \sin 4x + c \right]$

5. (a) $3 \cos^4 3t \sin^2 3t$ (b) $\frac{1}{2} \sin^4 2\theta \cos^2 2\theta$

(a) $[\frac{3}{16}(t - \frac{1}{12} \sin 12t + \frac{1}{9} \sin^3 6t) + c]$
(b) $[\frac{1}{32}(\theta - \frac{1}{8} \sin 8\theta - \frac{1}{6} \sin^3 4\theta) + c]$

Products of sines and cosines

6. (a) $\sin 3t \cos t$ (b) $2 \cos 4t \cos 2t$
    (a) $[-\frac{1}{8}(\cos 4t + 2 \cos 2t) + c]$ (b) $[\frac{1}{6} \sin 6t + \frac{1}{2} \sin 2t + c]$

7. (a) $3 \cos 6x \cos 2x$ (b) $4 \sin 5x \sin 2x$

(a) $[\frac{3}{16}(\sin 8x + 2 \sin 4x) + c]$
(b) $[2(\frac{1}{3} \sin 3x - \frac{1}{7} \sin 7x) + c]$

8. (a) $\frac{1}{2} \sin 9t \sin 3t$ (b) $2 \sin 4\theta \cos 3\theta$
    (a) $[\frac{1}{48}(2 \sin 6t - \sin 12t) + c]$ (b) $[-\frac{1}{7} \cos 7\theta - \cos \theta + c]$

9. (a) $9 \cos 5t \cos 4t$ (b) $\frac{3}{2} \cos 2x \sin x$
    (a) $[\frac{9}{2}(\frac{1}{9} \sin 9t + \sin t) + c]$ (b) $[\frac{3}{4}(\cos x - \frac{1}{3} \cos 3x) + c]$

'Sine θ' substitution

10. (a) $\dfrac{2}{\sqrt{(4 - x^2)}}$ (b) $\dfrac{1}{\sqrt{(9 - 4x^2)}}$

(a) $\left[2 \arcsin \dfrac{x}{2} + c \right]$ (b) $\left[\dfrac{1}{2} \arcsin \dfrac{2x}{3} + c \right]$

11. (a) $\sqrt{(16 - x^2)}$ (b) $\sqrt{(16 - 9x^2)}$

(a) $\left[8 \arcsin \dfrac{x}{4} + \dfrac{x}{2} \sqrt{(16 - x^2)} + c \right]$

(b) $\left[\dfrac{8}{3} \arcsin \dfrac{3x}{4} + \dfrac{x}{2} \sqrt{(16 - 9x^2)} + c \right]$

12. (a) $\dfrac{1}{\sqrt{(4 + 2x - x^2)}}$ (b) $\sqrt{(4 + 2x - x^2)}$

(a) $\left[\arcsin\dfrac{(x-1)}{\sqrt{5}} + c\right]$

(b) $\left[\dfrac{5}{2}\arcsin\dfrac{(x-1)}{\sqrt{5}} + \left(\dfrac{x-1}{2}\right)\sqrt{(4 + 2x - x^2)} + c\right]$

'Tan θ' substitution

13. (a) $\dfrac{2}{1 + x^2}$ (b) $\dfrac{3}{16 + x^2}$

(a) $[2\arctan x + c]$ (b) $\left[\dfrac{3}{4}\arctan\dfrac{x}{4} + c\right]$

14. (a) $\dfrac{1}{9 + 16x^2}$ (b) $\dfrac{1}{2x^2 + 4x + 18}$

(a) $\left[\dfrac{1}{12}\arctan\dfrac{4x}{3} + c\right]$ (b) $\left[\dfrac{1}{4\sqrt{2}}\arctan\dfrac{x+1}{2\sqrt{2}} + c\right]$

Definite integrals

In Problems 15 to 22 evaluate the definite integrals.

15. (a) $\displaystyle\int_0^{\pi/2} \sin^2\theta\, d\theta$ (b) $\displaystyle\int_0^1 2\cos^2\theta\, d\theta$

(a) $\left[\dfrac{\pi}{4}\text{ or } 0.785\,4\right]$ (b) $[1.454\,6]$

16. (a) $\displaystyle\int_0^{\pi/4} \sin^3 t \cos t\, dt$ (b) $\displaystyle\int_0^{\pi/3} \cos^2 2t\, dt$

(a) $[0.062\,5]$ (b) $[0.324\,8]$

17. (a) $\displaystyle\int_{\pi/4}^{\pi/2} 4\sin^2\theta\cos^2\theta\, d\theta$ (b) $\displaystyle\int_0^{\pi/2} 3\sin^2 2t\cos^4 2t\, dt$

(a) $\left[\dfrac{\pi}{8}\text{ or } 0.392\,7\right]$ (b) $\left[\dfrac{3\pi}{32}\text{ or } 0.294\,5\right]$

18. (a) $\displaystyle\int_0^{\pi/4} \sin 5\theta\cos 3\theta\, d\theta$ (b) $\displaystyle\int_0^{\pi/3} 4\cos 6t\sin 3t\, dt$

(a) $[\tfrac{1}{4}]$ (b) $[-\tfrac{8}{9}]$

19. (a) $\displaystyle\int_0^1 2\cos 7x\cos 2x\, dx$ (b) $\displaystyle\int_0^{\pi/2} \sin 8\alpha\sin 5\alpha\, d\alpha$

(a) $[-0.146\,0]$ (b) $[-0.205\,1]$

20. (a) $\displaystyle\int_0^3 \frac{1}{\sqrt{(9-x^2)}}\,dx$ (b) $\displaystyle\int_0^3 \sqrt{(9-x^2)}\,dx$

   (a) $\left[\dfrac{\pi}{2} \text{ or } 1.570\,8\right]$ (b) $\left[\dfrac{9\pi}{4} \text{ or } 7.068\,6\right]$

21. (a) $\displaystyle\int_0^{1/2} \frac{2}{\sqrt{(1-x^2)}}\,dx$ (b) $\displaystyle\int_0^{3/4} \sqrt{(9-16x^2)}\,dx$

   (a) $\left[\dfrac{\pi}{3} \text{ or } 1.047\,2\right]$ (b) $\left[\dfrac{9\pi}{16} \text{ or } 1.767\,1\right]$

22. (a) $\displaystyle\int_0^1 \frac{1}{1+x^2}\,dx$ (b) $\displaystyle\int_0^2 \frac{3}{4+x^2}\,dx$

   (a) $\left[\dfrac{\pi}{4} \text{ or } 0.785\,4\right]$ (b) $\left[\dfrac{3\pi}{8} \text{ or } 1.178\,1\right]$

Chapter 24

Integration using partial fractions

24.1   Introduction

The process of expressing a fraction in terms of simple fractions – called partial fractions – was explained in Chapter 6. Certain functions can only be integrated when they have been resolved into partial fractions. This is demonstrated in the following worked problems.

Worked problems on integration using partial fractions

(a) *Denominator containing linear factors*

Problem 1. Find $\displaystyle\int \frac{x-8}{x^2-x-2}\,dx$

It was shown on page 99 that

$$\frac{x-8}{x^2-x-2} \equiv \frac{3}{x+1} - \frac{2}{x-2}$$

Hence $\displaystyle\int \frac{x-8}{x^2-x-2}\,dx = \int \left(\frac{3}{x+1} - \frac{2}{x-2}\right)dx$

$$= 3\ln(x+1) - 2\ln(x-2) + c$$

$$\text{or } \ln\left\{\frac{(x+1)^3}{(x-2)^2}\right\} + c$$

Problem 2. Determine $\displaystyle\int \frac{6x^2+7x-25}{(x-1)(x+2)(x-3)}\,dx$

It was shown on page 101 that

$$\frac{6x^2+7x-25}{(x-1)(x+2)(x-3)} \equiv \frac{2}{(x-1)} - \frac{1}{(x+2)} + \frac{5}{(x-3)}$$

Hence $\displaystyle\int \frac{6x^2 + 7x - 25}{(x-1)(x+2)(x-3)}\,dx$

$$= \int\left(\frac{2}{(x-1)} - \frac{1}{(x+2)} + \frac{5}{(x-3)}\right)dx$$

$$= 2\ln(x-1) - \ln(x+2) + 5\ln(x-3) + c$$

$$\text{or } \ln\left\{\frac{(x-1)^2(x-3)^5}{(x+2)}\right\} + c$$

Problem 3. Evaluate $\displaystyle\int_3^4 \frac{x^3 - x^2 - 5x}{x^2 - 3x + 2}\,dx$

It was shown on page 101 that

$$\frac{x^3 - x^2 - 5x}{x^2 - 3x + 2} \equiv x + 2 + \frac{5}{(x-1)} - \frac{6}{(x-2)}$$

Hence $\displaystyle\int_3^4 \frac{x^3 - x^2 - 5x}{x^2 - 3x + 2}\,dx$

$$= \int_3^4 \left(x + 2 + \frac{5}{(x-1)} - \frac{6}{(x-2)}\right)dx$$

$$= \left[\frac{x^2}{2} + 2x + 5\ln(x-1) - 6\ln(x-2)\right]_3^4$$

$$= (8 + 8 + 5\ln 3 - 6\ln 2) - (\tfrac{9}{2} + 6 + 5\ln 2 - 6\ln 1)$$

$$= 3.368\ 4$$

(b)   *Denominator containing repeated linear factors*

Problem 4. Find $\displaystyle\int \frac{x+5}{(x+3)^2}\,dx$

It was shown on page 102 that $\dfrac{x+5}{(x+3)^2} \equiv \dfrac{1}{(x+3)} + \dfrac{2}{(x+3)^2}$

Hence $\displaystyle\int \frac{x+5}{(x+3)^2}\,dx = \int\left(\frac{1}{(x+3)} + \frac{2}{(x+3)^2}\right)dx$

$$= \ln(x+3) - \frac{2}{(x+3)} + c$$

Problem 5. Find $\displaystyle\int \frac{5x^2 - 19x + 3}{(x - 2)^2(x + 1)} \, dx$

It was shown on page 103 that

$$\frac{5x^2 - 19x + 3}{(x - 2)^2(x + 1)} \equiv \frac{2}{(x - 2)} - \frac{5}{(x - 2)^2} + \frac{3}{(x + 1)}$$

Hence $\displaystyle\int \frac{5x^2 - 19x + 3}{(x - 2)^2(x + 1)} \, dx$

$$= \int \left(\frac{2}{(x - 2)} - \frac{5}{(x - 2)^2} + \frac{3}{(x + 1)} \right) dx$$

$$= 2 \ln(x - 2) + \frac{5}{(x - 2)} + 3 \ln(x + 1) + c$$

$$= \ln(x - 2)^2(x + 1)^3 + \frac{5}{(x - 2)} + c$$

Problem 6. Evaluate $\displaystyle\int_5^6 \frac{2x^2 - 13x + 13}{(x - 4)^3} \, dx$

It was shown on page 104 that

$$\frac{2x^2 - 13x + 13}{(x - 4)^3} \equiv \frac{2}{(x - 4)} + \frac{3}{(x - 4)^2} - \frac{7}{(x - 4)^3}$$

Hence $\displaystyle\int_5^6 \frac{2x^2 - 13x + 13}{(x - 4)^3}$

$$= \int_5^6 \left(\frac{2}{(x - 4)} + \frac{3}{(x - 4)^2} - \frac{7}{(x - 4)^3} \right) dx$$

$$= \left[2 \ln(x - 4) - \frac{3}{(x - 4)} + \frac{7}{2(x - 4)^2} \right]_5^6$$

$$= (2 \ln 2 - \tfrac{3}{2} + \tfrac{7}{8}) - (2 \ln 1 - \tfrac{3}{1} + \tfrac{7}{2})$$

$$= 0.261\ 3$$

(c) Denominator containing a quadratic factor

Problem 7. Find $\displaystyle\int \frac{8x^2 - 3x + 19}{(x^2 + 3)(x - 1)} \, dx$

It was shown on page 105 that $\displaystyle\frac{8x^2 - 3x + 19}{(x^2 + 3)(x - 1)} \equiv \frac{2x - 1}{(x^2 + 3)} + \frac{6}{(x - 1)}$

Hence $\displaystyle\int \frac{8x^2 - 3x + 19}{(x^2 + 3)(x - 1)}\,dx$

$$= \int \left(\frac{2x - 1}{(x^2 + 3)} + \frac{6}{(x - 1)}\right) dx$$

$$= \int \left(\frac{2x}{(x^2 + 3)} - \frac{1}{(x^2 + 3)} + \frac{6}{(x - 1)}\right) dx$$

$$= \ln(x^2 + 3) - \frac{1}{\sqrt{3}} \arctan \frac{x}{\sqrt{3}} + 6 \ln(x - 1) + c$$

$$= \ln(x^2 + 3)(x - 1)^6 - \frac{1}{\sqrt{3}} \arctan \frac{x}{\sqrt{3}} + c$$

Problem 8. Find $\displaystyle\int \frac{2 + x + 6x^2 - 2x^3}{x^2(x^2 + 1)}\,dx$

It was shown on page 106 that

$$\frac{2 + x + 6x^2 - 2x^3}{x^2(x^2 + 1)} \equiv \frac{1}{x} + \frac{2}{x^2} + \frac{4 - 3x}{x^2 + 1}$$

Hence $\displaystyle\int \frac{2 + x + 6x^2 - 2x^3}{x^2(x^2 + 1)}\,dx$

$$= \int \left(\frac{1}{x} + \frac{2}{x^2} + \frac{4 - 3x}{x^2 + 1}\right) dx$$

$$= \int \left(\frac{1}{x} + \frac{2}{x^2} + \frac{4}{x^2 + 1} - \frac{3x}{x^2 + 1}\right) dx$$

$$= \ln x - \frac{2}{x} + 4 \arctan x - \frac{3}{2} \ln(x^2 + 1) + c$$

$$= \ln\left\{\frac{x}{(x^2 + 1)^{\frac{3}{2}}}\right\} - \frac{2}{x} + 4 \arctan x + c$$

Problem 9. Find $\displaystyle\int \frac{5(x^2 + x + 3)}{x(x^2 + 2x + 5)}\,dx$

Let $\displaystyle\frac{5(x^2 + x + 3)}{x(x^2 + 2x + 5)} \equiv \frac{A}{x} + \frac{Bx + C}{x^2 + 2x + 5}$

$$\equiv \frac{A(x^2 + 2x + 5) + (Bx + C)(x)}{x(x^2 + 2x + 5)}$$

Hence $5x^2 + 5x + 15 \equiv A(x^2 + 2x + 5) + (Bx + C)(x)$
by equating numerators.

Let $x = 0$, then $A = 3$

Equating the coefficients of x^2 gives $5 = A + B$, i.e. $B = 2$

Equating the coefficients of x gives: $5 = 2A + C$, i.e. $C = -1$

Hence $\displaystyle \int \frac{5(x^2 + x + 3)}{x(x^2 + 2x + 5)} \, dx = \int \left(\frac{3}{x} + \frac{2x - 1}{x^2 + 2x + 5} \right) dx$

Now $\displaystyle \int \frac{2x - 1}{x^2 + 2x + 5} \, dx = \int \frac{2x + 2}{(x^2 + 2x + 5)} \, dx - \int \frac{3}{(x^2 + 2x + 5)} \, dx$

i.e. the numerator of the first integral on the right-hand side has deliberately been made equal to the differential coefficient of the denominator so that the integral will integrate as $\ln(x^2 + 2x + 5)$.

$$\int \frac{3}{(x^2 + 2x + 5)} \, dx = \int \frac{3}{(x + 1)^2 + 4} \, dx = \int \frac{3}{(x + 1)^2 + (2)^2} \, dx$$

$$= \frac{3}{2} \arctan \frac{(x + 1)}{2}$$

Hence $\displaystyle \int \frac{5(x^2 + x + 3)}{x(x^2 + 2x + 5)} \, dx = 3 \ln x + \ln(x^2 + 2x + 5)$

$$- \frac{3}{2} \arctan \frac{(x + 1)}{2} + c$$

$$= \ln x^3 (x^2 + 2x + 5)$$

$$- \frac{3}{2} \arctan \frac{(x + 1)}{2} + c$$

(d) *Integrals of the form* $\displaystyle \int \frac{1}{x^2 - a^2} \, dx \text{ and } \int \frac{1}{a^2 - x^2} \, dx$

Problem 10. Determine (a) $\displaystyle \int \frac{1}{x^2 - a^2} \, dx$, (b) $\displaystyle \int_1^2 \frac{1}{(9 - x^2)} \, dx$

(a) Let $\displaystyle \frac{1}{x^2 - a^2} \equiv \frac{A}{(x - a)} + \frac{B}{(x + a)} \equiv \frac{A(x + a) + B(x - a)}{(x - a)(x + a)}$

Hence $1 = A(x + a) + B(x - a)$ by equating numerators

Let $x = a$, then $A = \dfrac{1}{2a}$. Let $x = -a$, then $B = -\dfrac{1}{2a}$

Hence $\displaystyle \int \frac{1}{x^2 - a^2} \, dx = \frac{1}{2a} \int \left(\frac{1}{(x - a)} - \frac{1}{(x + a)} \right) dx$

$$= \frac{1}{2a} [\ln(x - a) - \ln(x + a)] + c$$

$$= \frac{1}{2a} \ln \left(\frac{x - a}{x + a} \right) + c$$

Similarly, it may be shown that $\displaystyle\int \frac{1}{a^2 - x^2}\, dx = \frac{1}{2a}\ln\left(\frac{a+x}{a-x}\right) + c$

(b) $\displaystyle\int_1^2 \frac{1}{9 - x^2}\, dx = \int_1^2 \frac{1}{3^2 - x^2}\, dx = \frac{1}{2(3)}\left[\ln\left(\frac{3+x}{3-x}\right)\right]_1^2$ from (a)

$$= \frac{1}{6}\ln\frac{5}{2} \text{ or } 0.152\,7$$

Further problems on integration using partial fractions may be found in the following Section 24.2 (Problems 1 to 31).

24.2   Further problems

In Problems 1 to 20 integrate after resolving into partial fractions.

1. $\dfrac{8}{x^2 - 4}$ $\left[2\ln(x-2) - 2\ln(x+2) + c \text{ or } 2\ln\left(\dfrac{x-2}{x+2}\right) + c\right]$

2. $\dfrac{3x + 5}{x^2 + 2x - 3}$

$[2\ln(x-1) + \ln(x+3) + c \text{ or } \ln\{(x-1)^2(x+3)\} + c]$

3. $\dfrac{y - 13}{y^2 - y - 6}$

$\left[3\ln(y+2) - 2\ln(y-3) + c \text{ or } \ln\left\{\dfrac{(y+2)^3}{(y-3)^2}\right\} + c\right]$

4. $\dfrac{17x^2 - 21x - 6}{x(x+1)(x-3)}$

$[2\ln x + 8\ln(x+1) + 7\ln(x-3) + c \text{ or } \ln\{x^2(x+1)^8(x-3)^7\} + c]$

5. $\dfrac{6x^2 + 7x - 49}{(x-4)(x+1)(2x-3)}$

$\Big[3\ln(x-4) - 2\ln(x+1) + 2\ln(2x-3) + c \text{ or }$

$\ln\left\{\dfrac{(x-4)^3(2x-3)^2}{(x+1)^2}\right\} + c\Big]$

6. $\dfrac{x^2 + 2}{(x+4)(x-2)}$

$\left[x - 3\ln(x+4) + \ln(x-2) + c \text{ or } x + \ln\left\{\dfrac{(x-2)}{(x+4)^3}\right\} + c\right]$

7. $\dfrac{2x^2 + 4x + 19}{2(x-3)(x+4)}$

$$\left[x + \tfrac{7}{2}\ln(x-3) - \tfrac{5}{2}\ln(x+4) + c \text{ or } x + \ln\left\{\dfrac{(x-3)^{7/2}}{(x+4)^{5/2}}\right\} + c \right]$$

8. $\dfrac{2x^3 + 7x^2 - 2x - 27}{(x-1)(x+4)}$

$$\left[x^2 + x - 4\ln(x-1) + 7\ln(x+4) + c \text{ or} \right.$$

$$\left. x^2 + x + \ln\left\{\dfrac{(x+4)^7}{(x-1)^4}\right\} + c \right]$$

9. $\dfrac{2t-1}{(t+1)^2}$ $\qquad \left[2\ln(t+1) + \dfrac{3}{(t+1)} + c \right]$

10. $\dfrac{8x^2 + 12x - 3}{(x+2)^3}$ $\qquad \left[8\ln(x+2) + \dfrac{20}{(x+2)} - \dfrac{5}{2(x+2)^2} + c \right]$

11. $\dfrac{6x+1}{(2x+1)^2}$ $\qquad \left[\tfrac{3}{2}\ln(2x+1) + \dfrac{1}{(2x+1)} + c \right]$

12. $\dfrac{1}{x^2(x+2)}$

$$\left[-\dfrac{1}{2x} - \dfrac{1}{4}\ln x + \dfrac{1}{4}\ln(x+2) + c \text{ or } \dfrac{1}{4}\ln\left(\dfrac{x+2}{x}\right) - \dfrac{1}{2x} + c \right]$$

13. $\dfrac{9x^2 - 73x + 150}{(x-7)(x-3)^2}$

$$\left[5\ln(x-7) + 4\ln(x-3) + \dfrac{3}{(x-3)} + c \text{ or} \right.$$

$$\left. \ln\{(x-7)^5(x-3)^4\} + \dfrac{3}{x-3} + c \right]$$

14. $\dfrac{-(9x^2 + 4x + 4)}{x^2(x^2-4)}$

$$\left[\ln x - \dfrac{1}{x} + 2\ln(x+2) - 3\ln(x-2) + c \text{ or} \right.$$

$$\left. \ln\left\{\dfrac{x(x+2)^2}{(x-2)^3}\right\} - \dfrac{1}{x} + c \right]$$

15. $\dfrac{-(a^2 + 5a + 13)}{(a^2+5)(a-2)}$

$$\left[\ln(a^2+5)-\frac{1}{\sqrt{5}}\arctan\frac{a}{\sqrt{5}}-3\ln(a-2)+c \text{ or}\right.$$
$$\left.\ln\left\{\frac{(a^2+5)}{(a-2)^3}\right\}-\frac{1}{\sqrt{5}}\arctan\frac{a}{\sqrt{5}}+c\right]$$

16. $\dfrac{3-x}{(x^2+3)(x+3)}$

$$\left[\frac{1}{2\sqrt{3}}\arctan\frac{x}{\sqrt{3}}-\frac{1}{4}\ln(x^2+3)+\frac{1}{2}\ln(x+3)+c \text{ or}\right.$$
$$\left.\ln\left\{\frac{(x+3)^{1/2}}{(x^2+3)^{1/4}}\right\}+\frac{1}{2\sqrt{3}}\arctan\frac{x}{\sqrt{3}}+c\right]$$

17. $\dfrac{12-2x-5x^2}{(x^2+x+1)(3-x)}$

$$\left[\ln(x^2+x+1)+\frac{8}{\sqrt{3}}\arctan\frac{2(x+\frac{1}{2})}{\sqrt{3}}+3\ln(3-x)+c\right]$$

18. $\dfrac{x^3+7x^2+8x+10}{x(x^2+2x+5)}$

$$\left[x+2\ln x+\tfrac{3}{2}\ln(x^2+2x+5)-2\arctan\left(\frac{x+1}{2}\right)+c\right]$$

19. $\dfrac{5x^3-3x^2+41x-64}{(x^2+6)(x-1)^2}$

$$\left[\frac{2}{\sqrt{6}}\arctan\frac{x}{\sqrt{6}}-\frac{3}{2}\ln(x^2+6)+8\ln(x-1)+\frac{3}{(x-1)}+c\right]$$

20. $\dfrac{6x^3+5x^2+4x+3}{(x^2+x+1)(x^2-1)}$

$$\left[\ln(x^2+x+1)-\frac{4}{\sqrt{3}}\arctan\frac{2(x+\frac{1}{2})}{\sqrt{3}}\right.$$
$$\left.+3\ln(x-1)+\ln(x+1)+c\right]$$

In Problems 21 to 30 evaluate the definite integrals correct to four decimal places.

21. $\displaystyle\int_4^6\frac{x-7}{x^2-2x-3}\,dx$ $\quad[-0.425\,7]$

22. $\displaystyle\int_2^3\frac{x^3-2x^2-3x-2}{(x+2)(x-1)}\,dx$ $\quad[-0.993\,7]$

23. $\displaystyle\int_3^4 \frac{4x^2 + 15x - 1}{(x + 1)(x - 2)(x + 3)}\,dx$ [2.371 6]

24. $\displaystyle\int_3^5 \frac{x^2 + 1}{x^2 + x - 6}\,dx$ [2.523 2]

25. $\displaystyle\int_6^8 \frac{1}{(x^2 - 25)}\,dx$ [0.093 2]

26. $\displaystyle\int_2^3 \frac{1}{(16 - x^2)}\,dx$ [0.105 9]

27. $\displaystyle\int_1^2 \frac{2 + x + 6x^2 - 2x^3}{x^2(x^2 + 1)}\,dx$ [1.605 7]

28. $\displaystyle\int_2^3 \frac{2x^2 - x - 2}{(x - 1)^3}\,dx$ [2.511 3]

29. $\displaystyle\int_3^4 \frac{2x^3 + 4x^2 - 8x - 4}{x^2(x^2 - 4)}\,dx$ [1.041 8]

30. $\displaystyle\int_0^1 \frac{-(4x^2 + 9x + 8)}{(x + 1)^2(x + 2)}\,dx$ [-2.546 5]

31. The velocity constant k of a given chemical reaction is given by

$$kt = \int \frac{dx}{(3 - 0.4x)(2 - 0.6x)} \quad \text{where } x = 0 \text{ when } t = 0$$

Determine kt

$$\left[\ln\left\{\frac{2(3 - 0.4x)}{3(2 - 0.6x)}\right\}\right]$$

Chapter 25

The $t = \tan \dfrac{\theta}{2}$ substitution

25.1   Introduction

In the right-angled triangle shown in Fig. 25.1, let angle B be equal to $\dfrac{\theta}{2}$, with side $AB = 1$ and side $AC = t$.

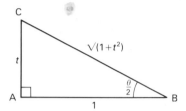

Fig. 25.1

Then $\tan \dfrac{\theta}{2} = \dfrac{t}{1} = t$, and by Pythagoras's theorem, side $BC = \sqrt{(1 + t^2)}$

Hence $\sin \dfrac{\theta}{2} = \dfrac{t}{\sqrt{(1 + t^2)}}$ and $\cos \dfrac{\theta}{2} = \dfrac{t}{\sqrt{(1 + t^2)}}$

From double-angle formulae (see Chapter 16),

$$\sin 2x = 2 \sin x \cos x$$

and it follows that $\sin \theta = 2 \sin \dfrac{\theta}{2} \cos \dfrac{\theta}{2}$

Hence, from above, $\sin \theta = 2\left(\dfrac{t}{\sqrt{(1 + t^2)}}\right)\left(\dfrac{1}{\sqrt{(1 + t^2)}}\right)$

i.e. $\sin \theta = \dfrac{2t}{(1 + t^2)}$ (1)

Also from double-angle formulae, $\cos 2x = \cos^2 x - \sin^2 x$ and it follows that $\cos \theta = \cos^2 \dfrac{\theta}{2} - \sin^2 \dfrac{\theta}{2}$

Hence, from above, $\cos \theta = \left[\dfrac{1}{\sqrt{(1+t^2)}}\right]^2 - \left[\dfrac{t}{\sqrt{(1+t^2)}}\right]^2$

$$= \frac{1}{(1+t^2)} - \frac{t^2}{(1+t^2)}$$

i.e. $$\cos \theta = \frac{(1-t^2)}{(1+t^2)} \qquad (2)$$

Since $t = \tan \dfrac{\theta}{2}$, then $\dfrac{dt}{d\theta} = \dfrac{1}{2} \sec^2 \dfrac{\theta}{2} = \dfrac{1}{2}\left(1 + \tan^2 \dfrac{\theta}{2}\right)$

from trigonometric identities

i.e. $\dfrac{dt}{d\theta} = \dfrac{1}{2}(1+t^2)$

from which,

$$d\theta = \frac{2}{(1+t^2)} dt \qquad (3)$$

Equations (1), (2) and (3) are used to determine integrals of the form $\displaystyle\int \dfrac{d\theta}{a\cos\theta + b\sin\theta + c}$, where a, b and c are constants (and may be zero) and are known as the $t = \tan \dfrac{\theta}{2}$ substitution. This is demonstrated in the following worked problems.

Worked problems on the $t = \tan \theta/2$ substitution

Problem 1. Find $\displaystyle\int \dfrac{d\theta}{\sin \theta}$

If $t = \tan \dfrac{\theta}{2}$ then, from equation (1), $\sin \theta = \dfrac{2t}{1+t^2}$ and from equation (3), $d\theta = \dfrac{2}{(1+t^2)} dt$

Thus $\displaystyle\int \dfrac{d\theta}{\sin \theta} = \int \dfrac{1}{\left(\dfrac{2t}{1+t^2}\right)} \cdot \dfrac{2}{(1+t^2)} dt = \int \dfrac{1}{t} dt = \ln t + c$

Hence $\displaystyle\int \dfrac{d\theta}{\sin \theta} = \ln\left(\tan \dfrac{\theta}{2}\right) + c$

Problem 2. Determine $\displaystyle\int \frac{d\theta}{1 + \cos\theta}$

If $t = \tan\dfrac{\theta}{2}$ then, from equation (2), $\cos\theta = \dfrac{(1 - t^2)}{(1 + t^2)}$ and from

equation (3), $d\theta = \dfrac{2}{(1 + t^2)}\,dt$

Thus $\displaystyle\int \frac{d\theta}{1 + \cos\theta} = \int \frac{1}{1 + \dfrac{(1 - t^2)}{(1 + t^2)}} \cdot \frac{2}{(1 + t^2)}\,dt$

$$= \int \frac{1}{\dfrac{(1 + t^2) + (1 - t^2)}{(1 + t^2)}} \cdot \frac{2}{(1 + t^2)}\,dt$$

$$= \int dt = t + c$$

Hence $\displaystyle\int \frac{d\theta}{1 + \cos\theta} = \tan\frac{\theta}{2} + c$

Problem 3. Determine $\displaystyle\int \frac{d\theta}{4 + 3\cos\theta}$

If $t = \tan\dfrac{\theta}{2}$, then $\cos\theta = \dfrac{(1 - t^2)}{(1 + t^2)}$ and $d\theta = \dfrac{2}{(1 + t^2)}\,dt$, from

$\qquad\qquad\qquad\qquad\qquad\qquad\qquad$ equations (2) and (3)

Thus $\displaystyle\int \frac{d\theta}{4 + 3\cos\theta} = \int \frac{1}{4 + 3\dfrac{(1 - t^2)}{(1 + t^2)}} \cdot \frac{2}{(1 + t^2)}\,dt$

$$= \int \frac{1}{\dfrac{4(1 + t^2) + 3(1 - t^2)}{(1 + t^2)}} \cdot \frac{2}{(1 + t^2)}\,dt = \int \frac{2}{7 + t^2}\,dt$$

$$= 2\int \frac{1}{(\sqrt{7})^2 + t^2}\,dt = 2\left(\frac{1}{\sqrt{7}} \arctan\frac{t}{\sqrt{7}}\right) + c$$

Hence $\displaystyle\int \frac{d\theta}{4 + 3\cos\theta} = \frac{2}{\sqrt{7}} \arctan\left(\frac{1}{\sqrt{7}} \tan\frac{\theta}{2}\right) + c$

Problem 4. Determine $\displaystyle\int \frac{d\theta}{\cos\theta}$

If $t = \tan\dfrac{\theta}{2}$, then from equation (2), $\cos\theta = \dfrac{(1-t^2)}{(1+t^2)}$ and from

equation (3), $d\theta = \dfrac{2}{(1+t^2)}\,dt$

Thus $\displaystyle\int\dfrac{d\theta}{\cos\theta} = \int\dfrac{1}{\dfrac{(1-t^2)}{(1+t^2)}}\cdot\dfrac{2}{(1+t^2)}\,dt = \int\dfrac{2}{(1-t^2)}\,dt$

$\dfrac{2}{(1-t^2)}$ may be resolved into partial fractions (see Chapter 6).

Let $\dfrac{2}{(1-t^2)} = \dfrac{2}{(1-t)(1+t)} \equiv \dfrac{A}{(1-t)} + \dfrac{B}{(1+t)}$

$$= \dfrac{A(1+t)+B(1-t)}{(1-t)(1+t)}$$

Hence $2 = A(1+t) + B(1-t)$

When $t = 1$, $2 = 2A$, from which, $A = 1$

When $t = -1$, $2 = 2B$, from which, $B = 1$

Hence $\displaystyle\int\dfrac{2}{(1-t^2)}\,dt = \int\dfrac{1}{(1-t)} + \dfrac{1}{(1+t)}\,dt$

$$= -\ln(1-t) + \ln(1+t) + c$$

$$= \ln\dfrac{(1+t)}{(1-t)} + c, \text{ by the laws of logarithms}$$

Thus $\displaystyle\int\dfrac{d\theta}{\cos\theta} = \ln\left\{\dfrac{1+\tan\dfrac{\theta}{2}}{1-\tan\dfrac{\theta}{2}}\right\} + c$

[Note that since $\tan\dfrac{\pi}{4} = 1$, the above result may be re-written as

$$\int\dfrac{d\theta}{\cos\theta} = \ln\left\{\dfrac{\tan\dfrac{\pi}{4} + \tan\dfrac{\theta}{2}}{\tan\dfrac{\pi}{4} - \tan\dfrac{\theta}{2}}\right\} + c$$

$$= \ln\left\{\tan\left(\dfrac{\pi}{4} + \dfrac{\theta}{2}\right)\right\} + c$$

from compound angles (see page 255)]

Problem 5. Find $\displaystyle\int \frac{d\theta}{\sin\theta + \cos\theta}$

If $t = \tan\dfrac{\theta}{2}$, then $\sin\theta = \dfrac{2t}{(1+t^2)}$, $\cos\theta = \dfrac{(1-t^2)}{(1+t^2)}$ and $d\theta = \dfrac{2}{(1+t^2)}\,dt$ from equations (1), (2) and (3).

Thus $\displaystyle\int \frac{d\theta}{\sin\theta + \cos\theta} = \int \frac{\dfrac{2}{(1+t^2)}\,dt}{\dfrac{2t}{(1+t^2)} + \dfrac{(1-t^2)}{(1+t^2)}}$

$$= \int \frac{\dfrac{2}{(1+t^2)}\,dt}{\dfrac{2t+1-t^2}{(1+t^2)}} = \int \frac{2\,dt}{1+2t-t^2}$$

$$= \int \frac{-2\,dt}{t^2-2t-1} = \int \frac{-2\,dt}{(t-1)^2-2}$$

$$= \int \frac{2\,dt}{2-(t-1)^2} = \int \frac{2\,dt}{(\sqrt{2})^2-(t-1)^2}$$

$$= 2\left[\frac{1}{2\sqrt{2}}\ln\left\{\frac{\sqrt{2}+(t-1)}{\sqrt{2}-(t-1)}\right\}\right] + c$$

from Problem 10(b), page 435, i.e.

$$\int \frac{d\theta}{\sin\theta + \cos\theta} = \frac{1}{\sqrt{2}}\ln\left\{\frac{\sqrt{2}-1+\tan\dfrac{\theta}{2}}{\sqrt{2}+1-\tan\dfrac{\theta}{2}}\right\} + c$$

Problem 6. Determine $\displaystyle\int \frac{dx}{3-3\sin x+2\cos x}$

Using equations (1), (2) and (3),

$$\int \frac{dx}{3-3\sin x+2\cos x} = \int \frac{\dfrac{2}{(1+t^2)}\,dt}{3-3\dfrac{(2t)}{(1+t^2)} + \dfrac{2(1-t^2)}{(1+t^2)}}$$

$$= \int \frac{\dfrac{2\,dt}{(1+t^2)}}{\dfrac{3(1+t^2) - 3(2t) + 2(1-t^2)}{(1+t^2)}} = \int \frac{2\,dt}{3 + 3t^2 - 6t + 2 - 2t^2}$$

$$= \int \frac{2\,dt}{t^2 - 6t + 5} = \int \frac{2\,dt}{(t-3)^2 - (2)^2}$$

$$= 2\left[\frac{1}{2(2)} \ln\left\{\frac{(t-3) - 2}{(t-3) + 2}\right\}\right] + c, \text{ from Problem 10}(a), \text{ page 434,}$$

$$= \frac{1}{2} \ln\left(\frac{t-5}{t-1}\right) + c$$

i.e. $\displaystyle \int \frac{dx}{3 - 3\sin x + 2\cos x} = \frac{1}{2} \ln\left\{ \frac{\tan\dfrac{x}{2} - 5}{\tan\dfrac{x}{2} - 1} \right\} + c$

Problem 7. Determine $\displaystyle \int \frac{d\theta}{3\sin\theta - 4\cos\theta}$

From equations (1), (2) and (3),

$$\int \frac{d\theta}{3\sin\theta - 4\cos\theta} = \int \frac{\dfrac{2}{(1+t^2)}\,dt}{3\left(\dfrac{2t}{1+t^2}\right) - 4\left(\dfrac{1-t^2}{1+t^2}\right)}$$

$$= \int \frac{\dfrac{2}{(1+t^2)}\,dt}{\dfrac{6t - 4 + 4t^2}{(1+t^2)}} = \int \frac{2\,dt}{4t^2 + 6t - 4}$$

$$= \int \frac{dt}{2t^2 + 3t - 2} = \frac{1}{2} \int \frac{dt}{t^2 + \dfrac{3}{2}t - 1}$$

$$= \frac{1}{2} \int \frac{dt}{\left(t + \dfrac{3}{4}\right)^2 - \dfrac{25}{16}}$$

$$= \frac{1}{2} \int \frac{dt}{\left(t + \dfrac{3}{4}\right)^2 - \left(\dfrac{5}{4}\right)^2}$$

$$= \frac{1}{2}\left[\frac{1}{2\left(\frac{5}{4}\right)} \ln\left\{\frac{\left(t + \frac{3}{4}\right) - \frac{5}{4}}{\left(t + \frac{3}{4}\right) + \frac{5}{4}}\right\}\right] + c$$

$$= \frac{1}{5} \ln\left\{\frac{t - \frac{1}{2}}{t + 2}\right\} + c$$

i.e. $\displaystyle\int \frac{d\theta}{3 \sin\theta - 4\cos\theta} = \frac{1}{5} \ln\left\{\frac{\tan\frac{\theta}{2} - \frac{1}{2}}{\tan\frac{\theta}{2} + 2}\right\} + c$

or

$$\frac{1}{5} \ln\left\{\frac{2\tan\frac{\theta}{2} - 1}{2\tan\frac{\theta}{2} + 4}\right\} + c$$

Further problems on the $t = \tan\dfrac{\theta}{2}$ *substitution may be found in the following Section 25.2 (Problems 1 to 12).*

25.2 Further problems

In Problems 1 to 9, integrate with respect to the variable.

1. $\displaystyle\int \frac{dx}{1 + \sin x}$ $\qquad \left[\dfrac{-2}{1 + \tan\dfrac{x}{2}} + c\right]$

2. $\displaystyle\int \frac{d\theta}{5 + 4\cos\theta}$ $\qquad \left[\dfrac{2}{3} \arctan\left(\dfrac{1}{3}\tan\dfrac{\theta}{2}\right) + c\right]$

3. $\displaystyle\int \frac{d\theta}{7 - 3\sin\theta + 6\cos\theta}$ $\qquad \left[\arctan\left(\dfrac{\tan\dfrac{\theta}{2} - 3}{2}\right) + c\right]$

4. $\displaystyle\int \frac{dx}{3 \sin x + 4 \cos x}$ $\left[\dfrac{1}{5} \ln \left\{\dfrac{1 + 2 \tan \dfrac{x}{2}}{4 - 2 \tan \dfrac{x}{2}}\right\} + c\right]$

5. $\displaystyle\int \frac{d\theta}{1 + \sin \theta - \cos \theta}$ $\left[\ln \left\{\dfrac{\tan \dfrac{\theta}{2}}{1 + \tan \dfrac{\theta}{2}}\right\} + c\right]$

6. $\displaystyle\int \frac{dx}{2 + \cos x}$ $\left[\dfrac{2}{\sqrt{3}} \arctan \left(\dfrac{\tan \dfrac{x}{3}}{\sqrt{3}}\right) + c\right]$

7. $\displaystyle\int \frac{d\theta}{4 \sin \theta - 3 \cos \theta}$ $\left[\dfrac{1}{5} \ln \left\{\dfrac{\tan \dfrac{\theta}{2} - \dfrac{1}{3}}{\tan \dfrac{\theta}{2} + 3}\right\} + c\right]$

8. $\displaystyle\int \frac{d\theta}{6 + 5 \sin \theta}$ $\left[\dfrac{2}{\sqrt{11}} \arctan \left(\dfrac{6 \tan \dfrac{\theta}{2} + 5}{\sqrt{11}}\right) + c\right]$

9. $\displaystyle\int \frac{d\alpha}{2 - 3 \sin \alpha}$ $\left[\dfrac{1}{\sqrt{5}} \ln \left\{\dfrac{2 \tan \dfrac{\alpha}{2} - 3 - \sqrt{5}}{2 \tan \dfrac{\alpha}{2} - 3 + \sqrt{5}}\right\} + c\right]$

10. Show that $\displaystyle\int_{0}^{\pi/6} \frac{3}{\cos x} \, dx = 1.648$, correct to 4 significant figures.

11. Show that $\displaystyle\int_{\pi/6}^{\pi/4} \frac{2}{\sin \theta} = 0.871\,2$, correct to 4 decimal places.

12. Show that $\displaystyle\int \frac{d\theta}{3 + 4 \cos \theta} = \dfrac{1}{\sqrt{7}} \ln \left\{\dfrac{\sqrt{7} + \tan \dfrac{\theta}{2}}{\sqrt{7} - \tan \dfrac{\theta}{2}}\right\} + c$

Chapter 26

Integration by parts

26.1   Introduction

When differentiating the product uv, where u and v are both functions of x, then:

$$\frac{d}{dx}(uv) = v\frac{du}{dx} + u\frac{dv}{dx}$$

This is the **product rule for differentiation.**
Rearranging this formula gives:

$$u\frac{dv}{dx} = \frac{d}{dx}(uv) - v\frac{du}{dx}$$

Integrating both sides with respect to x gives:

$$\int u\frac{dv}{dx}\,dx = \int \frac{d}{dx}(uv)\,dx - \int v\frac{du}{dx}\,dx$$

Since integration is the reversal of the differentiation process this becomes:

$$\int u\,dv = uv - \int v\,du$$

This formula enables products of certain simple functions to be integrated in cases where it is possible to evaluate $\int v\,du$. This is known as the **integration by parts formula** and is a useful method of integration enabling such integrals as $\int x\,e^x\,dx$, $\int x\cos x\,dx$, $\int t^2 \sin t\,dt$, $\int x^3 \ln x\,dx$, $\int \ln x\,dx$, $\int e^{ax}\sin bx\,dx$, etc., to be determined.

26.2   Application of the integration by parts formula

Problem 1. Find $\displaystyle\int x\,e^x\,dx$

In the integration by parts formula we must let one function of our product be equal to u and the other be equal to dv

Let $u = x$ and $dv = e^x\,dx$

Then $\dfrac{du}{dx} = 1$, i.e. $du = dx$ and $v = \int e^x \, dx = e^x$

There are now expressions for u, du, dv and v which are substituted into the formula:

$$\int u \; dv = u \; v - \int v \; du$$

$$\int x \; e^x \, dx = x \; e^x - \int e^x \; dx$$

$$= x e^x - e^x + c$$

Hence, $\int x \, e^x \, dx = e^x(x - 1) + c$

The following four points should be noted:

(i)   The above result may be checked by differentiation.

Thus $\dfrac{d}{dx} [e^x(x - 1) + c] = e^x(1) + (x - 1)e^x + 0$

$$= e^x + x e^x - e^x = x e^x$$

(ii) Given that $dv = e^x \, dx$ then $v = \int e^x \, dx$, which is strictly equal to e^x + a constant. However, the constant is omitted at this stage. If a constant, say k, were included, then:

$$\int x e^x \, dx = x(e^x + k) - \int (e^x + k) \, dx$$

$$= x e^x + xk - (e^x + kx + c)$$

$$= x e^x - e^x + c, \text{ as before}$$

Thus the constant k is an unnecessary addition. A constant is added only after the final integration (i.e. c in the above problem).

(iii) If instead of choosing to let $u = x$ and $dv = e^x \, dx$ we let

$u = e^x$ and $dv = x \, dx$ then $du = e^x \, dx$ and $v = \int x \, dx = \dfrac{x^2}{2}$

Hence $\int x e^x \, dx = (e^x)\left(\dfrac{x^2}{2}\right) - \int \left(\dfrac{x^2}{2}\right) e^x \, dx$

The integral on the far right-hand side is seen to be more complicated than the original, thus the original choice of letting $u = e^x$ and $dv = x \, dx$ was wrong. The choice must be such that the 'u part' becomes a constant after differentiation and the 'dv' part can be integrated easily. (It will be seen later that for the 'u part' to become a constant often requires more than one differentiation (see Problem 5).)

(iv)   If a product to be integrated contains an 'x' term then this term is chosen as the 'u part' and the other function as the 'dv part' except where a logarithmic term is involved (see Problems 6 and 7).

Problem 2. Determine $\int x \sin x \, dx$

Let $u = x$ and $dv = \sin x \, dx$
Then $du = dx$ and $v = \int \sin x \, dx = -\cos x$
Substituting into $\int u \, dv = uv - \int v \, du$ gives:

$$\int x \sin x \, dx = (x)(-\cos x) - \int (-\cos x) \, dx$$

$$= -x \cos x + \int \cos x \, dx$$

$$= -x \cos x + \sin x + c$$

This result can be checked by differentiating.

Thus

$$\frac{d}{dx}(-x \cos x + \sin x + c) = (-x)(-\sin x) + (\cos x)(-1) + \cos x$$

$$= x \sin x - \cos x + \cos x = x \sin x$$

Problem 3. Evaluate $\int_0^1 2x \, e^{3x} \, dx$

Let $u = 2x$ and $dv = e^{3x} \, dx$

Then $\dfrac{du}{dx} = 2$, i.e. $du = 2 \, dx$ and $v = \int e^{3x} \, dx = \dfrac{e^{3x}}{3}$

Substituting into $\int u \, dv = uv - \int v \, du$ gives:

$$\int 2x \, e^{3x} \, dx = (2x)\left(\frac{e^{3x}}{3}\right) - \int \left(\frac{e^{3x}}{3}\right)(2 \, dx)$$

$$= \frac{2}{3} x \, e^{3x} - \frac{2}{3} \int e^{3x} \, dx$$

$$= \frac{2}{3} x \, e^{3x} - \frac{2}{3}\left(\frac{e^{3x}}{3}\right) + c$$

$$= \frac{2}{3} x \, e^{3x} - \frac{2}{9} e^{3x} + c$$

Hence $\int_0^1 2x\, e^{3x}\, dx = \left[\dfrac{2}{3}x\, e^{3x} - \dfrac{2}{9}e^{3x}\right]_0^1$

$$= \left(\dfrac{2}{3}e^3 - \dfrac{2}{9}e^3\right) - \left(0 - \dfrac{2}{9}e^0\right)$$

$$= \dfrac{4}{9}e^3 + \dfrac{2}{9} = 8.926\,9 + 0.222\,2$$

$$= \textbf{9.149 correct to 3 decimal places}$$

Problem 4. Evaluate $\displaystyle\int_0^{\pi/2} 3t \cos 2t\, dt$

Let $u = 3t$ and $dv = \cos 2t\, dt$

Then $du = 3\, dt$ and $v = \displaystyle\int \cos 2t\, dt = \dfrac{1}{2}\sin 2t$

Substituting into $\int u\, dv = uv - \int v\, du$ gives:

$$\int 3t \cos 2t\, dt = (3t)\left(\dfrac{\sin 2t}{2}\right) - \int\left(\dfrac{\sin 2t}{2}\right)(3\, dt)$$

$$= \dfrac{3t}{2}\sin 2t - \dfrac{3}{2}\int \sin 2t\, dt$$

$$= \dfrac{3t}{2}\sin 2t - \dfrac{3}{2}\left(\dfrac{-\cos 2t}{2}\right) + c$$

$$= \dfrac{3t}{2}\sin 2t + \dfrac{3}{4}\cos 2t + c$$

Hence $\displaystyle\int_0^{\pi/2} 3t \cos 2t\, dt = \left[\dfrac{3t}{2}\sin 2t + \dfrac{3}{4}\cos 2t\right]_0^{\pi/2}$

$$= \left[\dfrac{3}{2}\left(\dfrac{\pi}{2}\right)\sin \pi + \dfrac{3}{4}\cos \pi\right] - \left[0 + \dfrac{3}{4}\cos 0\right]$$

$$= \left(0 + \dfrac{3}{4}(-1)\right) - \left(0 + \dfrac{3}{4}\right) = -\dfrac{3}{4} - \dfrac{3}{4} = -1\tfrac{1}{2}$$

Problem 5. Determine $\displaystyle\int x^2 \cos x\, dx$

Let $u = x^2$ and $dv = \cos x\, dx$
Then $du = 2x\, dx$ and $v = \int \cos x\, dx = \sin x$
Substituting into $\int u\, dv = uv - \int v\, du$ gives:

$$\int x^2 \cos x \, dx = (x^2)(\sin x) - \int (\sin x)(2x \, dx)$$

$$= x^2 \sin x \quad - 2 \left[\int x \sin x \, dx \right]$$

The integral on the right-hand side in the bracket is not a standard and cannot be determined 'on sight'. Since it is a product of two simple functions we may use integration by parts again.

Now $\int x \sin x \, dx = -x \cos x + \sin x$ (from Problem 2)

Thus $\int x^2 \cos x \, dx = x^2 \sin x - 2[-x \cos x + \sin x] + c$

$$= x^2 \sin x + 2x \cos x - 2 \sin x + c$$

$$= (x^2 - 2) \sin x + 2x \cos x + c$$

Generally, if the term in x is of power n then the integration by parts formula is applied n times, provided one of the functions is not $\ln x$, as in Problem 6 following.

Problem 6. $\int x^2 \ln x \, dx$

Whenever a product consists of a term in x and a logarithmic function, as in this problem, it is always the logarithmic function that is chosen as the 'u part'. The reason for this is that $\dfrac{d}{dx} (\ln x)$ is $\dfrac{1}{x}$, but $\int \ln x \, dx$ is not normally remembered as a standard integral.

Thus if $u = \ln x$ and $dv = x^2 \, dx$

then $du = \dfrac{1}{x} \, dx$ and $v = \int x^2 \, dx = \dfrac{x^3}{3}$

Substituting into $\int u \, dv = uv - \int v \, du$ gives:

$$\int x^2 \ln x \, dx = (\ln x)\left(\frac{x^3}{3}\right) - \int \left(\frac{x^3}{3}\right)\left(\frac{1}{x} \, dx\right)$$

$$= \frac{x^3}{3} \ln x - \frac{1}{3} \int x^2 \, dx$$

$$= \frac{x^3}{3} \ln x - \frac{1}{3}\left(\frac{x^3}{3}\right) + c$$

$$= \frac{x^3}{3} \ln x - \frac{x^3}{9} + c$$

$$= \frac{x^3}{9} (3 \ln x - 1) + c$$

Problem 7. Determine $\int \ln x \, dx$

In each of the previous problems the components of the product have been obvious. However, $\int \ln x \, dx$ is a special case for initially it appears not to be a product. However, $\int \ln x \, dx$ is the same as $\int 1 \times \ln x \, dx$

Let $u = \ln x$ and $dv = 1 \, dx$

Then $du = \dfrac{1}{x} \, dx$ and $v = \displaystyle\int 1 \, dx = x$

Hence $\displaystyle\int \ln x \, dx = (\ln x)(x) - \int (x)\left(\frac{1}{x} dx\right)$

$$= x \ln x - \int dx$$

$$= x \ln x - x + c$$

$$= x(\ln x - 1) + c$$

Problem 8. Find $\int e^{ax} \sin bx \, dx$

With an integral of a product of an exponential function and a sine or cosine function it does not matter which function is made equal to 'u'
Thus let $u = e^{ax}$ and $dv = \sin bx \, dx$

then $du = a e^{ax} \, dx$ and $v = \displaystyle\int \sin bx \, dx = \frac{-\cos bx}{b}$

Thus $\displaystyle\int e^{ax} \sin bx \, dx = (e^{ax})\left(\frac{-\cos bx}{b}\right) - \int \left(\frac{-\cos bx}{b}\right)(a e^{ax} \, dx)$

$$= -\frac{1}{b} e^{ax} \cos bx + \frac{a}{b}\left[\int e^{ax} \cos bx \, dx \right] \qquad (1)$$

It would seem that we are no nearer a solution of the initial integral. However, the integration by parts formula may be applied to the integral in the bracket.

Let $u = e^{ax}$ and $dv = \cos bx \, dx$

Then $du = ae^{ax} \, dx$ and $v = \displaystyle\int \cos bx \, dx = \dfrac{\sin bx}{b}$

Thus $\displaystyle\int e^{ax} \cos bx \, dx = (e^{ax})\left(\dfrac{\sin bx}{b}\right) - \int \left(\dfrac{\sin bx}{b}\right)(a\,e^{ax}\,dx)$

$$= \frac{1}{b} e^{ax} \sin bx - \frac{a}{b} \int e^{ax} \sin bx \, dx$$

Substituting this result into equation (1) gives:

$$\int e^{ax} \sin bx \, dx = -\frac{1}{b} e^{ax} \cos bx + \frac{a}{b} \left[\frac{1}{b} e^{ax} \sin bx - \frac{a}{b} \int e^{ax} \sin bx \, dx \right]$$

$$= \frac{-1}{b} e^{ax} \cos bx + \frac{a}{b^2} e^{ax} \sin bx - \frac{a^2}{b^2} \left[\int e^{ax} \sin bx \, dx \right]$$

The integral in the bracket on the right-hand side is the same as the integral on the left-hand side thus they may be combined on the left-hand side of the equation. Thus:

$$\int e^{ax} \sin bx \, dx + \frac{a^2}{b^2} \int e^{ax} \sin bx \, dx = \frac{-1}{b} e^{ax} \cos bx + \frac{a}{b^2} e^{ax} \sin bx$$

i.e. $\left(1 + \dfrac{a^2}{b^2}\right) \displaystyle\int e^{ax} \sin bx \, dx = \dfrac{e^{ax}}{b^2} (a \sin bx - b \cos bx)$

$\left(\dfrac{b^2 + a^2}{b^2}\right) \displaystyle\int e^{ax} \sin bx \, dx = \dfrac{e^{ax}}{b^2} (a \sin bx - b \cos bx)$

Hence $\displaystyle\int e^{ax} \sin bx \, dx = \left(\dfrac{b^2}{a^2 + b^2}\right) \dfrac{e^{ax}}{b^2} (a \sin bx - b \cos bx)$

$$= \left(\frac{e^{ax}}{a^2 + b^2}\right)(a \sin bx - b \cos bx) + c$$

A product of an exponential function and a sine or cosine function thus involves integration by parts twice. If, in the above problem, the exponential function is made equal to dv instead of u the same result is obtained. It is left as a student exercise to prove this.

By a similar method to above it may be shown that:

$$\int e^{ax} \cos bx \, dx = \left(\frac{e^{ax}}{a^2 + b^2}\right)(b \sin bx + a \cos bx) + c$$

Further problems on integration by parts may be found in the following Section 26.3 (Problems 1 to 27).

26.3   Further problems

Find the following integrals using integration by parts.

1. $\displaystyle\int x\,e^{3x}\,dx$ $\left[\dfrac{e^{3x}}{9}(3x-1)+c\right]$

2. $\displaystyle\int 3x\,e^{2x}\,dx$ $\left[\dfrac{3}{4}e^{2x}(2x-1)+c\right]$

3. $\displaystyle\int \dfrac{5x}{e^{4x}}\,dx$ $\left[\dfrac{-5}{16}e^{-4x}(4x+1)+c\right]$

4. $\displaystyle\int x\cos x\,dx$ $[x\sin x+\cos x+c]$

5. $\displaystyle\int x\ln x\,dx$ $\left[\dfrac{x^2}{4}(2\ln x-1)+c\right]$

6. $\displaystyle\int 3x\sin 2x\,dx$ $\left[\dfrac{-3x}{2}\cos 2x+\dfrac{3}{4}\sin 2x+c\right]$

7. $\displaystyle\int \ln 4x\,dx$ $[x(\ln 4x-1)+c]$

8. $\displaystyle\int \dfrac{\ln y\,dy}{y^2}$ $\left[-\dfrac{1}{y}(\ln y+1)+c\right]$

9. $\displaystyle\int 2x\cos 5x\,dx$ $\left[\dfrac{2}{25}(5x\sin 5x+\cos 5x)+c\right]$

10. $\displaystyle\int 2x^2\,e^x\,dx$ $[2e^x(x^2-2x+2)+c]$

11. $\displaystyle\int x^2\sin 2x\,dx$ $\left[\left(\dfrac{1}{4}-\dfrac{x^2}{2}\right)\cos 2x+\dfrac{x}{2}\sin 2x+c\right]$

12. $\displaystyle\int e^{2x}\cos x\,dx$ $\left[\dfrac{e^{2x}}{5}(\sin x+2\cos x)+c\right]$

13. $\displaystyle\int 4\theta^2\cos 3\theta\,d\theta$ $\left[4\left(\dfrac{\theta^2}{3}-\dfrac{2}{27}\right)\sin 3\theta+\dfrac{8}{9}\theta\cos 3\theta+c\right]$

14. $\displaystyle\int 3e^x\sin 2x\,dx$ $\left[\dfrac{3}{5}e^x(\sin 2x-2\cos 2x)+c\right]$

15. $\displaystyle\int 4x\sec^2 x\,dx$ $[4[x\tan x-\ln(\sec x)]+c]$

Evaluate the following integrals correct to 3 decimal places.

16. $\displaystyle\int_0^2 x\,e^x\,dx$ $[8.389]$

17. $\displaystyle\int_0^{\pi/2} x \cos x \, dx$ [0.571]

18. $\displaystyle\int_0^1 t \, e^{2t} \, dt$ [2.097]

19. $\displaystyle\int_0^{\pi/4} \phi \sin 2\phi \, d\phi$ [0.250]

20. $\displaystyle\int_1^2 \ln x^2 \, dx$ [0.773]

21. $\displaystyle\int_0^{\pi/2} 3x^2 \cos x \, dx$ [1.402]

22. $\displaystyle\int_0^1 x^2 \, e^x \, dx$ [0.718]

23. $\displaystyle\int_1^4 \sqrt{t} \ln t \, dt$ [4.282]

24. $\displaystyle\int_0^1 2e^t \cos 2t \, dt$ [1.125]

25. $\displaystyle\int_0^{\pi/2} e^\theta \sin \theta \, d\theta$ [2.905]

26. In the study of damped oscillations integrations of the following type are important:

$$C = \int_0^1 e^{-0.4\theta} \cos 1.2\theta \, d\theta$$

and $$S = \int_0^1 e^{-0.4\theta} \sin 1.2\theta \, d\theta$$

Determine C and S [$C = 0.66$, $S = 0.41$]

27. If a string is plucked at a point $x = \dfrac{l}{3}$ with an amplitude a and released, the equation of motion is

$$K = \frac{2}{l}\left\{ \int_0^{l/3} \frac{3a}{l} x \sin \frac{n\pi}{l} x \, dx + \int_{l/3}^l \frac{3a}{2l}(l - x) \sin \frac{n\pi}{l} x \, dx \right\}$$

where n is a constant.

Show that $K = \dfrac{9a}{\pi^2 n^2} \sin \dfrac{n\pi}{3}$

Chapter 27

Numerical integration

27.1 Introduction

In Chapter 21 it was shown that a number of functions, such as ax^n, $\sin ax$, $\cos ax$, e^{ax} and $\dfrac{1}{x}$, are termed **'standard integrals'**, and they may be integrated 'on sight'. It was also shown that with some other simple functions, such as $\cos(5x + 2)$, $(4t - 3)^7$ and $\dfrac{1}{7x + 2}$, an **algebraic substitution** may be used to change the function into a form which can be readily integrated. However, even with more advanced methods of integration, there are many mathematical functions which cannot be integrated by analytical methods and thus approximate methods have then to be used. Also, in some cases, only a set of observed tabulated numerical values of the function to be integrated may be available. Approximate values of definite integrals may be determined by what is termed **numerical integration**. It is shown in Chapter 28 that determining the value of a definite integral is, in fact, finding the area between a curve, the horizontal axis and the specified ordinates. Three methods of finding approximate areas under curves are the trapezoidal rule, the mid-ordinate rule and Simpson's rule and these rules are used as a basis for numerical integration.

27.2 The trapezoidal rule

Let a required definite integral be denoted by $\displaystyle\int_a^b y \, dx$ and be represented by the area under the graph of y between the limits $x = a$ and $x = b$, as shown in Fig. 27.1. Let the range of integration by divided into n equal intervals each of width d, such that $nd = b - a$, i.e. $d = \dfrac{b - a}{n}$.

Let the ordinates be labelled $y_1, y_2, \ldots y_{n+1}$, as shown. An approximation to the area under the curve is obtained by joining the tops of the ordinates by ystraight lines. Each strip of area is thus a trapezium and since the area

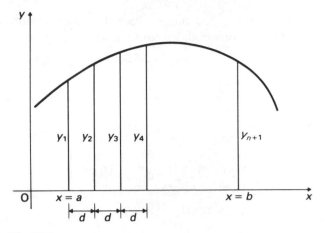

Fig. 27.1

of a trapezium is given by:

area of trapezium $= \dfrac{1}{2}$ (sum of the parallel sides) (perpendicular distance between them)

then

$$\int_a^b y \, \mathrm{d}x \approx \frac{1}{2}(y_1 + y_2)\mathrm{d} + \frac{1}{2}(y_2 + y_3)\mathrm{d} + \frac{1}{2}(y_3 + y_4)\mathrm{d} + \cdots$$

$$+ \frac{1}{2}(y_n + y_{n+1})\mathrm{d}$$

$$\approx \mathrm{d}\left[\frac{1}{2}y_1 + y_2 + y_3 + \cdots + y_n + \frac{1}{2}y_{n+1}\right]$$

i.e. $$\boxed{\int_a^b y \, \mathrm{d}x \approx \left(\begin{array}{c}\textbf{width of}\\ \textbf{interval}\end{array}\right)\left\{\frac{1}{2}\left(\begin{array}{c}\textbf{first + last}\\ \textbf{ordinate}\end{array}\right) + \begin{array}{c}\textbf{sum of remaining}\\ \textbf{ordinates}\end{array}\right\}}$$ (1)

To demonstrate this method of numerical integration let an integral be chosen whose value may be determined exactly using integration.

Evaluating $\displaystyle\int_1^3 \frac{1}{\sqrt{x}}\,\mathrm{d}x$ by integration gives $\left[\dfrac{x^{-\frac{1}{2}+1}}{-\frac{1}{2}+1}\right]_1^3 = \left[2\sqrt{x}\right]_1^3$

$$= 2(\sqrt{3} - \sqrt{1})$$

$$= \textbf{1.464}, \text{ correct}$$
$$\text{to 3 decimal places}$$

The range of integration is the difference between the upper and lower limits, i.e., $3 - 1 = 2$. Using the trapezoidal rule with, say, 4 intervals, gives an interval width d of $\dfrac{3-1}{4} = \dfrac{1}{2}$, and ordinates situated at 1.0, 1.5, 2.0, 2.5 and 3.0. Corresponding values of $\dfrac{1}{\sqrt{x}}$ are as shown in the table below, each given correct to 4 decimal places.

x	1.0	1.5	2.0	2.5	3.0
$\dfrac{1}{\sqrt{x}}$	1.0000	0.8165	0.7071	0.6325	0.5774

From equation (1):

$$\int_1^3 \frac{1}{\sqrt{x}}\,dx \approx \left(\frac{1}{2}\right)\left[\frac{1}{2}(1.0000 + 0.5774) + 0.8165 + 0.7071 + 0.6325\right]$$

$$= \mathbf{1.472}, \text{ correct to 3 decimal places}$$

Using the trapezoidal rule with, say, 8 intervals, each of width $\dfrac{3-1}{8}$, i.e. 0.25, gives ordinates situated at 1.00, 1.25, 1.50, 1.75, 2.00, 2.25, 2.50, 2.75 and 3.00. Corresponding values of $\dfrac{1}{\sqrt{x}}$ are as shown in the table below.

x	1.00	1.25	1.50	1.75	2.00	2.25	2.50	2.75	3.00
$\dfrac{1}{\sqrt{x}}$	1.0000	0.8944	0.8165	0.7559	0.7071	0.6667	0.6325	0.6030	0.5774

From equation (1):

$$\int_1^3 \frac{1}{\sqrt{x}}\,dx \approx (0.25)\left[\frac{1}{2}(1.0000 + 0.5774) + 0.8944 + 0.8165 + 0.7559\right.$$
$$\left. + 0.7071 + 0.6667 + 0.6325 + 0.6030\right]$$

$$= \mathbf{1.466}, \text{ correct to 3 decimal places}$$

The greater the number of intervals chosen (i.e. the smaller the interval width) the more accurate will be the value of the definite integral. The exact value is found when the number of intervals is infinite, i.e. when the interval width d tends to zero, and this is of course what the process of integration is based upon.

27.3   The mid-ordinate rule

Let a required definite integral be denoted again by $\int_a^b y\,dx$ and represented by the area under the graph of y between the limits $x = a$ and $x = b$, as shown in Fig. 27.2

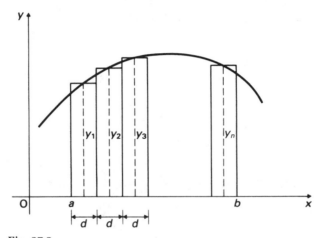

Fig. 27.2

With this rule each strip of width d is assumed to be replaced by a rectangle of height equal to the ordinates at the middle point of each interval, shown as $y_1, y_2, y_3, \ldots, y_n$ in Fig. 27.2. Thus,

$$\int_a^b y\,dx \approx dy_1 + dy_2 + dy_3 + \cdots + dy_n$$

$$\approx d(y_1 + y_2 + y_3 + \cdots + y_n)$$

i.e.

$$\boxed{\int_a^b y\,dx \approx \textbf{(width of interval)(sum of mid-ordinates)}} \qquad (2)$$

The more intervals chosen the more accurate will be the value of the definite integral.

Applying this rule to evaluating $\int_1^3 \dfrac{1}{\sqrt{x}}\,dx$ with, say, 4 intervals means that the width interval d is $\dfrac{3-1}{4}$, i.e. $\dfrac{1}{2}$ and that ordinates exist at 1.0, 1.5, 2.0, 2.5 and 3.0. Hence mid-ordinates y_1, y_2, y_3 and y_4 occur at 1.25, 1.75, 2.25 and 2.75. Corresponding values of $\dfrac{1}{\sqrt{x}}$ are shown in the following table:

x	1.25	1.75	2.25	2.75
$\dfrac{1}{\sqrt{x}}$	0.894 4	0.755 9	0.666 7	0.603 0

From equation (2):

$$\int_1^3 \frac{1}{\sqrt{x}}\, dx \approx \left(\frac{1}{2}\right)[0.894\,4 + 0.755\,9 + 0.666\,7 + 0.603\,0]$$

$$= \mathbf{1.460},\text{ correct to 3 decimal places}$$

Using the mid-ordinate rule with, say, 8 intervals, each of width 0.25, gives ordinates at 1.00, 1.25, 1.50, 1.75... and thus mid-ordinates at 1.125, 1.375, 1.625, 1.875, ... Corresponding values of $\dfrac{1}{\sqrt{x}}$ are shown in the following table.

x	1.125	1.375	1.625	1.875	2.125	2.375	2.625	2.875
$\dfrac{1}{\sqrt{x}}$	0.942 8	0.852 8	0.784 5	0.730 3	0.686 0	0.648 9	0.617 2	0.589 8

From equation (2):

$$\int_1^3 \frac{1}{\sqrt{x}}\, dx \approx (0.25)\Big[0.942\,8 + 0.852\,8 + 0.784\,5 + 0.730\,3 + 0.686\,0$$

$$+\, 0.648\,9 + 0.617\,2 + 0.589\,8 \Big]$$

$$= \mathbf{1.463},\text{ correct to 3 decimal places}$$

As before, the greater the number of values chosen the nearer the result will be to the true one.

27.4   Simpson's rule

In Section 27.2, it is shown that the approximation made with the trapezoidal rule is to join the tops of two successive ordinates by a straight line, i.e. by using a linear approximation of the form $a + bx$. With Simpson's rule, the approximation made is to join the tops of three successive ordinates by a parabola, i.e. by using a quadratic approximation of the form $a + bx + cx^2$.

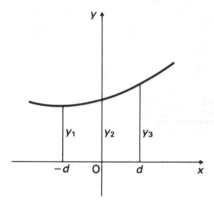

Fig. 27.3

Figure 27.3 shows three ordinates, y_1, y_2 and y_3 of a parabola $y = a + bx + cx^2$ at $x = -d$, $x = 0$ and $x = d$ respectively. Thus the width of each of the two intervals is d. The area under the parabola from $x = -d$ to $x = d$ is given by:

$$\int_{-d}^{d} (a + bx + cx^2)\,dx = \left[ax + \frac{bx^2}{2} + \frac{cx^3}{3} \right]_{-d}^{d}$$

$$= \left(ad + \frac{bd^2}{2} + \frac{cd^3}{3} \right) - \left(-ad + \frac{bd^2}{2} - \frac{cd^3}{3} \right)$$

$$= 2ad + \frac{2}{3}cd^3 = \frac{1}{3}d(6a + 2cd^2) \qquad (3)$$

Since $y = a + bx + cx^2$, at $x = -d$, $y_1 = a - bd + cd^2$,

$$\text{at } x = 0,\ y_2 = a,$$

and $\qquad\qquad$ at $x = d$, $y_3 = a + bd + cd^2$

Hence $\quad y_1 + y_3 = 2a + 2cd^2$

and $\quad y_1 + 4y_2 + y_3 = 6a + 2cd^2 \qquad\qquad (4)$

Thus the area under the parabola between $x = -d$ and $x = d$ in Fig. 27.3

is (from equations (3) and (4)):

$$\frac{1}{3} d(y_1 + 4y_2 + y_3)$$

and this result can be seen to be independent of the position of the origin.

Let a definite integral be denoted by $\int_a^b y\, dx$ and represented by the area under the graph of y between the limits $x = a$ and $x = b$, as shown in Fig. 27.4. The range of integration, $b - a$, is divided into an **even** number of intervals, say, $2n$, each of width d.

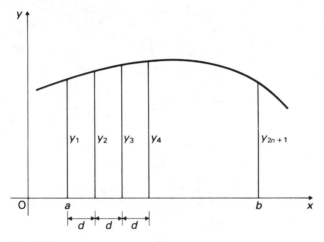

Fig. 27.4

Since an even number of intervals is specified, an odd number of ordinates, $2n + 1$, exists. Let an approximation to the curve over the first two intervals be a parabola of the form $y = a + bx + cx^2$ which passes through the tops of the three ordinates y_1, y_2 and y_3. Similarly, let an approximation to the curve over the next two intervals be the parabola which passes through the tops of the ordinates y_3, y_4 and y_5, and so on.

Then
$$\int_a^b y\, dx \approx \frac{1}{3} d(y_1 + 4y_2 + y_3) + \frac{1}{3} d(y_3 + 4y_4 + y_5) + \cdots$$

$$+ \frac{1}{3} d(y_{2n-1} + 4y_{2n} + y_{2n+1})$$

$$\approx \frac{1}{3} d[(y_1 + y_{2n+1}) + 4(y_2 + y_4 + \cdots + y_{2n})$$

$$+ 2(y_3 + y_5 + \cdots + y_{2n-1})]$$

i.e.

$$\int_a^b y\,dx \approx \frac{1}{3}\begin{pmatrix}\text{width of}\\\text{interval}\end{pmatrix}\left\{\begin{pmatrix}\text{first + last}\\\text{ordinate}\end{pmatrix} + 4\begin{pmatrix}\text{sum of even}\\\text{ordinates}\end{pmatrix}\right.$$
$$\left. + 2\begin{pmatrix}\text{sum of remaining}\\\text{odd ordinates}\end{pmatrix}\right\} \qquad (5)$$

Again, the more intervals chosen the more accurate will be the value of the definite integral.

Applying this rule to evaluate $\int_1^3 \frac{1}{\sqrt{x}}\,dx$ with, say, 4 intervals, using the table of values produced in Section 27.2 and using equation (5) gives:

$$\int_1^3 \frac{1}{\sqrt{x}}\,dx \approx \frac{1}{3}(0.5)\{(1.000\,0 + 0.577\,4) + 4(0.816\,5 + 0.632\,5) + 2(0.707\,1)\}$$

$$\approx \frac{1}{3}(0.5)\{1.577\,4 + 5.796\,0 + 1.414\,2\}$$

$$= \mathbf{1.465}, \text{ correct to 3 decimal places}$$

Using 8 intervals, the table of values produced in Section 27.2, and equation (5) gives:

$$\int_1^3 \frac{1}{\sqrt{x}}\,dx \approx \frac{1}{3}(0.25)\{(1.000\,0 + 0.577\,4) + 4(0.894\,4 + 0.755\,9 + 0.666\,7 + 0.603\,0) + 2(0.816\,5 + 0.707\,1 + 0.632\,5)\}$$

$$\approx \frac{1}{3}(0.25)\{1.577\,4 + 11.680\,0 + 4.312\,2\}$$

$$= \mathbf{1.465}, \text{ correct to 3 decimal places, which is the same value as obtained by integration}$$

Simpson's rule is generally regarded as the most accurate of the three approximate methods used in numerical integration.

Worked problems on numerical integration

Problem 1. Evaluate $\int_0^1 \frac{2}{1+x^2}\,dx$, using the trapezoidal rule with 8 intervals, giving the answer correct to 3 decimal places.

The range of integration is from 0 to 1 and hence if 8 intervals are chosen then each will be of width $\dfrac{1-0}{8}$, i.e. d = 0.125. Thus ordinates occur at 0, 0.125, 0.250, 0.375 ... and for each of these values $\dfrac{2}{1+x^2}$ may be evaluated. The results are shown in the following table.

x	0	0.125	0.250	0.375	0.500	0.625	0.750	0.875	1.000
$\dfrac{2}{1+x^2}$	2.000 0	1.969 2	1.882 4	1.753 4	1.600 0	1.438 2	1.280 0	1.132 7	1.000 0

(Note that since 3-decimal-place accuracy is required in the final answer, values in the table are taken correct to one decimal place more than this, i.e. correct to 4 decimal places.)
From equation (1), using the trapezoidal rule with 8 intervals:

$$\int_0^1 \frac{2}{1+x^2}\,dx \approx d\left[\frac{1}{2}\left(\begin{array}{c}\text{first + last} \\ \text{ordinate}\end{array}\right) + \text{sum of remaining ordinates}\right]$$

$$\approx (0.125)\left[\frac{1}{2}(2.000\,0 + 1.000\,0) + 1.969\,2 + 1.882\,4\right.$$

$$+ 1753\,4 + 1.600\,0 + 1.438\,2$$

$$\left. + 1.280\,0 + 1.132\,7\right]$$

$$= \mathbf{1.569}, \text{ correct to 3 decimal places}$$

Problem 2. Determine the value of $\displaystyle\int_1^5 \ln x\,dx$ using the mid-ordinate rule with (a) 4 intervals, (b) 8 intervals. Give the answers correct to 4 significant figures.

(a)   The range of integration is from 1 to 5 and hence if 4 intervals are chosen then each will be of width $\dfrac{5-1}{4}$, i.e. d = 1

Hence ordinates occur at 1, 2, 3, 4 and 5 and thus mid-ordinates occur at 1.5, 2.5, 3.5 and 4.5. Corresponding values of $\ln x$ are shown in the following table.

x	1.5	2.5	3.5	4.5
ln x	0.405 47	0.916 29	1.252 76	1.504 08

From equation (2), using the mid-ordinate rule with 4 intervals:

$$\int_1^5 \ln x \, dx \approx (\text{width of interval})(\text{sum of mid-ordinates})$$

$$\approx (1)(0.405\,47 + 0.916\,29 + 1.252\,76 + 1.504\,08)$$

$$= \textbf{4.079}, \text{ correct to 4 significant figures}$$

(b) With 8 intervals, width d $= \dfrac{5-1}{8} = 0.5$, ordinates occur at

1.0, 1.5, 2.0, 2.5, ..., 5, and thus mid-ordinates occur at 1.25,
1.75, 2.25, ..., 4.75
Hence from equation (2):

$$\int_1^5 \ln x \, dx \approx (0.5)[\ln 1.25 + \ln 1.75 + \ln 2.25 + \ln 2.75$$

$$+ \ln 3.25 + \ln 3.75 + \ln 4.25 + \ln 4.75]$$

$$= \textbf{4.055}, \text{ correct to 4 significant figures}$$

Problem 3. Evaluate $\displaystyle\int_0^{\pi/2} \dfrac{1}{1 + \frac{1}{2}\sin^2\theta}\, d\theta$, using Simpson's rule
with 6 intervals, correct to 3 decimal places.

With 6 intervals chosen, each will have a width of $\dfrac{\frac{\pi}{2} - 0}{6}$, i.e.

$\dfrac{\pi}{12}$ rad (or 15°) and the ordinates occur at 0, $\dfrac{\pi}{12}, \dfrac{\pi}{6}, \dfrac{\pi}{4}, \dfrac{\pi}{3}, \dfrac{5\pi}{12}$ and $\dfrac{\pi}{2}$

Corresponding values of $\dfrac{1}{1 + \frac{1}{2}\sin^2\theta}$ are evaluated and are shown
in the table below.

θ	0	$\dfrac{\pi}{12}$ (or 15°)	$\dfrac{\pi}{6}$ (or 30°)	$\dfrac{\pi}{4}$ (or 45°)	$\dfrac{\pi}{3}$ (or 60°)	$\dfrac{5\pi}{12}$ (or 75°)	$\dfrac{\pi}{2}$ (or 90°)
$\dfrac{1}{1 + \frac{1}{2}\sin^2\theta}$	1.0000	0.9676	0.8889	0.8000	0.7273	0.6819	0.6667

From equation (5), using Simpson's rule with 6 intervals:

$$\int_0^{\pi/2} \frac{1}{1 + \frac{1}{2}\sin^2 \theta}\, d\theta$$

$$\approx \frac{1}{3}\left(\begin{array}{c}\text{width of}\\ \text{intervals}\end{array}\right)\left\{\begin{array}{c}\left(\begin{array}{c}\text{first + last}\\ \text{ordinate}\end{array}\right) + 4\left(\begin{array}{c}\text{sum of even}\\ \text{ordinates}\end{array}\right)\\ + 2\left(\begin{array}{c}\text{sum of remaining}\\ \text{odd ordinates}\end{array}\right)\end{array}\right\}$$

$$\approx \frac{1}{3}\left(\frac{\pi}{12}\right)[(1.000\,0 + 0.666\,7)$$

$$+ 4(0.967\,6 + 0.800\,0 + 0.681\,9)$$

$$+ 2(0.888\,9 + 0.727\,3)]$$

$$\approx \frac{\pi}{36}\{1.666\,7 + 9.798\,0 + 3.232\,4\}$$

$$\approx 0.408\,3\pi = \mathbf{1.283}, \text{ correct to 3 decimal places}$$

Problem 4. The velocity v of a car has the following values for corresponding values of time t from $t = 0$ to $t = 8$ s

v m/s	0	0.6	1.7	2.8	4.9	7.0	9.2	10.8	12.0
t s	0	1	2	3	4	5	6	7	8

The distance travelled by the car in 8 s is given by $\displaystyle\int_0^8 v\, dt$

Determine the approximate distance travelled by using (*a*) the trapezoidal rule, and (*b*) Simpson's rule, using 8 intervals in each case.

Since 8 intervals are chosen, each has a width of 1 s

(*a*) From equation (1), using the trapezoidal rule with 8 intervals:

$$\int_0^8 v\, dt \approx (1)\left[\frac{1}{2}(0 + 12.0) + 0.6 + 1.7 + 2.8 + 4.9 + 7.0 \right.$$
$$\left. + 9.2 + 10.8\right]$$

$$= \mathbf{43\ m}$$

(b) From equation (5), using Simpson's rule with 8 intervals:

$$\int_0^8 v \, dt \approx \frac{1}{3}(1)[(0 + 12.0) + 4(0.6 + 2.8 + 7.0 + 10.8) \\ + 2(1.7 + 4.9 + 9.2)]$$

$$\approx \frac{1}{3}(1)[12.0 + 84.8 + 31.6]$$

$$= \textbf{42.8 m}$$

Problem 5. Evaluate $\int_0^{1.2} e^{-x^2/2} \, dx$ using (a) the trapezoidal rule, (b) the mid-ordinate rule and (c) Simpson's rule. Use 6 intervals in each case and give answers correct to 3 significant figures.

Since 6 intervals are chosen then each is of width $\dfrac{1.2 - 0}{6}$, i.e. 0.2

Hence ordinates occur at 0, 0.2, 0.4, 0.6, 0.8, 1.0 and 1.2
Corresponding values of $e^{-x^2/2}$ are evaluated and shown in the following table.

x	0	0.2	0.4	0.6	0.8	1.0	1.2
$e^{-x^2/2}$	1.0000	0.9802	0.9231	0.8353	0.7261	0.6065	0.4868

(a) From equation (1), using the trapezoidal rule with 6 intervals:

$$\int_0^{1.2} e^{-x^2/2} \, dx \approx (0.2)\left[\frac{1}{2}(1.0000 + 0.4868) + 0.9802 + 0.9231 \\ + 0.8353 + 0.7261 + 0.6065\right]$$

$$= \textbf{0.963}, \text{ correct to 3 significant figures}$$

(b) Mid-ordinates occur at 0.1, 0.3, 0.5, 0.7, 0.9 and 1.1 and corresponding values of $e^{-x^2/2}$ are evaluated and shown in the following table.

x	0.1	0.3	0.5	0.7	0.9	1.1
$e^{-x^2/2}$	0.9950	0.9560	0.8825	0.7827	0.6670	0.5461

From equation (2), using the mid-ordinate rule with 6 intervals:

$$\int_0^{1.2} e^{-x^2/2}\, dx \approx (0.2)(0.995\,0 + 0.956\,0 + 0.882\,5 + 0.782\,7$$
$$+ 0.667\,0 + 0.546\,1)$$

$$= \mathbf{0.966}, \text{ correct to 3 significant figures}$$

(c) From equation (5), using Simpson's rule with 6 intervals and the table of values in part (a) above:

$$\int_0^{1.2} e^{-x^2/2}\, dx \approx \frac{1}{3}(0.2)[(1.000\,0 + 0.486\,8) + 4(0.9802 + 0.835\,3$$
$$+ 0.606\,5) + 2(0.923\,1 + 0.726\,1)]$$

$$\approx \frac{1}{3}(0.2)[1.486\,8 + 9.688\,0 + 3.298\,4]$$

$$= \mathbf{0.965}, \text{ correct to 3 significant figures}$$

Problem 6. Evaluate, correct to 3 decimal places, $\int_1^3 \dfrac{5}{x}\, dx$ using

(a) integration,
(b) the trapezoidal rule, with (i) 4 intervals, (ii) 8 intervals,
(c) Simpson's rule with (i) 4 intervals, (ii) 8 intervals.
(d) Determine the percentage error in parts (b) and (c) compared with the value obtained in part (a)

(a) $\int_1^3 \dfrac{5}{x}\, dx = 5[\ln x]_1^3 = 5[\ln 3 - \ln 1] = 5\ln 3 = \mathbf{5.493}$,
correct to 3
decimal places

(b) (i) With 4 intervals each interval is of width $\dfrac{3-1}{4} = 0.5$ and ordinates occur at 1.0, 1.5, 2.0, 2.5 and 3.0. Corresponding values of $\dfrac{5}{x}$ are evaluated and shown in the following table.

x	1.0	1.5	2.0	2.5	3.0
$\dfrac{5}{x}$	5.000 0	3.333 3	2.500 0	2.000 0	1.666 7

From equation (1), using the trapezoidal rule with 4 intervals:

$$\int_1^3 \frac{5}{x} \, dx \approx (0.5)\left[\frac{1}{2}(5.000\,0 + 1.666\,7) + 3.333\,3 + 2.500\,0 + 2.000\,0\right]$$

= **5.583**, correct to 3 decimal places

(ii) With 8 intervals each interval is of width $\dfrac{3-1}{8} = 0.25$ and

ordinates occur at 1.00, 1.25, 1.50, 1.75, ..., 3.00

Corresponding values of $\dfrac{5}{x}$ are evaluated and shown in the

following table.

x	1.00	1.25	1.50	1.75	2.00	2.25	2.50	2.75	3.00
$\dfrac{5}{x}$	5.000 0	4.000 0	3.333 3	2.857 1	2.500 0	2.222 2	2.000 0	1.818 2	1.666 7

From equation (1), using the trapezoidal rule with 8 intervals:

$$\int_1^3 \frac{5}{x} \, dx \approx (0.25)\left[\frac{1}{2}(5.000\,0 + 1.666\,7) + 4.000\,0 + 3.333\,3 + 2.857\,1\right.$$
$$\left. + 2.500\,0 + 2.222\,2 + 2.000\,0 + 1.818\,2\right]$$

= **5.516**, correct to 3 decimal places

(c) (i) From equation (5), using Simpson's rule with 4 intervals and the table of values in part (b) (i) above:

$$\int_1^3 \frac{5}{x} \, dx \approx \frac{1}{3}(0.5)[(5.000\,0 + 1.666\,7) + 4(3.333\,3 + 2.000\,0) + 2(2.500\,0)]$$

= **5.500**, correct to 3 decimal places

(ii) From equation (5), using Simpson's rule with 8 intervals and the table of values in part (b) (ii) above:

$$\int_1^3 \frac{5}{x} \, dx \approx \frac{1}{3}(0.25)[(5.000\,0 + 1.666\,7) + 4(4.000\,0 + 2.857\,1$$
$$+ 2.222\,2 + 1.818\,2) + 2(3.333\,3$$
$$+ 2.500\,0 + 2.000\,0)]$$

$$\approx \frac{1}{3}(0.25)[6.666\,7 + 43.590\,0 + 15.666\,6]$$

= **5.494**, correct to 3 decimal places

(d) Percentage error $= \left(\dfrac{\text{approximate value} - \text{true value}}{\text{true value}}\right) \times 100\%$

where true value = 5.493 from part (a)

With the trapezoidal rule using 4 intervals,

$$\text{percentage error} = \left(\frac{5.583 - 5.493}{5.493}\right) \times 100\% = \mathbf{1.638\%}$$

and using 8 intervals,

$$\text{percentage error} = \left(\frac{5.516 - 5.493}{5.493}\right) \times 100\% = \mathbf{0.419\%}$$

With Simpson's rule using 4 intervals,

$$\text{percentage error} = \left(\frac{5.500 - 5.493}{5.493}\right) \times 100\% = \mathbf{0.127\%}$$

and using 8 intervals,

$$\text{percentage error} = \left(\frac{5.494 - 5.493}{5.493}\right) \times 100\% = \mathbf{0.018\%}$$

Thus when evaluating $\int_1^3 \dfrac{5}{x}\,dx$ the following conclusions may be drawn from above:

(i) the larger the number of intervals chosen the more accurate is the result, and
(ii) Simpson's rule is more accurate than the trapezoidal rule when the same number of intervals are chosen.

Further problems on numerical integration may be found in the following Section 27.5 (Problems 1 to 28).

27.5   Further problems

In Problems 1 to 5, evaluate the definite integrals using the trapezoidal rule, giving the answers correct to 3 decimal places.

1. $\displaystyle\int_0^2 \frac{1}{1 + \theta^2}\,d\theta$ (Use 8 intervals)   [1.106]

2. $\displaystyle\int_1^4 3 \ln 2x \, dx$ (Use 6 intervals)   [13.827]

3. $\displaystyle\int_0^{\pi/2} \frac{1}{1 + \sin x}\,dx$ (Use 6 intervals)   [1.006]

4. $\int_{0}^{\pi/2} \sqrt{(\sin x)}\, dx$ (Use 5 intervals)   [1.162]

5. $\int_{0}^{\pi} t \sin t\, dt$ (Use 8 intervals)   [3.101]

In Problems 6 to 10, evaluate the definite integrals using the mid-ordinate rule, giving the answers correct to 3 decimal places.

6. $\int_{1}^{4} \sqrt{(x^2 - 1)}\, dx$ (Use 6 intervals)   [6.735]

7. $\int_{0}^{\pi/4} \sqrt{(\cos^3 \theta)}\, d\theta$ (Use 9 intervals)   [0.674]

8. $\int_{0}^{1.5} e^{-(1/3)x^2}\, dx$ (Use 6 intervals)   [1.197]

9. $\int_{0.4}^{2} \dfrac{dx}{1 + x^4}$ (Use 8 intervals)   [0.672]

10. $\int_{0}^{\pi/2} \dfrac{1}{1 + \cos x}\, dx$ (Use 6 intervals)   [0.997]

In Problems 11 to 16, evaluate the definite integrals using Simpson's rule, giving the answers correct to 3 decimal places.

11. $\int_{\pi/6}^{\pi/3} \tan \theta\, d\theta$ (Use 6 intervals)   [0.549]

12. $\int_{0}^{2} \dfrac{1}{1 + x^3}\, dx$ (Use 8 intervals)   [1.090]

13. $\int_{0}^{\pi/3} \sqrt{\left(1 - \dfrac{1}{3}\sin^2 x\right)}\, dx$ (Use 6 intervals)   [0.994]

14. $\int_{1}^{3} \dfrac{\ln x}{x}\, dx$ (Use 10 intervals)   [0.603]

15. $\int_{0}^{0.4} \dfrac{\sin \theta}{\theta}\, d\theta$ (Use 8 intervals)   [0.380]

16. $\int_{0}^{\pi/4} \sqrt{(\sec x)}\, dx$ (Use 6 intervals)   [0.831]

In Problems 17 and 18 evaluate the definite integrals using
(*a*) integration, (*b*) the trapezoidal rule, (*c*) the mid-ordinate rule
and (*d*) Simpson's rule. In each of the approximate methods give
the answers correct to 3 decimal places.

17. $\int_{1}^{3} \dfrac{9}{x^2} \, dx$ (Use 8 intervals)

[(*a*) 6 (*b*) 6.089 (*c*) 5.956 (*d*) 6.004]

18. $\int_{0}^{5} \sqrt{(3x + 1)} \, dx$ (Use 10 intervals)

[(*a*) 14 (*b*) 13.977 (*c*) 14.011 (*d*) 13.998]

In Problems 19 to 21, evaluate the definite integrals using
(*a*) the trapezoidal rule, (*b*) the mid-ordinate rule and
(*c*) Simpson's rule. Use 6 intervals in each case and give answers
correct to 3 decimal places.

19. $\int_{0}^{0.9} \sqrt{(1 - x^2)} \, dx$ [(*a*) 0.752 (*b*) 0.758 (*c*) 0.756]

20. $\int_{0.6}^{2.4} \sqrt{(1 + x^3)} \, dx$ [(*a*) 4.006 (*b*) 3.986 (*c*) 3.992]

21. $\int_{0}^{\pi/2} \dfrac{1}{\sqrt{\left(1 - \dfrac{1}{2}\sin^2 \theta\right)}} \, d\theta$ [(*a*) 1.854 (*b*) 1.854 (*c*) 1.854]

22.   A curve is given by the following values:

x	0	1.0	2.0	3.0	4.0	5.0	6.0
y	3	6	12	20	30	42	56

The area under the curve between $x = 0$ and $x = 6.0$ is given by
$\int_{0}^{6.0} y \, dx$. Determine the approximate value of this definite
integral, correct to 4 significant figures, using Simpson's
rule.   [138.3]

23.   A function of x, $f(x)$, has the following values for
corresponding values of x.

x	0 0.1	0.2	0.3	0.4	0.5	0.6
$f(x)$	0 0.099 5	0.196 0	0.286 6	0.368 4	0.438 8	0.495 2

Evaluate $\displaystyle\int_0^{0.6} f(x)\,dx$ using (a) the trapezoidal rule, and (b) Simpson's rule, giving answers correct to 3 decimal places.
[(a) 0.164 (b) 0.164]

24. Use Simpson's rule to estimate $\displaystyle\int_1^3 y\,dx$ for the following pairs of (x, y) values.

x	1.00	1.25	1.50	1.75	2.00	2.25	2.50	2.75	3.00
y	0	0.278 9	0.608 2	0.979 3	1.386 3	1.824 6	2.290 7	2.781 9	3.295 8

[2.944]

25. A vehicle starts from rest and its velocity is measured every second for 6.0 seconds, with values as follows:

time t (s)	0 1.0 2.0 3.0 4.0 5.0 6.0
velocity v (m/s)	0 1.2 2.4 3.7 5.2 6.0 9.2

The distance travelled in 6.0 seconds is given by $\displaystyle\int_0^{6.0} v\,dt$.

Estimate this distance using Simpson's rule giving the answer correct to 3 significant figures.   [22.7 m]

26. An alternating current i has the following values at equal intervals of 2.0×10^{-3} seconds.

time $\times 10^{-3}$ (s)	0 2.0 4.0 6.0 8.0 10.0 12.0
current i (A)	0 1.7 3.5 5.0 3.7 2.0 0

Charge q, in coulombs, is given by $q = \displaystyle\int_0^{12.0 \times 10^{-3}} i\,dt$. Use Simpson's rule to determine the approximate charge in the 12.0×10^{-3} second period.   [32.8×10^{-3} C]

27. The velocity v of a body moving in a straight line at time t is given in the table below.

t (s)	0 0.5 1.0 1.5 2.0 2.5 3.0 3.5 4.0
v (m/s)	0 0.07 0.13 0.22 0.27 0.32 0.34 0.31 0

The total distance travelled is given by $\int_0^{4.0} v \, dt$. Estimate the total distance travelled in 4.0 s using (a) the trapezoidal rule, and (b) Simpson's rule, giving the answers in centimetres.   [(a) 83 cm (b) 86 cm]

28. Determine the value of $\int_0^2 \dfrac{x}{\sqrt{(2x^2 + 1)}} \, dx$ using integral calculus. Find also the percentage error introduced by estimating the definite integral by (a) the trapezoidal rule, (b) the mid-ordinate rule and (c) Simpson's rule, using 4 intervals in each case. Give answers correct to 3 decimal places. [1; (a) -2.073%, (b) 1.063%, (c) 0.213%]

Chapter 28

Areas under and between curves

28.1 The area between a curve, the x-axis and given ordinates

There are several instances in branches of engineering and science where the area under a curve is required to be accurately determined. For example, the areas, between given limits, of:

(a) velocity/time graphs give distances travelled,
(b) force/distance graphs give work done,
(c) acceleration/time graphs give velocities,
(d) voltage/current graphs give power,
(e) pressure/volume graphs give work done,
(f) normal distribution curves give frequencies.

Provided there is a known relationship [e.g. $y = f(x)$] between the variables forming the axes of the above graphs then the areas may be calculated exactly using integral calculus. If a relationship between variables is not known then areas have to be approximately determined using such techniques as the trapezoidal rule, the mid-ordinate rule or Simpson's rule (see Chapter 27).

Let A be the area enclosed between the curve $y = f(x)$, the x-axis and the ordinates $x = a$ and $x = b$. Also let A be subdivided into a number of elemental strips each of width δx as shown in Fig. 28.1.

One such strip is shown as PQRBA, with point P having coordinates (x, y) and point Q having coordinates $(x + \delta x, y + \delta y)$. Let the area PQRBA be δA, which can be seen from Fig. 28.1 to consist of a rectangle PRBA, of area $y\delta x$, and PQR, which approximates to a triangle of area $\frac{1}{2}\delta x\delta y$,

i.e. $\delta A \approx y\delta x + \frac{1}{2}\delta x\delta y$

Dividing both sides by δx gives:

$$\frac{\delta A}{\delta x} \approx y + \frac{1}{2}\delta y$$

As δx is made smaller and smaller, the number of rectangles increases and all such areas as PQR become smaller and smaller. Also δy becomes smaller

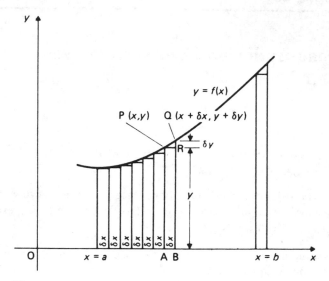

Fig. 28.1

and in the limit as δx approaches zero, $\dfrac{\delta A}{\delta x}$ becomes the differential coefficient $\dfrac{\mathrm{d}A}{\mathrm{d}x}$ and δy becomes zero,

i.e. $\lim\limits_{\delta x \to 0}\left(\dfrac{\delta A}{\delta x}\right) = \dfrac{\mathrm{d}A}{\mathrm{d}x} = y + \frac{1}{2}(0) = y$

Hence $\dfrac{\mathrm{d}A}{\mathrm{d}x} = y$ \hfill (1)

This shows that when a limiting value is taken, all such areas as PQR become zero. Hence the area beneath the curve is given by the sum of all such rectangles as PRBA,

i.e.   Area $= \Sigma y \delta x$

Between the limits $x = a$ and $x = b$,

Area $A = \lim\limits_{\delta x \to 0} \sum\limits_{x=a}^{x=b} y \delta x$ \hfill (2)

From equation (1), $\dfrac{\mathrm{d}A}{\mathrm{d}x} = y$ and by integration:

$$\int \frac{\mathrm{d}A}{\mathrm{d}x}\,\mathrm{d}x = \int y \,\mathrm{d}x$$

Hence $A = \int y \, dx$

The ordinates $x = a$ and $x = b$ limit the area and such ordinate values are shown as limits.

Thus $A = \int_a^b y \, dx$ $\qquad\qquad\qquad\qquad\qquad$ (3)

Equations (2) and (3) show that:

Area $A = \lim_{\delta x \to 0} \sum_{x=a}^{x=b} y \delta x = \int_a^b y \, dx$

This statement that the limiting value of a sum is equal to the integral between the same limits forms a fundamental theorem of integration. This can be illustrated by considering simple shapes of known areas. For example, Fig. 28.2(a) shows a rectangle bounded by the line $y = h$, ordinates $x = a$ and $x = b$ and the x-axis.

Let the rectangle be divided into n equal vertical strips of width δx. The area of strip PQAB is $h\delta x$ and since there are n strips making up the total area the total area $= nh\delta x$. The base length of the rectangle, i.e. $(b - a)$, is made up of n strips, each δx in width, hence $n\delta x = (b - a)$. Therefore the total area $= h(b - a)$

The total area is also obtained by adding the areas of all such strips as PQAB and is independent of the value of n, that is, n can be infinitely large.

Hence total area $= \lim_{\delta x \to 0} \sum_{x=a}^{x=b} h\delta x = h(b - a)$ $\qquad\qquad$ (4)

Also the total area is given by $\int_a^b y \, dx = \int_a^b h \, dx$

$$= [hx]_a^b = h(b - a) \qquad\qquad (5)$$

But this is the area obtained from equation (4).

Hence $\lim_{\delta x \to 0} \sum_{x=a}^{x=b} h\delta x = \int_a^b h \, dx$

Similarly for, say, a trapezium bounded by the line $y = x$, the ordinates $x = a$ and $x = b$ and the x-axis (as shown in Fig. 28.2(b)), the total area is given by:

(half the sum of the parallel sides)(perpendicular distance between these sides)

i.e. $\quad \frac{1}{2}(a + b)(b - a)$ or $\frac{1}{2}(b^2 - a^2)$ $\qquad\qquad\qquad$ (6)

Also, the total area will be given by the sum of all areas such as PQAB which each have an area of $y\delta x$ provided δx is infinitely small,

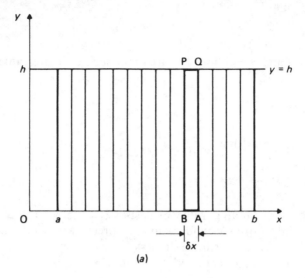

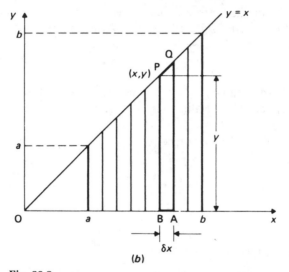

Fig. 28.2

i.e.   total area $= \lim\limits_{\delta x \to 0} \sum\limits_{x=a}^{x=b} y\delta x = \frac{1}{2}(b^2 - a^2)$ from above   (7)

Also, the total area $= \int_a^b y \, dx = \int_a^b x \, dx = \left[\dfrac{x^2}{2}\right]_a^b = \frac{1}{2}(b^2 - a^2)$ (8)

Equations (7) and (8) give further evidence that

$$\lim\limits_{\delta x \to 0} \sum\limits_{x=a}^{x=b} y\delta x = \int_a^b y \, dx$$ (9)

The two simple illustrations used above show that equation (9) is valid in these two cases and we will assume that it is generally true, although a more rigorous proof is beyond the scope of this book.

If the area between a curve $x = f(y)$, the y-axis and ordinates $y = m$ and $y = n$ is required, then by similar reasoning to the above:

$$\text{Area} = \int_m^n x \, dy$$

Thus finding the area beneath a curve is the same as determining the value of a definite integral as previously discussed in Chapter 21.

A part of the curve $y = 2x^2 + 3$ is shown in Fig. 28.3, which is produced from the table of values shown below.

x	-2	-1	0	1	2	3
$y = 2x^2 + 3$	11	5	3	5	11	21

The area between the curve, the x-axis and the ordinates $x = -1$ and $x = 3$ is shown shaded. This area is given by:

$$\text{Area} = \int_{-1}^{3} y \, dx = \int_{-1}^{3} (2x^2 + 3) \, dx$$

$$= \left[\frac{2x^3}{3} + 3x \right]_{-1}^{3}$$

$$= \left[\frac{2(3)^3}{3} + 3(3) \right] - \left[\frac{2(-1)^3}{3} + 3(-1) \right]$$

$$= 30\tfrac{2}{3} \text{ square units}$$

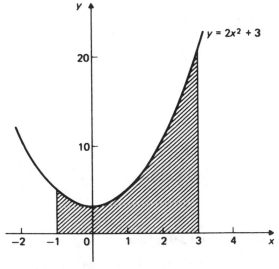

Fig. 28.3   Graph of $y = 2x^2 + 3$

With the curve $y = 2x^2 + 3$ shown in Fig. 28.3 all values of y are positive. Hence all the terms in $\Sigma y \delta x$ are positive and $\int_a^b y \, dx$ is positive. However, if a curve should drop below the x-axis, then y becomes negative, all terms in $\Sigma y \delta x$ become negative and $\int_a^b y \, dx$ is negative.

In Fig. 28.4 the total area between the curve $y = f(x)$, the x-axis and the ordinates $x = a$ and $x = b$ is given by

$$\text{area P}\left(\text{i.e. } \int_a^c f(x) \, dx\right) + \text{area Q}\left(\text{i.e. } - \int_c^d f(x) \, dx\right)$$

$$+ \text{area R}\left(\text{i.e. } \int_d^b f(x) \, dx\right)$$

i.e. $$\int_a^c f(x) \, dx - \int_c^d f(x) \, dx + \int_d^b f(x) \, dx$$

This is **not** the same as the value given by $\int_a^b f(x) \, dx$

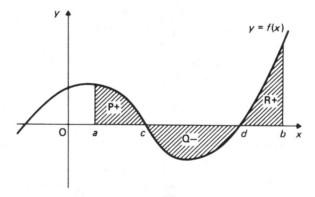

Fig. 28.4

For this reason, if there is any doubt about the shape of the graph of a function or any possibility of all or part of it lying below the x-axis, a sketch should be made over the required limits to determine if any part of the curve lies below the x-axis.

28.2   The area between two curves

Let the graphs of the functions $y = f_1(x)$ and $y = f_2(x)$ intersect at points A ($x = a$) and B ($x = b$) as shown in Fig. 28.5

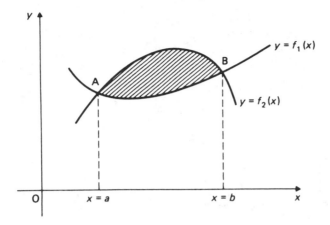

Fig. 28.5

At the points of intersection $f_1(x) = f_2(x)$.

The area enclosed between the curve $y = f_2(x)$, the x-axis and the ordinates $x = a$ and $x = b$ is given by $\displaystyle\int_a^b f_2(x)\,\mathrm{d}x$

The area enclosed between the curve $y = f_1(x)$, the x-axis and the ordinates $x = a$ and $x = b$ is given by $\displaystyle\int_a^b f_1(x)\,\mathrm{d}x$

It follows that the area enclosed between the two curves (shown shaded in Fig. 28.5) is given by:

$$\textbf{Shaded area} = \int_a^b f_2(x)\,\mathrm{d}x - \int_a^b f_1(x)\,\mathrm{d}x$$

$$= \int_a^b [f_2(x) - f_1(x)]\,\mathrm{d}x$$

Worked problems on finding areas under and between curves

Problem 1. Sketch the curves and find the areas enclosed by the given curves, the x-axis and the given ordinates: (a) $y = \sin 2x$, $x = 0$, $x = \dfrac{\pi}{2}$; (b) $y = 3\cos\dfrac{1}{2}x$, $x = 0$, $x = \dfrac{2\pi}{3}$

(a) A sketch of $y = \sin 2x$ in the range $x = 0$ to $x = \pi$ is shown in Fig. 28.6(a)

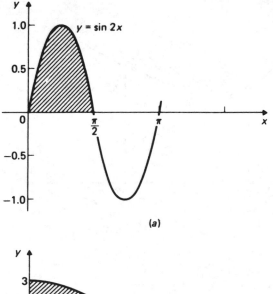

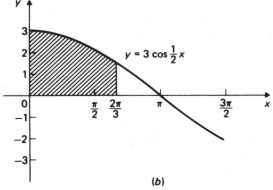

Fig. 28.6   Graphs of $y = \sin 2x$ and $y = 3 \cos \frac{1}{2}x$

The area shown shaded is given by:

$$\text{Area} = \int_0^{\pi/2} \sin 2x \, dx$$

$$= \left[-\frac{\cos 2x}{2} \right]_0^{\pi/2} = \left(\frac{-\cos 2(\pi/2)}{2} \right) - \left(\frac{-\cos 0}{2} \right)$$

$$= \left(\frac{-\cos \pi}{2} \right) - \left(\frac{-\cos 0}{2} \right) = (- -\tfrac{1}{2}) - (-\tfrac{1}{2})$$

$$= 1 \text{ square unit}$$

(b) A sketch of $y = 3 \cos \frac{1}{2}x$ in the range $x = 0$ to $x = \dfrac{2\pi}{3}$ is shown in Fig. 28.6(b)

The area shown shaded is given by:

$$\text{Area} = \int_0^{2\pi/3} 3 \cos \tfrac{1}{2}x \, dx$$

$$= [6 \sin \tfrac{1}{2}x]_0^{2\pi/3} = \left(6 \sin \frac{\pi}{3}\right) - (6 \sin 0)$$

$$= 6 \sin 60° = \textbf{5.196 square units}$$

Problem 2. Find the area enclosed by the curve $y = 2x^2 - x + 3$, the x-axis and the ordinates $x = -1$ and $x = 2$

A table of values is produced as shown below.

x	-1	0	1	2
y	6	3	4	9

The area between the curve, the x-axis and the ordinates $x = -1$ and $x = 2$ is wholly above the x-axis, since all values of y in the table are positive. Thus the area is positive. In such cases as this it is unnecessary to actually draw the graph.

$$\text{Area} = \int_{-1}^{2} (2x^2 - x + 3) \, dx$$

$$= \left[\frac{2x^3}{3} - \frac{x^2}{2} + 3x\right]_{-1}^{2}$$

$$= (\tfrac{16}{3} - 2 + 6) - (-\tfrac{2}{3} - \tfrac{1}{2} - 3)$$

$$= (9\tfrac{1}{3}) - (-4\tfrac{1}{6}) = \textbf{13}\tfrac{1}{2} \textbf{ square units}$$

Problem 3. Calculate the area of the figure bounded by the curve $y = 2e^{t/2}$, the t-axis and ordinates $t = -1$ and $t = 3$

A table of values is produced as shown below.

t	-1	0	1	2	3
$y = 2e^{t/2}$	1.213	2.000	3.297	5.437	8.963

Since all the values of y are positive, the area required is wholly above the t-axis. Hence the area enclosed by the curve, the t-axis and the ordinates $t = -1$ and $t = 3$ is given by:

$$\text{Area} = \int_{-1}^{3} 2e^{t/2} \, dt$$

$$= [4e^{t/2}]_{-1}^{3} = 4[e^{\frac{3}{2}} - e^{-\frac{1}{2}}]$$

$$= 4[4.481\,7 - 0.606\,5]$$

$$= \textbf{15.50 square units}$$

Problem 4. Find the area enclosed by the curve $y = x^2 + 3$, the x-axis and the ordinates $x = 0$ and $x = 3$. Sketch the curve within these limits. Find also, using integration, the area enclosed by the curve and the y-axis, between the same limits.

A table of values is produced as shown below.

x	0	1	2	3
y	3	4	7	12

(a) Part of the curve $y = x^2 + 3$ is shown in Fig. 28.7

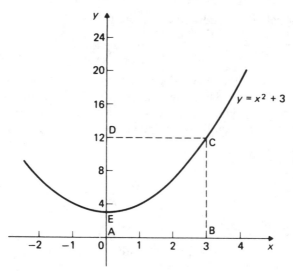

Fig. 28.7   Graph of $y = x^2 + 3$

The area enclosed by the curve, the x-axis and ordinates $x = 0$ and $x = 3$ (i.e. area ECBA of Fig. 28.7) is given by:

$$\text{Area} = \int_0^3 (x^2 + 3)\, dx = \left[\frac{x^3}{3} + 3x \right]_0^3$$

$$= \textbf{18 square units}$$

(b) When $x = 3$, $y = x^2 + 3 = 12$
     when $x = 0$, $y = 3$

If, $y = x^2 + 3$ then $x^2 = y - 3$ and $x = \sqrt{(y - 3)}$
Hence the area enclosed by the curve $y = x^2 + 3$ (i.e. the curve $x = \sqrt{(y - 3)}$), the y-axis and the ordinates $y = 3$ and $y = 12$ (i.e. area EDC of Fig. 28.7) is given by:

$$\text{Area} = \int_{y=3}^{y=12} x\, dy = \int_3^{12} \sqrt{(y - 3)}\, dy$$

Let $u = y - 3$

then $\dfrac{du}{dy} = 1$, i.e. $dy = du$

Hence $\displaystyle\int (y - 3)^{\frac{1}{2}}\, dy = \int u^{\frac{1}{2}}\, du = \dfrac{2u^{\frac{3}{2}}}{3}$

Since $u = y - 3$ then

$$\text{Area} = \int_{3}^{12} \sqrt{(y - 3)}\, dy = \left[\tfrac{2}{3}(y - 3)^{\frac{3}{2}}\right]_{3}^{12}$$

$$= \tfrac{2}{3}\left[\sqrt{9^3} - 0\right]$$

$$= 18 \text{ square units}$$

The sum of the areas in parts (a) and (b) is 36 square units, which is equal to the area of the rectangle DCBA

Problem 5. Calculate the area between the curve $y = x^3 - x^2 - 6x$ and the x-axis

$y = x^3 - x^2 - 6x = x(x^2 - x - 6)$

$\qquad\qquad\qquad = x(x - 3)(x + 2)$

Thus when $y = 0$, $x = 0$ or $(x - 3) = 0$ or $(x + 2) = 0$, i.e. $x = 0$, $x = 3$ or $x = -2$

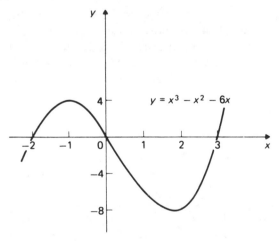

Fig. 28.8   Graph of $y = x^3 - x^2 - 6x$

Hence the curve cuts the x-axis at $x = 0$, 3 and -2. Since the curve is a continuous function, only one other value needs to be calculated before a sketch of the curve can be produced. For example, when $x = 1$, $y = -6$, which shows that the portion of the curve between ordinates $x = 0$ and $x = 3$ is negative. Hence the portion of the curve between ordinates $x = 0$ and $x = -2$ must be positive.

A sketch of part of the curve $y = x^3 - x^2 - 6x$ is shown in Fig. 28.8

If $y = f(x)$ had not factorised as above, then a table of values could have been produced and the graph sketched in the usual manner.

The sketch shows that the area needs to be calculated in two parts, one part being positive and the other negative, as shown in the second integral below.

The area between the curve and the x-axis is given by:

$$\text{Area} = \int_{-2}^{0} (x^3 - x^2 - 6x)\, dx - \int_{0}^{3} (x^3 - x^2 - 6x)\, dx$$

$$= \left[\frac{x^4}{4} - \frac{x^3}{3} - 3x^2 \right]_{-2}^{0} - \left[\frac{x^4}{4} - \frac{x^3}{3} - 3x^2 \right]_{0}^{3}$$

$$= (5\tfrac{1}{3}) - (-15\tfrac{3}{4})$$

$$= 21\tfrac{1}{12} \text{ square units}$$

Problem 6. Find the area enclosed between the curves $y = x^2 + 2$ and $y + x = 14$

The first step is to find the points of intersection of the two curves. This will enable us to limit the range of values when drawing up a table of values in order to sketch the curves. At the points of intersection the curves are equal (i.e. their coordinates are the same). Since $y = x^2 + 2$ and $y + x = 14$ (i.e. $y = 14 - x$) then $x^2 + 2 = 14 - x$ at the points of intersection,

i.e. $x^2 + x - 12 = 0$

$\qquad (x - 3)(x + 4) = 0$

Hence $x = 3$ and $x = -4$ at the points of intersection.
Tables of values may now be produced as shown below.

x		-4	-3	-2	-1	0	1	2	3
$y = x^2 + 2$		18	11	6	3	2	3	6	11

x		-4	0	3
$y = 14 - x$		18	14	11

$y = 14 - x$ is a straight line thus only two points are needed (plus one more to check).

A sketch of the two curves is shown in Fig. 28.9

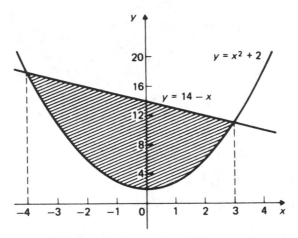

Fig. 28.9   Graphs of $y = x^2 + 2$ and $y = 14 - x$

The area between the two curves (shown shaded) is given by:

Shaded area $= \displaystyle\int_{-4}^{3} (14 - x)\,dx - \int_{-4}^{3} (x^2 + 2)\,dx$

$$= \left[14x - \frac{x^2}{2} \right]_{-4}^{3} - \left[\frac{x^3}{3} + 2x \right]_{-4}^{3}$$

$$= 101\tfrac{1}{2} - 44\tfrac{1}{3}$$

$$= 57\tfrac{1}{6} \text{ square units}$$

Problem 7. Find the points of intersection of the two curves $x^2 = 2y$ and $\dfrac{y^2}{16} = x$. Sketch the two curves and calculate the area enclosed by them.

$x^2 = 2y$, i.e. $y = \dfrac{x^2}{2}$ or $y^2 = \dfrac{x^4}{4}$

$\dfrac{y^2}{16} = x$, i.e. $y^2 = 16x$

At the points of intersection, $\dfrac{x^4}{4} = 16x$

i.e. $x^4 = 64x$

Hence $x^4 - 64x = 0$

$$x(x^3 - 64) = 0$$

i.e. $x = 0$ or $x^3 - 64 = 0$

Hence at the points of intersection $x = 0$ and $x = 4$

Using $y = \dfrac{x^2}{2}$, when $x = 0$, $y = 0$

when $x = 4$, $y = \dfrac{(4)^2}{2} = 8$

[Check, using $y^2 = 16x$. When $x = 0$, $y = 0$

When $x = 4$, $y^2 = 64$, $y = 8$]

Hence the points of intersection of the two curves $x^2 = 2y$ and $\dfrac{y^2}{16} = x$ are (0, 0) and (4, 8)

A sketch of the two curves (given the special name of parabolas) is shown in Fig. 28.10
The area enclosed by the two curves, i.e. OABC (shown shaded), is given by:

$$\text{Area} = \int_0^4 4\sqrt{x}\,dx - \int_0^4 \frac{x^2}{2}\,dx$$

(Note that for one curve $y = \pm 4\sqrt{x}$. The $-4\sqrt{x}$ is neglected since the shaded area required is above the x-axis, and hence positive.)

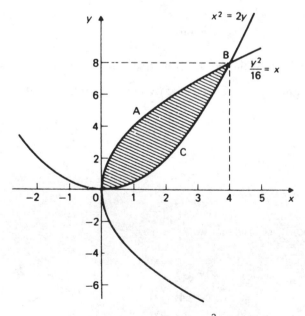

Fig. 28.10   Graphs of $x^2 = 2y$ and $\dfrac{y^2}{16} = x$

$$\text{Area} = 4[\tfrac{2}{3}x^{\frac{3}{2}}]_0^4 - \frac{1}{2}\left[\frac{x^3}{3}\right]_0^4$$

$$= 21\tfrac{1}{3} - 10\tfrac{2}{3}$$

$$= 10\tfrac{2}{3} \textbf{ square units}$$

Further problems on areas under and between curves may be found in the following Section 28.3 (Problems 1 to 48).

28.3 Further problems

All answers are in square units.

In Problems 1 to 22 find the area enclosed between the given curve, the horizontal axis and the given ordinates. Sketch the curve in the given range for each.

1. $y = 2x$; $x = 0$, $x = 5$ [25]
2. $y = x^2 - x + 2$; $x = -1$, $x = 2$ $[7\tfrac{1}{2}]$

3. $y = \dfrac{x}{2}$; $x = 3$, $x = 7$ [10]

4. $y = p - 1$; $p = 1$, $p = 5$ [8]
5. $F = 8S - 2S^2$; $S = 0$, $S = 2$ $[10\tfrac{2}{3}]$
6. $y = (x - 1)(x - 2)$; $x = 0$, $x = 3$ $[1\tfrac{5}{6}]$
7. $y = 8 + 2x - x^2$; $x = -2$, $x = 4$ [36]
8. $u = 2(4 - t^2)$; $t = -2$, $t = 2$ $[21\tfrac{1}{3}]$
9. $y = x(x - 1)(x + 3)$; $x = -2$, $x = 1$ $[7\tfrac{11}{12}]$
10. $x = 4a^3$; $a = -2$, $a = 2$ [32]
11. $y = x(x - 1)(x - 3)$; $x = 0$, $x = 3$ $[3\tfrac{1}{12}]$
12. $a = t^3 + t^2 - 4t - 4$; $t = -3$, $t = 3$ $[24\tfrac{1}{3}]$

13. $y = \sin \theta$; $\theta = 0$, $\theta = \dfrac{\pi}{2}$ [1]

14. $y = \cos x$; $x = \dfrac{\pi}{4}$, $x = \dfrac{\pi}{2}$ [0.292 9]

15. $y = 3 \sin 2\beta$; $\beta = 0$, $\beta = \dfrac{\pi}{4}$ $[1\tfrac{1}{2}]$

16. $y = 5 \cos 3\alpha$; $\alpha = 0$, $\alpha = \dfrac{\pi}{6}$ $[1\tfrac{2}{3}]$

17. $y = \sin x - \cos x$; $x = 0$, $x = \dfrac{\pi}{4}$ [0.414 2]

18. $2y^2 = x; x = 0, x = 2$ $[2\frac{2}{3}]$
19. $5 = xy; x = 2, x = 5$ [4.581]
20. $y = 2e^{2t}; t = 0, t = 2$ [53.60]
21. $ye^{4x} = 3; x = 1, x = 3$ [0.013 7]
22. $y = 2x + e^x; x = 0, x = 3$ [28.09]
23. Find the area between the curve $y = 3x - x^2$ and the x-axis.
    $[4\frac{1}{2}]$
24. Calculate the area enclosed between the curve
    $y = 12 - x - x^2$ and the x-axis using integration. $[57\frac{1}{6}]$
25. Sketch the curve $y = \sec^2 2x$ from $x = 0$ to $x = \dfrac{\pi}{4}$ and
    calculate the area enclosed between the curve, the x-axis and
    the ordinates $x = 0$ and $x = \dfrac{\pi}{6}$ [0.866]
26. Find the area of the template enclosed between the curve
    $y = \dfrac{1}{x - 2}$, the x-axis and the ordinates $x = 3$ cm and $x = 5$ cm.
    [1.098 6 cm²]
27. Sketch the curves $y = x^2 + 4$ and $y + x = 10$ and find the
    area enclosed by them. $[20\frac{5}{6}]$
28. Calculate the area enclosed between the curves $y = \sin \theta$ and
    $y = \cos \theta$ and the y-axis between the limits $\theta = 0$ and
    $\theta = \dfrac{\pi}{4}$ [0.414 2]
29. Find the area between the two parabolas $9y^2 = 16x$ and
    $x^2 = 6y$ $[3\frac{5}{9}]$
30. Calculate the area of the metal plate enclosed between
    $y = x(x - 4)$ and the x-axis where x is in metres. $[10\frac{2}{3}$ m²]
31. Sketch the curve $x^2 - y = 3x + 10$ and find the area
    enclosed between it and the x-axis. $[57\frac{1}{6}]$
32. Find the area enclosed by the curve $y = 4(x^2 - 1)$, the x-axis
    and the ordinates $x = 0$ and $x = 2$. Find also the area
    enclosed by the curve and the y-axis between the same
    limits. $[8, 21\frac{1}{3}]$
33. Calculate the area between the curve $y = x(x^2 - 2x - 3)$ and
    the x-axis using integration. $[11\frac{5}{6}]$
34. Find the area of the figure bounded by the curve $y = 3e^{2x}$,
    the x-axis and the ordinates $x = -2$ and $x = 2$.   [81.87]
35. Find the area enclosed between the curves $y = x^2 - 3x + 5$
    and $y - 1 = 2x$ $[4\frac{1}{2}]$

36. Find the points of intersection of the two curves $\dfrac{x^2}{2} = \sqrt{2y}$
   and $y^2 = 8x$ and calculate the area enclosed by them.
   [(0, 0), (4, 5.657); 7.542]

37. Calculate the area bounded by the curve $y = x^2 + x + 4$ and
   the line $y = 2(x + 5)$ $[20\frac{5}{6}]$

38. Find the area enclosed between the curves $y = x^2$ and
   $y = 8 - x^2$ $[21\frac{1}{3}]$

39. Calculate the area between the curve $y = 3x^3$ and the line
   $\dfrac{y}{12} = x$ in the first quadrant.   [12]

40. Find the area bounded by the three straight lines
   $y = 4(2 - x)$, $y = 4x$ and $3y = 4x$ [2]

41. A vehicle has an acceleration a of $(30 + 2t)$ metres per
   second after t seconds. If the vehicle starts from rest find its
   velocity after 10 seconds. $\left(\text{Velocity} = \displaystyle\int_{t_1}^{t_2} a\, dt \right)$ [400 m/s]

42. A car has a velocity v of $(3 + 4t)$ metres per second after t
   seconds. How far does it move in the first 4 seconds? Find the
   distance travelled in the fifth second.
   $\left(\text{Distance travelled} = \displaystyle\int_{t_1}^{t_2} v\, dt \right)$ [44 m; 21 m]

43. A gas expands according to the law $pv = $ constant. When the
   volume is 2 m^3 the pressure is 200 kPa. Find the work done
   as the gas expands from 2 m^3 to a volume of 5 m^3
   $\left(\text{Work done} = \displaystyle\int_{v_1}^{v_2} p\, dv \right)$ [367 kJ]

44. The brakes are applied to a train and the velocity v at any
   time t seconds after applying the brakes is given by
   $(16 - 2.5t)$ m/s. Calculate the distance travelled in 8 seconds.
   $\left(\text{Distance travelled} = \displaystyle\int_{t_1}^{t_2} v\, dt \right)$ [48 m]

45. The force F newtons acting on a body at a distance x metres
   from a fixed point is given by $F = 3x + \dfrac{1}{x^2}$. Find the work
   done when the body moves from the position where $x = 1$ m
   to that where $x = 3$ m $\left(\text{Work done} = \displaystyle\int_{x_1}^{x_2} F\, dx \right)$
   $[12\frac{2}{3}$ newton metres]

46. The velocity v of a body t seconds after a certain instant is
   $(4t^2 + 3)$ m/s. Find how far it moves in the interval from $t = 2$ s

to $t = 6$ s $\left(\text{Distance travelled} = \int_{t_1}^{t_2} v \, dt \right)$ $[289\frac{1}{3}$ m]

47. The heat required to raise the temperature of carbon dioxide from 300 K to 600 K is determined from the area formed when the heat capacity (C_p) is plotted against the temperature (T) between 300 K and 600 K. If $C_p = 27 + 42 \times 10^{-3}T - 14.22 \times 10^{-6}T^2$, determine the area by integration.   [12 870]

48. The entropy required to raise hydrogen sulphide from 400 K to 500 K is determined from the area formed when C_p is plotted against the temperature (T) between 400 K and 500 K. Given that $C_p = 37 + 0.008T$, determine the area by integration.   [4060]

Chapter 29

Mean and root mean square values

29.1   Mean or average values

Figure 29.1 shows the positive half cycle of a periodic waveform of an alternating quantity. If the negative half cycle is the same shape as the positive half cycle then every positive value is balanced by a corresponding negative value and thus the average value of the complete cycle is zero, i.e. the average or mean value over a complete cycle of a symmetrically alternating quantity is zero. However, over half a cycle it has a non zero value.

Let the area of the waveform in Fig. 29.1 representing the positive half cycle be divided into, say, 7 strips each of width d, with ordinates at the mid-point of each strip (mid-ordinates) represented by y_1, y_2, y_3 and so on. Let EF be drawn parallel to base OG such that the area under the curve between O and G is equal to the area of rectangle OEFG. Then OE represents the mean or average height of the waveform, i.e. the average height of the y ordinates.

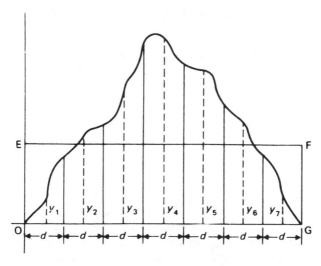

Fig. 29.1

From the mid-ordinate rule (see Chapter 27):

Area under curve $= d(y_1 + y_2 + y_3 + y_4 + y_5 + y_6 + y_7)$

Also, area of rectangle OEFG $=$ (OE)(OG)

$=$ (average value)(7d)

But area under curve $=$ area of rectangle OEFG.

Thus $d(y_1 + y_2 + y_3 + y_4 + y_5 + y_6 + y_7) =$ (average value)(7d)
Let the average value be denoted by $\bar{y}$ (pronounced y bar)

Then $\bar{y} = \dfrac{d(y_1 + y_2 + y_3 + y_4 + y_5 + y_6 + y_7)}{7d}$

i.e. $\bar{y} = \dfrac{\textbf{area under curve}}{\textbf{length of base}}$

In the example shown in Fig. 29.1 the mid-ordinate rule is used to find the area under the curve, although other approximate methods such as Simpson's rule or the trapezoidal rule (see Chapter 27) could equally well have been used.

An exact method of finding areas under curves is that of integration discussed in Chapter 28, although this is only possible if (a) there is an equation relating the variables, and (b) the equation can be integrated. Figure 29.2 shows part of a curve $y = f(x)$. The mean value, $\bar{y}$, of the curve between the limits $x = a$ and $x = b$ is given by:

$$\bar{y} = \frac{\text{area under curve}}{\text{length of base}} = \frac{\text{area PSRQ}}{b - a}$$

From Chapter 28, area under the curve $y = f(x)$ between the limits $x = a$ and $x = b$ is given by:

$$\int_a^b f(x)\, dx$$

Hence $\bar{y} = \dfrac{\int_a^b f(x)\, dx}{b - a} = \dfrac{1}{b - a} \int_a^b y\, dx$

29.2   Root mean square values

The root mean square value of a quantity is 'the square root of the average value of the squared values of the quantity' taken over an interval. In many scientific applications – particularly those involving periodic waveforms – mean values, when determined, are found to be zero because there are equal numbers of positive and negative values which cancel each other out. In such cases the root mean square (r.m.s.) values can be valuable, e.g.:

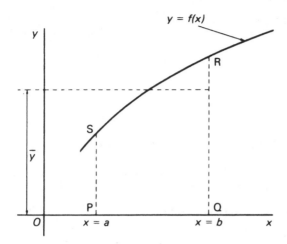

Fig. 29.2

(a)   the average rate of the heating effect of an electric current (i.e. proportional to current²),
(b)   the standard deviation, used in statistics to estimate the spread or scatter of a set of data (i.e. proportional to distance²), and
(c)   the average linear velocity of a particle in a body which is rotating about an axis (i.e. proportional to velocity²).

Each of these applications depend upon **square values**, which do not cancel. For example, a direct current I amperes passing through a resistor R ohms for t seconds produces a heating effect given by I^2Rt joules. When an alternating current i amperes is passed through the same resistor R for the same time, the instantaneous value of the heating effect, i.e. i^2Rt, varies. In order to give the same heating effect as the equivalent direct current, i^2 is replaced by the **mean square value**. Then the r.m.s. value of an alternating current is defined as that current which will give the same heating effect as the equivalent direct current, I amperes.

The r.m.s. value obtained without using integration

Referring to Fig. 29.1,

Average or mean value, $\bar{y} = \dfrac{\text{area under curve}}{\text{length of base}}$

$$= \frac{d(y_1 + y_2 + y_3 + y_4 + y_5 + y_6 + y_7)}{7d}$$

$$= \frac{y_1 + y_2 + y_3 + y_4 + y_5 + y_6 + y_7}{7}$$

Average value of the squares of the function $= \dfrac{y_1{}^2 + y_2{}^2 + y_3{}^2 + y_4{}^2 + y_5{}^2 + y_6{}^2 + y_7{}^2}{7}$

The square root of the average value of the squares of the function, i.e. the r.m.s. value $= \sqrt{\left(\dfrac{y_1{}^2 + y_2{}^2 + y_3{}^2 + y_4{}^2 + y_5{}^2 + y_6{}^2 + y_7{}^2}{7}\right)}$

The r.m.s. value obtained using integration

The average value $\bar{y} = \dfrac{1}{b-a} \displaystyle\int_a^b y \, \mathrm{d}x$ (from Section 29.1)

The average value of the square of the function $= \dfrac{1}{b-a} \displaystyle\int_a^b y^2 \, \mathrm{d}x$

The square root of the average value of the squares of the function, i.e. the r.m.s. value $= \sqrt{\left[\dfrac{1}{b-a} \displaystyle\int_a^b y^2 \, \mathrm{d}x\right]}$

One of the principal applications of r.m.s. values is in alternating currents and voltages in electrical engineering. Alternating waveforms are frequently of the form $i = I_m \sin \theta$ or $v = V_m \sin \theta$, and, when determining r.m.s. values, integrals of the form $\int (I_m \sin \theta)^2 \, \mathrm{d}\theta$ result, i.e. it is necessary to be able to integrate $\sin^2 \theta$ and also $\cos^2 \theta$.

From Section 16.3, equation (10) (page 269):

$$\cos 2A = 1 - 2 \sin^2 A$$

Thus $\sin^2 A = \dfrac{1 - \cos 2A}{2}$

Hence $\displaystyle\int \sin^2 A \, \mathrm{d}A = \int \dfrac{1 - \cos 2A}{2} \, \mathrm{d}A = \dfrac{1}{2}\left[A - \dfrac{\sin 2A}{2}\right] + c$, as shown in Chapter 23.

From Section 16.3, equation (9) (page 269)

$$\cos 2A = 2 \cos^2 A - 1$$

Thus $\cos^2 A = \dfrac{1 + \cos 2A}{2}$

Hence $\displaystyle\int \cos^2 A \, \mathrm{d}A = \int \dfrac{1 + \cos 2A}{2} \, \mathrm{d}A = \dfrac{1}{2}\left[A + \dfrac{\sin 2A}{2}\right] + c$, as shown in Chapter 23.

Worked problems on mean and r.m.s. values

Problem 1. Determine (a) the mean value and (b) the r.m.s. value of $y = 3x^2$ between $x = 1$ and $x = 3$, using integration.

(a)   Mean value, $\bar{y} = \dfrac{1}{3-1} \displaystyle\int_{1}^{3} 3x^2 \, dx$

$= \dfrac{1}{2} [x^3]_1^3 = \dfrac{1}{2}(27 - 1) = \mathbf{13}$

(b)   r.m.s. value $= \sqrt{\left[\dfrac{1}{3-1} \displaystyle\int_{1}^{3} (3x^2)^2 \, dx \right]}$

$= \sqrt{\left[\dfrac{1}{2} \displaystyle\int_{1}^{3} 9x^4 \, dx \right]}$

$= \sqrt{\left\{ \dfrac{9}{2} \left[\dfrac{x^5}{5} \right]_1^3 \right\}}$

$= \sqrt{\left\{ \dfrac{9}{10} [243 - 1] \right\}} = \sqrt{\left[\dfrac{9(242)}{10} \right]}$

$= \sqrt{217.8}$ $= \mathbf{14.76}$

Problem 2. A sinusoidal electrical current is given by $i = 10.0 \sin \theta$ amperes. Determine the mean value of the current over half a cycle using integration.

Average or mean value, $\bar{y} = \dfrac{1}{\pi - 0} \displaystyle\int_{0}^{\pi} 10.0 \sin \theta \, d\theta$

$= \dfrac{10.0}{\pi} [-\cos \theta]_0^\pi$

$= \dfrac{10.0}{\pi} [(-\cos \pi) - (-\cos 0)]$

$= \dfrac{10.0}{\pi} [(--1) - (-1)] = \dfrac{10.0}{\pi} (2)$

$= \dfrac{2}{\pi} \times 10.0 = \mathbf{6.366 \ amperes}$

Note that for a sine wave, the mean value $= \dfrac{2}{\pi} \times$ maximum value.

Problem 3. Using the current given in Problem 2, determine the r.m.s. value using integration.

r.m.s. value $= \sqrt{\left\{ \dfrac{1}{\pi} \displaystyle\int_{0}^{\pi} (10.0 \sin \theta)^2 \, d\theta \right\}}$

$= \sqrt{\left\{ \dfrac{100.0}{\pi} \displaystyle\int_{0}^{\pi} \sin^2 \theta \, d\theta \right\}}$

From Section 16.3, $\cos 2\theta = 1 - 2\sin^2\theta$

$$\text{from which } \sin^2\theta = \frac{1 - \cos 2\theta}{2}$$

$$\text{Hence r.m.s. value} = \sqrt{\left\{\frac{100.0}{\pi}\int_0^\pi \frac{1 - \cos 2\theta}{2}\,d\theta\right\}}$$

$$= \sqrt{\left\{\left(\frac{100.0}{\pi}\right)\frac{1}{2}\left[\theta - \frac{\sin 2\theta}{2}\right]_0^\pi\right\}}$$

$$= \sqrt{\left\{\left(\frac{100.0}{\pi}\right)\frac{1}{2}\left[\left(\pi - \frac{\sin 2\pi}{2}\right)\right.\right.}$$

$$\left.\left. - \left(0 - \frac{\sin 2(0)}{2}\right)\right]\right\}$$

$$= \sqrt{\left\{\frac{100.0}{\pi}\left(\frac{\pi}{2}\right)\right\}} = \frac{10.0}{\sqrt{2}} = \textbf{7.071 amperes}$$

Note that for a sine wave, the r.m.s. value $= \dfrac{1}{\sqrt{2}} \times$ maximum value.

Problem 4. Find the area bounded by the curve $y = 6x - x^2$ and the x-axis for values of x from 0 to 6. Determine also the mean value and the r.m.s. value of y over the same range.

A table of values is drawn up as shown below.

$y = 6x - x^2$	x	0	1	2	3	4	5	6
	y	0	5	8	9	8	5	0

Since all the values of y are positive the area required is wholly above the x-axis. Hence the area enclosed by the curve, the x-axis and the ordinates $x = 0$ and $x = 6$ is given by:

$$\text{Area} = \int_0^6 y\,dx = \int_0^6 (6x - x^2)\,dx$$

$$= \left[\frac{6x^2}{2} - \frac{x^3}{3}\right]_0^6 = (108 - 72) - (0 - 0) = \textbf{36 square units}$$

$$\text{Mean or average value} = \frac{\text{area under curve}}{\text{length of base}} = \frac{36}{6} = \textbf{6}$$

$$\text{r.m.s. value} = \sqrt{\left\{\frac{1}{6 - 0}\int_0^6 (6x - x^2)^2\,dx\right\}}$$

$$= \sqrt{\left\{\frac{1}{6}\int_0^6 (36x^2 - 12x^3 + x^4)\,dx\right\}}$$

$$= \sqrt{\left\{\frac{1}{6}\left[\frac{36x^3}{3} - \frac{12x^4}{4} + \frac{x^5}{5}\right]_0^6\right\}}$$

$$= \sqrt{\left\{\frac{1}{6}\left[\left(12(6)^3 - 3(6)^4 + \frac{(6)^5}{5}\right) - (0)\right]\right\}}$$

$$= \sqrt{\left\{\frac{1}{6}(2\,592 - 3\,888 + 1\,555.2)\right\}}$$

$$= \sqrt{\left\{\frac{1}{6}(259.2)\right\}} = \sqrt{43.2} = \mathbf{6.573}$$

Further problems on mean and r.m.s. values may be found in the following Section 29.3 (Problems 1 to 35).

29.3 Further problems

In Problems 1 to 7 find the mean values over the ranges stated.

1. $y = 2\sqrt{x}$ from $x = 0$ to $x = 4$ $[2\frac{2}{3}]$
2. $y = t(2 - t)$ from $t = 0$ to $t = 2$ $[\frac{2}{3}]$
3. $y = \sin\theta$ from $\theta = 0$ to $\theta = 2\pi$ $[0]$

4. $y = \sin\theta$ from $\theta = 0$ to $\theta = \pi$ $\left[\dfrac{2}{\pi}\text{ or } 0.637\right]$

5. $y = 2\cos 2x$ from $x = 0$ to $x = \dfrac{\pi}{4}$ $\left[\dfrac{4}{\pi}\text{ or } 1.273\right]$

6. $y = 2e^x$ from $x = 1$ to $x = 4$ $[34.59]$

7. $y = \dfrac{2}{x}$ from $x = 1$ to $x = 3$ $[1.099]$

8. Determine the mean value of the curve $y = t - t^2 + 2$ which lies above the t-axis by the mid-ordinate rule and check your result using integration. $[1\frac{1}{2}]$

9. The velocity v of a piston moving with simple harmonic motion at any time t is given by $v = k\sin\omega t$. Find the mean velocity between $t = 0$ and $t = \dfrac{\pi}{\omega}$ $\left[\dfrac{2k}{\pi}\right]$

10. Calculate the mean value of $y = 3x - x^2$ in the range $x = 0$ to $x = 3$ by integration $[1.5]$

11. If the speed v m/s of a car is given by $v = 3t + 5$, where t is the time in seconds, find the mean value of the speed from $t = 2$ s to $t = 5$ s $[15\frac{1}{2}$ m/s$]$

12. The number of atoms N remaining in a mass of material during radioactive decay after time t seconds is given by

$N = N_0\,e^{-\lambda t}$, where N_0 and λ are constants. Determine the mean number of atoms in the mass of material for the time period $t = 0$ to $t = \dfrac{1}{\lambda}$ [$0.632N_0$]

13.   A force $9\sqrt{x}$ newtons acts on a body whilst it moves from $x = 0$ to $x = 4$ metres. Find the mean value of the force with respect to distance x [12 N]

14.   The rotor of an electric motor has a tangential velocity v (given by $v = (9 - t^2)$) metres per second after t seconds. Find how far a point on the circumference of the rotor moves in 3 seconds from $t = 0$ and the average velocity during this time.   [18 m, 6 m/s]

15.   The vertical height y kilometres of a rocket fired from a launcher varies with the horizontal distance x kilometres and is given by $y = 6x - x^2$. Determine the mean height of the rocket from $x = 0$ to $x = 6$ kilometres.   [6 km]

In Problems 16 to 23 find the r.m.s. values over the ranges stated.

16. $y = 2x$ from $x = 0$ to $x = 4$ [4.619]
17. $y = x^2$ from $x = 1$ to $x = 3$ [4.919]

18. $y = \sin t$ from $t = 0$ to $t = 2\pi$ $\left[\dfrac{1}{\sqrt{2}} \text{ or } 0.707\right]$

19. $y = \sin t$ from $t = 0$ to $t = \pi$ $\left[\dfrac{1}{\sqrt{2}} \text{ to } 0.707\right]$

20. $y = 4 + 2\cos x$ from $x = 0$ to $x = 2\pi$ [4.243]

21. $y = \sin 3\theta$ from $\theta = 0$ to $\theta = \dfrac{\pi}{6}$ $\left[\dfrac{1}{\sqrt{2}} \text{ or } 0.707\right]$

22. $y = 1 + \sin t$ from $t = 0$ to $t = 2\pi$ [1.225]

23. $y = \cos\theta - \sin\theta$ from $\theta = 0$ to $\theta = \dfrac{\pi}{4}$ [0.603]

24.   Determine: (a) the average value; and (b) the r.m.s. value of a sine wave of maximum value 5.0 for:
(i) a half cycle, and
(ii) one cycle   (a) (i) [3.18]   (ii) [0]
(b) (i) [3.54]   (ii) [3.54]

25.   The distances of points, y, from the mean value of a frequency distribution are related to the variate, x, by the equation $y = x + \dfrac{1}{x}$. Determine the standard deviation (i.e. the r.m.s. value), correct to 4 significant figures, for values of x from 1 to 2   [2.198]

26. Show that the ratio of the r.m.s. value to the mean value of
$y = \sin x$ over the period $x = 0$ to $x = \pi$ is given by $\dfrac{\pi}{2\sqrt{2}}$

27. Draw the graph of $4t - t^2$ for values of t from 0 to 4.
Determine the area bounded by the curve and the t axis by
integration. Find also the mean and r.m.s. values over the
same range.
[Area $= 10\frac{2}{3}$ square units; mean value $= 2.67$; r.m.s.
value $= 2.92$]

28. An alternating voltage is given by $v = 20.0 \cos 50\pi t$ volts.
Find: (a) the mean value; and (b) the r.m.s. value over the
interval from $t = 0$ to $t = 0.01$ seconds   [12.73 V, 14.14 V]

29. A voltage, $v = 24 \sin 50\pi t$ volts is applied across an
electrical circuit. Find its mean and r.m.s. values over the
range $t = 0$ to $t = 10$ ms, each correct to 4 significant figures.
[15.28 V, 16.97 V]

30. A sinusoidal voltage has a maximum value of 150 volts.
Calculate its r.m.s. and mean values.   [106.1 V, 95.49 V]

31. In a frequency distribution the average distance from the
mean, p, is related to the variable, q, by the equation
$p = 3q^2 - 2$. Determine the r.m.s. deviation from the mean
for values of q from -2 to $+3$, correct to 3 significant figures.
[8.66]

32. If the dipolar coupling (y) between two parallel magnetic
dipoles in a liquid is given by $y = 1 - 3x^2$ determine the
average value of y between $x = 1$ and -1 [0]

33. Determine the average heat capacity $\bar{c}_p$ of magnesium
between 300 K and 400 K given that:
$c_p = 6.2 + 1.3 \times 10^{-3} T - 6.8 \times 10^4 \, T^{-2}$ [6.09]

34. If the rate of a chemical reaction (r) is given by:
$r = 2.5(3.2 - x)(3 - x)$,
where x is the moles of the product, determine the average
rate for x to increase from 0 to 1 mole.   [17.1]

35. Find the average velocity ($\bar{v}$) of a chemical change during the
first 5 minutes of reaction when $v = e^{-3t}$, where t is the time
in minutes. $\left[\dfrac{1 - e^{-15}}{15} \right]$

Chapter 30

Volumes of solids of revolution

30.1   Introduction to volumes of solids of revolution

Figure 30.1(a) shows a plane area ABCD bounded by the curve $y = f(x)$, the x-axis and the ordinates $x = a$ and $x = b$. If this area is rotated 360° about the x-axis, then a volume known as a **solid of revolution** is produced, as shown in Fig. 30.1(b).

Let the area ABCD be divided into a large number of strips, each of width δx. A typical strip is shown shaded. When the area ABCD is rotated 360° about the x-axis, each strip produces a solid of revolution which approximates to a circular disc of radius y and thickness δx. The smaller δx becomes, the more accurately the solid of revolution is represented by the disc.

The volume of one such disc = (circular cross-sectional area)(thickness)

$$= (\pi y^2)(\delta x)$$

The total volume of the solid of revolution between ordinates $x = a$ and $x = b$ is given by the sum of all such elemental strips as δx approaches zero.

i.e.   Total volume $= \lim\limits_{\delta x \to 0} \sum\limits_{x=a}^{x=b} \pi y^2 \delta x$

When dealing with areas, it was shown that the limiting value of a sum between limits is equal to the integral between the same limits.

i.e. $\lim\limits_{\delta x \to 0} \sum\limits_{x=a}^{x=b} \pi y^2 \delta x = \int_a^b \pi y^2 \, dx$

Hence, when the curve $y = f(x)$ is rotated one revolution about the x-axis between the limits $x = a$ and $x = b$, the volume V generated is given by:

Volume, $V = \int_a^b \pi y^2 \, dx$

Similarly, if a curve $x = f(y)$ is rotated about the y-axis between the limits $y = c$ and $y = d$, as shown in Fig. 30.2, then the volume generated is given by:

volume $= \lim\limits_{\delta y \to 0} \sum\limits_{y=c}^{y=d} \pi x^2 \delta y = \int_c^d \pi x^2 \, dy$

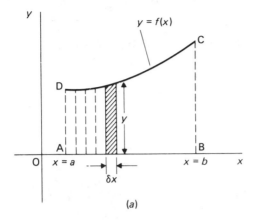

(a)

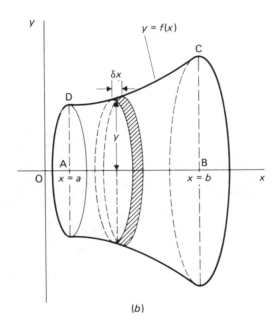

(b)

Fig. 30.1

Worked problems on volumes of solids of revolution

Problem 1. Find the volume of the solid of revolution between the limits $x = 0$ and $x = 4$ when the following curves are rotated about the x-axis: (a) $y = 3$ (b) $y = x$

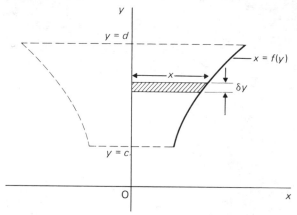

Fig. 30.2

(a) If the area bounded by $y = 3$, the x-axis and the limits of $x = 0$ and $x = 4$ is rotated about the x-axis (see Fig. 30.3(a)), a cylinder of radius 3 units and height 4 units is produced.

$$\text{Volume generated} = \int_0^4 \pi y^2 \, dx = \int_0^4 \pi(3)^2 \, dx$$

$$= 9\pi \int_0^4 dx = 9\pi[x]_0^4$$

$$= \textbf{36}\pi \text{ cubic units}$$

(b) If the area bounded by $y = x$, the x-axis and limits $x = 0$ and $x = 4$ is rotated about the x-axis (see Fig. 30.3(b)), a cone of base radius 4 units and perpendicular height 4 units is produced.

$$\text{Volume generated} = \int_0^4 \pi y^2 \, dx = \int_0^4 \pi x^2 \, dx$$

$$= \pi \left[\frac{x^3}{3} \right]_0^4$$

$$= \frac{\textbf{64}\pi}{\textbf{3}} \text{ cubic units}$$

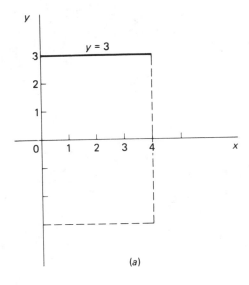

(a)

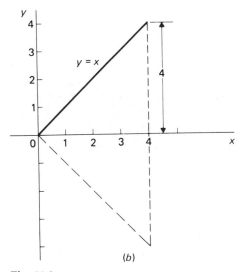

(b)

Fig. 30.3

Problem 2. The curve $y = 2x^2 + 3$ is rotated 360° about: (a) the x-axis, between the limits of $x = 1$ and $x = 3$; and (b) the y-axis, between the same limits. Find the volume of the solid of revolution produced in each case.

The relevant portion of the curve $y = 2x^2 + 3$ is shown in Fig. 30.4

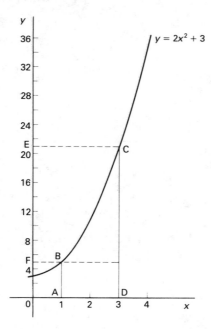

Fig. 30.4

(a) The volume produced when the curve $y = 2x^2 + 3$ is rotated about the x-axis between $x = 1$ and $x = 3$ (i.e. rotating area ABCD) is given by:

$$\text{volume} = \int_1^3 \pi y^2 \, dx$$

$$= \int_1^3 \pi(2x^2 + 3)^2 \, dx = \pi \int_1^3 (4x^4 + 12x^2 + 9) \, dx$$

$$= \pi \left[\frac{4x^5}{5} + \frac{12x^3}{3} + 9x \right]_1^3$$

$$= \pi[(194.4 + 108 + 27) - (0.8 + 4 + 9)]$$

$$= \textbf{315.6}\pi \textbf{ cubic units}$$

(b) When $x = 1$, $y = 2(1)^2 + 3 = 5$
When $x = 3$, $y = 2(3)^2 + 3 = 21$

Since $y = 2x^2 + 3$, then $x^2 = \dfrac{y - 3}{2}$

The volume produced when the curve $y = 2x^2 + 3$ is rotated about the y-axis between $y = 5$ and $y = 21$ (i.e. rotating area BCEF) is given by:

$$\text{volume} = \int_5^{21} \pi x^2 \, dy = \int_5^{21} \pi \frac{(y-3)}{2} \, dy$$

$$= \frac{\pi}{2} \left[\frac{y^2}{2} - 3y \right]_5^{21}$$

$$= \frac{\pi}{2} \left[\left\{ \frac{(21)^2}{2} - 3(21) \right\} - \left\{ \frac{(5)^2}{2} - 3(5) \right\} \right]$$

$$= 80.0\pi \textbf{ cubic units}$$

Problem 3. Find the volume generated when the area above the x-axis bounded by the curve $x^2 + y^2 = 4$ and the ordinates $x = 2$ and $x = -2$ is rotated about the x-axis.

Figure 30.5 shows the part of the curve $x^2 + y^2 = 4$ lying above the x-axis.

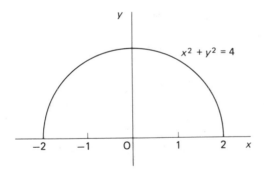

Fig. 30.5

The curve $x^2 + y^2 = 4$ is a circle, centre O and radius 2. (In general $x^2 + y^2 = r^2$ represents a circle, centre O and radius r.)

If the semicircle shown in Fig. 30.5 is rotated about the x-axis then the volume generated is given by:

$$\text{volume} = \int_{-2}^{2} \pi y^2 \, dx = \int_{-2}^{2} \pi (4 - x^2) \, dx$$

$$= \pi \left[4x - \frac{x^3}{3} \right]_{-2}^{2} = \frac{32\pi}{3} \textbf{ cubic units}$$

Problem 4. Calculate the volume of the frustum of a sphere of radius 5 cm which lies between two parallel planes at 1 cm and 3 cm from the centre and on the same side of it.

The volume of the frustum of the sphere may be found by rotating the curve $x^2 + y^2 = 5^2$ (i.e. a circle, centre O, radius 5) about the x-axis between the limits $x = 1$ and $x = 3$ (i.e. rotating the area shown shaded in Fig. 30.6).

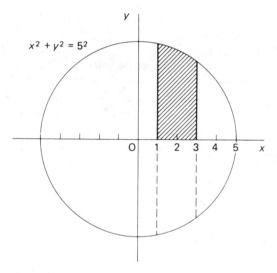

Fig. 30.6

$$\text{Volume of frustrum} = \int_1^3 \pi y^2 \, dx = \pi \int_1^3 (5^2 - x^2) \, dx$$

$$= \pi \left[25x - \frac{x^3}{3} \right]_1^3$$

$$= \pi[(75 - 9) - (25 - \tfrac{1}{3})]$$

$$= 41\tfrac{1}{3}\pi \text{ cm}^3$$

Problem 5. The curve $y = 3 \sec \dfrac{x}{2}$ is rotated about the x-axis between the limits $x = 0$ and $x = \dfrac{\pi}{3}$. Find the volume of the solid formed.

Using a calculator the following table of values is produced.

x	0	$\dfrac{\pi}{6}$	$\dfrac{\pi}{3}$
$y = 3\sec\dfrac{x}{2}$	3.000	3.106	3.464

A part of the curve $y = 3\sec\dfrac{x}{2}$ is shown in Fig. 30.7. If the area

shown shaded $\left(\text{i.e. between limits } x = 0 \text{ and } x = \dfrac{\pi}{3}\right)$ is rotated

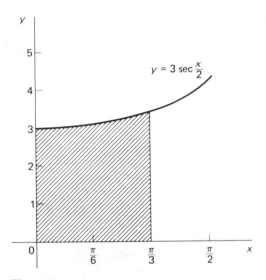

Fig. 30.7

about the x-axis then the volume formed is given by:

$$\text{volume} = \int_0^{\pi/3} \pi y^2 \, dx = \int_0^{\pi/3} \pi \left(3\sec\frac{x}{2}\right)^2 dx$$

$$= 9\pi \int_0^{\pi/3} \sec^2 \frac{x}{2} \, dx$$

$$= 9\pi \left[2 \tan \frac{x}{2}\right]_0^{\pi/3}$$

$$= 18\pi \left[\tan \frac{\pi}{6} - \tan 0\right] = \frac{18\pi}{\sqrt{3}}$$

$$= 6\sqrt{3}\pi \text{ or } \textbf{32.65 cubic units}$$

Problem 6. Find the volume of the solid formed by revolving the area enclosed between the curve $y = \dfrac{1}{x^2}$ and the lines $y = 2$ and $y = 4$ about the y-axis.

Since $y = \dfrac{1}{x^2}$, as $x \to 0$, $y \to \infty$ and as $x \to \infty$, $y \to 0$

When $x = \frac{1}{2}$, $y = 4$ and when $x = 1$, $y = 1$. Using these values a sketch of $y = \dfrac{1}{x^2}$ may be drawn as shown in Fig. 30.8

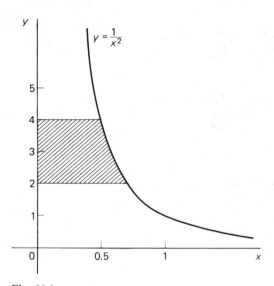

Fig. 30.8

If the area bounded by the curve $y = \dfrac{1}{x^2}$ and the lines $y = 2$ and $y = 4$ (shown shaded in Fig. 30.8) is rotated about the y-axis the volume is given by:

$$\text{volume} = \int_2^4 \pi x^2 \, dy$$

Since $y = \dfrac{1}{x^2}$, then $x^2 = \dfrac{1}{y}$

Hence volume $= \displaystyle\int_{2}^{4} \pi \frac{1}{y}\,dy$

$= \pi[\ln y]_{2}^{4}$

$= \pi[\ln 4 - \ln 2] = \pi \ln \frac{4}{2}$

$= \pi \ln 2$ or 0.693π **cubic units**

Problem 7. The area enclosed by the curve $y = 2e^{x/2}$, the x-axis and ordinates $x = -2$ and $x = 3$ is rotated about the x-axis. Calculate the volume generated.

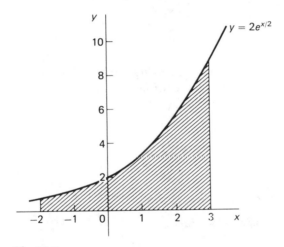

Fig. 30.9

If the area shown shaded in Fig. 30.9 is rotated about the x-axis, the volume generated is given by:

$$\text{volume} = \int_{-2}^{3} \pi y^2\,dx = \int_{-2}^{3} \pi(2e^{x/2})^2\,dx$$

$$= 4\pi \int_{-2}^{3} e^x\,dx = 4\pi[e^x]_{-2}^{3}$$

$$= 4\pi[e^3 - e^{-2}] = 79.8\pi \ \text{cubic units}$$

Problem 8. The area enclosed between the two parabolas $x^2 = 2y$ and $y^2 = 16x$ is rotated about the x-axis. Find the volume of the solid produced.

At the points of intersection the curves are equal.

Curve 1: $x^2 = 2y$ or $y = \dfrac{x^2}{2}$, i.e. $y^2 = \dfrac{x^4}{4}$

Curve 2: $y^2 = 16x$

At the points of intersection $\dfrac{x^4}{4} = 16x$

$$\frac{x^4}{4} - 16x = 0$$

$$x\left(\frac{x^3}{4} - 16\right) = 0$$

Hence $x = 0$ or $\dfrac{x^3}{4} - 16 = 0$, i.e. $x = 4$

The area enclosed by the two curves is shown in Fig. 28.10, page 488.

The volume produced by revolving the shaded area about the x-axis is given by: (the volume formed by revolving OAB) − (the volume formed by revolving OCB).

i.e. Volume $= \displaystyle\int_0^4 \pi(16x)\,dx - \int_0^4 \pi\left(\frac{x^4}{4}\right)dx$

$= \pi \displaystyle\int_0^4 \left(16x - \frac{x^4}{4}\right)dx = \pi\left[\frac{16x^2}{2} - \frac{x^5}{20}\right]_0^4$

$= \pi[(128 - 51.2) - (0)]$

$= \mathbf{76.8\pi}$ **cubic units**

Further problems on volumes of solids of revolution may be found in the following Section 30.2 (Problems 1 to 36).

30.2   Further problems

(All answers are in cubic units and are left in terms of π.)
In Problems 1 to 15 find the volume of the solids of revolution formed by revolving the areas enclosed by the given curves, the x-axis and the given ordinates through 360° about the x-axis.

1. $y = 2$; $x = 1$, $x = 5$ $[16\pi]$
2. $y = 2x$; $x = 0$, $x = 6$ $[288\pi]$
3. $y = x^2$; $x = -1$, $x = 2$ $[6\frac{3}{5}\pi]$
4. $y = x + 1$; $x = 1$, $x = 2$ $[6\frac{1}{3}\pi]$
5. $y + 1 = x^2$; $x = -1$, $x = 1$ $[1\frac{1}{15}\pi]$
6. $y = 3x^2 + 4$; $x = 0$, $x = 2$ $[153\frac{3}{5}\pi]$

7. $\dfrac{y^2}{3} = x; \; x = 1, \; x = 6$ $[52\frac{1}{2}\pi]$

8. $y = \dfrac{2}{x}; \; x = 2, \; x = 4$ $[\pi]$

9. $3xy = 5; \; x = 1, \; x = 3$ $[1\frac{23}{27}\pi]$
10. $x^2 + y^2 = 16; \; x = -4, \; x = 4$ $[85\frac{1}{3}\pi]$
11. $x = \sqrt{(9 - y^2)}; \; x = 0, \; x = 3$ $[18\pi]$

12. $y = 3e^x; \; x = 0, \; x = 2$ $\left[\dfrac{9e^4\pi}{2} \text{ or } 245.7\pi\right]$

13. $y = 2\sec x; \; x = 0, \; x = \dfrac{\pi}{4}$ $[4\pi]$

14. $y = 3\operatorname{cosec} x; \; x = \dfrac{\pi}{6}, \; x = \dfrac{\pi}{3}$ $[10.39\pi]$

15. $\dfrac{y}{6} = \sqrt{x^3}; \; x = 2, \; x = 4$ $[2\,160\pi]$

In Problems 16 to 24 find the volume of the solids of revolution formed by revolving the areas enclosed by the given curves, the y-axis and the given ordinates through 360° about the y-axis.

16. $y = x^2; \; y = 1, \; y = 4$ $[7\frac{1}{2}\pi]$
17. $y = 2x^2 - 3; \; y = 0, \; y = 2$ $[4\pi]$
18. $2y = x^4; \; y = 1, \; y = 4$ $[6.60\pi]$

19. $y = \dfrac{3}{x}; \; y = 2, \; y = 3$ $[1.5\pi]$

20. $x^2 + y^2 = r^2; \; y = 0, \; y = r$ $[\frac{2}{3}\pi r^3]$
21. $y = \sqrt{(25 - x^2)}; \; y = -5, \; y = 5$ $[166\frac{2}{3}\pi]$

22. $y = \sqrt{x^3}; \; y = 0, \; y = 1$ $\left[\dfrac{3\pi}{7}\right]$

23. $x\sqrt{y} = 1; \; y = 2, \; y = 3$ $[\pi \ln\frac{3}{2} \text{ or } 0.405\pi]$
24. $\sqrt{x} = (\sqrt{3})y; \; y = 0, \; y = 2$ $[57.6\pi]$
25. The curve $y = 3x^2 - 4$ is rotated about: (a) the x-axis between the limits $x = 0$ and $x = 2$, and (b) the y-axis between the same limits. Find the volume generated in each case.   (a) $[25.6\pi]$ (b) $[24\pi]$
26. Find the volume of a pressure vessel generated when the area above the x-axis bounded by the curve $x^2 + y^2 = 36$ and the ordinates $x = 6$ m and $x = -6$ m is rotated about the x-axis. $[288\pi \text{ m}^3]$

27. Calculate the volume of the plug formed by the frustum of a sphere of radius 7 cm which lies between two parallel planes at 2 cm and 5 cm from the centre and on the same side of it. $[108\pi \text{ cm}^3]$

28. The area enclosed between the two curves $x^2 = 4y$ and $y^2 = 4x$ is rotated about the x-axis. Find the volume of the solid formed. $[19\frac{1}{5}\pi]$

29. Calculate the volume of the solid gun-mounting formed by revolving the area between the curve $y = \dfrac{3}{2x^2}$ and the lines $y = 1$ m and $y = 4$ m about the y-axis. $[2.079\pi \text{ m}^3]$

30. The curve $y = x^2 + 2x + 1$ is rotated about the x-axis. Find the volume generated between the limits $x = 0$ and $x = 2$ $[48\frac{2}{5}\pi]$

31. Find the volume of the solid obtained by rotating about the x-axis the part of the curve $y = 4x - x^2$ lying above the x-axis. $[34\frac{2}{15}\pi]$

32. The portion of the curve $y = x^2 + \dfrac{2}{x}$ lying between $x = 1$ cm and $x = 2$ cm is revolved about the x-axis. Calculate the volume of the solid for the component formed. $[14.2\pi \text{ cm}^3]$

33. The curve $y = \dfrac{5}{x + 1}$ is rotated about the x-axis between the limits $x = 0$ and $x = 4$ m. Find the volume of the podium generated. $[20\pi \text{ m}^3]$

34. Calculate the volume of a frustum of a sphere of radius 8 cm which lies between two parallel planes at 3 cm and 4 cm from the centre and on opposite sides of it. $[417\frac{2}{3}\pi \text{ cm}^3]$

35. The area enclosed between the curves $x^2 = (2\sqrt{2})y$ and $\dfrac{y^2}{8} = x$ is rotated 360° about the x-axis. Find the volume produced. $[38.4\pi]$

36. The area between $\dfrac{y}{x^2} = 1$ and $y + x^2 = 8$ is rotated through 4 right angles about the x-axis. Calculate the volume generated. $[170\frac{2}{3}\pi]$

Chapter 31

Centroids of simple shapes

31.1   Centroids of plane areas

Centroid

A **lamina** is a thin sheet of uniform thickness. If, when supported at a particular point, a lamina balances perfectly, then this point is called the **centre of gravity**. It is through the centre of gravity that the mass of the lamina is considered to act. If a lamina of negligible thickness, and hence negligible mass, is considered, then the term centre of gravity is inappropriate. As we are now dealing only with a shape or area, the term **centre of area** or simply **centroid** is used for the point where the centre of gravity of a lamina of that shape would lie.

The first moment of area

A 'moment' (in mechanics) is the measure of the power of a force in causing rotation (i.e. the 'moment of a force' is the product of the force and the perpendicular distance from a fixed point).

i.e.   moment = force × distance

But   force = mass × acceleration,

and in the case of a thin uniform lamina,

   mass $\propto$ area

Hence force $\propto$ area × acceleration

Since acceleration due to gravity can be taken to be a constant, then:

   force $\propto$ area

The first moment of area is defined as the product of the area and the perpendicular distance of its centroid from a given axis in the plane of the area.

In Fig. 31.1 any area A is shown with its centroid at point C. XX is any axis in the same plane as A.

The first moment of area about the axis XX (which is at a perpendicular

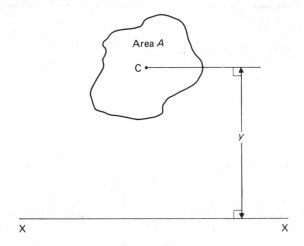

Fig. 31.1

distance y from the centroid C) is given by (area A)(distance y)

i.e. **first moment of area = Ay cubic units**

When dealing with centroids, the expression 'taking moments about XX' is often used for 'first moment of area about axis XX'.

It is found that the centroid of: (a) a rectangle lies on the intersection of the diagonals, (b) a triangle, of perpendicular height h, lies at a point $\dfrac{h}{3}$ from the base, (c) a circle lies at its centre, (d) a semicircle, of radius r, lies on the centre line at a distance $\dfrac{4r}{3\pi}$ from the diameter.

If the centroid of an area between a curve and given limits is required then integration may be used.

Figure 31.2 shows a plane area EFGH bounded by the curve $y = f(x)$, the x-axis and ordinates $x = a$ and $x = b$. Let this area be divided into a large number of strips each of width δx. A typical strip is shown shaded. The centroid of the shaded strip is at its centre, i.e. at coordinates $\left(x, \dfrac{y}{2}\right)$.

The area of the strip is given by $y\delta x$

Therefore the first moment of area of the strip about axis OY

= (area of strip)(perpendicular distance between the centroid and axis OY)

= $(y\delta x)x = xy\delta x$

The total first moment of area EFGH about axis OY is given by the sum of the first moments of area of all such strips, i.e. in the limit $\displaystyle\sum_{x=a}^{x=b} xy\delta x = \int_{a}^{b} xy\,\mathrm{d}x$

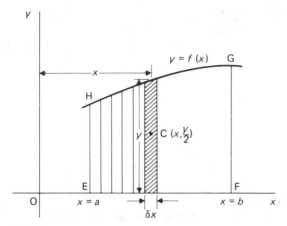

Fig. 31.2

The first moment of area of the strip about axis OX

= (area of strip)(perpendicular distance between the centroid and axis OX)

$$= (y\delta x)\frac{y}{2} = \tfrac{1}{2}y^2\delta x$$

The total first moment of area EFGH about axis OX is given by the sum of the first moments of area of all such strips, i.e. in the limit

$$\sum_{x=a}^{x=b} \tfrac{1}{2}y^2\delta x = \frac{1}{2}\int_a^b y^2\,\mathrm{d}x$$

The area A bounded by the curve $y = f(x)$, the x-axis and the limits $x = a$ and $x = b$, i.e. the area EFGH, is given by:

$$A = \int_a^b y\,\mathrm{d}x$$

If $\bar{x}$ and $\bar{y}$ (pronounced 'x bar' and 'y bar' respectively) are the distances of the centroid of the area A about axes OY and OX respectively then:

$$(\bar{x})(\text{total area } A) = \sum_{x=a}^{x=b} x(y\delta x)$$

i.e. in the limit
$$A\bar{x} = \int_a^b xy\,\mathrm{d}x$$

$$\boxed{\bar{x} = \frac{\displaystyle\int_a^b xy\,\mathrm{d}x}{\displaystyle\int_a^b y\,\mathrm{d}x}}$$

Similarly, $(\bar{y})$(total area A) = $\displaystyle\sum_{x=a}^{x=b} \frac{y}{2}(y\delta x)$

i.e. in the limit $A\bar{y} = \dfrac{1}{2}\displaystyle\int_a^b y^2 \, dx$

$$\bar{y} = \frac{\dfrac{1}{2}\displaystyle\int_a^b y^2 \, dx}{\displaystyle\int_a^b y \, dx}$$

Centroid of area between a curve and the y-axis

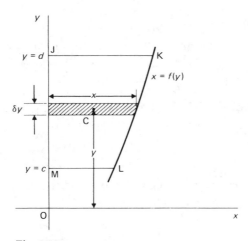

Fig. 31.3

If the position of the centroid of the area between a curve $(x = f(y))$ and the y-axis is required, say area JKLM of Fig. 31.3, then with similar reasoning to above:

$(\bar{x})$(total area) = $\displaystyle\sum_{y=c}^{y=d} \frac{x}{2}(x\delta y)$

i.e. $$\bar{x} = \frac{\dfrac{1}{2}\displaystyle\int_c^d x^2 \, dy}{\displaystyle\int_c^d y \, dy}$$

Similarly, $(\bar{y})$(total area) $= \displaystyle\sum_{y=c}^{y=d} y(x\delta y)$

i.e.

$$\bar{y} = \frac{\displaystyle\int_c^d xy\,dy}{\displaystyle\int_c^d x\,dy}$$

The expressions for $\bar{x}$ and $\bar{y}$ are the same as those used for finding the position of the centroid of the area between a curve and the x-axis except that the x's and y's have been interchanged. (See Problem 4 below.)

Theorem of Pappus

The theorem of Pappus states:

If a plane area is rotated about an axis in its own plane but not intersecting it, the volume of the solid formed is given by the product of the area and the distance moved through by the centroid of the area.

i.e. volume generated = area × distance moved through by the centroid.

The theorem essentially enables volumes of solids to be calculated. However, if the volume is known then the centroid of an area may be calculated (see Problems 8 and 9 below).

Let A be any area whose centroid is at C (see Fig. 31.4).

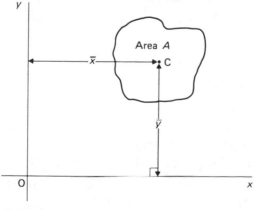

Fig. 31.4

If the distance of the centroid C from the axis OX is $\bar{y}$ then the distance moved through by the centroid when the area A is rotated $360°$ about the

x-axis is $2\pi\bar{y}$ (i.e. the circumference of a circle of radius $\bar{y}$) and the volume V of the solid produced is given by:

volume generated = area + $2\pi\bar{y}$

i.e. $V = 2\pi A\bar{y}$ **cubic units**

or $\bar{y} = \dfrac{V}{2\pi A}$ **units**

Similarly, if $\bar{x}$ is the perpendicular distance of the centroid C from the y-axis then rotation of area A about the y-axis will generate a volume V given by

$V = A(2\pi\bar{x})$, i.e. $\bar{x} = \dfrac{V}{2\pi A}$ **units**

Worked problems on centroids

Problem 1. Find the first moment of area about axis XX for each of the shapes shown in Fig. 31.5

The first moment of area about axis XX = (area of shape) × (perpendicular distance between the centroid of the shape and the axis XX).

(a) The centroid of a rectangle lies at the intersection of the diagonals. First moment of area of rectangle

$$= (3 \times 6)(8 + 1\tfrac{1}{2})$$
$$= (18)(9\tfrac{1}{2}) = \textbf{171 cm}^3$$

(b) The centroid of a circle lies at its centre.

First moment of area of circle = $[\pi(4)^2][6 + 4]$
$$= (16\pi)(10) = \textbf{160}\pi \textbf{ cm}^3$$

(c) The centroid of a triangle, of perpendicular height h, lies at a point $\dfrac{h}{3}$ from the base. The perpendicular height of the triangle is $5 \sin 60°$ or 4.330 cm. Since there are 180° in any triangle, the third angle of the triangle is 60°. The triangle is thus equilateral, each side being 5 cm in length.

First moment of area of triangle = $[\tfrac{1}{2}(5)(4.330)][9 + \tfrac{1}{3}(4.330)]$
$$= \tfrac{1}{2}(5)(4.330)(10.443)$$
$$= \textbf{113.0 cm}^3$$

(d) The centroid of a semicircle, of radius r, lies on the centre line at a distance $\dfrac{4r}{3\pi}$ from the diameter.

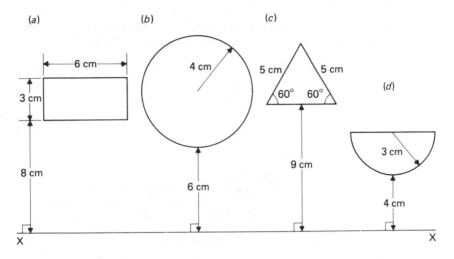

Fig. 31.5

First moment of area of semicircle $= [\frac{1}{2}\pi(3)^2]\left[4 + 3 - \left(\dfrac{4(3)}{3\pi}\right)\right]$

$$= \tfrac{1}{2}(9)\pi[4 + 3 - (1.273)]$$

$$= \mathbf{80.96 \ cm^3}$$

Problem 2. Find the position of the centroid of the area bounded by the curve $y = 2x^2$, the x-axis and the ordinates $x = 0$ and $x = 3$

$$\bar{x} = \frac{\displaystyle\int_0^3 xy \, dx}{\displaystyle\int_0^3 y \, dx} = \frac{\displaystyle\int_0^3 x(2x^2) \, dx}{\displaystyle\int_0^3 2x^2 \, dx} = \frac{\displaystyle\int_0^3 2x^3 \, dx}{\displaystyle\int_0^3 2x^2 \, dx}$$

$$= \frac{\left[\dfrac{2x^4}{4}\right]_0^3}{\left[\dfrac{2x^3}{3}\right]_0^3} = \frac{\frac{81}{2}}{18} = 2\tfrac{1}{4}$$

$$\bar{y} = \frac{\dfrac{1}{2}\displaystyle\int_0^3 y^2 \, dx}{\displaystyle\int_0^3 y \, dx} = \frac{\dfrac{1}{2}\displaystyle\int_0^3 (2x^2)^2 \, dx}{18} = \frac{1}{36}\displaystyle\int_0^3 4x^4 \, dx$$

$$= \frac{1}{9}\left[\dfrac{x^5}{5}\right]_0^3 = \tfrac{1}{9}(\tfrac{243}{5}) = 5\tfrac{2}{5}$$

Hence **the centroid of the area bounded by the curve $y = 2x^2$, the x-axis and the ordinates $x = 0$ and $x = 3$ is at $(2\frac{1}{4}, 5\frac{2}{5})$**

[Note that functions within an integral in a numerator must not be 'cancelled' with a function within an integral in the denominator. For example, from above, $\bar{x} = \dfrac{\displaystyle\int_0^3 2x^3\,dx}{\displaystyle\int_0^3 2x^2\,dx}$. Whereas

constants (which do not affect integration) may be cancelled, the x^2 in the denominator **must not** be cancelled with the x^3 in the numerator to produce $\bar{x} = \displaystyle\int_0^3 x\,dx$. This is incorrect; each integration must be evaluated separately.]

Problem 3. Calculate the coordinates of the centroid of the area lying between the curve $y = 4x - x^2$ and the x-axis.

It is necessary to obtain the limits of integration.

$$y = 4x - x^2 = x(4 - x)$$

When $y = 0$ (i.e. the x-axis), $x = 0$ or $(4 - x) = 0$, i.e. $x = 4$. Hence the curve cuts the x-axis at $x = 0$ and $x = 4$ as shown in Fig. 31.6

$$\bar{x} = \frac{\displaystyle\int_0^4 xy\,dx}{\displaystyle\int_0^4 y\,dx} = \frac{\displaystyle\int_0^4 x(4x - x^2)\,dx}{\displaystyle\int_0^4 (4x - x^2)\,dx} = \frac{\displaystyle\int_0^4 (4x^2 - x^3)\,dx}{\displaystyle\int_0^4 (4x - x^2)\,dx}$$

$$= \frac{\left[\dfrac{4x^3}{3} - \dfrac{x^4}{4}\right]_0^4}{\left[\dfrac{4x^2}{2} - \dfrac{x^3}{3}\right]_0^4} = \frac{(\frac{256}{3} - 64)}{(32 - \frac{64}{3})} = \frac{21\frac{1}{3}}{10\frac{2}{3}} = 2$$

$$\bar{y} = \frac{\dfrac{1}{2}\displaystyle\int_0^4 y^2\,dx}{\displaystyle\int_0^4 y\,dx} = \frac{\dfrac{1}{2}\displaystyle\int_0^4 (4x - x^2)^2\,dx}{10\frac{2}{3}} = \frac{3}{64}\displaystyle\int_0^4 (16x^2 - 8x^3 + x^4)\,dx$$

$$= \frac{3}{64}\left[\frac{16x^3}{3} - \frac{8x^4}{4} + \frac{x^5}{5}\right]_0^4 = \frac{3}{64}\left[\frac{16(64)}{3} - 512 + \frac{16(64)}{5}\right]$$

$$= \mathbf{1.6}$$

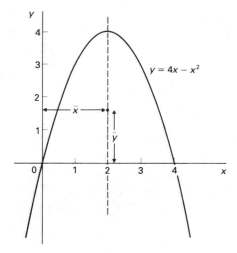

Fig. 31.6

Hence **the centroid of the area lying between $y = 4x - x^2$ and the x-axis is at (2, 1.6)**

From the sketch shown in Fig. 31.6 it can be seen that the curve is symmetrical about $x = 2$. Thus the calculation for $\bar{x}$ need not have been made. However, if there is any doubt as to whether an axis of symmetry exists, the above calculations should be made.

Problem 4. Determine the position of the centroid of the area enclosed by the curve $y = 4x^2$, the y-axis and ordinates $y = 1$ and $y = 9$, correct to 3 decimal places.

$$\bar{x} = \frac{\dfrac{1}{2}\displaystyle\int_1^9 x^2\, dy}{\displaystyle\int_1^9 x\, dy} = \frac{\dfrac{1}{2}\displaystyle\int_1^9 \dfrac{y}{4}\, dy}{\displaystyle\int_1^9 \sqrt{\dfrac{y}{4}}\, dy} = \frac{\dfrac{1}{2}\left[\dfrac{y^2}{8}\right]_1^9}{\dfrac{1}{2}\left[\dfrac{2y^{\frac{3}{2}}}{3}\right]_1^9}$$

$$= \frac{\dfrac{1}{2}\left[\dfrac{81}{8} - \dfrac{1}{8}\right]}{\dfrac{1}{2}\left[\dfrac{2(27)}{3} - \dfrac{2}{3}\right]} = \frac{5}{8\frac{2}{3}} = \mathbf{0.577}$$

$$\bar{y} = \frac{\displaystyle\int_1^9 xy\, dy}{\displaystyle\int_1^9 x\, dy} = \frac{\displaystyle\int_1^9 \left(\sqrt{\dfrac{y}{4}}\right) y\, dy}{8\frac{2}{3}} = \frac{\dfrac{1}{2}\displaystyle\int_1^9 y^{\frac{3}{2}}\, dy}{8\frac{2}{3}}$$

$$= \frac{\frac{1}{2}\left[\frac{2y^{\frac{5}{2}}}{5}\right]_1^9}{8\frac{2}{3}} = \frac{\frac{1}{5}[243 - 1]}{8\frac{2}{3}} = \mathbf{5.585}$$

Hence **the position of the centroid of the area is at (0.577, 5.585)**

Problem 5. Find the position of the centroid of the area enclosed between the curve $y = 9 - x^2$ and the x-axis.

$y = 9 - x^2$. When $y = 0$ (i.e. the x-axis) then $9 - x^2 = 0$ or $9 = x^2$
Hence $x = \pm 3$
Thus the curve $y = 9 - x^2$ cuts the x-axis at $x = -3$ and $x = +3$.
Also, when $x = 0$, $y = 9$. A sketch of a part of $y = 9 - x^2$ is shown in Fig. 31.7

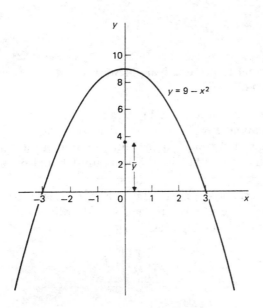

Fig. 31.7

It may be seen from Fig. 31.7 that the area is symmetrical about the y-axis. Hence $\bar{x} = 0$.

$$\bar{y} = \frac{\frac{1}{2}\int_{-3}^{3} y^2\, dx}{\int_{-3}^{3} y\, dx} = \frac{\frac{1}{2}\int_{-3}^{3} (9 - x^2)^2\, dx}{\int_{-3}^{3} (9 - x^2)\, dx} = \frac{\frac{1}{2}\int_{-3}^{3} (81 - 18x^2 + x^4)\, dx}{\int_{-3}^{3} (9 - x^2)\, dx}$$

$$= \frac{\dfrac{1}{2}\left[81x - \dfrac{18x^3}{3} + \dfrac{x^5}{5}\right]_{-3}^{3}}{\left[9x - \dfrac{x^3}{3}\right]_{-3}^{3}}$$

$$= \frac{\frac{1}{2}[(243 - 162 + \frac{243}{5}) - (-243 + 162 - \frac{243}{5})]}{[(27 - 9) - (-27 + 9)]}$$

$$= \frac{\frac{1}{2}[(129.6) - (-129.6)]}{[(18) - (-18)]}$$

$$= \frac{128.6}{36} = 3.6$$

Hence **the centroid of the area enclosed by $y = 9 - x^2$ and the x-axis lies at (0, 3.6)**

Problem 6. Prove by integration that the centroid of a triangle of perpendicular height h and base length b lies at a point $\dfrac{h}{3}$ from the base.

The equation $y = mx + c$ represents a straight line of slope m and y-axis intercept c. Hence the equation $y = \dfrac{b}{h} x$ represents a straight line of slope $\dfrac{b}{h}$ and intercept O.

The area bounded by $y = \dfrac{b}{h} x$, the x-axis and ordinates $x = 0$ and $x = h$ forms a right-angled triangle OAB as shown in Fig. 31.8

$$\bar{x} = \frac{\displaystyle\int_0^h xy\,dx}{\displaystyle\int_0^h y\,dx} = \frac{\displaystyle\int_0^h x\left(\dfrac{b}{h} x\right) dx}{\text{area of triangle OAB}} = \frac{\dfrac{b}{h}\left[\dfrac{x^3}{3}\right]_0^h}{\frac{1}{2}bh}$$

$$= \frac{\dfrac{b}{h}\left(\dfrac{h^3}{3}\right)}{\dfrac{bh}{2}} = \frac{\dfrac{bh^2}{3}}{\dfrac{bh}{2}} = \frac{2h}{3}$$

$\bar{x}$ is the perpendicular distance of the centroid from the vertex of

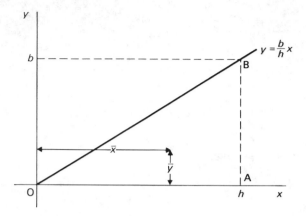

Fig. 31.8

triangle OAB. Hence **the distance of the centroid from the base AB is $\frac{1}{3}h$.**

$$\text{[Also } \bar{y} = \frac{\dfrac{1}{2}\displaystyle\int_0^h y^2\, dx}{\displaystyle\int_0^h y\, dx} = \frac{\dfrac{1}{2}\left(\dfrac{b}{h}\right)^2\displaystyle\int_0^h x^2\, dx}{\dfrac{bh}{2}}$$

$$= \frac{\dfrac{1}{2}\left(\dfrac{b}{h}\right)^2\left[\dfrac{x^3}{3}\right]_0^h}{\dfrac{bh}{2}} = \frac{\dfrac{1}{2}\left(\dfrac{b}{h}\right)^2\left(\dfrac{h^3}{3}\right)}{\dfrac{bh}{2}} = \frac{b}{3}$$

Hence if OA is considered as the base of triangle OAB then the centroid is at a distance of $\dfrac{b}{3}$ from it.]

Problem 7. Locate the position of the centroid of the area enclosed by the curves $x = \dfrac{y^2}{4}$ and $y = \dfrac{x^2}{4}$

Figure 31.9 shows the two curves intersecting at $(0, 0)$ and $(4, 4)$ and enclosing a shaded area, the centroid of which is required.

$$\bar{x} = \frac{\displaystyle\int_0^4 xy\, dx}{\displaystyle\int_0^4 y\, dx}$$

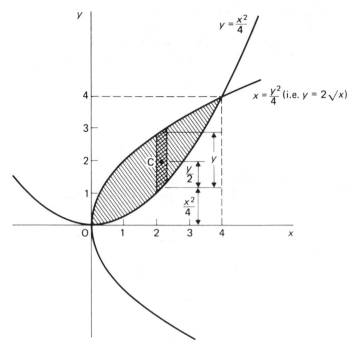

Fig. 31.9

The value of y in this integral is given by the height of the typical strip shown, i.e. $\left(2\sqrt{x} - \dfrac{x^2}{4}\right)$

Hence $\bar{x} = \dfrac{\displaystyle\int_0^4 x\left(2\sqrt{x} - \dfrac{x^2}{4}\right)dx}{\displaystyle\int_0^4 \left(2\sqrt{x} - \dfrac{x^2}{4}\right)dx} = \dfrac{\displaystyle\int_0^4 \left(2x^{\frac{3}{2}} - \dfrac{x^3}{4}\right)dx}{\displaystyle\int_0^4 \left(2x^{\frac{1}{2}} - \dfrac{x^2}{4}\right)dx}$

$= \dfrac{\left[\dfrac{4x^{\frac{5}{2}}}{5} - \dfrac{x^4}{16}\right]_0^4}{\left[\dfrac{4x^{\frac{3}{2}}}{3} - \dfrac{x^3}{12}\right]_0^4} = \dfrac{[\frac{4}{5}(32) - 16]}{[\frac{4}{3}(8) - \frac{64}{12}]}$

$= \dfrac{9\frac{3}{5}}{5\frac{1}{3}} = \mathbf{1.8}$

Care must be taken when finding $\bar{y}$ since the centroid of the typical strip is not now at a distance $\dfrac{y}{2}$ from the x-axis.

Taking moments about OX gives:

$$(\text{Total area})(\bar{y}) = \sum_{x=0}^{x=4} (\text{area of strip})(\text{perpendicular distance of the centroid of the strip to axis OX})$$

$y = 2\sqrt{x} - \dfrac{x^2}{4}$. Therefore $\dfrac{y}{2} = \dfrac{1}{2}\left(2\sqrt{x} - \dfrac{x^2}{4}\right)$

The perpendicular distance from the centroid C of the strip to the axis OX is given by $\dfrac{1}{2}\left(2\sqrt{x} - \dfrac{x^2}{4}\right) + \dfrac{x^2}{4}$

Hence $(\text{area})(\bar{y}) = \displaystyle\int_0^4 \left(2\sqrt{x} - \dfrac{x^2}{4}\right)\left[\dfrac{1}{2}\left(2\sqrt{x} - \dfrac{x^2}{4}\right) + \dfrac{x^2}{4}\right] dx$

$$5\tfrac{1}{3}(\bar{y}) = \int_0^4 \left(2\sqrt{x} - \dfrac{x^2}{4}\right)\left(\sqrt{x} + \dfrac{x^2}{8}\right) dx$$

$$\bar{y} = \dfrac{\displaystyle\int_0^4 \left(2x - \dfrac{x^4}{32}\right) dx}{5\tfrac{1}{3}}$$

$$= \dfrac{\left[x^2 - \dfrac{x^5}{5(32)}\right]_0^4}{5\tfrac{1}{3}} = \dfrac{16 - 6\tfrac{2}{5}}{5\tfrac{1}{3}}$$

$$= 1.8$$

Hence **the position of the centroid of the area enclosed by the curves** $x = \dfrac{y^2}{4}$ **and** $y = \dfrac{x^2}{4}$ **is at (1.8, 1.8)**

Thus when finding centroids of areas enclosed by two curves, say $y_1 = f(x)$ and $y_2 = g(x)$, then the formula derived for $\bar{x}$ $\left(\text{i.e.} \dfrac{\int xy\,dx}{\int y\,dx}\right)$ may be used, where y represents the difference between y_1 and y_2. However, the formula derived for $\bar{y}$ $\left(\text{i.e.} \dfrac{\tfrac{1}{2}\int y^2\,dx}{\int y\,dx}\right)$ must not be used. $\bar{y}$ is obtained from first principles as shown above.

Problem 8. Calculate the position of the centroid of a semicircle of radius r: (a) by using the theorem of Pappus, and (b) by using integration (given that the equation of a circle, centre O and radius r, is $x^2 + y^2 = r^2$)

Figure 31.10 shows a semicircle with its diameter lying on the x-axis, with its centre at the origin.

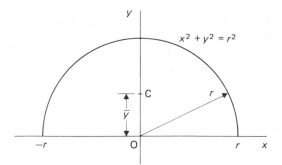

Fig. 31.10

(a) *Using the theorem of Pappus*
If the semicircular area is rotated about the x-axis then the
volume of the solid generated is that of a sphere, i.e. $\frac{4}{3}\pi r^3$. The area
of the semicircle is $\dfrac{\pi r^2}{2}$

The centroid lies on the axis of symmetry OY. Pappus's theorem
states:

volume generated = area × distance moved through by centroid

i.e. $\frac{4}{3}\pi r^3 = \left(\dfrac{\pi r^2}{2}\right)(2\pi\bar{y})$

$$\bar{y} = \dfrac{\frac{4}{3}\pi r^3}{\left(\dfrac{\pi r^2}{2}\right)(2\pi)} = \dfrac{4r}{3\pi}$$

(b) *Using integration*
The curve $x^2 + y^2 = r^2$ is symmetrical about $x = 0$. Hence the
centroid lies on the y-axis, i.e. $\bar{x} = 0$

$$\bar{y} = \dfrac{\dfrac{1}{2}\displaystyle\int_{-r}^{r} y^2\, dx}{\text{area}} = \dfrac{\dfrac{1}{2}\displaystyle\int_{-r}^{r} (r^2 - x^2)\, dx}{\dfrac{\pi r^2}{2}} = \dfrac{\dfrac{1}{2}\left[r^2 x - \dfrac{x^3}{3}\right]_{-r}^{r}}{\dfrac{\pi r^2}{2}}$$

$$= \dfrac{1}{\pi r^2}\left[\left(r^3 - \dfrac{r^3}{3}\right) - \left(-r^3 + \dfrac{r^3}{3}\right)\right] = \dfrac{1}{\pi r^2}\left[2\left(r^3 - \dfrac{r^3}{3}\right)\right]$$

$$= \dfrac{2}{\pi r^2}\left(\tfrac{2}{3}r^3\right) = \dfrac{4r}{3\pi}$$

Hence **the centroid lies on the axis of symmetry at a distance of**
$\dfrac{4r}{3\pi}$ **from the diameter.**

Problem 9. (*a*) Find the area bounded by the curve $y = 3x^2$, the
x-axis and ordinates $x = 0$ and $x = 2$
(*b*) If this area is revolved: (i) about the x-axis, (ii) about the
y-axis, find the volume of the solid produced in each case.
(*c*) Find the position of the centroid of the area: (i) by using
integration, (ii) by using the theorem of Pappus.

(*a*) The relevant area is shown shaded in Fig. 31.11

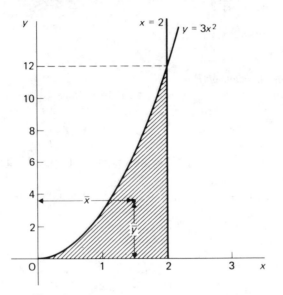

Fig. 31.11

$$\textbf{Area} = \int_0^2 y \, dx = \int_0^2 3x^2 \, dx = [x^3]_0^2 = \textbf{8 square units}$$

(*b*) (i) When revolved about the x-axis, volume of shaded

$$\text{area} = \int_0^2 \pi y^2 \, dx$$

i.e. $\textbf{volume}_{\textbf{O}x} = \pi \int_0^2 9x^4 \, dx = 9\pi \left[\dfrac{x^5}{5} \right]_0^2 = \textbf{57.6}\boldsymbol{\pi}$ **cubic units**

(ii) When revolved about the y-axis the limits are $y = 0$ and
$y = 12$

Since $y = 3x^2$ then $x^2 = \dfrac{y}{3}$

The volume generated by the shaded area when revolved about
the y-axis is given by: (volume generated by $x = 2$) − (volume
generated by $y = 3x^2$)

When a curve is revolved about the y-axis to produce a solid of revolution, the volume produced is given by:

$$\int_{y_1}^{y_2} \pi x^2 \, dy$$

Thus, **volume**$_{OY} = \int_0^{12} \pi(2)^2 \, dy - \int_0^{12} \pi \left(\frac{y}{3}\right) dy$

$$= \pi \int_0^{12} \left[(2)^2 - \frac{y}{3}\right] dy = \pi \left[4y - \frac{y^2}{6}\right]_0^{12}$$

$$= \pi[48 - \tfrac{144}{6}] = \pi[48 - 24] = \mathbf{24\pi \text{ cubic units}}$$

(c) (i) Using integration:

$$\bar{x} = \frac{\int_0^2 xy \, dx}{\int_0^2 y \, dx} = \frac{\int_0^2 3x^3 \, dx}{8} = \frac{\left[\dfrac{3x^4}{4}\right]_0^2}{8}$$

$$= \frac{(3)\frac{16}{4}}{8} = \mathbf{1.5 \text{ units}}$$

$$\bar{y} = \frac{\dfrac{1}{2}\int_0^2 y^2 \, dx}{\int_0^2 y \, dx} = \frac{\dfrac{1}{2}\int_0^2 9x^4 \, dx}{8}$$

$$= \frac{\dfrac{9}{2}\left[\dfrac{x^5}{5}\right]_0^2}{8} = \frac{9}{16}\left(\frac{32}{5}\right) = \mathbf{3.6 \text{ units}}$$

(ii) Using the theorem of Pappus:
Volume generated when revolved about OY $= (\text{area})(2\pi\bar{x})$

$$24\pi = (8)(2\pi\bar{x})$$

$$\bar{x} = \frac{24\pi}{8(2\pi)} = \mathbf{1.5 \text{ units}}$$

Volume generated when revolved about OX $= (\text{area})(2\pi\bar{y})$

$$57.6\pi = (8)(2\pi\bar{y})$$

$$\bar{y} = \frac{57.6\pi}{8(2\pi)} = \mathbf{3.6 \text{ units}}$$

Hence **the centroid of the shaded area is at (1.5, 3.6)**

Further problems on centroids may be found in the following Section 31.2 (Problems 1 to 24).

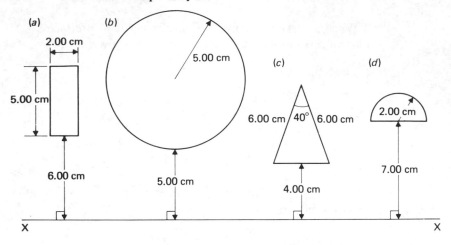

Fig. 31.12

31.2   Further problems

1. Find the first moment of area about the axis XX for the
    shapes shown in Fig. 31.12
    (a) [85 cm³]   (b) [250π cm³]   (c) [68.0 cm³]   (d) [49.3 cm³]

In Problems 2 to 6 find the positions of the centroids of the areas
bounded by the given curves, the x-axis and the given ordinates.

2. $y = 3x$; $x = 0$, $x = 4$ $[(2\frac{3}{4}, 4)]$
3. $y = 2x + 1$; $x = 0$, $x = 2$ $[(1\frac{2}{9}, 1\frac{13}{18})]$
4. $y = 4x^2$; $x = 1$, $x = 3$ $[(2.308, 11.17)]$
5. $y = x^3$; $x = 0$, $x = 1$ $[(\frac{4}{5}, \frac{2}{7})]$
6. $y = 2x(x + 1)$; $x = -1$, $x = 0$ $[(-\frac{1}{2}, -\frac{1}{5})]$
7. Calculate the position of the centroid of the sheet of metal
    formed by the x-axis and the part of the curve $y = 5x - x^2$
    which lies above the x-axis. $[(2\frac{1}{2}, 2\frac{1}{2})]$
8. Find the coordinates of the centroid of the plate which lies
    between the curve $\frac{y}{x} = x - 3$ and the x-axis. $[(1.5, -0.9)]$
9. Calculate the position of the centroid of the area of the
    metal plate bounded by the axes and the part of the curve
    $y = 4 - x^2$ which lies in the first quadrant. $[(0.75, 1.60)]$
10. A portion of the curve $y = 16 - x^2$ lies above the x-axis.
    Calculate the coordinates of the centroid of the area formed
    by the curve and the x-axis. $[(0, 6.4)]$

11. Find the position of the centroid of the area lying between $y = 3x^2$, the y-axis and the ordinates $y = 0$ and $y = 4$ [(0.433, 2.4)]

12. Determine the position of the centroid of the area enclosed by the curve $y = 3\sqrt{x}$, the x-axis and the ordinate $x = 4$ [(2.4, 2.25)]

13. Sketch the area enclosed by the curve $\dfrac{y}{\sqrt{3}} = \sqrt{x}$, the y-axis and the ordinate $y = 5$. Calculate the coordinates of the centroid of this area.   [(2.5, 3.75)]

14. Sketch the curve $y^2 = 4x$ between limits of $x = 0$ and $x = 5$. Find the position of the centroid of this area.   [(3, 0)]

15. Determine the position of the centroid of a metal template in the shape of a quadrant of a circle of radius r, given that the equation of a circle, centre O and radius r, is $x^2 + y^2 = r^2$. Check your answers by using the theorem of Pappus.
[On the centre line, distance $\dfrac{4\sqrt{2}}{3\pi}$ (i.e. 0.6r) from the centre]

16. Find the coordinates of the centroid of the wooden lamina enclosed by the curve $y = 3 - \sqrt{x}$, the x-axis and the ordinate $x = 4$ [(1.68, 0.9)]

17. Find the points of intersection of the curves $y = 2x^2$ and $y = 4x$. Determine the coordinates of the centroid of the area enclosed by the two curves.   [(0, 0), (2, 8); (1, 3.2)]

18. Determine the position of the centre of area of the metal plate enclosed between the curves $x^2 = y$ and $y^2 = x$ [(0.45, 0.45)]

19. Calculate the points of intersection of the curves $\dfrac{x^2}{2} = y$ and $y^2 = 2x$ and find the position of the centroid of the area enclosed by them.   [(0, 0), (2, 2); (0.9, 0.9)]

20. Find the area of the sheet of thin cardboard bounded by the curve $y = 4x^2$ and the x-axis, between the limits $x = 0$ and $x = 3$. If this area is revolved about: (a) the x-axis; and (b) the y-axis, find the volume of the solid of revolution produced in each case. Determine the coordinates of the centroid of the cardboard: (i) by using integration; (ii) by using the theorem of Pappus.
[36 square units; (a) 777.6π cubic units; (b) 162π cubic units; (2.25, 10.8)]

21. Sketch the loop $y^2 = x(3 - x)^2$ which is the shape of a rudder. Calculate the coordinates of the centroid of the area enclosed by the loop.   [(1.286, 0)]

22. Determine the position of the centroid of the sheet of paper enclosed by the curve $y = 3x^2 + 2$, the y-axis and ordinates $y = 2$ and $y = 5$ [(0.375, 3.80)]

23. Sketch the part of the curve $(x - 1)(4 - x)$ which lies above the x-axis. Find the area enclosed by the curve and the x-axis. If the area is revolved completely about the x-axis find the volume generated. Determine the position of the centroid of the area.
    $[4\frac{1}{2}$ square units; 8.1π cubic units; (2.5, 0.9)]

24. Sketch the curves $y = 3x^2 + 2x - 1$ and $y + 5 = x(2x + 7)$ and determine the points of intersection. Calculate the coordinates of the centroid of the area enclosed by the two curves. $[(1, 4), (4, 55); (2\frac{1}{2}, 25\frac{4}{7})]$

Chapter 32

Second moments of areas of regular sections

32.1 Second moment of area

Consider three small bodies rotating at an angular velocity ω rad/s around an axis DD at distances r_1, r_2 and r_3 as shown in Fig. 32.1(a). Let the bodies have cross-sectional areas a_1, a_2 and a_3, and corresponding masses of m_1, m_2 and m_3, moving with tangential velocities v_1, v_2 and v_3 respectively as shown.

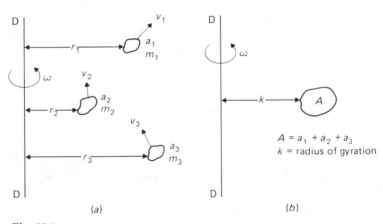

Fig. 32.1

The kinetic energy of the system $= \frac{1}{2}m_1v_1^2 + \frac{1}{2}m_2v_2^2 + \frac{1}{2}m_3v_3^2$

Since $v = \omega r$, then $v_1 = \omega r_1$, $v_2 = \omega r_2$ and $v_3 = \omega r_3$

Hence the kinetic energy $= \frac{1}{2}m_1(\omega r_1)^2 + \frac{1}{2}m_2(\omega r_2)^2 + \frac{1}{2}m_3(\omega r_3)^2$

$$= \frac{1}{2}\omega^2(m_1r_1^2 + m_2r_2^2 + m_3r_3^2)$$

If such a system contained many bodies then, in general,

kinetic energy $= \frac{1}{2}\omega^2 \Sigma mr^2$

The expression Σmr^2 is known as the **moment of inertia** and is defined as the second moment of the total mass about axis DD of the system. It is a measure of the amount of work done to give the system an angular velocity of ω rad/s, or the amount of work which can be done by a system

535

turning at ω rad/s. When mass is proportional to area then the moment of inertia (i.e. Σmr^2) becomes Σar^2. The term Σar^2 is called the **second moment of area** and it is a quantity much used, particularly in mechanical engineering, for example, in the theory of bending of beams (where the beam equation is $\dfrac{M}{I} = \dfrac{\sigma}{y} = \dfrac{E}{R}$, and I denotes the second moment of area) and in torsion of shafts, and also in naval architecture, for example, in calculations involving water planes and centres of pressure. Whereas the first moment of area is given by Σar, the second moment of area is given by Σar^2 (i.e. the distance is '*squared*' when finding the '*second*' moment).

If the areas a_1, a_2 and a_3 of Fig. 32.1(a) are replaced by a single area A, such that $A = a_1 + a_2 + a_3$, at a distance k from axis DD such that $Ak^2 = \Sigma ar^2$, then k is called the **radius of gyration** of area A about axis DD (see Fig. 32.1(b)).

In order to find the second moment of area of regular sections about a given axis, the second moment of area of a typical element is first determined and then the sum of all such second moment of areas found by integrating between appropriate limits. This is shown in the following derivations for rectangles, triangles, circles and semicircles.

(i)   *Second moment of area of a rectangle*

Consider the rectangle shown in Fig. 32.2(a), of length l and breadth b.

(a) *To find the second moment of area of the rectangle about axis DD:*
Consider an elemental strip, width δx, parallel to, and distance x from, axis DD as shown.

Area of elemental strip $= b\delta x$

Second moment of area of strip about axis DD $= x^2(b\delta x)$

Hence the total second moment of area of the rectangle about axis DD

$$= \sum_{x=0}^{x=l} x^2 b\delta x$$

$$= \int_0^l x^2 b \, dx = \left[\frac{x^3}{3}\right]_0^l = \frac{bl^3}{3}$$

But the total area A of the rectangle $= lb$

Hence **the second moment of area about axis DD** $= lb\left(\dfrac{l^2}{3}\right)$

$$= A\frac{l^2}{3}$$

(a)

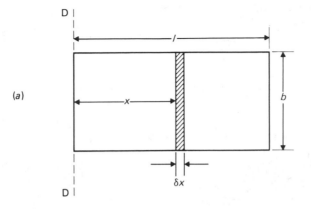

(b)

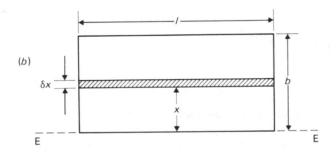

(c)

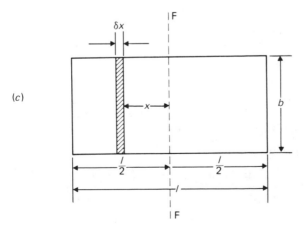

Fig. 32.2

Since the second moment of area $= Ak_{DD}{}^2$

$$k_{DD}{}^2 = \frac{l^2}{3}$$

i.e. **the radius of gyration about axis DD, $k_{DD} = \dfrac{l}{\sqrt{3}}$**

(b) *To find the second moment of area of the rectangle about axis EE:*
Consider an elemental strip, width δx, parallel to, and distance x from axis EE as shown in Fig. 32.2(b).

Area of elemental strip $= l\delta x$

Second moment of area of strip about EE $= x^2(l\delta x)$

Hence the total second moment of area of the rectangle about axis EE

$$= \sum_{x=0}^{x=b} x^2 l\delta x = \int_0^b x^2 l \, dx$$

$$= l\left[\frac{x^3}{3}\right]_0^b = \frac{lb^3}{3}$$

But the total area of the rectangle $= lb$

Hence **the second moment of area about axis EE $= lb\left[\dfrac{b^2}{3}\right]$**

$$= A\frac{b^2}{3}$$

Since $Ak_{EE}{}^2 = A\dfrac{b^2}{3}$

then $\quad k_{EE}{}^2 = \dfrac{b^2}{3}$ and $k_{EE} = \dfrac{b}{\sqrt{3}}$

(c) *To find the second moment of area of the rectangle about axis FF:*
FF is an axis passing through the centre of the rectangle parallel to the breadth.
Consider, again, an elemental strip, width δx, parallel to, and distance x from axis FF as shown in Fig. 32.2(c).

Area of elemental strip $= b\delta x$

Second moment of area of the strip about FF $= x^2(b\delta x)$

Hence the total second moment of area of the rectangle about axis FF

$$= \sum_{x=-l/2}^{x=l/2} x^2 b\delta x = \int_{-l/2}^{l/2} x^2 b\delta x$$

The only difference between this case and that about axis DD in case (a) is in the limits of integration.

$$\int_{-l/2}^{l/2} x^2 b \, dx = b\left[\frac{x^3}{3}\right]_{-l/2}^{l/2} = b\left[\left(\frac{l^3}{24}\right) - \left(-\frac{l^3}{24}\right)\right] = \frac{bl^3}{12}$$

Hence **the second moment of area of the rectangle about FF** $= A\dfrac{l^2}{12}$

$$(\text{radius of gyration})^2 = \frac{\text{second moment of area}}{\text{area}}$$

i.e.
$$k_{FF}^2 = \frac{A\dfrac{l^2}{12}}{A} = \frac{l^2}{12}$$

$$k_{FF} = \frac{l}{\sqrt{12}} \text{ or } \frac{l}{2\sqrt{3}}$$

(ii)   *Second moment of area of a triangle about an edge*

Consider triangle DEF shown in Fig. 32.3(a). Let the edge EF be the base, of length b, and let the perpendicular height be h.

To find the second moment of area of the triangle about edge EF:

Consider an elemental strip, width δx, parallel to and distance x from EF as shown in Fig. 32.3(a). Let the length of the strip be y.

Area of strip $= y\delta x$

Second moment of area of the strip about EF $= x^2(y\delta x)$

Second moment of area of triangle DEF about EF $= \displaystyle\sum_{x=0}^{x=h} x^2 y \delta x$

$$= \int_0^h x^2 y \, dx$$

In order to obtain an expression for y in terms of x, similar triangles are used. From Fig. 32.3(b) it may be deduced that:

$$\frac{h-x}{y} = \frac{h}{b},$$

from which $y = \dfrac{b}{h}(h-x)$

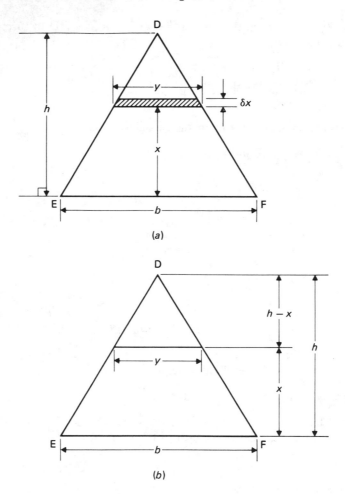

Fig. 32.3

Thus $\displaystyle\int_0^h x^2 y \, \mathrm{d}x = \int_0^h x^2 \frac{b}{h}(h-x)\,\mathrm{d}x$

$\qquad\qquad\qquad = \dfrac{b}{h}\displaystyle\int_0^h (x^2 h - x^3)\,\mathrm{d}x$

$\qquad\qquad\qquad = \dfrac{b}{h}\left[\dfrac{x^3}{3}h - \dfrac{x^4}{4}\right]_0^h = \dfrac{bh^4}{12h} = \dfrac{\boldsymbol{bh^3}}{\boldsymbol{12}}$

The area A of triangle DEF is given by $\frac{1}{2}bh$

Hence **the second moment of area of the triangle about EF**

$$= (\tfrac{1}{2}bh)\left(\frac{h^2}{6}\right) = A\left(\frac{h^2}{6}\right)$$

Hence $k_{EF}^2 = \dfrac{h^2}{6}$ and $\boldsymbol{k_{EF}} = \dfrac{\boldsymbol{h}}{\boldsymbol{\sqrt{6}}}$

(*iii*)   *Second moment of area of a circle, of radius r, about its polar axis*

A polar axis is an axis through the centre of the circle perpendicular to the plane of the circle. This is shown as axis BB in Fig. 32.4(*a*) and a plan view of the circle is shown in Fig. 32.4(*b*).

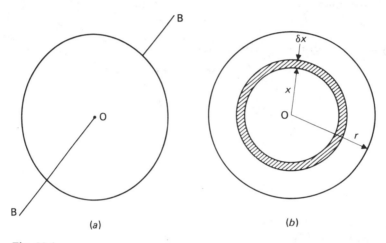

(*a*)   (*b*)

Fig. 32.4

Consider an elemental annulus of width δx (shown shaded in Fig. 32.4(*b*)) and radius x from the centre of the circle (i.e. distance x from the polar axis BB). If δx is very small then the area of the elemental annulus equals the circumference times the width.

i.e.   area of annulus = $(2\pi x)(\delta x)$

The second moment of area of the annulus about the centre of the circle

$$= (x^2)(2\pi x \delta x)$$

The total second moment of area of the circle about an axis through its centre perpendicular to the plane

$$= \sum_{x=0}^{x=r} x^2 (2\pi x \delta x)$$

$$= \int_0^r 2\pi x^3 \, dx = 2\pi \left[\frac{x^4}{4} \right]_0^r = \frac{\pi r^4}{2}$$

But the total area A of the circle is given by πr^2

Hence **the second moment of area of a circle about its polar axis**

$$= (\pi r^2)\left(\frac{r^2}{2}\right) = A\,\frac{r^2}{2}$$

Hence $k_{BB}{}^2 = \dfrac{r^2}{2}$ and $k_{BB} = \dfrac{r}{\sqrt{2}}$

There are two important theorems associated with second moments of areas which enable the second moment of area and the radius of gyration to be obtained about axes other than those used above.

32.2   Parallel-axis theorem

Let C be the centroid of any area A as shown in Fig. 32.5

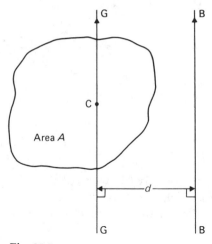

Fig. 32.5

Let the second moment of area of A about an axis GG, passing through the centroid, be denoted by $Ak_{GG}{}^2$. Let BB be an axis parallel to axis GG and at perpendicular distance d from it. Axes GG and BB are in the same plane as area A. The parallel-axis theorem states:

$$Ak_{BB}{}^2 = Ak_{GG}{}^2 + Ad^2$$

or $\qquad k_{BB}{}^2 = k_{GG}{}^2 + d^2$

As an example, this may be shown to be true from the results obtained for the rectangle of length l, breadth b and area A.

The second moment of area of a rectangle about an axis through its centroid, parallel to the breadth (i.e. about axis FF of Fig. 32.2(c)), is given by $A \dfrac{l^2}{12}$

Let the second moment of area of the rectangle about an axis coinciding with the breadth (i.e. axis DD of Fig. 32.2(a)) be denoted by $Ak_{DD}{}^2$. The distance between the parallel axes FF and DD is $\dfrac{l}{2}$. Hence, by the parallel-axis theorem:

$$Ak_{DD}{}^2 = Ak_{FF}{}^2 + Ad^2$$

$$= A\frac{l^2}{12} + A\left(\frac{l}{2}\right)^2$$

$$= A\left[\frac{l^2}{12} + \frac{l^2}{4}\right] = A\frac{l^2}{3}$$

This, of course, is the result previously obtained on page 536.

Second moment of area of a triangle about an axis through its centroid, parallel to the base, using the parallel-axis theorem

It was shown earlier that the second moment of area of a triangle about its base is $A\dfrac{h^2}{6}$, where h is the perpendicular height of the triangle. The centroid of a triangle is situated at a distance $\dfrac{h}{3}$ from the base.

Let the second moment of area of a triangle about an axis through its centroid, parallel to the base, be denoted by $Ak_{CC}{}^2$

Thus, using the parallel-axis theorem:

$$A\frac{h^2}{6} = Ak_{CC}{}^2 + A\left(\frac{h}{3}\right)^2$$

i.e. $Ak_{CC}{}^2 = A\dfrac{h^2}{6} - A\dfrac{h^2}{9} = A\dfrac{h^2}{18}$

Since $A = \frac{1}{2}bh$ then $Ak_{CC}{}^2 = (\frac{1}{2}bh)\dfrac{h^2}{18}$ or $\dfrac{bh^3}{36}$

Hence $k_{CC}{}^2 = \dfrac{h^2}{18}$ and $k_{CC} = \dfrac{h}{\sqrt{18}}$ or $\dfrac{h}{3\sqrt{2}}$

Second moment of area of a triangle about an axis through its vertex, parallel with the base, using the parallel-axis theorem

If an axis is drawn through the vertex D of Fig. 32.3(*a*), parallel to base EF, then:

$$Ak_D{}^2 = Ak_{CC}{}^2 + Ad^2$$

where d = distance between the centroid and the vertex of the triangle, i.e. $d = \frac{2}{3}h$.

Hence $$Ak_D{}^2 = A\,\frac{h^2}{18} + A(\tfrac{2}{3}h)^2$$

$$= A\left[\frac{h^2}{18} + \frac{4}{9}h^2\right] = A\,\frac{h^2}{3} \text{ or } \frac{bh^3}{4}$$

Thus $$k_D{}^2 = \frac{h^2}{2} \text{ and } k_D = \frac{h}{\sqrt{2}}$$

32.3   Perpendicular-axis theorem

Consider a plane area A having three mutually perpendicular axes OX, OY and OZ, as shown in Fig. 32.6, with axes OX and OY lying in the plane of area A.

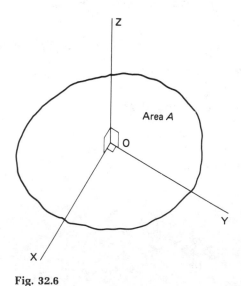

Area A

Fig. 32.6

The perpendicular axis-theorem states:

$$Ak_{OZ}{}^2 = Ak_{OX}{}^2 + Ak_{OY}{}^2$$

or $$k_{OZ}{}^2 = k_{OX}{}^2 + k_{OY}{}^2$$

The second moment of area of a circle about a diameter,
using the perpendicular-axis theorem

Consider a circle of radius r having three mutually perpendicular axes OX, OY and ZZ as shown in Fig. 32.7

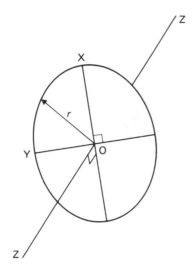

Fig. 32.7

It was shown earlier that the second moment of area of a circle about its polar axis (i.e. axis ZZ of Fig. 32.7) is $A\dfrac{r^2}{2}$

By symmetry, the second moment of area of the circle about axis OX will equal that about axis OY,

i.e. $Ak_{OX}^2 = Ak_{OY}^2$

By the perpendicular-axis theorem:

$$Ak_{ZZ}^2 = Ak_{OX}^2 + Ak_{OY}^2$$
$$= 2Ak_{OX}^2$$

i.e. $A\dfrac{r^2}{2} = 2Ak_{OX}^2$

$$Ak_{OX}^2 = A\dfrac{r^2}{4}$$

Hence **the second moment of area of a circle about a diameter** $= A\dfrac{r^2}{4}$ or $\dfrac{\pi r^4}{4}$

Thus $k_{OX}^2 = \dfrac{r^2}{4}$ and $k_{OX} = \dfrac{r}{2}$

Second moment of area of a circle about a tangent using the parallel-axis theorem

Consider a circle of radius r having a tangent BB parallel to diameter XX as shown in Fig. 32.8

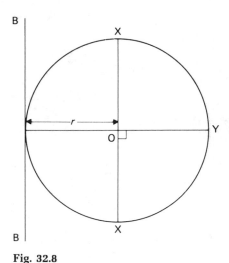

Fig. 32.8

By the parallel-axis theorem:

$$Ak_{BB}^2 = Ak_{XX}^2 + Ar^2$$

But $Ak_{XX}^2 = A\dfrac{r^2}{4}$

Hence $Ak_{BB}^2 = A\dfrac{r^2}{4} + Ar^2 = A\left(\dfrac{r^2}{4} + r^2\right)$

Thus **the second moment of area of a circle about a tangent** $= \frac{5}{4}Ar^2$

$$k_{BB}^2 = \tfrac{5}{4}r^2 \text{ and } k_{BB} = \dfrac{\sqrt{5}}{2}r$$

The second moment of area of a semicircle about its diameter

The second moment of area of a semicircle about its diameter is one-half of that for a circle about its diameter,

i.e. $\dfrac{1}{2}\left(\dfrac{\pi r^4}{4}\right)$ or $\dfrac{\pi r^3}{8}$

An alternative method of finding the second moment of area of a circle about a diameter is to take an elemental strip of width δx, distance x from axis OY, as shown in Fig. 32.9

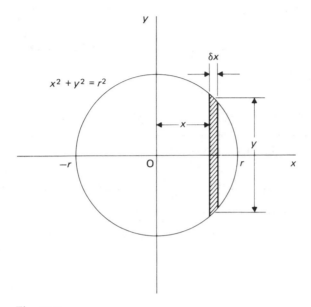

Fig. 32.9

The second moment of area of the strip about OY $= x^2 y \delta x$

The second moment of area of the circle about OY

$$= \sum_{x=-r}^{x=r} x^2 y \delta x = \int_{-r}^{r} x^2 y \, dx \qquad (1)$$

If now the same procedure is adopted for a semicircle, the second moment of area of the semicircle about axis OY

$$= \sum_{x=0}^{x=r} x^2 y \delta x = \int_{0}^{r} x^2 y \, dx \qquad (2)$$

(Note that the evaluation of the integrals in equations (1) and (2) is too advanced for this text.)

Since $\displaystyle\int_{-r}^{r} x^2 y \, dx = 2 \int_{0}^{r} x^2 y \, dx$ it follows that the second moment of area

of a semicircle about its diameter is half of that for a circle about its diameter,

i.e. $\dfrac{\pi r^4}{8}$

Since area A of a semicircle $= \dfrac{\pi r^2}{2}$, the second moment of area about the

diameter $= \left(\dfrac{\pi r^2}{2}\right)\left(\dfrac{r^2}{4}\right)$ or $A\,\dfrac{r^2}{4}$

Thus **the radius of gyration k for a semicircle about its diameter is $\dfrac{r}{2}$**, the same value as for the circle.

32.4   Summary of derived standard results

Shape	Position of axis	Second moment of area	Radius of gyration, k
Rectangle length l breadth b area A	1. Coinciding with b	$\dfrac{bl^3}{3}$ or $A\,\dfrac{l^2}{3}$	$\dfrac{l}{\sqrt{3}}$
	2. Coinciding with l	$\dfrac{lb^3}{3}$ or $A\,\dfrac{b^2}{3}$	$\dfrac{b}{\sqrt{3}}$
	3. Through centroid, parallel to b	$\dfrac{bl^3}{12}$ or $A\,\dfrac{l^2}{12}$	$\dfrac{l}{\sqrt{12}}$ or $\dfrac{l}{2\sqrt{3}}$
Triangle perpendicular height h base b area A	1. Coinciding with base	$\dfrac{bh^3}{12}$ or $A\,\dfrac{h^2}{6}$	$\dfrac{h}{\sqrt{6}}$
	2. Through centroid, parallel to base	$\dfrac{bh^3}{36}$ or $A\,\dfrac{h^2}{18}$	$\dfrac{h}{\sqrt{18}}$ or $\dfrac{h}{3\sqrt{2}}$
	3. Through vertex, parallel to base	$\dfrac{bh^3}{4}$ or $A\,\dfrac{h^2}{2}$	$\dfrac{h}{\sqrt{2}}$
Circle radius r area A	1. Through centre, perpendicular to plane (i.e. polar axis)	$\dfrac{\pi r^4}{2}$ or $A\,\dfrac{r^2}{2}$	$\dfrac{r}{\sqrt{2}}$
	2. Coinciding with diameter	$\dfrac{\pi r^4}{4}$ or $A\,\dfrac{r^2}{4}$	$\dfrac{r}{2}$
	3. About a tangent	$\dfrac{5\pi}{4}r^4$ or $\dfrac{5}{4}Ar^2$	$\dfrac{\sqrt{5}}{2}r$
Semicircle radius r area A	Coinciding with diameter	$\dfrac{\pi r^4}{8}$ or $A\,\dfrac{r^2}{4}$	$\dfrac{r}{2}$

Worked problems on second moments of area

Problem 1. For each of the shapes shown in Fig. 32.10 find the second moment of area and the radius of gyration about the given axes.

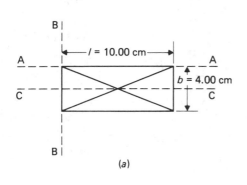

(a)

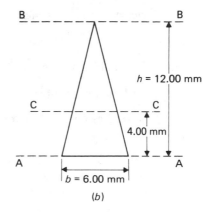

(b)

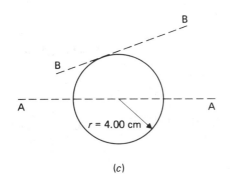

(c)

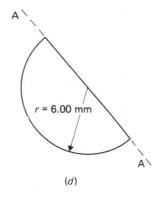

(d)

Fig. 32.10

(a) For the rectangle:

Second moment of area about AA $= \dfrac{lb^3}{3} = \dfrac{10.00(4.00)^3}{3} = \mathbf{213.3\ cm^4}$

Radius of gyration, $k_{AA} = \dfrac{b}{\sqrt{3}} = \dfrac{4.00}{\sqrt{3}} = \mathbf{2.309\ cm}$

Second moment of area about BB $= \dfrac{bl^3}{3} = \dfrac{4.00(10.00)^3}{3} = \mathbf{1\,333\ cm^4}$

Radius of gyration, $k_{BB} = \dfrac{l}{\sqrt{3}} = \dfrac{10.00}{\sqrt{3}} = \mathbf{5.774\ cm}$

The second moment of area about the centroid of a rectangle is $\dfrac{bl^3}{12}$ when the axis through the centroid is parallel with the breadth. In this case, the axis through the centroid is parallel with the length.

Hence the second moment of area about $CC = \dfrac{lb^3}{12} = \dfrac{10.00(4.00)^3}{12}$

$$= 53.33 \text{ cm}^4$$

Radius of gyration, $k_{CC} = \dfrac{b}{2\sqrt{3}} = \dfrac{4.00}{2\sqrt{3}} = 1.155 \text{ cm}$

(b) For the triangle:

Second moment of area about $AA = \dfrac{bh^3}{12}$

$$= \dfrac{(6.00)(12.00)^3}{12} = 864 \text{ mm}^4$$

Radius of gyration, $k_{AA} = \dfrac{h}{\sqrt{6}} = \dfrac{12.00}{\sqrt{6}} = 4.899 \text{ mm}$

Second moment of area about $BB = \dfrac{bh^3}{4}$

$$= \dfrac{6.00(12.00)^3}{4} = 2\,592 \text{ mm}^4$$

Radius of gyration, $k_{BB} = \dfrac{h}{\sqrt{2}} = \dfrac{12.00}{\sqrt{2}} = 8.485 \text{ mm}$

Second moment of area about CC (i.e. through centroid) $= \dfrac{bh^3}{36}$

$$= \dfrac{6.00(12.00)^3}{36} = 288.0 \text{ mm}^4$$

Radius of gyration, $k_{CC} = \dfrac{h}{3\sqrt{2}} = \dfrac{12.00}{3\sqrt{2}} = 2.828 \text{ mm}$

(c) For the circle:

Second moment of area about AA (i.e. diameter) $= \dfrac{\pi r^4}{4} = \dfrac{\pi(4.00)^4}{4}$

$$= 201.1 \text{ cm}^4$$

Radius of gyration, $k_{AA} = \dfrac{r}{2} = \dfrac{4.00}{2} = 2.000 \text{ cm}$

Second moment of area about BB (i.e. tangent) $= \dfrac{5}{4}\pi r^4 = \dfrac{5}{4}\pi(4.00)^4$

$$= 1\,005 \text{ cm}^4$$

Radius of gyration, $k_{BB} = \dfrac{\sqrt{5}}{2}r = \dfrac{\sqrt{5}}{2}(4.00) = 4.472 \text{ cm}$

(*d*) For the semicircle:

Second moment of area about AA (i.e. diameter) $= \dfrac{\pi r^4}{8} = \dfrac{\pi(6.00)^4}{8}$

$$= 508.9 \text{ mm}^4$$

Radius of gyration, $k_{\text{AA}} = \dfrac{r}{2} = \dfrac{6.00}{2} = 3.000 \text{ mm}$

Problem 2. Find the second moment of area of the rectangle of length *l* and breadth *b* shown in Fig. 32.11 about axis YY.

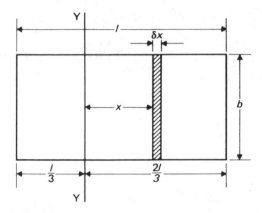

Fig. 32.11

There are two methods of finding the second moment of area about axis YY.

Method 1. From first principles
Consider an elemental strip, width δx, parallel to and at distance *x* from axis YY as shown in Fig. 32.11

Area of strip $= b\delta x$

Second moment of area of strip about YY $= x^2(b\delta x)$

Second moment of area of rectangle about YY $= \displaystyle\sum_{x=-l/3}^{x=2l/3} x^2 b\delta x$

$$= \int_{-l/3}^{2l/3} x^2 b \; dx$$

$$= b\left[\frac{x^3}{3}\right]_{-l/3}^{2l/3}$$

$$= b \left[\frac{\left(\frac{2l}{3}\right)^3}{3} - \frac{\left(-\frac{l}{3}\right)^3}{3} \right]$$

$$= b \left[\frac{8l^3}{81} + \frac{l^3}{81} \right]$$

$$= \frac{bl^3}{9} \left(= A \frac{l^2}{9} \right)$$

Method 2. Using the parallel-axis theorem
The parallel-axis theorem states:

$$Ak_{YY}^2 = Ak_{CC}^2 + Ad^2$$

where $Ak_{CC}^2 =$ second moment of area about the centroid, parallel with b, $= \frac{bl^3}{12}$, and $d =$ perpendicular distance between axis YY and the centroid, i.e.

$$\frac{l}{2} - \frac{l}{3}, \text{ or } \frac{l}{6}$$

Hence $Ak_{YY}^2 = \frac{bl^3}{12} + bl\left(\frac{l}{6}\right)^2$

$$= \frac{bl^3}{12} + \frac{bl^3}{36} = \frac{bl^3}{9}$$

Problem 3. Find the second moment of area and the radius of gyration about axis XX for each of the shapes shown in Fig. 32.12. The broken lines indicate an axis through the centroid of the shape parallel to axis XX.

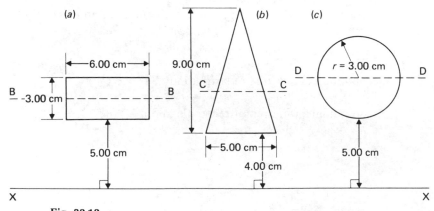

Fig. 32.12

(*a*) For the rectangle:

Second moment of area about BB $= \dfrac{bl^3}{12}$ (where $l = 3.00$ cm, i.e.

the length of the side at right angles to axis BB, and $b = 6.00$ cm)

$$= \frac{(6.00)(3.00)^3}{12}$$

$$= 13.50 \text{ cm}^4$$

Using the parallel-axis theorem:

$$Ak_{XX}^2 = Ak_{BB}^2 + Ad^2$$

where d = distance between BB and XX $= 5.00 + 1.50 = 6.50$ cm

Hence the second moment of area of the rectangle about XX,

$$Ak_{XX}^2 = 13.50 + (18.00)(6.50)^2$$

$$= 13.50 + 760.5 = \textbf{774.0 cm}^4$$

$$k_{XX}^2 = \frac{774.0}{\text{area}} = \frac{774.0}{18.00} = 43.00$$

Hence radius of gyration, $k_{XX} = \sqrt{43.00} = \textbf{6.557 cm}$

(*b*) For the triangle:

The second moment of area about CC $= \dfrac{bh^3}{36} = \dfrac{(5.00)(9.00)^3}{36}$

$$= 101.3 \text{ cm}^4$$

Using the parallel-axis theorem:

$$Ak_{XX}^2 = Ak_{CC}^2 + Ad^2$$

where $d = 4.00 + \frac{1}{3}(9.00) = 7.00$ cm

Hence the second moment of area of the triangle about XX,

$$Ak_{XX}^2 = 101.3 + [\tfrac{1}{2}(5.00)(9.00)](7.00)^2$$

$$= 101.3 + 1\,102.5 = \textbf{1 204 cm}^4$$

$$k_{XX}^2 = \frac{1\,204}{\text{area}} = \frac{1\,204}{\frac{1}{2}(5.00)(9.00)} = 53.51$$

Hence radius of gyration, $k_{XX} = \sqrt{53.51} = \textbf{7.315 cm}$

(*c*) For the circle:

Second moment of area about DD $= \dfrac{\pi r^4}{4} = \dfrac{\pi(3.00)^4}{4} = 63.62 \text{ cm}^4$

Using the parallel-axis theorem:

$$Ak_{XX}^2 = Ak_{DD}^2 + Ad^2$$

where $d = 5.00 + 3.00 = 8.00$ cm

Hence the second moment of area of the circle about XX,

$Ak_{XX}^2 = 63.62 + [\pi(3.00)^2][8.00]^2$

$\qquad = 63.62 + 1\,810 = \mathbf{1\,874\ cm^4}$

$k_{XX}^2 = \dfrac{1\,874}{\text{area}} = \dfrac{1\,874}{\pi(3.00)^2} = 66.28$

Hence radius of gyration, $k_{XX} = \sqrt{66.28} = \mathbf{8.141\ cm}$

Problem 4. Calculate the second moment of area and the square of the radius of gyration of a rectangular lamina of length 56.0 cm and width 22.0 cm about an axis through one corner, perpendicular to the plane of the lamina.

Figure 32.13 shows the rectangle, having three mutually perpendicular axes at one corner.

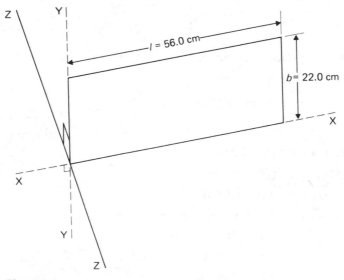

Fig. 32.13

The perpendicular-axis theorem states:

$\qquad Ak_{ZZ}^2 = Ak_{XX}^2 + Ak_{YY}^2$

or $k_{ZZ}^2 = k_{XX}^2 + k_{YY}^2$

$k_{XX}^2 = \dfrac{b^2}{3} = \dfrac{(22.0)^2}{3} = 161.3\ cm^2$

$k_{YY}^2 = \dfrac{l^2}{3} = \dfrac{(56.0)^2}{3} = 1\,045\ cm^2$

Hence $k_{ZZ}^2 = 161.3 + 1\,045 = \mathbf{1206\ cm^2}$

$Ak_{ZZ}^2 = (56.0 \times 22.0)(1\,206) = \mathbf{1\,486\,000\ cm^4}$ or $\mathbf{1.486 \times 10^6\ cm^4}$

Problem 5. A circular lamina, centre O, has a radius of 12.00 cm. A hole of radius 4.00 cm and centre A, where OA = 5.00 cm, is cut in the lamina. Find the second moment of area and the radius of gyration of the remainder about a diameter through O perpendicular to OA.

The circular lamina is shown in Fig. 32.14

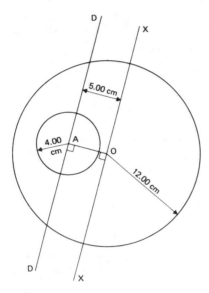

Fig. 32.14

For the 12.00-cm-radius circle:

Second moment of area about XX (i.e. diameter) $= \dfrac{\pi r^4}{4} = \dfrac{\pi (12.00)^4}{4}$

$$= 16\,290\ cm^4$$

For the 4.00-cm-radius circle:

Second moment of area about diameter (i.e. axis DD), parallel to XX,

$$= \frac{\pi r^4}{4} = \frac{\pi (4.00)^4}{4}$$

$$= 201.1\ cm^4$$

Using the parallel-axis theorem, the second moment of area of the smaller circle about axis XX (denoted by Ak_{XX}^2) is given by:

$$Ak_{XX}^2 = Ak_{DD}^2 + \text{(area of smaller circle)(perpendicular distance}$$
$$\text{between DD and XX)}^2$$

$$= 201.1 + [\pi(4.00)^2][5.00]^2$$

$$= 1\,458 \text{ cm}^4$$

Hence **the second moment of area of the lamina remaining about its diameter XX is** $16\,290 - 1458$, i.e. **14 832 cm^4**

(Note that the second moment of area of the smaller circle is subtracted since that area has been removed from the lamina.)

Now $14\,832 = Ak_{XX}^2$

Area of lamina, $A = \pi(12.00)^2 - \pi(4.00)^2 = 402.1 \text{ cm}^2$

Hence $k_{XX}^2 = \dfrac{14\,832}{\text{area}} = \dfrac{14\,832}{402.1}$

and **radius of gyration about XX**, $k_{XX} = \sqrt{\left(\dfrac{14\,832}{402.1}\right)}$

$$= 6.073 \text{ cm}$$

Problem 6. Determine the second moment of area about axis XX for the composite area shown in Fig. 32.15

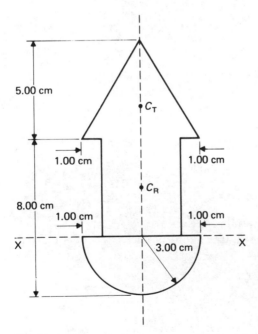

Fig. 32.15

The composite area of Fig. 32.15 comprises:

a triangle, perpendicular height 5.00 cm, base 6.00 cm, centroid C_T,

a rectangle, length 5.00 cm, breadth 4.00 cm, centroid C_R and

a semicircle, radius 3.00 cm

Second moment of area of triangle about an axis through C_T parallel to XX

$$= \frac{bh^3}{36} = \frac{(6.00)(5.00)^3}{36} = 20.83 \text{ cm}^4$$

By the parallel-axis theorem, the second moment of area of the triangle about XX $= 20.83 + [\frac{1}{2}(6.00)(5.00)][5.00 + \frac{1}{3}(5.00)]^2$

$$= \mathbf{687.5} \textbf{ cm}^4$$

Second moment of area of rectangle about XX $= \dfrac{bl^3}{3} = \dfrac{(4.00)(5.00)^3}{3}$

$$= \mathbf{166.7} \textbf{ cm}^4$$

Second moment of area of semicircle about XX $= \dfrac{\pi r^4}{8} = \dfrac{\pi(3.00)^4}{8}$

$$= \mathbf{31.81} \textbf{ cm}^4$$

Total second moment of area about XX $= 687.5 + 166.7 + 31.81$

$$= \mathbf{886.0} \textbf{ cm}^4$$

Problem 7. (*a*) Find the second moment of area and the radius of gyration about axis XX for the I-section beam shown in Fig. 32.16 (*b*) Determine the position of the centroid of the I-section. (*c*) Calculate the second moment of area and the radius of gyration about an axis NN through the centroid of the section, parallel to axis XX.

The I-section is divided into three rectangles P, Q and R as shown in Fig. 32.16 with the centroid of each denoted by C_P, C_Q and C_R respectively.

(*a*) **For rectangle P:**
Second moment of area about C_P (an axis through C_P parallel to XX),

$$Ak_{C_P}^2 = \frac{bl^3}{12} \text{ (where } l = 2.00 \text{ cm and } b = 8.00 \text{ cm)}$$

$$= \frac{(8.00)(2.00)^3}{12} = 5.333 \text{ cm}^4$$

Using the parallel-axis theorem:
$$Ak_{XX}^2 = Ak_{C_P}^2 + Ad^2$$

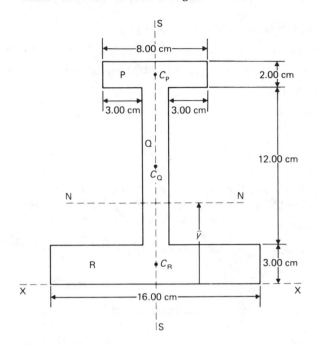

Fig. 32.16

where A = area of rectangle P = 8.00 × 2.00 = 16.00 cm^2 and d = perpendicular distance between C_P and XX = 16.00 cm.

Hence $Ak_{XX}^2 = 5.333 + (16.00)(16.00)^2 = \mathbf{4\,101 \ cm^4}$

For rectangle Q:

Similarly to above, $Ak_{C_Q}^2 = \dfrac{bl^3}{12} = \dfrac{(2.00)(12.00)^3}{12} = 288.0 \ cm^4$

$$Ak_{XX}^2 = 288.0 + [(2.00)(12.00)](9.00)^2$$

$$= \mathbf{2\,232 \ cm^4}$$

For rectangle R:

$$Ak_{XX}^2 = \frac{lb^3}{3} = \frac{(16.00)(3.00)^3}{3} = \mathbf{144.0 \ cm^4}$$

Total second moment of area for the I-section beam about axis XX

$$= 4\,101 + 2\,232 + 144.0 = \mathbf{6\,477 \ cm^4}$$

Total area of I-section = (8.00)(2.00) + (2.00)(12.00) + (16.00)(3.00)

$$= 88.00 \ cm^2$$

$$Ak_{XX}^2 = 6\,477 \ cm^4$$

Hence $k_{XX}^2 = \dfrac{6\,477}{\text{area}} = \dfrac{6\,477}{88.00}$

and radius of gyration about axis XX, $k_{XX} = \sqrt{\left(\dfrac{6\,477}{88.00}\right)} = \textbf{8.579 cm}$

(b) The centroid will lie on the axis of symmetry (shown as SS in Fig. 32.16).
Hence $\bar{x} = 0$

Part	Area (a cm^2)	Distance of centroid from XX (i.e. y cm)	Moment about XX (i.e. ay cm^3)
P	16.00	16.00	256.00
Q	24.00	9.00	216.00
R	48.00	1.50	72.00

$\Sigma a = A = 88.00$ $\qquad\qquad\qquad$ $\Sigma ay = 544.00$

$A\bar{y} = \Sigma ay$

Thus $\bar{y} = \dfrac{\Sigma ay}{A} = \dfrac{544.0}{88.00} = 6.182$ cm

The centroid is thus positioned on the axis of symmetry 6.182 cm from axis XX.

(c) $Ak_{XX}^2 = Ak_{NN}^2 + Ad^2$ from the parallel-axis theorem
$6\,477 = Ak_{NN}^2 + (88.00)(6.182)^2$
Hence $Ak_{NN}^2 = 6\,477 - 3\,363 = 3\,114$ cm^4

$$k_{NN}^2 = \dfrac{3\,114}{\text{area}} = \dfrac{3\,114}{88.00}\,\text{cm}^2$$

$$k_{NN} = \sqrt{\left(\dfrac{3\,114}{88.00}\right)} = 5.949\text{ cm}$$

Thus **the second moment of area about the centroid is $3\,114$ cm^4 and its radius of gyration is 5.949 cm**

Further problems on second moments of area may be found in the following Section 32.5 (Problems 1 to 28).

32.5   Further problems

1. For the rectangle shown in Fig. 32.17(a), find the second moment of area and the radius of gyration: (i) about axis AA; (ii) about axis BB; and (iii) about axis CC.
    (i) [1 080 cm^4, 3.464 cm]   (ii) [6 750 cm^4, 8.660 cm]
    (iii) [1 688 cm^4, 4.331 cm]

2. For the triangle shown in Fig. 32.17(*b*), find the second moment of area and the radius of gyration: (i) about axis AA; (ii) about axis BB; and (iii) about axis CC.
    (i) [425.3 mm^4, 3.674 mm]   (ii) [1 276 mm^4, 6.365 mm]
    (iii) [141.8 mm^4, 2.122 mm]
3. For the circle shown in Fig. 32.17(*c*), find the second moment of area and the radius of gyration: (i) about axis AA; (ii) about axis BB; and (iii) about axis CC.
    (i) [490.9 cm^4, 2.500 cm]   (ii) [2 454 cm^4, 5.590 cm]
    (iii) [2 454 cm^4, 5.590 cm]
4. For the semicircle shown in Fig. 32.17(*d*), find the second moment of area and the radius of gyration about axis AA.
    [100.5 cm^4, 2.000 cm]

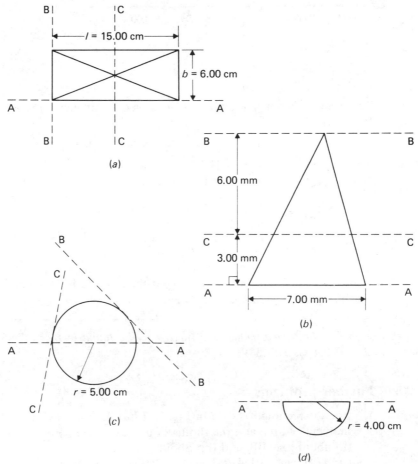

Fig. 32.17

5. Find the second moment of area of a rectangular wooden lamina, having dimensions l and b, about an axis parallel to b at a distance $\dfrac{l}{4}$ from one end. $[\frac{7}{48}bl^3]$

6. Show that the radius of gyration of a semicircle of radius r about an axis through its centroid parallel with its diameter is given by $0.264\,3r$

7. For each of the rectangles shown in Fig. 32.18, find the second moment of area and the radius of gyration about axis XX by using the parallel-axis theorem.
   (a) [210.7 cm⁴, 5.132 cm]   (b) [20 200 cm⁴, 12.66 cm]

8. For each of the triangles shown in Fig. 32.19, find the second moment of area and the radius of gyration about axis YY by using the parallel-axis theorem.
   (a) [12 250 cm⁴, 13.47 cm]   (b) [1 695 cm⁴, 8.069 cm]

9. For each of the circles shown in Fig. 32.20, find the second moment of area and the radius of gyration about axis AA by using the parallel-axis theorem.
   (a) [53 410 mm⁴, 11.85 mm]   (b) [29 730 mm⁴, 21.62 mm]

10. Calculate the second moment of area and the radius of gyration of a rectangular cover of length 32.0 mm and width 12.0 mm about an axis through one corner perpendicular to the plane of the lamina.   [149 500 mm⁴, 19.73 mm]

11. Find the second moment of area and the radius of gyration of a plate in the shape of a quadrilateral PQRS about the side PQ, given that the lengths of PQ, PS and SR are 8.00, 10.00 and 12.00 cm respectively and that the angles QPS and PSR are right angles.   [3 667 cm⁴, 6.056 cm]

(a)   (b)

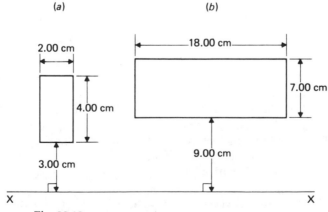

Fig. 32.18

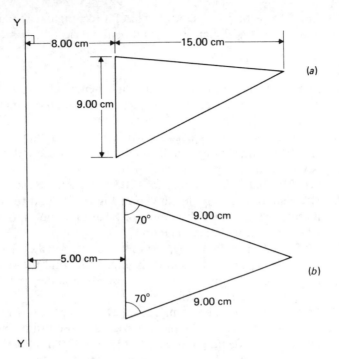

Fig. 32.19

12. Determine the second moment of area and the square of the radius of gyration of a template in the shape of an isosceles triangle having sides of 5.00, 6.00 and 6.00 cm about an axis through its centroid and parallel to the 5.00-cm side. [38.14 cm⁴, 2.347 cm²]

13. Calculate the second moment of area and the radius of gyration of a circular metal cover of diameter 25.0 mm about an axis perpendicular to the plane of the cover and passing through the centre.   [38.350 mm⁴, 8.839 mm]

14. For each of the composite shapes shown in Fig. 32.21, find the second moment of area and the radius of gyration about the axes AA
    (a) [285.8 cm⁴, 2.210 cm]   (b) [14 890 mm⁴, 6.055 mm]
    (c) [2 960 cm⁴, 3.854 cm]

15. Determine the second moment of area of an equilateral triangular metal plate of side 8.00 cm about a pole passing through its centroid and perpendicular to the plane of the plate. [254.4 cm⁴]

16. A circular cover, centre O, has a radius of 48.0 mm. A hole of radius 15.0 mm and centre B, where OB = 25.0 mm, is cut in

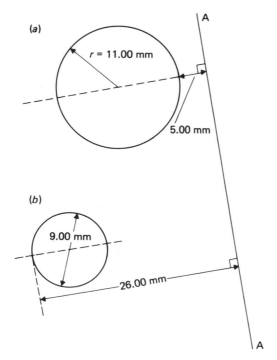

(a)

A

r = 11.00 mm

5.00 mm

(b)

9.00 mm

26.00 mm

A

Fig. 32.20

the cover. Calculate the second moment of area (in cm⁴) and
the radius of gyration (in cm) of the remainder about a
diameter through O perpendicular to OB
[368.8 cm⁴, 2.376 cm]

17. Determine in terms of l, the second moment of area of the
lamina shown in Fig. 32.22 about PQ, given that
PQ = QR = ST = l and UT = $\frac{1}{3}$SR $[\frac{10}{9}l^4]$

18. For each of the sections shown in Fig. 32.23, find the second
moment of area and the radius of gyration about the axes AA.
(a) [3 184 cm⁴, 8.145 cm]   (b) [272 300 mm⁴, 19.23 mm]

19. Calculate the second moment of area and the radius of
gyration about the axes BB for each of the composite shapes
shown in Fig. 32.24
(a) [108 300 cm⁴, 17.06 cm]   (b) [1 394 cm⁴, 4.224 cm]

20. Determine the second moment of area and the square of the
radius of gyration about the axes CC for each of the sections
shown in Fig. 32.25
(a) [14 530 cm⁴, 113.5 cm²]   (b) [4 186 cm⁴, 27.72 cm²]

21. Calculate the radius of gyration of a rectangular door 2 m

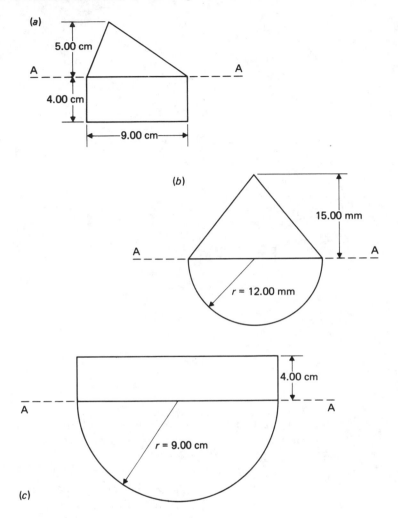

Fig. 32.21

high by 1.4 m wide about a vertical axis through its hinge.
[0.808 m]

22. For the area shown in Fig. 32.26 find the second moment of
area and the radius of gyration: (i) about axis DD, and
(ii) about axis EE. (Note that the circular area has been
removed)
(i) [2 302 cm⁴, 4.460 cm]   (ii) [10 710 cm⁴, 9.620 cm]

23. Show that the radius of gyration of a square plate of side x
about a diagonal is given by $\dfrac{x}{2\sqrt{3}}$

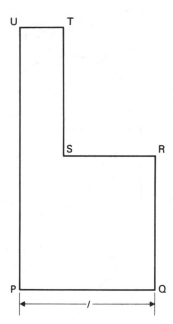

Fig. 32.22

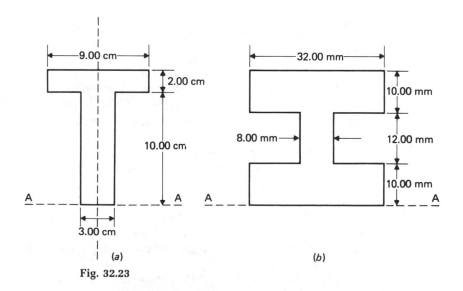

(a)

(b)

Fig. 32.23

24. A circular door is hinged so that it turns about a tangent. If its diameter is 0.9 m find its second moment of area and radius of gyration about the hinge. [0.161 m⁴, 0.503 m]

25. A uniform rectangular template of dimensions 32.0 cm by

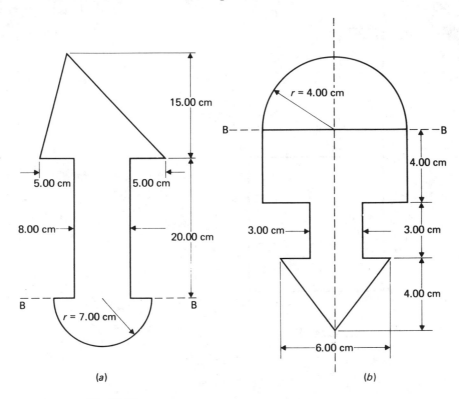

(a)   (b)

Fig. 32.24

24.0 cm has a circular hole of radius 10.0 cm removed. The
centre of the hole coincides with the centroid of the
rectangle. Calculate the second moment of area and the
radius of gyration about an axis through the centroid
perpendicular to the template.   [86 690 cm⁴, 13.82 cm]

26. (a) Find the second moment of area and the radius of
     gyration about the axis XX for the beam section shown in
     Fig. 32.27
     (b) Determine the position of the centroid of the section.
     (c) Calculate the second moment of area and the radius of
     gyration about an axis through the centroid, parallel to the
     axis XX.
     (a) [22 020 cm⁴, 11.67 cm]   (b) [8.944 cm from XX on the
     axis of symmetry]   (c) [9 111 cm⁴, 7.499 cm]

27. For the H-beam section shown in Fig. 32.28, find the second
     moment of area and the radius of gyration: (a) about axis
     XX, (b) about axis YY, and (c) about an axis through its
     centroid parallel to axis XX.

(*a*) [66 410 cm⁴, 14.14 cm]   (*b*) [48 970 cm⁴, 12.14 cm]
(*c*) [15 690 cm⁴, 6.875 cm]

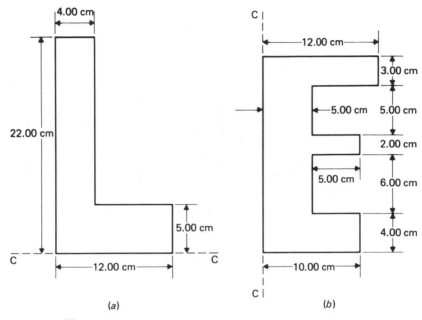

(*a*)

(*b*)

Fig. 32.25

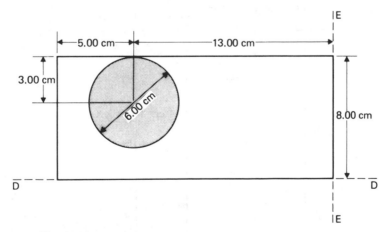

Fig. 32.26

28. For the I-beam section shown in Fig. 32.29 calculate the second moment of area and the square of the radius of gyration: (i) about axis AA, (ii) about axis BB, (iii) about the

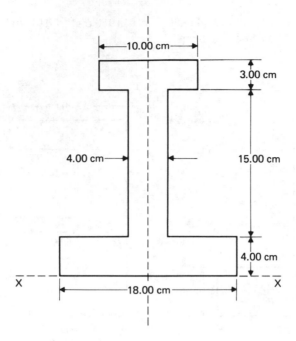

Fig. 32.27

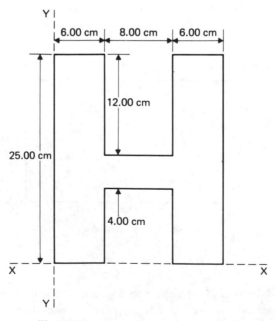

Fig. 32.28

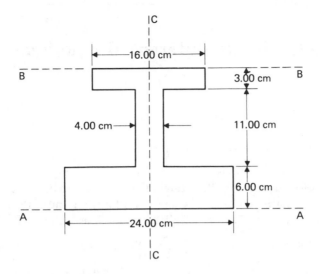

Fig. 32.29

axis of symmetry CC, (iv) about an axis through the centroid
parallel to axis AA, (v) about an axis through the centroid
perpendicular to the plane of the section.
(i) [24 450 cm^4, 103.6 cm^2] (ii) [45 810 cm^4, 194.1 cm^2]
(iii) [7 995 cm^4, 33.88 cm^2]   (iv) [10 320 cm^4, 43.73 cm^2]
(v) [18 315 cm^4, 77.61 cm^2]

Chapter 33

Introduction to differential equations

33.1   Families of curves

A graph depicting the equations $y = 3x + 1$, $y = 3x + 2$ and $y = 3x - 4$ is shown in Fig. 33.1, and three parallel straight lines are seen to be the result.

Equations of the form $y = 3x + c$, where c can have any numerical value, will produce an infinite number of parallel straight lines called a **family of curves**. A few of these can be seen in Fig. 33.1. Since $y = 3x + c$, $\dfrac{dy}{dx} = 3$, that is, the slope of every member of the family is 3. When additional information is given, for example both the **general equation** $y = 3x + c$ and the member

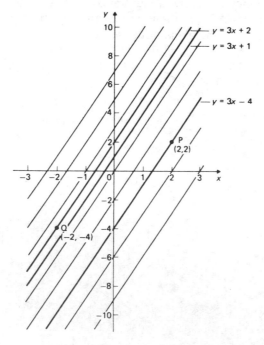

Fig. 33.1   Some members of the family of curves satisfying the equation $y = 3x + c$

of the family passing through the point (2, 2), shown as P in Fig. 33.1, then one particular member of the family is identified. The only line meeting both these conditions is the line $y = 3x - 4$. This is established by substituting $x = 2$ and $y = 2$ in the general equation $y = 3x + c$ and determining the value of c. Then, $2 = 3(2) + c$, giving $c = -6 + 2$, i.e. -4. Thus the **particular solution** is $y = 3x - 4$. Similarly, at point Q having coordinates $(-2, -4)$, the particular member of the family meeting both the conditions that it belongs to the family $y = 3x + c$ and that it passes through Q is the line $y = 3x + 2$. The additional information given, to enable a particular member of a family to be selected, is called the **boundary conditions**.

The equation $\dfrac{dy}{dx} = 3$ is called a **differential equation** since it contains a differential coefficient. It is also called a **first-order** differential equation, since it contains the first differential coefficient only, and has no differential coefficients such as $\dfrac{d^2y}{dx^2}$ or higher orders.

Another family of an infinite number of curves is produced by drawing a graph depicting the equations $y = 2x^2 + c$. Two of the curves in the family are $y = 2x^2$ (when $c = 0$) and $y = 2x^2 - 12$ (when $c = -12$) and these curves, together with others belonging to the family, are shown in Fig. 33.2.

The slope at any point of these curves is found by differentiating $y = 2x^2 + c$ and is given by $\dfrac{dy}{dx} = 4x$, i.e. the gradient of all of the curves is given by 4 times the value of the abscissa at every point. When boundary conditions are stated, particular curves can be identified. For example, the curve belonging to the family of curves $y = 2x^2 + c$ and which passes through point P, having coordinates (2, 8), is obtained by substituting $x = 2$ and $y = 8$ in the general equation. This gives $8 = 2(2)^2 + c$, i.e. $c = 0$, and hence the curve is $y = 2x^2$. Similarly, the curve satisfying the general equation and passing through the point $Q = (-2, -4)$ is $y = 2x^2 - 12$.

33.2 The solution of differential equations of the form $\dfrac{dy}{dx} = f(x)$

Differential equations are used extensively in science and engineering.

There are many types of differential equation and possibly the simplest type is of the form $\dfrac{dy}{dx} = f(x)$.

The solution of any first-order differential equation of the form $\dfrac{dy}{dx} = f(x)$ involves eliminating the differential coefficient, i.e., forming an equation

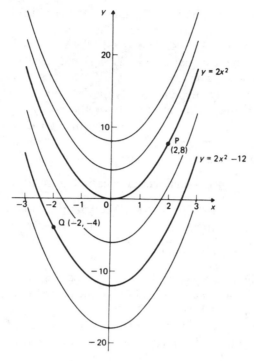

Fig. 33.2 Some members of the family of curves satisfying the equation
$y = 2x^2 + c$

which does not contain $\dfrac{dy}{dx}$. This is achieved by integrating, because $\displaystyle\int \dfrac{dy}{dx}\,dx$
is equal to y.

For example, when $\dfrac{dy}{dx} = 2x + 3$, by integrating:

$$\int \frac{dy}{dx} = dx = \int (2x + 3)\,dx$$

giving $y = x^2 + 3x + c$

It was shown in Section 33.1 that the solution of an equation of this form produces a family of curves. A solution to a differential equation containing an arbitrary constant of integration is called the **general solution** of the differential equation and is representative of a family of curves.

When additional information is given, so that a particular curve from the family can be identified, the solution is called a **particular solution** of the differential equation or just a **particular integral**.

For example, when determining the particular solution to the equation

$\dfrac{dy}{dx} = x^2 + 5$, given the boundary conditions that x is equal to 2 when y is

equal to 5, the first step is to integrate to obtain the general solution.

When $\dfrac{dy}{dx} = x^2 + 5$, then $y = \dfrac{x^3}{3} + 5x + c$, the general solution.

Using the information given, i.e. substituting for x and y in the general solution, gives:

$$5 = \frac{(2)^3}{3} + 5(2) + c$$

and so $c = 5 - \frac{8}{3} - 10 = -7\frac{2}{3}$

Hence the particular solution is

$$y = \frac{x^3}{3} + 5x - 7\frac{2}{3}$$

Worked problems on the solution of differential equations of the form $\dfrac{dy}{dx} = f(x)$

Problem 1. Find the general solutions to the equations:

(a) $\dfrac{dy}{dx} + \dfrac{5}{x} = 4x$

(b) $x\dfrac{dy}{dx} = 3x^3 - 4x^2 + 5x$

(c) $5\dfrac{dM}{d\theta} = 3e^\theta - 4e^{-\theta}$

(d) $\dfrac{ds}{dt} = u + at$, where u and a are constants

(e) $\dfrac{di}{dt} = \omega I_m \cos \omega t$, where ω and I_m are constants

Each of these equations is of the form $\dfrac{dy}{dx} = f(x)$ and the general solution can be obtained by integration.

(a) $\dfrac{dy}{dx} = 4x - \dfrac{5}{x}$

Integrating: $\int \dfrac{dy}{dx}\,dx = \int \left(4x - \dfrac{5}{x} \right) dx$

$$y = \dfrac{4x^2}{2} - 5 \ln x + c$$

i.e. $y = 2x^2 - 5 \ln x + c$

(b) $x \dfrac{dy}{dx} = 3x^3 - 4x^2 + 5x$

Dividing throughout by x gives:

$$\dfrac{dy}{dx} = 3x^2 - 4x + 5$$

Integrating gives: $y = \dfrac{3x^3}{3} - \dfrac{4x^2}{2} + 5x + c$

i.e. $y = x^3 - 2x^2 + 5x + c$

(c) $5 \dfrac{dM}{d\theta} = 3e^\theta - 4e^{-\theta}$

$$\dfrac{dM}{d\theta} = \tfrac{1}{5}(3e^\theta - 4e^{-\theta})$$

Integrating gives: $M = \tfrac{1}{5}(3e^\theta + 4e^{-\theta}) + c$

(d) $\dfrac{ds}{dt} = u + at$

Integrating gives: $s = ut + \tfrac{1}{2}at^2 + c$

(e) $\dfrac{di}{dt} = \omega I_m \cos \omega t$

Integrating gives: $i = \dfrac{\omega I_m}{\omega} \sin \omega t + c$

i.e. $i = I_m \sin \omega t = c$

Problem 2. Find the particular solutions of the following equations satisfying the given boundary conditions:

(a) $\dfrac{dy}{dx} + x = 2$ and $y = 3$ when $x = 1$

(b) $x \dfrac{dy}{dx} = 3 - x^3$ and $y = 3\tfrac{2}{3}$ when $x = 1$

(c) $3 \dfrac{dr}{d\theta} + \cos \theta = 0$ and $r = 5$ when $\theta = \dfrac{\pi}{2}$

(d) $\dfrac{dy}{dx} = e^x - 2 \sin 2x$ and $y = 2$ when $x = \dfrac{\pi}{4}$

(a) $\dfrac{dy}{dx} = 2 - x$

Integrating gives: $y = 2x - \dfrac{x^2}{2} + c$, the general solution.

Substituting the boundary conditions $y = 3$ and $x = 1$ to evaluate c gives:

$$3 = 2(1) - \frac{(1)^2}{2} + c$$

i.e. $c = 1\tfrac{1}{2}$

Hence the particular solution is $y = 2x - \dfrac{x^2}{2} + 1\tfrac{1}{2}$

(b) $x \dfrac{dy}{dx} = 3 - x^3$

Dividing throughout by x to express the equation in the form

$\dfrac{dy}{dx} = f(x)$ gives:

$\dfrac{dy}{dx} = \dfrac{3}{x} - x^2$

Integrating gives: $y = 3 \ln x - \dfrac{x^3}{3} + c$, the general solution.

Substituting the boundary conditions gives: $3\tfrac{2}{3} = 3 \ln 1 - \dfrac{(1)^3}{3} + c$

i.e. $c = 4$

Hence the particular solution is $y = 3 \ln x - \dfrac{x^3}{3} + 4$

(c) $3 \dfrac{dr}{d\theta} + \cos \theta = 0$

$\dfrac{dr}{d\theta} = -\tfrac{1}{3} \cos \theta$

Integrating gives: $r = -\tfrac{1}{3} \sin \theta + c$, the general solution.

Substituting the boundary conditions gives: $5 = -\tfrac{1}{3} \sin \dfrac{\pi}{2} + c$

i.e. $c = 5\tfrac{1}{3}$

Hence the particular solution is: $r = -\tfrac{1}{3} \sin \theta + 5\tfrac{1}{3}$

(d) $\dfrac{dy}{dx} = e^x - 2 \sin 2x$

Integrating gives: $y = e^x + \cos 2x + c$, the general solution.

Substituting the boundary conditions gives: $2 = e^{\pi/4} + \cos\dfrac{\pi}{2} + c$

i.e. $\qquad\qquad\qquad\qquad\qquad\qquad c = 2 - e^{\pi/4}$

Hence the particular solution is $y = e^x + \cos 2x + 2 - e^{\pi/4}$
Expressed in this form, the true value of y is stated. The value of
$e^{\pi/4}$ is 2.193 3 correct to 4 decimal places and the result can be
expressed as $y = e^x + \cos 2x - 0.193\,3$, correct to 4 decimal
places. However, when a result can be accurately expressed in
terms of e or π, then it is usually better to leave the result in this
form, unless the problem specifies the accuracy required.

Further problems on the solution of differential equations of the form $\dfrac{dy}{dx} = f(x)$
may be found in Section 33.4, Problems 1 to 30, page 582.

33.3 The solution of differential equations of the form $\dfrac{dQ}{dt} = kQ$

The natural laws of growth and decay are of the form $y = Ae^{kx}$, where A
and k are constants. For such a law to apply, the rate of change of a variable
must be proportional to the variable itself. This can be shown by differentiation. Since

$$y = Ae^{kx}$$

$$\frac{dy}{dx} = Ake^{kx}, \text{ i.e. } \frac{dy}{dx} = kAe^{kx}$$

But Ae^{kx} is equal to y. Hence, $\dfrac{dy}{dx} = ky$

Three of the natural laws are shown below and all such laws can be shown
to be of a similar form.

(i) For linear expansion, the amount by which a rod expands when heated
depends on the length of the rod, that is, the increase of length with
respect to temperature is proportional to the length of the rod. Thus,
mathematically:

$$\frac{dl}{d\theta} = kl \text{ and the law is } l = l_0 e^{k\theta}$$

(ii) For Newton's law of cooling, the fall of temperature with respect to
time is proportional to the excess of its temperature above that of its

surroundings, i.e.

$$\frac{d\theta}{dt} = -k\theta \text{ and the law is } \theta = \theta_0 e^{-kt}$$

(iii)   In electrical work, when current decays in a circuit containing resistance and inductance connected in series, the change of current with respect to time is proportional to the current flowing at any instant,

i.e. $\dfrac{di}{dt} = ki$ and the law is $i = Ae^{kt}$ where $k = -\dfrac{1}{T}$ and T is the time constant of the circuit.

These are just some of many examples of natural or exponential laws. For more examples see Chapter 3. In general, differential equations of the form $\dfrac{dQ}{dt} = kQ$ depict natural laws and the solutions are always of the form $Q = Ae^{kt}$. This can be shown as follows:

Since $\dfrac{dQ}{dt} = kQ$, $\dfrac{dQ}{Q} = k \, dt$

Integrating: $\displaystyle\int \frac{dQ}{Q} = \int k \, dt$

i.e. $\ln Q = kt + c$

By the definition of a logarithm, if $y = e^x$ then $x = \ln y$, i.e. if $\ln y = x$, then $y = e^x$. It follows that when $\ln Q = kt + c$

$$Q = e^{kt+c}$$

By the laws of indices, $e^a e^b = e^{a+b}$ and applying this principle gives:

$$Q = e^{kt} e^c$$

But e^c is a constant, say A, thus

$$Q = Ae^{kt}$$

Checking by differentiation:

when $Q = Ae^{kt}$

$$\frac{dQ}{dt} = kAe^{kt} = kQ$$

Hence Ae^{kt} is a solution to the differential equation $\dfrac{dQ}{dt} = kQ$

Thus the general solution of any differential equation of the form

$$\frac{dQ}{dt} = kQ \text{ is } Q = Ae^{kt}$$

and when boundary conditions are given, the particular solution can be obtained, as shown in the worked problems following.

Worked problems on the solution of equations of the form $\frac{dQ}{dt} = kQ$

Problem 1. Solve the equation $\frac{dy}{dx} = 6y$ given that $y = 3$ when $x = 0.5$

Since $\frac{dy}{dx} = 6y$ is of the form $\frac{dQ}{dt} = kQ$, the solution to the general equation will be of the form $Q = Ae^{kt}$, i.e. $y = Ae^{6x}$
Substituting the boundary conditions gives: $3 = Ae^{6(0.5)}$

i.e.
$$A = \frac{3}{e^3} = 0.149\,4$$

Hence the particular solution is $y = \dfrac{3}{e^3} e^{6x} = 3e^{3(2x-1)}$ **or** $\mathbf{0.149\,4e^{6x}}$

Problem 2. Determine the particular solutions of the following equations and their given boundary conditions, expressing the values of the constants correct to 4 significant figures:

(a) $\dfrac{dM}{di} - 4M = 0$ and $M = 5$ when $i = 1$

(b) $\dfrac{1}{15}\dfrac{dl}{dm} + \dfrac{l}{4} = 0$ and $l = 15.41$ when $m = 0.714\,3$

(a) Rearranging the equation into the form $\dfrac{dQ}{dt} = kQ$ gives:

$$\frac{dM}{di} = 4M$$

The general solution is of the form $Q = Ae^{kt}$, giving

$$M = Ae^{4i}$$

Substituting the boundary conditions gives: $5 = Ae^{(4)(1)}$

i.e. $\qquad\qquad\qquad\qquad\qquad\qquad\qquad A = \dfrac{5}{e^4} = 0.091\,58$

Hence the particular solution is $M = \mathbf{0.091\,58}\ e^{4i}$

(b) Writing the equation in the form $\dfrac{dQ}{dt} = kQ$ gives:

$$\frac{dl}{dm} = -\frac{15}{4}l$$

The general solution is $l = Ae^{-\frac{15}{4}m}$

Substituting the boundary conditions gives: $15.41 = Ae^{(-\frac{15}{4})(0.714\,3)}$

i.e. $\qquad\qquad\qquad\qquad\qquad\qquad\qquad A = 224.4$

Hence the particular solution is $l = \mathbf{224.4}\,e^{-3.750m}$

Problem 3. The decay of current in an electrical circuit containing resistance R ohms and inductance L henrys in series is given by $L\dfrac{di}{dt} + Ri = 0$, where i is the current flowing at time t seconds. Determine the general solution of the equation. In such a circuit, R is 5 kΩ, L is 3 henrys and the current falls to 5 A in 0.7 ms. Determine how long it will take for the current to fall to 2 amperes. Express your answer correct to 2 significant figures.

Since $\dfrac{di}{dt} = -\dfrac{R}{L}i$, then the general solution of the equation is

$i = Ae^{-Rt/L}$

By substituting the given values of R, L, i and t, the value of constant A is determined, i.e.

$$5 = Ae^{(-5 \times 10^3 \times 0.7 \times 10^{-3})/3}$$

$$= Ae^{-\frac{3.5}{3}}$$

giving $\quad A = 16.056$

To determine the time for i to fall to 2 A, substituting in the general solution for i, A, R and L gives

$$2 = 16.056\,e^{(-5 \times 10^3 \times t)/3}$$

$$= 16.056\,e^{-5t/3} \text{ when } t \text{ is stated in milliseconds.}$$

Thus, $\quad e^{-5t/3} = \dfrac{2}{16.056} = 0.124\,56,$

and taking natural logarithms gives:

$$-\frac{5t}{3}\ln e = \ln 0.124\,56$$

But ln $e = 1$, hence $t = -\frac{3}{5}\ln 0.124\,56 = 1.25$ ms

i.e. **the time for i to fall to 2 A is 1.3 ms,** correct to 2 significant figures.

Problem 4. A copper conductor heats up to 50°C when carrying a current of 200 A. If the temperature coefficient of linear expansion, α_0, for copper is 17×10^{-6}/°C at 0°C and the equation relating temperature θ with length l is $\dfrac{dl}{d\theta} = \alpha l$, find the increase in length of the conductor at 50°C correct to the nearest centimetre, when l is 1 000 m at 0°C

Since $\dfrac{dl}{d\theta} = \alpha l$, then $l = Ae^{\alpha\theta}$ is the general solution.

But l is 1 000 when θ is 0°C, and substituting these values in the general solution of the equation gives:

$1\,000 = Ae^0$, i.e. $A = 1\,000$

Substituting for A, α and θ in the general equation gives

$l = 1\,000\,e^{(17 \times 10^{-6} \times 50)} = 1\,000.850$ m

i.e. **the increase in length of the conductor at 50°C is 85 cm,** correct to the nearest centimetre.

Problem 5. The rate of cooling of a body is proportional to the excess of its temperature above that of its surrounding, θ °C.

The equation is $\dfrac{d\theta}{dt} = k\theta$, where k is a constant.

A body cools from 90°C to 70°C in 3.0 minutes at a surrounding temperature of 15°C. Determine how long it will take for the body to cool to 50°C.

The general solution of the equation $\dfrac{d\theta}{dt} = k\theta$ is $\theta = Ae^{kt}$

Letting the temperature 90°C correspond to a time t of zero gives an excess of body temperature above the surroundings of $(90 - 15)$

Hence, $(90 - 15) = Ae^{(k)(0)}$, i.e. $A = 75$

3.0 minutes later, the general solution becomes:

$(70 - 15) = 75\,e^{(k)(3)}$

i.e. $e^{3k} = \dfrac{55}{75}$

Taking natural logarithms,

$$3k = \ln \frac{55}{75}, \text{ and } k = \frac{1}{3} \ln \frac{55}{75}$$

i.e. $k = -0.103\,38$

At 50°C, $(50 - 15) = 75\,e^{-0.103\,38t}$

$$\frac{35}{75} = e^{-0.103\,38t}$$

Taking natural logarithms gives:

$$t = -\frac{1}{0.103\,38} \ln \frac{35}{75}$$

$$= 7.37$$

That is, **the time for the body to cool to 50°C is 7.37 minutes**, or 7 minutes 22 seconds, correct to the nearest second.

Problem 6. The rate of decay of a radioactive material is given by $\dfrac{dN}{dt} = -\lambda N$ where λ is the decay constant and N the number of radioactive atoms disintegrating per second. Determine the half-life of a zinc isotope, taking the decay constant as 2.22×10^{-4} atoms per second.

The half-life of an element is the time for N to become one-half of its original value. Since $\dfrac{dN}{dt} = -\lambda N$, then applying the general solution to this equation gives:
$N = Ae^{-\lambda t}$, where the constant A represents the original number of radioactive atoms present since $N = A$ when $t = 0$. For half-life conditions, the ratio $\dfrac{N}{A}$ is $\dfrac{1}{2}$, hence

$$\tfrac{1}{2} = e^{-\lambda t} = e^{-2.22 \times 10^{-4}t}$$

Thus, $\ln \tfrac{1}{2} = -2.22 \times 10^{-4}t$

i.e. $t = -\dfrac{1}{2.22 \times 10^{-4}} \ln 0.5$

$$= 3122 \text{ seconds or 52 minutes, 2 seconds}$$

Thus, **the half-life is 52 minutes**, correct to the nearest minute.

Further problems on the solution of equations of the form $\dfrac{dQ}{dt} = kQ$ *may be found in the following Section 33.4 (Problems 31 to 40), page 584.*

33.4   Further problems

Solution of equations of the form $\dfrac{dy}{dx} = f(x)$

In Problems 1 to 15, find the general solutions of the equations.

1. $\dfrac{dy}{dx} = 3x - \dfrac{4}{x^2}$ $\left[y = \dfrac{3x^2}{2} + \dfrac{4}{x} + c \right]$

2. $\dfrac{dy}{dx} + 3 = 4x^2$ $\left[y = \dfrac{4x^3}{3} - 3x + c \right]$

3. $3\dfrac{dy}{dx} + \dfrac{2}{\sqrt{x}} = 5\sqrt{x}$ $[y = \tfrac{2}{3}\sqrt{x}(\tfrac{5}{3}x - 2) + c]$

4. $\dfrac{du}{dV} - \dfrac{1}{V} = 4$ $[u = 4V + \ln V + c]$

5. $6 - 5\dfrac{dy}{dx} = \dfrac{1}{x - 2}$ $[y = \tfrac{1}{5}(6x - \ln(x - 2)) + c]$

6. $2x^2 - \dfrac{3}{x} + 4\dfrac{dy}{dx} = 0$ $\left[y = \dfrac{1}{4}\left(3 \ln x - 2\dfrac{x^3}{3} \right) + c \right]$

7. $\dfrac{di}{d\theta} = \cos \theta$ $[i = \sin \theta + c]$

8. $6\dfrac{dV}{dt} = 4 \sin\left(100t + \dfrac{\pi}{6} \right)$ $\left[V = -\dfrac{1}{150}\cos\left(100t + \dfrac{\pi}{6} \right) + c \right]$

9. $\dfrac{di}{dt} - \dfrac{t}{10} + 140 = 0$ $\left[i = \dfrac{t^2}{20} - 140t + c \right]$

10. $3\dfrac{dv}{dt} + 0.7t^2 - 1.4 = 0$ $[v = \tfrac{1}{9}(4.2t - 0.7t^3) + c]$

11. $\dfrac{dy}{d\theta} = 3e^\theta - \dfrac{4}{e^{2\theta}}$ $\left[y = 3e^\theta + \dfrac{2}{e^{2\theta}} + c \right]$

12. $\dfrac{dV}{dx} = 3x - \dfrac{5}{x} - \sec^2 x$ $\left[V = \dfrac{3x^2}{2} - 5 \ln x - \tan x + c \right]$

13. $\dfrac{1}{2}\dfrac{dy}{dx} + 2x^{\frac{1}{2}} = e^{x/2}$ $\left[y = 4\left(e^{x/2} - \dfrac{2x^{\frac{3}{2}}}{3} \right) + c \right]$

14. $x\dfrac{dy}{dx} = 2 - 3x^2$ $\left[y = 2 \ln x - \dfrac{3x^2}{2} + c \right]$

15. $\dfrac{dM}{d\theta} = \tfrac{1}{2} \sin 3\theta - \tfrac{1}{3} \cos 2\theta$ $[M = -\tfrac{1}{6}(\cos 3\theta + \sin 2\theta) + c]$

In Problems 16 to 25, determine the particular solutions of the differential equations for the boundary conditions given.

16. $x\dfrac{dy}{dx} - 2 = x^3$ and $y = 1$ when $x = 1$

$$\left[y = 2 \ln x + \frac{x^3}{3} + \frac{2}{3} \right]$$

17. $x\left(x - \dfrac{dy}{dx} \right) = 3$ and $y = 2$ when $x = 1$

$$\left[y = \frac{x^2}{2} - 3 \ln x + 1\tfrac{1}{2} \right]$$

18. $\dfrac{ds}{dt} - 4t^2 = 9$ and $s = 27$ when $t = 3$ $\qquad \left[s = 9t + \dfrac{4t^3}{3} - 36 \right]$

19. $e^{-p}\dfrac{dq}{dp} = 5$ and $q = 2.718$ when $p = 0$ $\qquad [q = 5e^p - 2.282]$

20. $3 - \dfrac{dy}{dx} = e^{2x} - 2e^x$ and $y = 7$ when $x = 0$

$[y = 3x - \tfrac{1}{2}e^{2x} + 2e^x + 5\tfrac{1}{2}]$

21. $\dfrac{dy}{d\theta} - \sin 3\theta = 5$ and $y = \dfrac{5\pi}{6}$ when $\theta = \dfrac{\pi}{6}$

$[y = 5\theta - \tfrac{1}{3}\cos 3\theta]$

22. $3 \sin\left(2\theta - \dfrac{\pi}{3} \right) + 4\dfrac{dv}{d\theta} = 0$ and $v = 3.7$ when $\theta = \dfrac{2\pi}{3}$

$$\left[v = \frac{3}{8}\left(\cos\left(2\theta - \frac{\pi}{3} \right) \right) + 4.075 \right]$$

23. $\dfrac{1}{6}\dfrac{dM}{d\theta} + 1 = \sin \theta$ and $M = 3$ when $\theta = \pi$

$[M = 3\{(2\pi - 1) - 2\cos\theta - 2\theta\}]$

24. $\dfrac{2}{(u + 1)^2} = 4 - \dfrac{dz}{du}$ and $z = 14$ when $u = 5$

$$\left[z = 4u + \frac{2}{u + 1} - 6\tfrac{1}{3} \right]$$

25. $\dfrac{1}{2e^x} + 4 = x - 3\dfrac{dy}{dx}$ and $y = 3$ when $x = 0$

$$\left[y = \frac{1}{3}\left\{ \frac{x^2}{2} + \frac{1}{2e^x} - 4x \right\} + 2\tfrac{5}{6} \right]$$

26. The bending moment of a beam, M, and shear force F are related by the equation $\dfrac{dM}{dx} = F$, where x is the distance from one end of the beam. Determine M in terms of x when $F = -w(l - x)$ where w and l are constants, and $M = \frac{1}{2}wl^2$ when $x = 0$ $\qquad [M = \frac{1}{2}w(l - x)^2]$

27. The angular velocity ω of a flywheel of moment of inertia I is given by $I\dfrac{d\omega}{dt} + N = 0$, where N is a constant. Determine ω in terms of t given that $\omega = \omega_0$ when $t = 0$

$$\left[\omega = \omega_0 - \frac{Nt}{I}\right]$$

28. The gradient of a curve is given by $\dfrac{dy}{dx} = 2x - \dfrac{x^2}{3}$. Determine the equation of the curve if it passes through the point $x = 3$, $y = 4$ $\qquad \left[y = x^2 - \dfrac{x^2}{9} - 2\right]$

29. The acceleration of a body f is equal to its rate of change of velocity, $\dfrac{dv}{dt}$. Determine an equation for v in terms of t given that the velocity is 5 when $t = 0$ $\qquad [v = 5 + ft]$

30. The velocity of a body v is equal to its rate of change of distance, $\dfrac{dx}{dt}$. Determine an equation for x in terms of t given $v = u + at$, where u and a are constants and $x = 0$ when $t = 0$ $[x = ut + \frac{1}{2}at^2]$

Solution of equations of the form $\dfrac{dQ}{dt} = kQ$

In Problems 31 to 33 determine the general solutions to the equations.

31. $\dfrac{dp}{dq} = 9p$ $\qquad [p = Ae^{9q}]$

32. $\dfrac{dm}{dn} + 5m = 0$ $\qquad [m = Ae^{-5n}]$

33. $\dfrac{1}{6}\dfrac{dw}{dx} + \frac{3}{5}w = 0$ $\qquad [w = Ae^{-\frac{18}{5}x}]$

In Problems 34 to 36 determine the particular solutions to the equations, expressing the values of the constants correct to 3 significant figures.

34. $\dfrac{dQ}{dt} = 15.0Q$ and $Q = 7.3$ when $t = 0.015$ $\quad [Q = 5.83\,e^{15.0t}]$

35. $\dfrac{1}{7}\dfrac{dl}{dm} - \tfrac{1}{3}l = 0$ and $l = 1.7 \times 10^4$ when $m = 3.4 \times 10^{-2}$
$[l = 1.57 \times 10^4\,e^{2.33m}]$

36. $0.741\dfrac{dy}{dx} + 0.071\,y = 0$ and $y = 73.4$ when $x = 15.7$
$[y = 330\,e^{-0.0958x}]$

37. The difference in tension, T newtons, between two sides of a belt when in contact with a pulley over an angle of θ radians and when it is on the point of slipping is given by $\dfrac{dT}{d\theta} = \mu T$, where μ is the coefficient of friction between the material of the belt and that of the pulley at the point of slipping. When $\theta = 0$ radians, the tension is 170 N and the coefficient of friction as slipping starts is 0.31. Determine the tension at the point of slipping when θ is $\dfrac{5\pi}{6}$ radians. Also determine the angle of lap in degrees, to give a tension of 340 N just before slipping starts. [383 N, 128°]

38. The charge Q coulombs at time t seconds for a capacitor of capacitance C farads when discharging through a resistance of R ohms is given by:

$$R\dfrac{dQ}{dt} + \dfrac{Q}{C} = 0$$

A circuit contains a resistance of 500 kilohms and a capacitance of 8.7 microfarads, and after 147 milliseconds the charge falls to 7.5 coulombs. Determine the initial charge and the charge after one second, correct to 3 significant figures. [7.76 C, 6.17 C]

39. The variation of resistance, R ohms, of a copper conductor with temperature, $\theta\,°C$, is given by $\dfrac{dR}{d\theta} = \alpha R$, where α is the temperature coefficient of resistance of copper. Taking α as 39×10^{-4} per °C, determine the resistance of a copper conductor at 30°C, correct to 4 significant figures, when its resistance at 80°C is 57.4 ohms. [47.23 ohms]

40. The rate of growth of bacteria is directly proportional to the amount of bacteria present. Form a differential equation for the rate of growth when n is the number of bacteria at time t seconds. If the number of bacteria present at $t = 0$ is n_0, solve the equation. When the number of bacteria doubles in one hour, determine by how many times it will have increased in twelve hours. $[n = n_0 e^{kt}, 2^{12}]$

Chapter 34

First order differential equations by separation of variables

34.1 Introduction

First order differential equations of the form $\dfrac{dy}{dx} = f(x)$, $\dfrac{dy}{dx} = f(y)$ and $\dfrac{dy}{dx} = f(x) \cdot f(y)$ can all be solved by direct integration. In each case it is possible to separate the y terms to one side of the equation and the x terms to the other side. Solving such equations is therefore known as solution by **separation of variables**.

34.2 Solution of differential equations of the form $\dfrac{dy}{dx} = f(x)$

An equation of the form $\dfrac{dy}{dx} = f(x)$ may be solved directly by integration. The solution is $y = \int f(x)\, dx$, as shown in Chapter 33. Below are some further examples of this important form.

Worked problems on solving differential equations of the form $\dfrac{dy}{dx} = f(x)$

Problem 1. Find the general solutions of the following differential equations.

(a) $\dfrac{dy}{dx} = 5x^2 + \cos 3x$ (b) $x\dfrac{dy}{dx} = 3 - 2x^2$

(a) If $\dfrac{dy}{dx} = 5x^2 + \cos 3x$, then $y = \displaystyle\int (5x^2 + \cos 3x)\, dx$

i.e.
$$y = \frac{5x^3}{3} + \frac{1}{3}\sin 3x + c$$

(b) If $x\dfrac{dy}{dx} = 3 - 2x^2$, then $\dfrac{dy}{dx} = \dfrac{3 - 2x^2}{x} = \dfrac{3}{x} - 2x$

Hence
$$y = \int \left(\frac{3}{x} - 2x\right) dx$$

i.e.
$$y = 3\ln x - x^2 + c$$

Problem 2. Find the particular solutions of the following differential equations satisfying the given boundary conditions:

(a) $3\dfrac{dy}{dx} + x = 6$ $\qquad$ and $y = 5\frac{1}{2}$ when $x = 3$

(b) $2\dfrac{dr}{d\theta} + \sin 2\theta = 0$ and $r = 2$ when $\theta = \dfrac{\pi}{2}$

(a) $\quad$ If $3\dfrac{dy}{dx} + x = 6$, then $\dfrac{dy}{dx} = \dfrac{6 - x}{3} = 2 - \dfrac{x}{3}$

Hence
$$y = \int \left(2 - \frac{x}{3}\right) dx$$

i.e.
$$y = 2x - \frac{x^2}{6} + c$$

(This is the general solution.)

Substituting the boundary conditions, $y = 5\frac{1}{2}$ wnen $x = 3$ to evaluate c gives:

$$5\tfrac{1}{2} = 6 - 1\tfrac{1}{2} + c$$

i.e. $\quad c = 1$

Hence the particular solution is $y = 2x - \dfrac{x^2}{6} + 1$

(b) $\quad$ If $2\dfrac{dr}{d\theta} + \sin 2\theta = 0$, then $\dfrac{dr}{d\theta} = \dfrac{-\sin 2\theta}{2}$

Hence
$$r = \int \frac{-\sin 2\theta}{2}\, d\theta$$

i.e.
$$r = \frac{1}{4}\cos 2\theta + c$$

Substituting the boundary conditions $r = 2$ when $\theta = \dfrac{\pi}{2}$ to evaluate c gives:

$$2 = \frac{1}{4}\cos \pi + c$$

i.e. $c = 2 - \left(-\dfrac{1}{4}\right) = 2\dfrac{1}{4}$

Hence the particular solution is

$$r = \frac{1}{4}\cos 2\theta + 2\frac{1}{4} = \frac{1}{4}(\cos 2\theta + 9)$$

Further problems on solving differential equations of the form $\dfrac{dy}{dx} = f(x)$ *may be found in Section 34.5 (Problems 1 to 14), page 595.*

34.3 Solution of differential equations of the form $\dfrac{dy}{dx} = f(y)$

An equation of the form $\dfrac{dy}{dx} = f(y)$ may be rearranged to give:

$$\frac{dx}{dy} = \frac{1}{f(y)}$$

i.e. $\qquad\qquad dx = \dfrac{dy}{f(y)}$

Integrating both sides gives: $\displaystyle\int dx = \int \frac{dy}{f(y)}$

Hence the solution may be obtained by direct integration.

Worked problems on solving differential equations of the form $\dfrac{dy}{dx} = f(y)$

Problem 1. Solve the differential equations:

(a) $\dfrac{dy}{dx} = 2 + y$ (b) $3\dfrac{dy}{dx} = \sec 2y$

(a)   Rearranging $\dfrac{dy}{dx} = 2 + y$ gives $dx = \dfrac{dy}{(2 + y)}$

Integrating both sides gives: $\displaystyle\int dx = \int \dfrac{dy}{(2 + y)}$

i.e. $\hspace{4cm} x = \ln(2 + y) + c \hspace{2cm} (1)$

The general solution of differential equations can sometimes be rearranged. In this case, for example, if $C = \ln D$, where D is a constant, then:

$$x = \ln(2 + y) + \ln D$$

i.e. $\hspace{1cm} x = \ln D(2 + y)$, from the law of logarithms,

from which $e^x = D(2 + y) \hspace{3cm} (2)$

Equations (1) and (2) are both acceptable general solutions of the differential equation $\dfrac{dy}{dx} = 2 + y$

(b)   Rearranging $3\dfrac{dy}{dx} = \sec 2y$ gives $dx = \dfrac{3}{\sec 2y} dy = 3\cos 2y\, dy$

Integrating both sides gives: $\displaystyle\int dx = \int 3\cos 2y\, dy$

Hence the general solution is $\hspace{1cm} x = \dfrac{3}{2}\sin 2y + c$

Problem 2. The rate at which a body cools is given by the equation $\dfrac{d\theta}{dt} = -k\theta$, where θ is the temperature of the body above its surroundings and k is a constant. Solve the equation for θ given that at $t = 0$, $\theta = \theta_0$

$\dfrac{d\theta}{dt} = -k\theta$ is of the form $\dfrac{dy}{dx} = f(y)$

Rearranging gives $dt = \dfrac{-1}{k\theta} d\theta$

Integrating both sides gives: $\displaystyle\int dt = \dfrac{-1}{k} \int \dfrac{d\theta}{\theta}$

i.e. $\hspace{4cm} t = \dfrac{-1}{k}\ln\theta + c \hspace{2cm} (1)$

Substituting the boundary conditions $t = 0$ and $\theta = \theta_0$ to find c

gives:

$$0 = \frac{-1}{k} \ln \theta_0 + c$$

i.e. $c = \frac{1}{k} \ln \theta_0$

Substituting $c = \frac{1}{k} \ln \theta_0$ in equation (1) gives:

$$t = -\frac{1}{k} \ln \theta + \frac{1}{k} \ln \theta_0$$

$$t = \frac{1}{k} (\ln \theta_0 - \ln \theta) = \frac{1}{k} \ln \left(\frac{\theta_0}{\theta} \right)$$

$$kt = \ln \left(\frac{\theta_0}{\theta} \right)$$

$$e^{kt} = \frac{\theta_0}{\theta}$$

$$e^{-kt} = \frac{\theta}{\theta_0}$$

Hence $\theta = \theta_0 \, e^{-kt}$

Further problems on solving differential equations of the form $\dfrac{dy}{dx} = f(y)$ may be found in Section 34.5 (Problems 15 to 29), page 596.

34.4 Solution of 'variable separable' type of differential equations

An equation of the form $\dfrac{dy}{dx} = f(x) \cdot f(y)$, where $f(x)$ is a function of x only and $f(y)$ is a function of y only, may be rearranged thus:

$$\frac{dy}{f(y)} = f(x) \, dx$$

Integrating both sides gives $\displaystyle \int \frac{dy}{f(y)} = \int f(x) \, dx$, i.e. the left-hand side is the integral of a function of y with respect to y and the right-hand side is the integral of a function of x with respect to x.

When two variables can be rearranged into two separate groups as shown above, each consisting of only one variable, the variables are said to be separable.

The equations of the type $\dfrac{dy}{dx} = f(x)$ and $\dfrac{dy}{dx} = f(y)$ discussed in Sections 34.2 and 34.3 are, in fact, merely special simple cases of 'separating the variables'.

Worked problems on solving differential equations of the form $\dfrac{dy}{dx} = f(x) \cdot f(y)$

Problem 1. Solve the differential equations:

(a) $\dfrac{dy}{dx} = \dfrac{3x^2 - 2}{2y - 1}$ (b) $2xy\dfrac{dy}{dx} = 1 + y^2$

(a) $\dfrac{dy}{dx} = \dfrac{3x^2 - 2}{2y - 1}$

Separating the variables gives $(2y - 1)\, dy = (3x^2 - 2)\, dx$

Integrating both sides gives $\displaystyle\int (2y - 1)\, dy = \int (3x^2 - 2)\, dx$

Hence the general solution is $y^2 - y = x^3 - 2x + C$

Note that when integrating both sides of an equation there is no need to put an arbitrary constant on both sides of the result. In this case, if this was done, then:

$$y^2 - y + A = x^3 - 2x + B$$
$$\text{and } y^2 - y \quad\;\; = x^3 - 2x + C, \text{ where } C = B - A$$

(b) $2xy\dfrac{dy}{dx} = 1 + y^2$

Separating the variables gives: $\dfrac{2y}{1 + y^2}\, dy = \dfrac{1}{x}\, dx$

Integrating both sides gives: $\displaystyle\int \dfrac{2y}{(1 + y^2)}\, dy = \int \dfrac{1}{x}\, dx$

Hence the general solution is $\ln(1 + y^2) = \ln x + C$ (3)

or $\ln(1 + y^2) - \ln x = C$

from which $\ln\left(\dfrac{1 + y^2}{x}\right) = C$

and $\dfrac{1 + y^2}{x} = e^C$ (4)

Also, if in equation (1), $C = \ln A$, we have $\ln(1 + y^2) = \ln x + \ln A$

$$\ln(1 + y^2) = \ln(Ax)$$

i.e. $$1 + y^2 = Ax \qquad (5)$$

Equations (3), (4) and (5) are all valid general solutions to the differential equation $2xy\dfrac{dy}{dx} = 1 + y^2$, none of them being any more correct than the others. Thus, by manipulation, it is possible to obtain several general solutions to a differential equation.

Problem 2. Find the particular solution of $\dfrac{dy}{dx} = 3e^{2x-3y}$ given that $y = 0$ when $x = 0$

$\dfrac{dy}{dx} = 3e^{2x-3y} = 3(e^{2x})(e^{-3y})$ by the laws of indices.

Separating the variables gives $\dfrac{dy}{e^{-3y}} = 3e^{2x}\,dx$

i.e. $$e^{3y}\,dy = 3e^{2x}\,dx$$

Integrating both sides gives $\displaystyle\int e^{3y}\,dy = 3\int e^{2x}\,dx$

$$\frac{1}{3}e^{3y} = \frac{3}{2}e^{2x} + C$$

(This is the general solution.)

When $y = 0$, $x = 0$, thus $\dfrac{1}{3}e^0 = \dfrac{3}{2}e^0 + C$

$$C = \frac{1}{3} - \frac{3}{2} = -\frac{7}{6}$$

Hence the particular solution is $\dfrac{1}{3}e^{3y} = \dfrac{3}{2}e^{2x} - \dfrac{7}{6}$

or $$2e^{3y} = 9e^{2x} - 7$$

Problem 3. An electrical circuit contains inductance L and resistance R connected to a constant voltage source E. The current i is given by the differential equation $E - L\dfrac{di}{dt} = Ri$, where L and R are constants. Find the current in terms of time t

given that when $t = 0$, $i = 0$

$$E - L\frac{di}{dt} = Ri. \quad \text{Rearranging gives} \quad \frac{di}{dt} = \frac{E - Ri}{L}$$

and

$$\frac{di}{E - Ri} = \frac{dt}{L}$$

Integrating both sides gives: $\displaystyle\int \frac{di}{E - Ri} = \int \frac{dt}{L}$

$$-\frac{1}{R}\ln(E - Ri) = \frac{t}{L} + c$$

(This is the general solution.)

$t = 0$ when $i = 0$ hence $\displaystyle -\frac{1}{R}\ln E = c$

Thus, $\displaystyle -\frac{1}{R}\ln(E - Ri) = \frac{t}{L} - \frac{1}{R}\ln E$

This particular solution must now be transposed to find i:

$$-\frac{1}{R}\ln(E - Ri) + \frac{1}{R}\ln E = \frac{t}{L}$$

$$\frac{1}{R}(\ln E - \ln(E - Ri)) = \frac{t}{L}$$

$$\ln\left(\frac{E}{E - Ri}\right) = \frac{Rt}{L}$$

$$\frac{E}{E - Ri} = e^{Rt/L}$$

$$\frac{E - Ri}{E} = e^{-Rt/L}$$

$$Ri = E - Ee^{-Rt/L}$$

Hence current $i = \dfrac{E}{R}(1 - e^{-RtL})$

This expression for current represents the natural law of growth of current in an inductive circuit.

Further problems on solving 'variables separable' types of differential equations may be found in the following Section 34.5 (Problems 30 to 46), page 598.

34.5   Further problems

Differential equations of the form $\dfrac{dy}{dx} = f(x)$

In Problems 1 to 9 solve the differential equations.

1. $\dfrac{dy}{dx} = 2x^4$ $\left[y = \dfrac{2}{5}x^5 + c \right]$

2. $\dfrac{dy}{dx} = 5x + \sin x$ $\left[y = \dfrac{5}{2}x^2 - \cos x + c \right]$

3. $x\dfrac{dy}{dx} = 4 - x^2$ $\left[y = 4 \ln x - \dfrac{x^2}{2} + c \right]$

4. $\dfrac{dy}{dx} - 2x^3 = e^{3x}$ $[6y = 2e^{3x} + 3x^4 + c]$

5. $x^2\dfrac{dy}{dx} = 2 + x$ $\left[y = \ln x - \dfrac{2}{x} + c \right]$

6. $\dfrac{dy}{dx} + x = 2$ and $y = 3$ when $x = 2$ $\left[y = 2x - \dfrac{x^2}{2} + 1 \right]$

7. $2\dfrac{dr}{d\theta} + \cos\theta = 0$ and $r = \dfrac{5}{2}$ when $\theta = \dfrac{\pi}{2}$ $\left[r = 3 - \dfrac{1}{2}\sin\theta \right]$

8. $x\left(x - \dfrac{dy}{dx} \right) = 3$ and $y = 1$ when $x = 1$

   $[2y = x^2 - 6 \ln x + 1]$

9. $\dfrac{1}{2e^t} + 4 = t - 3\dfrac{d\theta}{dt}$ and $\theta = \dfrac{1}{6}$ when $t = 0$

   $\left[\theta = \dfrac{1}{3}\left(\dfrac{t^2}{2} + \dfrac{e^{-t}}{2} - 4t \right) \right]$

10. The acceleration of a body a is equal to its rate of change of velocity, $\dfrac{dv}{dt}$. Determine an equation for v in terms of t given that the velocity is u when $t = 0$ $[v = u + at]$

11. The velocity of a body, v, is equal to its rate of change of distance, $\dfrac{ds}{dt}$. Determine an equation for s in terms of t, given $v = 5 + 2t$ and $s = 0$ when $t = 0$ $[s = 5t + t^2]$

12. The gradient of a curve is given by $\dfrac{dy}{dx} = 4x - \dfrac{x^3}{6}$. Determine

the equation of the curve if it passes through the point
$\left(2, 3\frac{1}{3}\right)$ $\left[y = 2x^2 - \dfrac{x^4}{24} - 4\right]$

13.   An object is thrown vertically upwards with an initial velocity, u, of 30 m/s. The motion of the object follows the differential equation $\dfrac{ds}{dt} = u - gt$, where s is the height in metres at time t seconds and $g = 9.81\,\text{m/s}^2$. Determine the height of the object after 4 seconds if $s = 0$ when $t = 0$.
[41.52 m]

14.   The angular velocity ω of a flywheel of moment of inertia I is given by $I\dfrac{d\omega}{dt} + k = 0$, where k is a constant. Determine ω in terms of t given that $\omega = \omega_0$ when $t = 0$
$\left[\omega = \omega_0 - \dfrac{kt}{I}\right]$

Differential equations of the form $\dfrac{dy}{dx} = f(y)$

In Problems 15 to 22 solve the differential equations.

15. $\dfrac{dy}{dx} = 3 + 2y$ $\left[\dfrac{1}{2}\ln(3 + 2y) = x + c\right]$

16. $5\dfrac{dy}{dx} = \cot 2y$ $\left[\dfrac{5}{2}\ln(\sec 2y) = x + c\right]$

17. $y\dfrac{dy}{dx} = 3 - y^2$ $\left[-\dfrac{1}{2}\ln(3 - y^2) = x + c\right]$

18. $2\dfrac{dy}{dx} + 3y = 4$ $\left[-\dfrac{2}{3}\ln(4 - 3y) = x + c\right]$

19. $\dfrac{dy}{dx} = 2\tan y.$ $[\arcsin(e^{2x+c})]$

20. $y\dfrac{dy}{dx} = 1 - y$, and $y = 0$ when $x = 1$
$[x + y + \ln(1 - y) = 1]$

21. $(y^2 + 1)\dfrac{dy}{dx} = 2y$ and $y = 1$ when $x = \dfrac{1}{4}$

$[y^2 + 2\ln y = 4x]$

22. $\sqrt{y}\dfrac{dy}{dx} - 1 = 0$ and $y = 4$ when $x = \dfrac{1}{3}$ $\qquad \left[\dfrac{2}{3}\sqrt{y^3} = x + 5\right]$

23. An equation of motion may be represented by the equation

$\dfrac{dv}{dt} + kv^2 = 0$ where v is the velocity of a body travelling in a

restraining medium. Show that $v = \dfrac{v_0}{1 + ktv_0}$ given that

$v = v_0$ when $t = 0$

24. The current in an electric circuit is given by $L\dfrac{di}{dt} + Ri = 0$

where L and R are constants. Solve for i given that $i = I$
when $t = 0$ $\qquad [i = I\,e^{-Rt/L}]$

25. The difference in tension, T newtons, between two sides of a
belt when in contact with a pulley over an angle of θ radians

and when it is on the point of slipping is given by $\dfrac{dT}{d\theta} = \mu T$,

where μ is the coefficient of friction between the material of
the belt and that of the pulley at the point of slipping. When
$\theta = 0$ radians, the tension is 150 N and $\mu = 0.29$ as slipping

starts. Find the tension at the point of slipping when $\theta = \dfrac{2\pi}{3}$

radians. Also determine the angle of lap correct to the
nearest degree to give a tension of 300 N just before slipping
starts. $\qquad$ [275.3 N, 137°]

26. The charge Q coulomb at time t seconds for a capacitor of
capacitance C farads when discharging through a resistance

of R ohms is given by $R\dfrac{dQ}{dt} + \dfrac{Q}{C} = 0$. Solve for Q given that

$Q = Q_0$ when $t = 0$. A circuit contains a resistance of 400
kilohms and a capacitance of 7.3 microfarads, and after 225
milliseconds the charge falls to 7.0 coulombs. Find the initial
charge and the charge after 2 seconds, correct to 3 significant
figures. $\qquad [Q = Q_0\,e^{-t/CR}$, 7.65 C; 3.81 C]

27. In a chemical reaction in which x is the amount transformed
in time t the velocity of the reaction is given by

$\dfrac{dx}{dt} = k(a - x)$ where k is a constant and a is the

concentration at time $t = 0$ when $x = 0$. Find x in terms of
of t $\qquad [x = a(1 - e^{-kt})]$

28. The rate of decay of a radioactive substance is given by $\dfrac{dN}{dt} = -\lambda N$, where λ is the decay constant and λN is the number of radioactive atoms disintegrating per second. Determine the half-life of radium in years (i.e. the time for N to become one-half of its original value) taking the decay constant for radium as 1.36×10^{-11} atoms per second and assuming a '365-day' year.   [1 616 years]

29. The variation of resistance R ohms of a copper conductor with temperature, $\theta\,°C$, is given by $\dfrac{dR}{d\theta} = \alpha R$, where α is the temperature coefficient of resistance of copper. If $R = R_0$ at $\theta = 0°C$, solve the equation for R. Taking α as 39×10^{-4} per $°C$, find the resistance of a copper conductor at $20°C$, correct to 4 significant figures, when its resistance at $100°C$ is 62.0 ohms. $[R = R_0\,e^{\alpha\theta};\ 45.42\text{ ohms}]$

'Variable separable' types of differential equations

In Problems 30 to 36 solve the differential equations.

30. $\dfrac{dy}{dx} = (2y)(x^2)$ $\left[\begin{array}{l} \dfrac{1}{2}\ln y = \dfrac{x^3}{3} + c \\[2mm] \text{or } \dfrac{1}{2}\ln 2y = \dfrac{x^3}{3} + k \end{array}\right]$

31. $\dfrac{dy}{dx} = y\cos x$ $[\ln y = \sin x + c]$

32. $(x + 2)\dfrac{dy}{dx} = (1 - y)$ $[\ln(x + 2)(1 - y) = c]$

33. $\dfrac{dy}{dx} = \dfrac{2x^2 - 1}{3y + 2}$ and $x = 0$ when $y = 0$

$\left[\dfrac{3y^2}{2} + 2y = \dfrac{2x^3}{3} - x\right]$

34. $\dfrac{dy}{dx} = e^{x-2y}$ and $x = 0$ when $y = 0$ $[e^{2y} = 2e^x - 1]$

35. $\dfrac{1}{2}\dfrac{dy}{dx} = e^{x+3y}$ and $x = 0$ when $y = 0$ $[7 - e^{-3y} = 6e^x]$

36. $y(1 + x) + x(1 - y)\dfrac{dy}{dx} = 0$ and $x = 1$ when $y = 1$

$[\ln(xy) = y - x]$

37.   Show that the solution of the equation $xy\dfrac{dy}{dx} = 1 + 2y^2$ may

be of the form $y = \sqrt{\left(\dfrac{x^4 k - 1}{2}\right)}$, where k is a constant.

38.   Show that the solution of the differential equation

$\dfrac{y^2 + 2}{x^2 + 2} = \dfrac{y}{x}\dfrac{dy}{dx}$ is of the form $\sqrt{\left(\dfrac{x^2 + 2}{y^2 + 2}\right)} = \text{constant}$.

39.   Prove that $y = x$ is a solution of the equation

$x\sqrt{(y^2 - 1)} - y\sqrt{(x^2 - 1)}\dfrac{dy}{dx} = 0$ when $x = 1$ and $y = 1$

40.   Solve $xy = (1 + x^2)\dfrac{dy}{dx}$ for y $\left[y = \dfrac{1}{k}\sqrt{(1 + x^2)}\right]$

41.   Find the curve which satisfies the equation $2xy\dfrac{dy}{dx} = x^2 + 1$

and which passes through the point $(1, 2)$
$[2y^2 = x^2 + 2\ln x + 7]$

42.   Solve the equation $y\cos^2 x\dfrac{dy}{dx} = \tan x + 2$ given that $y = 2$

when $x = \dfrac{\pi}{4}$ $[y^2 = \tan^2 x + 4\tan x - 1]$

43.   A capacitor C is charged by applying a steady voltage E
through a resistance R. The p.d. between the plates, V, is
given by the differential equation $CR\dfrac{dV}{dt} + V = E$. Solve the
equation for V given that $V = 0$ when $t = 0$ and evaluate V
when $E = 20$ volts, $C = 25$ microfarads, $R = 300$ kilohms
and $t = 2$ seconds. $[V = E(1 - e^{-t/CR}); 4.681$ volts$]$

44.   Find an adiabatic expansion of a gas $C_p\dfrac{dv}{V} + C_v\dfrac{dp}{P} = 0$,

where C_p and C_v are constants. Show that $pv^n = \text{constant}$,

where $n = \dfrac{C_p}{C_v}$

45.   The streamlines of a cylinder of radius a in a stream of
liquid of ambient velocity v are given by the equation:
$\dfrac{dr}{d\theta} = \dfrac{r(a^2 - r^2)}{(a^2 + r^2)}\cot\theta$

r is the distance of the centre of the cylinder from the applied force, the line joining them being at an angle θ to the direction of the liquid flow. Solve the equation. (Hint: let

$$a^2 + r^2 = a^2 - r^2 + 2r^2) \qquad \left[\left(\frac{a^2}{r} - r\right)\sin\theta = C\right]$$

46. The equilibrium constant (K) of a chemical reaction varies with temperature (T) according to the equation:

$$\frac{d(\ln K)}{dT} = \frac{\Delta H}{RT^2}.$$ If $\Delta H = 10^4$, $R = 8.3$ and $K = 4$ when $T = 600$, solve the equation completely.

$$\left[\ln K = \frac{-\Delta H}{RT} + 3.39\right]$$

Chapter 35

Probability

35.1   Definitions and simple probability

Probability

The probability of an event occurring means the likelihood or chance of it occurring. It is measured on a scale extending from a minimum of zero to a maximum of one. On this scale, zero corresponds to an absolute impossibility and one corresponds to an absolute certainty. When one red and one black marble are concealed in a bag and one of these is drawn from it, the probability of drawing a red marble is $\frac{1}{2}$ (or 1 in 2) and that of drawing a black marble is $\frac{1}{2}$ (or 1 in 2). When three red marbles and one black marble are placed in the bag, the probability of drawing one red marble is $\frac{3}{4}$ (or 3 in 4) and that of drawing one black marble is $\frac{1}{4}$ (or 1 in 4).

For a bag containing m red marbles and n black marbles, the probability p of drawing a red marble is given by the ratio:

$$p = \frac{\text{number of red marbles}}{\text{the total number of marbles}}, \text{ i.e. } p = \frac{m}{m+n}$$

Also the probability q of not drawing a red marble, i.e. drawing a black marble, is the ratio:

$$q = \frac{\text{number of black marbles}}{\text{the total number of marbles}}, \text{ i.e. } q = \frac{n}{m+n}$$

In general, when p is the probability of an event happening and q is the probability of an event not happening, then

$$q = 1 - p \text{ and the total probability } p + q \text{ is unity.}$$

Expectation

Generally, expectation can be defined as the product of the probability of success and the number of attempts. Thus if I buy 10 tickets in a raffle and the probability of winning a prize is $\frac{1}{200}$, the expectation of success, i.e. winning a prize, is $10 \times \frac{1}{200}$ or $\frac{1}{20}$. Expectation of success can be considered also in the following way.

If p is the probability of success and M is the reward for success, then the product is defined as the expectation of success, i.e.

$$\text{expectation, } E = pM$$

For example, if the probability p of winning a prize is $\frac{1}{50}$ and the prize, M, is £100, the expectation is pM, i.e. $\frac{1}{50} \times £100$ or £2

Independent and dependent events

When the occurrence of one event does not affect the probability of the occurrence of another, then the events are called **independent**. For example, a bag contains six red and four white balls. A ball is drawn at random from the bag and then replaced (a process termed **'with replacement'**), after its colour is noted. The probability of drawing a red ball is $\frac{6}{10}$. A second ball is now drawn from the bag and its colour noted, and again the probability of drawing a red ball is $\frac{6}{10}$. Since the probability of drawing a red ball on the second draw is in no way affected by the first draw, these two events are independent.

Conversely, when the probability of one event occurring does affect the probability of another event occurring, then the events are called **dependent**. For the bag containing six red and four white balls, on the first draw the probability of drawing a red ball is $\frac{6}{10}$. If another ball is withdrawn without putting the first ball back (a process termed **'without replacement'**), then the probability of drawing a red ball on the second draw is either $\frac{5}{9}$, when a red ball is removed as a result of the first draw, or $\frac{6}{9}$, when a white ball is removed as a result of the first draw.

Mutually exclusive events

Two or more events are called **mutually exclusive** when the occurrence of one of them excludes the occurrence of the others, i.e. when not more than one event can happen at the same time. An example is provided by the tossing of a coin, since the appearance of a head mutually excludes the appearance of a tail.

Worked problems on simple probability

Problem 1. Determine the probabilities of: (a) drawing a black ball from a bag containing 7 red and 20 black balls, (b) selecting at random a male from a group of 15 males and 29 females, (c) winning a prize in a raffle by buying 5 tickets when a total of 450 tickets are sold, and (d) winning a prize in a raffle by buying 10 tickets when there are 6 prizes and a total of 800 tickets are solid.

(a) Applying $p = \dfrac{n}{m+n}$, where p is the probability of drawing a black ball, m is the number of red balls and n the number of black balls, gives:

$$p = \frac{20}{7+20} = \frac{20}{27}$$

i.e. **the probability of drawing a black ball is $\dfrac{20}{27}$**

(b) Applying $p = \dfrac{m}{m+n}$, where p is the probability of selecting a male, m is the number of males and n is the number of females, gives:

$$p = \frac{15}{15+29} = \frac{15}{44}$$

i.e. **the probability of selecting a male is $\dfrac{15}{44}$**

(c) Applying $p = \dfrac{m}{m+n}$, where p is the probability of winning a prize, m is the number of tickets bought and $m+n$ the total number of tickets sold, gives:

$$p = \frac{m}{m+n} = \frac{5}{450} = \frac{1}{90}$$

i.e. **the probability of winning a prize is** $\dfrac{1}{90}$

(*d*) The probability of winning a prize by buying one ticket is the ratio:

$$\frac{\text{number of prizes}}{\text{number of tickets sold}}, \text{ i.e. } \frac{6}{800} \text{ or } \frac{3}{400}$$

That is, the probability of success is $\dfrac{3}{400}$. Hence the probability of
winning a prize by buying ten tickets is (the probability of success) × (the number of attempts),

i.e. $10 \times \dfrac{3}{400}$ or $\dfrac{3}{40}$

Thus, **the probability of winning a prize is** $\dfrac{3}{40}$

Problem 2. (*a*) The probability of an event happening is $\dfrac{3}{19}$.
Determine the probability of it not happening. (*b*) A bench can seat 7 people. Determine the probability of any one person: (i) sitting in the middle, and (ii) sitting at the end. (*c*) A person's chance of winning a raffle is $\dfrac{1}{18}$. If he buys 15 tickets determine the total number sold. (*d*) A bag contains 8 red, 5 blue and 4 white balls. Determine the probabilities of drawing: (i) a red ball, (ii) a blue ball, and (iii) a white ball.

(*a*) When p is the probability of an event happening and q the probability of the event not happening, then $p + q = 1$. Hence $q = 1 - p$. The value of p is given as $\dfrac{3}{19}$, hence

$$q = 1 - \frac{3}{19} = \frac{16}{19}$$

i.e. **the probability of the event not happening is** $\dfrac{16}{19}$

(b) There are 7 seats on the bench and the probability of one person sitting on any particular seat is $\frac{1}{7}$. Hence:

(i) the probability of one person sitting in the middle of the bench is $\frac{1}{7}$; and (ii) the probability of one person sitting on either of the two ends of the bench is $\frac{1}{7}$ for the left end and $\frac{1}{7}$ for the right end, giving a total probability of $\frac{2}{7}$

(c) Let p be the probability of the person winning the raffle, m be the number of tickets the person bought and $m + n$ the total number of tickets sold. Then:

$$p = \frac{m}{m+n}, \text{ i.e. } m+n = \frac{m}{p}$$

$$m+n = \frac{15}{\frac{1}{18}} = 270$$

i.e. **the total number of tickets sold is 270**

(d) Let p_R be the probability of drawing a red ball, given by the ratio:

$$p_R = \frac{\text{number of red balls}}{\text{the total number of balls}}$$

i.e. $p_R = \dfrac{8}{8+5+4} = \dfrac{8}{17}$

Similarly, $p_B = \dfrac{5}{17}$ and $p_W = \dfrac{4}{17}$

i.e. **the probability of drawing a red ball is** $\dfrac{8}{17}$**, a blue ball** $\dfrac{5}{17}$ **and a white ball** $\dfrac{4}{17}$

Further problems on simple probability may be found in Section 35.3 (Problems 1 to 8), page 612.

35.2 Laws of probability

The addition law of probability

This law applies to mutually exclusive events and is usually recognised by the words '**either ... or**'. It states that when two events are mutually exclusive, if the probability of the first event happening is p_1 and the probability of the second event happening is p_2, then the probability of **either** the first **or** the second event happening is $p_1 + p_2$. For example, fifteen cards are marked from 1 to 15 and one is drawn at random. It is required to determine the probability of the card selected being a multiple of 2 or 3. The probability of the card being a multiple of 2 is $\dfrac{7}{15}$ (given by the numbers 2, 4, 6, 8, 10, 12 and 14). The probability of the card being a multiple of 3 is $\dfrac{3}{15}$ (given by the numbers 3, 9 and 15, the numbers 6 and 12 already having been selected). Hence the probability of the card being a multiple of **either 2 or 3** is

$$\frac{7}{15} + \frac{3}{15}, \text{ that is, } \frac{10}{15} \text{ or } \frac{2}{5}$$

Similarly for n mutually exclusive events which can be related by the addition law, where n is a positive integer, the total probability is $p_1 + p_2 + p_3 + \cdots + p_n$, the events being linked by the words **or**.

The multiplication law of probability

This law applies to both dependent and independent events and is usually recognised by the words '**both ... and**'. It states that when the probability of one event happening is p_1 and the probability of a second event happening is p_2, then the probability of **both** the first **and** the second events happening is $p_1 \times p_2$. This law is derived as follows. Assuming the first event can happen in a total of n_1 ways, of which a_1 are successful, and the second event can happen in a total of n_2 ways, of which a_2 are successful, then the probability of both the first and the second events happening is given by combining each successful first event with each successful second event, i.e. $a_1 \times a_2$. For example, if there are two successful first events, say $a_1 = 1$ and 2, and three successful second events, say $a_2 = 3$, 4 and 5, then the probability of both the first and second events happening is given by the combinations 1 and 3, 1 and 4, 1 and 5, 2 and 3, 2 and 4, 2 and 5, i.e. 6 combinations given by $a_1 \times a_2$. Also, the total possible number of occurrences is $n_1 \times n_2$, i.e. each of the number of ways the first event can happen combined with each of the number of ways the second event can happen. Hence, the probability of both the first and the second event happening is

$$p = \frac{a_1 \times a_2}{n_1 \times n_2} = p_1 \times p_2$$

As an example of the multiplication law, it is required to determine the probability of drawing two white balls in succession from a bag containing six red and four white balls without replacement. Since the problem can be expressed as drawing **both** one white ball **and** another white ball, the multiplication law is indicated. The probability p_1 of drawing one white ball on the first draw is

$$p_1 = \frac{\text{number of white balls}}{\text{the total number of balls}} = \frac{4}{10} = \frac{2}{5}$$

The number of white balls is now reduced by 1, to 3, and the total number of balls reduced by 1, to 9. If p_2 is the probability of drawing one white ball on the second draw, then

$$p_2 = \frac{\text{number of white balls}}{\text{the total number of balls}} = \frac{3}{9} = \frac{1}{3}$$

Applying the multiplication law gives:

$$\text{probability of drawing two white balls} = p_1 \times p_2 = \frac{2}{5} \times \frac{1}{3} = \frac{2}{15}$$

Similarly for n events which can be related by the multiplication law, the total probability is $p_1 \times p_2 \times p_3 \times \cdots \times p_n$, the events being linked by the words **and**.

Worked problems on the laws of probability

Problem 1. Two balls are drawn in turn with replacement from a bag containing 8 red balls, 15 white balls, 24 black balls and 17 orange balls. Determine the probabilities of having:
(a) two red balls,
(b) a red and a white ball,
(c) no orange balls,
(d) a black and red or a black and orange ball,
(e) at least one black ball,
(f) at most one orange ball, and
(g) a white ball on the first draw but the second ball not white.

(a) The probability of drawing a red ball on the first draw is the ratio:

$$\frac{\text{number of red balls}}{\text{total number of balls}}, \text{ i.e. } \frac{8}{8 + 15 + 24 + 17} = \frac{8}{64} = \frac{1}{8}$$

Since drawing is with replacement, the red ball is now returned to the bag and a second draw made. The probability of drawing a red ball on the second draw is again $\frac{1}{8}$. Thus the probability of drawing **both** a red ball on the first draw **and** a red ball on the second draw is $\dfrac{1}{8} \times \dfrac{1}{8} = \dfrac{1}{64}$, the '**both ... and**' indicating the multiplication law. That is, the probability of selecting two red balls is $\dfrac{1}{64}$, with replacement.

(b) The probability of drawing a red ball on the first draw is $\dfrac{1}{8}$ (see part (a)). The probability of drawing a white ball on the second draw is

$$\dfrac{15 \text{ (white balls)}}{64 \text{ (total number of balls)}}$$

Hence the probability of drawing **both** a red ball on the first draw **and** a white ball on the second draw is $\dfrac{1}{8} \times \dfrac{15}{64} = \dfrac{15}{512}$

The probability of having a red and white ball can also be achieved by drawing a white ball first and a red ball second. Hence the probability of drawing **both** a white ball on the first draw **and** a red ball on the second draw is $\dfrac{15}{64} \times \dfrac{1}{8} = \dfrac{15}{512}$

The probability of drawing **either** a red then white ball **or** a white then red ball is $\dfrac{15}{512} + \dfrac{15}{512}$ (the 'either ... or' indicating the addition law), i.e. $\dfrac{15}{256}$. That is, the probability of having a red ball and a white ball is $\dfrac{15}{256}$, with replacement.

(c) The probability of having no orange balls really means the probability of drawing a red, white or black ball on both the first and second draws. The probability of drawing a red, white or black ball is $\dfrac{8 + 15 + 24}{64}$, that is $\dfrac{47}{64}$

Hence the probability of drawing red, white and black balls on **both** the first **and** the second draws is $\dfrac{47}{64} \times \dfrac{47}{64}$, i.e. $\dfrac{2\,209}{4\,096}$ since drawing is with replacement. That is, the probability of drawing no orange balls is $\dfrac{2\,209}{4\,096}$, with replacement.

(d) The probability of having a black and red is $\dfrac{24}{64} \times \dfrac{8}{64} + \dfrac{8}{64} \times \dfrac{24}{64}$, that is, $\dfrac{192}{2\,048}$ (see part (b)), or $\dfrac{3}{32}$

The probability of having a black and orange is
$\frac{24}{64} \times \frac{17}{64} + \frac{17}{64} \times \frac{24}{64} = \frac{408}{2\,048}$ (see part (b)), or $\frac{51}{256}$

Hence the probability of having **either** a black and red **or** a black and orange is $\frac{3}{32} + \frac{51}{256}$, i.e. $\frac{75}{256}$. Thus, the probability of having a black and red or a black and orange ball is $\mathbf{\frac{75}{256}}$, with replacement.

(e) The outcome of at least one black ball can be achieved by drawing a black and a non-black ball, a non-black ball and a black ball, or by drawing two black balls.

The probability of drawing a black and a non-black is $\frac{24}{64} \times \frac{40}{64}$, i.e. $\frac{15}{64}$. Since drawing is with replacement, the probability of drawing a non-black and a black ball is also $\frac{15}{64}$

The probability of drawing **either** a black and non-black **or** a non-black and black ball is $\frac{15}{64} + \frac{15}{64}$, i.e. $\frac{15}{32}$. The probability of drawing two black balls is $\frac{24}{64} \times \frac{24}{64}$, i.e. $\frac{9}{64}$. Thus the probability of drawing **either** (a black and non-black or a non-black and black) **or** two black balls is $\frac{15}{32} + \frac{9}{64}$, i.e. $\frac{39}{64}$. Thus, the probability of drawing at least one black ball is $\mathbf{\frac{39}{64}}$, with replacement.

(f) The possibilities for at most one orange ball are: (i) an orange ball on the first draw and a non-orange ball on the second draw, (ii) a non-orange ball on the first draw and an orange ball on the second draw, or (iii) no orange balls at all.

The probability of (i) is $\frac{17}{64} \times \frac{47}{64}$, i.e. $\frac{799}{4\,096}$

The probability of (ii) is $\frac{47}{64} \times \frac{17}{64}$, i.e. $\frac{799}{4\,096}$

The probability of (iii) is $\frac{47}{64} \times \frac{47}{64}$, i.e. $\frac{2\,209}{4\,096}$

The probability of (i) **or** (ii) **or** (iii) is $\frac{799 + 799 + 2\,209}{4\,096}$, i.e.

$\dfrac{3\,807}{4\,096}$. That is, the probability of having at most one orange ball

is $\dfrac{\mathbf{3\,807}}{\mathbf{4\,096}}$, with replacement.

(*g*) The probability of having a white ball on the first draw and one non-white ball on the second is $\dfrac{15}{64} \times \dfrac{49}{64}$, i.e. $\dfrac{\mathbf{735}}{\mathbf{4\,096}}$

Problem 2. The probability of three events happening are $\dfrac{1}{8}$ for event A, $\dfrac{1}{5}$ for event B, and $\dfrac{2}{7}$ for event C. Determine:

(*a*) the probability of all three events happening,
(*b*) the probability of event A and B but not C happening,
(*c*) the probability of only event B happening, and
(*d*) the probability of event A or event B happening but not event C

Let p_A, p_B and p_C be the probabilities of events A, B and C happening respectively. Let $\overline{p_A}$, $\overline{p_B}$ and $\overline{p_C}$ be the probabilities of those events not happening. Then:

$$p_A = \frac{1}{8},\ p_B = \frac{1}{5},\ p_C = \frac{2}{7},\ \overline{p_A} = \frac{7}{8},\ \overline{p_B} = \frac{4}{5}\text{ and }\overline{p_C} = \frac{5}{7},$$

since the probability of an event happening plus the probability of it not happening must be unity, i.e. $p_A + \overline{p_A} = 1$, etc.

(*a*) The probability of all three events happening is the same as the probability of A **and** B **and** C, i.e. the multiplication law is indicated. Thus:

$$p_A \times p_B \times p_C = \frac{1}{8} \times \frac{1}{5} \times \frac{2}{7} = \frac{1}{140}$$

That is, the probability that all three events will happen is $\dfrac{1}{140}$.

An example of this is that if there are three commonly occurring faults during the manufacture of an article and these occur with the probabilities shown as p_A, p_B and p_C, then the probability of all three of these faults occurring in any one article is $\dfrac{1}{140}$

(*b*) The probability of **both** A **and** B happening is

$$p_A \times p_B = \frac{1}{8} \times \frac{1}{5} = \frac{1}{40}$$

The probability of **both** A and B happening **and** C not happening is $p_A \times p_B \times \overline{p_C}$, i.e. $\dfrac{1}{8} \times \dfrac{1}{5} \times \dfrac{5}{7}$, i.e. $\dfrac{1}{56}$

That is, the probability of A and B but not C happening is $\dfrac{1}{56}$

(c) The probability of only event B happening means that **both** event A and event C are not happening **and** event B is happening.
    The probability of this occurring is given by $\overline{p_A} \times p_B \times \overline{p_C}$, i.e.
$\dfrac{7}{8} \times \dfrac{1}{5} \times \dfrac{5}{7}$, that is, $\dfrac{1}{8}$

Thus, the probability of only event B happening is $\dfrac{1}{8}$

(d) The probability of event A **or** event B happening is $(p_A + p_B)$.
    The probability of (event A or event B happening) **and** event C not happening is $(p_A + p_B) \times \overline{p_C}$. That is, $\overline{p_C}(p_A + p_B)$. Then,

$$\overline{p_C}(p_A + p_B) = \frac{5}{7}\left(\frac{1}{8} + \frac{1}{5}\right)$$
$$= \frac{5}{7} \times \frac{13}{40}$$
$$= \frac{13}{56}$$

i.e. the probability of event A or event B but not event C happening is $\dfrac{13}{56}$

Problem 3. One bag contains 3 red and 5 black marbles and a second bag contains 4 green and 7 white marbles. One marble is drawn from the first bag and two marbles from the second bag, without replacement. Determine the probability of having:
(a) one red and two white marbles,
(b) no green marbles, and
(c) either one black and two green or one black and two white marbles.

(a) The probability of having one red marble on the first draw is $\dfrac{3}{8}$

The probability of having one white marble on the second draw is $\dfrac{7}{11}$

The probability of having one white marble on the third draw is $\dfrac{6}{10}$, since there is no replacement. Hence the probability of having one red **and** one white marble on successive draws is
$$\frac{3}{8} \times \frac{7}{11} \times \frac{6}{10} = \frac{63}{440}$$

i.e. the probability of having one red and two white marbles is $\dfrac{63}{440}$, without replacement.

(*b*) The probability of having no green marbles on the first draw is unity, since the first bag contains no green marbles.

The probability of drawing no green marbles on the second draw is the probability of drawing a white marble (since the second bag only contains green and white marbles), i.e. $\dfrac{7}{11}$

The probability of drawing no green marbles on the third draw is again the probability of drawing a white marble, i.e. $\dfrac{6}{10}$, since there is no replacement.

Hence the probability of drawing no green marbles on the first draw **and** no green marbles on the second draw **and** no green marbles on the third draw is $\dfrac{1}{1} \times \dfrac{7}{11} \times \dfrac{6}{10}$, i.e. $\dfrac{21}{55}$

That is, the probability of having no green marbles is $\dfrac{21}{55}$, without replacement.

(*c*) The probability of one black and two green marbles, without replacement, is $\dfrac{5}{8} \times \dfrac{4}{11} \times \dfrac{3}{10}$ (see part (*a*)), i.e. $\dfrac{3}{44}$

The probability of one black and two white marbles, without replacement, is $\dfrac{5}{8} \times \dfrac{7}{11} \times \dfrac{6}{10}$ (see part (*a*)), i.e. $\dfrac{21}{88}$

Thus, the probability of **either** one black and two green marbles **or** one black and two white is $\dfrac{3}{44} + \dfrac{21}{88}$, i.e. $\dfrac{27}{88}$

Further problems on the laws of probability may be found in the following Section 35.3 (Problems 9 to 21), page 613.

35.3   Further problems

Simple probability

1.   A box contains 132 rivets of which 23 are undersized, 47 are oversized and 62 are satisfactory. Determine the probability of drawing at random: (*a*) one undersized; (*b*) one oversized; and

(*c*) one satisfactory rivet from the box.

(*a*) $\left[\dfrac{23}{132}\right]$ (*b*) $\left[\dfrac{47}{132}\right]$ (*c*) $\left[\dfrac{31}{66}\right]$

2. Four hundred resistors are examined and 6 per cent are found to be defective. Determine the probability that one selected at random will be defective and also the probability that it will not be defective. $\left[\dfrac{3}{50}, \dfrac{47}{50}\right]$

3. A purse contains 7 copper and 13 silver coins. Determine the probability of selecting a copper coin when one is taken at random. $\left[\dfrac{7}{20}\right]$

4. Determine the probability of winning a prize in a raffle by buying 3 tickets, when there are 7 prizes and a total of 450 tickets are sold. $\left[\dfrac{7}{150}\right]$

5. Determine the probability of drawing a multiple of 3 from 50 cards marked from 1 to 50. $\left[\dfrac{8}{25}\right]$

6. Determine the probability of an event not happening when the probability of it happening is $\dfrac{7}{93}$ $\left[\dfrac{86}{93}\right]$

7. The probability of winning a prize in a raffle by buying 4 tickets is $\dfrac{1}{300}$. Find out how many tickets were sold in a raffle having 5 prizes. [6 000]

8. A batch of 700 components contains 16 defective ones. Determine the probability of selecting at random one defective item. $\left[\dfrac{4}{175}\right]$

Laws of probability

Problems 9 to 14 refer to a box that contains 131 similar transistors, of which 70 are satisfactory, 43 gives too high a gain under normal operating conditions and 18 give too low a gain. Determine the probabilities stated.

9. The probability when drawing two transistors in turn, at random, with replacement, of having: (*a*) two satisfactory, (*b*) none with low gain, (*c*) one satisfactory and one with high gain, (*d*) one with low gain and none satisfactory.

$(a) \left[\dfrac{4\,900}{17\,161}\right]$ $(b) \left[\dfrac{12\,769}{17\,161}\right]$ $(c) \left[\dfrac{6\,020}{17\,161}\right]$ $(d) \left[\dfrac{2\,196}{17\,161}\right]$

10. Determine the probabilities required in Problem 9, but with no replacement between draws.

$(a) \left[\dfrac{483}{1\,703}\right]$ $(b) \left[\dfrac{6\,328}{8\,515}\right]$ $(c) \left[\dfrac{602}{1\,703}\right]$ $(d) \left[\dfrac{1\,089}{8\,515} \text{ or } \dfrac{2\,117}{17\,030}\right]$

11. The probability, when drawing two transistors at random, with replacement, of having: (a) two satisfactory or two with high gain, (b) the second draw being not low gain, and (c) at least one with high gain.

$(a) \left[\dfrac{6\,749}{17\,161}\right]$ $(b) \left[\dfrac{113}{131}\right]$ $(c) \left[\dfrac{9\,417}{17\,161}\right]$

12. Determine the probabilities required in Problem 11, but with no replacement between draws. $(a) \left[\dfrac{3\,318}{8\,515}\right]$

$(b) \left[\text{either } \dfrac{112}{130} \text{ or } \dfrac{113}{130}, \text{ depending on the gain of the first transistor drawn}\right]$ $(c) \left[\dfrac{4\,687}{8\,515}\right]$

13. The probability, when drawing two transistors at random, with replacement, of having: (a) two satisfactory or one with high gain and one with low gain, (b) at most one with low gain, and (c) one satisfactory and one with low gain or one satisfactory and one with high gain.

$(a) \left[\dfrac{6\,448}{17\,161}\right]$ $(b) \left[\dfrac{16\,837}{17\,161}\right]$ $(c) \left[\dfrac{8\,540}{17\,161}\right]$

14. Determine the probabilities required in Problem 13, but with no replacement between draws.

$(a) \left[\dfrac{3\,189}{8\,515}\right]$ $(b) \left[\dfrac{8\,362}{8\,515}\right]$ $(c) \left[\dfrac{854}{1\,703}\right]$

15. A box contains 13 tickets numbered from 1 to 13. Two tickets are drawn from the box, one at a time, with replacement. Determine the probabilities that they are: (a) odd then even, and (b) even then odd numbers.

$(a) \left[\dfrac{42}{169}\right]$ $(b) \left[\dfrac{42}{169}\right]$

16. Determine the probabilities required in Problem 15, but with no replacement between draws. $(a) \left[\dfrac{7}{26}\right]$ $(b) \left[\dfrac{7}{26}\right]$

17. The probabilities of an engine failing are given by: p_1, failure due to overheating; p_2, failure due to ignition problems; p_3, failure due to fuel blockage. When $p_1 = \dfrac{1}{7}$, $p_2 = \dfrac{2}{9}$ and

$p_3 = \dfrac{3}{11}$, determine the probabilities of:

(a) both p_1 and p_2 happening,
(b) either p_2 or p_3 happening, and
(c) both p_1 and either p_2 or p_3 happening.

(a) $\left[\dfrac{2}{63} \right]$ (b) $\left[\dfrac{49}{99} \right]$ (c) $\left[\dfrac{7}{99} \right]$

18. Actuarial tables show that the life expectancy of three men, A, B and C, over a twenty-year period depends on their age and is given by $p_A = \dfrac{4}{15}$, $p_B = \dfrac{11}{15}$ and $p_C = \dfrac{14}{15}$. Determine the probabilities that in twenty years:
(a) all three men will be alive,
(b) A will be alive but B and C will be dead,
(c) at least one man will be alive.

(a) $\left[\dfrac{616}{3\,375} \right]$ (b) $\left[\dfrac{16}{3\,375} \right]$ (c) $\left[\dfrac{3\,331}{3\,375} \right]$

19. Bag A contains 13 white and 15 black marbles. Bag B contains 17 green and 19 red marbles. Three marbles are drawn at random, one from bag A and two from bag B, with replacement. Determine the probabilities of having:
(a) one white, one green and one red marble,
(b) no black marbles and two green marbles,
(c) either one white and two green or one white, one red and one green marble.

(a) $\left[\dfrac{8\,398}{36\,288} \right]$ (b) $\left[\dfrac{3\,757}{36\,288} \right]$ (c) $\left[\dfrac{12\,155}{36\,288} \right]$

20. Determine the probabilities required in Problem 19, but with no replacement between draws.

(a) $\left[\dfrac{8\,398}{35\,280} \right]$ (b) $\left[\dfrac{221}{2\,205} \right]$ (c) $\left[\dfrac{5\,967}{17\,640} \right]$

21. Three types of seed are planted with a chance of growth of $\dfrac{1}{8}, \dfrac{1}{5}, \dfrac{1}{10}$. What are the probabilities that they all grow, that only one type grows, and that at least one type grows?

$\left[\dfrac{1}{400}, \dfrac{127}{400}, \dfrac{37}{100} \right]$

Chapter 36

The binomial and Poisson probability distributions

36.1 Probability distributions

Frequency distributions are discussed in *Technician Mathematics 1*, Chapter 14, in which it is shown how data can be grouped into classes containing the variable and its frequency. These distributions can be readily analysed to give a better understanding of the data by means of histograms, frequency polygons and numerical values, like measures of central tendency and deviation. The relative frequency of a class is the ratio of the frequency of the class to the total frequency of all classes, and hence probability and relative frequency are closely linked. It follows that information on probabilities can be presented in a similar way to information on frequencies. These data then form a **probability distribution** and contain information on the variable and its probability. A probability distribution can also be represented pictorially by means of a histogram and frequency polygon. A **discrete** probability distribution is one in which the values of probability can have certain values only and will have 'steps' in any pictorial representation. A **continuous** probability distribution can have any value between certain limits and will be a smooth curve when represented pictorially.

Three of the principal theoretical probability distributions are introduced in this book. Two discrete probability distributions called the binomial distribution and the Poisson distribution are introduced in this chapter and a continuous distribution called the normal distribution is introduced in Chapter 37.

36.2 The binomial distribution

As the name implies, the binomial distribution deals with 'two numbers' only and these are often taken as the probability that an event will happen, p (called the probability of success), and the probability that an event will not happen, q (called the probability of failure). In this context, $p + q$ must be equal to unity. The binomial distribution is the basis of much of the statistical work done in industrial inspection and in research.

Suppose that a large number of balls are placed in a bag, 10 per cent of them being red and the remainder black. Let the number be large enough so that if drawing takes place with replacement or without replacement, the result is almost the same. The probability p of drawing a red ball is $\dfrac{1}{10}$. The probability of drawing two red balls (**both** a red ball **and** another red ball) is given by the multiplication law of probability and is $\dfrac{1}{10} \times \dfrac{1}{10}$, i.e. $\dfrac{1}{100}$.

It follows that the probability of drawing n red balls is $\left(\dfrac{1}{10}\right)^n$. Similarly, the probability of drawing n black balls is $\left(\dfrac{9}{10}\right)^n$. These probabilities can be readily determined, but to determine the probability that when 7 balls are drawn, 3 will be red and the remainder black is far more difficult and it is this type of problem which can be solved using the binomial distribution.

Let R signify a red ball and B a black ball. The various possibilities which exist when drawing two balls from the bag are: two reds, a red and a black, a black and red and two blacks, i.e.

R R, R B, B R, B B

The probability of drawing two red balls is p^2, a red and black ball is pq, and so on, and summarising the results when two balls are drawn at random:

Result	R R	R B or B R	B B
Probability	p^2	$2pq$	q^2

When three balls are drawn at random from the bag, the various possibilities are as shown below, together with their probabilities.

R R R	p^3
R R B	p^2q
R B R	p^2q
R B B	pq^2
B R R	p^2q
B R B	pq^2
B B R	pq^2
B B B	q^3

Summary:

Result	3R	2R, 1B	1R, 2B	3B
Probability	p^3	$3p^2q$	$3pq^2$	q^3

When the same procedure is carried out for four balls drawn at random, the summary of results is:

Result	4R	3R, 1B	2R, 2B	1R, 3B	4B
Probability	p^4	$4p^3q$	$6p^2q^2$	$4pq^3$	q^4

Comparing these summaries with the results obtained in Chapter 1 (the binomial expansion) for expansions of $(a + b)^2$, $(a + b)^3$ and $(a + b)^4$, we can see that the terms obtained in the probability summaries are of the same form as those obtained when carrying out a binomial expansion.

In practical statistics, for example, sampling in industry for inspection purposes, red balls in a bag correspond to defective items in a large batch of items and the black balls correspond to items which are not defective. In this case the binomial probability distribution is defined as:

The probability that 0, 1, 2, 3, ... defective items in a sample of n items drawn at random from a large population, whose probability of defective items is p and whose probability of non-defective items is q, is given by the successive terms of the expansion of $(q + p)^n$, taking terms in succession from left to right.

For example, a certain machine is producing, say, 1 000 components per hour, and batches of 10 are taken at random for inspection purposes every 30 minutes. By inspecting these components, it is possible to predict the defect rate for the machine over a period of time. Let this be, say, 15 per cent, then p is $\dfrac{15}{100}$ or $\dfrac{3}{20}$. The probability q of a non-defective item is $1 - \dfrac{3}{20}$, i.e. $\dfrac{17}{20}$. If a sample of, say, 4 components is selected at random from the output of the machine, the probability of having 0, 1, 2, 3 or 4 defective components is given by the successive terms of the expansion of $(q + p)^4$ taken from left to right. Applying the binomial expansion $(q + p)^4$ gives:

$$(q + p)^4 = q^4 + 4q^3p + 6q^2p^2 + 4qp^3 + p^4,$$

and substituting $p = \dfrac{3}{20}$ and $q = \dfrac{17}{20}$ gives:

$$\left(\frac{17}{20} + \frac{3}{20}\right)^4 = \left(\frac{17}{20}\right)^4 + 4\left(\frac{17}{20}\right)^3\left(\frac{3}{20}\right) + 6\left(\frac{17}{20}\right)^2\left(\frac{3}{20}\right)^2$$

$$+ 4\left(\frac{17}{20}\right)\left(\frac{3}{20}\right)^3 + \left(\frac{3}{20}\right)^4$$

$$= 0.522 + 0.368 + 0.098 + 0.011 + 0.001$$

when these values are determined correct to 3-decimal-place accuracy. This result shows that:

Number defective	0	1	2	3	4
Probability	0.522	0.368	0.098	0.011	0.001

This result means that when drawing random samples of 4 components from the output of the machine when it has a defect production rate of 15 per cent, it is likely that, when drawing say 100 such samples, in 52 samples there will be no defective items, in 37 samples, there will be one defective item, in 10 samples there will be two defective items, and so on. Providing these probabilities remain reasonably constant, it can be predicted that the production defect rate for the machine is remaining reasonably constant.

A more general statement of the binomial probability distribution is:

If p is the probability that an event will happen and q the probability that it will not happen, then the probability that the event will happen 0, 1, 2, 3, ... times in n trials is given by the successive terms of the expansion of $(q + p)^n$, taken from left to right.

It is shown in Chapter 1 that the terms comprising the expansion of $(q + p)^n$ can be obtained either using Pascal's triangle or by using the general binomial expansion:

$$(q + p)^n = q^n + nq^{n-1}p + \frac{n(n-1)}{2!}q^{n-2}p^2 + \cdots$$

An example of this more general application is to determine the probability of a three-child family having, say, 2 boys, assuming the probability of birth of a boy and girl are the same. The probability of a boy being born is $\frac{1}{2}$ and the probability of a girl being born is also $\frac{1}{2}$. The probability of 0, 1, 2 or 3 boys being born to a family of 3 children is given by the successive terms of the expansion of $(q + p)^3$ taken from left to right.

$$(q + p)^3 = q^3 + 3q^2p + 3qp^2 + p^3$$

and substituting $q = \frac{1}{2}$ and $p = \frac{1}{2}$ gives:

$$\left(\frac{1}{2} + \frac{1}{2}\right)^3 = \left(\frac{1}{2}\right)^3 + 3\left(\frac{1}{2}\right)^2\left(\frac{1}{2}\right) + 3\left(\frac{1}{2}\right)\left(\frac{1}{2}\right)^2 + \left(\frac{1}{2}\right)^3$$

$$= \frac{1}{8} + \frac{3}{8} + \frac{3}{8} + \frac{1}{8}$$

and these terms
give the   0 boys   1 boy   2 boys   3 boys
probabilities of

i.e. the probability of there being 2 boys in a 3-child family is $\frac{3}{8}$

The results of a binomial probability distribution can be represented pictorially by drawing a histogram, and the histograms for the machine having 15 per cent defective items and for the number of boys in a 3-child family are shown in Figs 36.1(a) and 36.1(b) respectively.

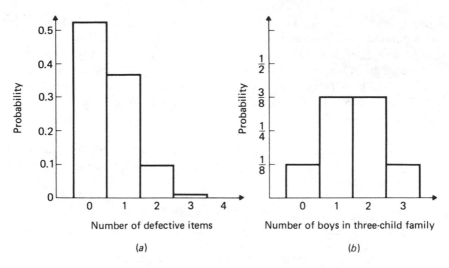

Fig. 36.1   Histograms of binomial probability distribution

Worked problems on the binomial distribution

Problem 1. A bag contains a large number of marbles of which 25 per cent are white and the remainder green. Five marbles are drawn from the bag at random. Determine the probability of having:

(*a*)   2 white and 3 green marbles,
(*b*)   at least 3 white marbles, and
(*c*)   not more than 3 green marbles.

Let p be the probability of having a white marble and q be the probability of not having a white marble, i.e. the probability of having a green marble. The sample number n is 5. Then the probability of having 0, 1, 2, 3, 4 or 5 white marbles is given by the successive terms of the expansion of $(q + p)^5$ taken from left to right. Using Pascal's triangle to obtain the expansion of $(q + p)^5$ (see Chapter 1) gives:

$$(q + p)^5 = q^5 + 5q^4p + 10q^3p^2 + 10q^2p^3 + 5qp^4 + p^5$$

Substituting $q = \dfrac{3}{4}$ and $p = \dfrac{1}{4}$ gives:

$$\left(\frac{3}{4} + \frac{1}{4}\right)^5 = \left(\frac{3}{4}\right)^5 + 5\left(\frac{3}{4}\right)^4\left(\frac{1}{4}\right) + 10\left(\frac{3}{4}\right)^3\left(\frac{1}{4}\right)^2 + 10\left(\frac{3}{4}\right)^2\left(\frac{1}{4}\right)^3$$

$$+ 5\left(\frac{3}{4}\right)\left(\frac{1}{4}\right)^4 + \left(\frac{1}{4}\right)^5$$

and summarising:

Term	1st	2nd	3rd	4th	5th	6th
Marbles	0W, 5G	1W, 4G	2W, 3G	3W, 2G	4W, 1G	5W, 0G
Probability	$\left(\dfrac{3}{4}\right)^5$	$5\left(\dfrac{3}{4}\right)^4\left(\dfrac{1}{4}\right)$	$10\left(\dfrac{3}{4}\right)^3\left(\dfrac{1}{4}\right)^2$	$10\left(\dfrac{3}{4}\right)^2\left(\dfrac{1}{4}\right)^3$	$5\left(\dfrac{3}{4}\right)\left(\dfrac{1}{4}\right)^4$	$\left(\dfrac{1}{4}\right)^5$

(a) The probability of having 2 white and 3 green marbles is given by the 3rd term of the expansion,

i.e. $10\left(\dfrac{3}{4}\right)^3\left(\dfrac{1}{4}\right)^2$, i.e. $\dfrac{10 \times 3^3}{4^5}$ or $\dfrac{\mathbf{135}}{\mathbf{512}}$

(b) The probability of having at least 3 white marbles is made up of the sum of the 4th, 5th and 6th terms.

The 4th term $= 10\left(\dfrac{3}{4}\right)^2\left(\dfrac{1}{4}\right)^3 = \dfrac{10 \times 3^2}{4^5} = \dfrac{90}{1\,024}$

The 5th term $= 5\left(\dfrac{3}{4}\right)\left(\dfrac{1}{4}\right)^4 = \dfrac{5 \times 3}{4^5} = \dfrac{15}{1\,024}$

The 6th term $= \left(\dfrac{1}{4}\right)^5 = \dfrac{1}{1\,024}$

Hence the sum of these term is $\dfrac{90 + 15 + 1}{1\,024} = \dfrac{106}{1\,024} = \dfrac{\mathbf{53}}{\mathbf{512}}$

(c) The probability of having not more than 3 green marbles is given by the sum of the 3rd, 4th, 5th and 6th terms. Since the sum of all the terms is unity and the values of the 3rd and subsequent terms have been previously calculated, then the required value is $\dfrac{53}{512} + \dfrac{135}{512}$, these terms being the solutions to parts (b) and (a),

i.e. $\dfrac{188}{512}$ or $\dfrac{\mathbf{47}}{\mathbf{128}}$

Problem 2. A machine produces 20 per cent defective components. In a sample of 6, drawn at random, determine the probability that:
(a) there will be 4 defective items,
(b) there will be not more than 3 defective items, and
(c) all the items will be non-defective.

Let p be the probability of a component being defective, then $p = \dfrac{1}{5}$. Also, let q be the probability of a component not being defective, then $q = \dfrac{4}{5}$. The probability of 0, 1, 2, 3, 4, 5 or 6 defective items in a random sample of 6 items is given by the successive terms of the expansion of $(q + p)^6$, taken from left to right. Using Pascal's triangle to expand $(q + p)^6$ gives:

$$(q + p)^6 = q^6 + 6q^5p + 15q^4p^2 + 20q^3p^3 + 15q^2p^4 + 6qp^5 + p^6$$

and corresponds to:

0 defectives, 1 defective, 2 defectives, 3 defectives, 4 defectives, 5 defectives, 6 defectives

The problem requires only information on the first 5 terms of the expansion, so only these will be considered.

Substituting for q and p gives:

$$\left(\frac{4}{5} + \frac{1}{5}\right)^6 = \left(\frac{4}{5}\right)^6 + 6\left(\frac{4}{5}\right)^5\left(\frac{1}{5}\right) + 15\left(\frac{4}{5}\right)^4\left(\frac{1}{5}\right)^2 + 20\left(\frac{4}{5}\right)^3\left(\frac{1}{5}\right)^3$$
$$+ 15\left(\frac{4}{5}\right)^2\left(\frac{1}{5}\right)^4 + \cdots$$

Summarising:

Term	1st	2nd	3rd	4th	5th
Defectives	0	1	2	3	4
Probability	$\left(\frac{4}{5}\right)^6$	$6\left(\frac{4}{5}\right)^5\left(\frac{1}{5}\right)$	$15\left(\frac{4}{5}\right)^4\left(\frac{1}{5}\right)^2$	$20\left(\frac{4}{5}\right)^3\left(\frac{1}{5}\right)^3$	$15\left(\frac{4}{5}\right)^2\left(\frac{1}{5}\right)^4$

(a) The probability of 4 defective items is given by the 5th term and is

$$15\left(\frac{4}{5}\right)^2\left(\frac{1}{5}\right)^4, \text{ i.e. } \frac{15 \times 4^2}{5^6} \text{ or } \frac{240}{15\,625} \text{ or } \frac{48}{3\,125}$$

(b) The probability of not more than 3 defective items is given by the sum of the first 4 terms.

$$\text{1st term} = \left(\frac{4}{5}\right)^6 = \frac{4\,096}{15\,625}$$

$$\text{2nd term} = 6\left(\frac{4}{5}\right)^5\left(\frac{1}{5}\right) = \frac{6 \times 4^5}{5^6} = \frac{6\,144}{15\,625}$$

$$\text{3rd term} = 15\left(\frac{4}{5}\right)^4\left(\frac{1}{5}\right)^2 = \frac{3\,840}{15\,625}$$

$$\text{4th term} = 20\left(\frac{4}{5}\right)^3\left(\frac{1}{5}\right)^3 = \frac{1\,280}{15\,625}$$

Hence the sum of the first 4 terms is $\dfrac{4\,096 + 6\,144 + 3\,840 + 1\,280}{15\,625}$

that is, $\dfrac{15\,360}{15\,625}$, or $\dfrac{3\,072}{3\,125}$

(c) The probability of all items non-defective is given by the first term, and is $\dfrac{4\,096}{15\,625}$ from part (b)

Problem 3. Four hundred families have 4 children each. Assuming equal probabilities for boy and girl births, determine how many families will have: (a) 3 boys, (b) 2 girls, and (c) either 2 boys and 2 girls or 3 boys and 1 girl.

Let the probability of a boy being born be p and that of a girl being born be q. Then $p = q = \dfrac{1}{2}$. The probability of 0, 1, 2, 3 or 4 boys in a family having 4 children is given by the successive terms of the expansion of $(q + p)^4$, taken from left to right.

$$(q + p)^4 = q^4 + 4q^3p + 6q^2p^2 + 4qp^3 + p^4$$

Number of boys   0   1   2   3   4

(a) The probability of having 3 boys is given by the 4th term.

$$4qp^3 = 4\left(\frac{1}{2}\right)\left(\frac{1}{2}\right)^3 = \frac{4}{16} = \frac{1}{4}$$

(b) The probability of having 2 girls is the same as the probability of having 2 boys and is given by the 3rd term of the expansion.

$$6q^2p^2 = 6\left(\frac{1}{2}\right)^2\left(\frac{1}{2}\right)^2 = \frac{6}{16} = \frac{3}{8}$$

(c) The probability of having 2 boys and 2 girls or 3 boys and 1 girl will be the sum of the 3rd and 4th terms. From parts (a) and (b), this is $\dfrac{1}{4} + \dfrac{3}{8}$, i.e. $\dfrac{5}{8}$

Further problems on the binomial probability distribution may be found in Section 36.4 (Problems 1 to 15), page 629.

36.3   The Poisson distribution

The calculations associated with the binomial distribution become very laborious when the sample number n becomes larger than about 10. When n

is large and the probability p is small, so that the expectation np is less than 5, then a very good approximation to the binomial distribution can be obtained by using another probability distribution called the Poisson distribution, in which the calculations are normally far easier. In addition, the Poisson distribution is also used in its own right to determine probabilities associated with events which cannot be resolved by using a binomial distribution.

The binomial expansion of $(q + p)^n$ is:

$$q^n + nq^{n-1}p + \frac{n(n-1)}{2!} q^{n-2}p^2 + \frac{n(n-1)(n-2)}{3!} q^{n-3}p^3 + \cdots$$

Also in the binomial distribution $q = 1 - p$. Hence when p is small $q \approx 1$. Making this approximation, the binomial expansion of $(q + p)^n$ becomes:

$$1 + np + \frac{n(n-1)}{2!} p^2 + \frac{n(n-1)(n-2)}{3!} p^3 + \cdots$$

Also, when n is large, terms such as $(n-1)$, $(n-2)$, $(n-3), \ldots$ are approximately equal to n, and applying this approximation to the binomial expansion of $(q + p)^n$ gives:

$$1 + np + \frac{n^2 p^2}{2!} + \frac{n^3 p^3}{3!} + \cdots$$

Writing the expectation, np, as λ gives:

$$1 + \lambda + \frac{\lambda^2}{2!} + \frac{\lambda^3}{3!} + \cdots$$

but it is stated in Chapter 2 that the power series for e^λ is

$$1 + \lambda + \frac{\lambda^2}{2!} + \frac{\lambda^3}{3!} + \cdots$$

One of the requirements of any probability distribution is that the sum of all the probabilities is equal to unity. Since n and p are in no way related, this is not necessarily the case for the terms making up the expansion of e^λ. To meet this condition, both sides of the equation are divided by e^λ, giving:

$$\frac{e^\lambda}{e^\lambda} = 1 = \frac{1}{e^\lambda}\left(1 + \lambda + \frac{\lambda^2}{2!} + \frac{\lambda^3}{3!} + \cdots\right)$$

The successive terms are $e^{-\lambda}$, $\lambda e^{-\lambda}$, $\dfrac{\lambda^2 e^{-\lambda}}{2!}$, $\dfrac{\lambda^3 e^{-\lambda}}{3!}, \ldots$

These successive terms taken from left to right give a very good approximation to the binomial distribution when n is large, p is small and λ is the expectation np.

For example, a machine produces 3 per cent defective items. We can determine the probability that there will be, say, two defective items in a sample of 15 items selected at random from the output of the machine by using either: (*a*) the binomial distribution, or (*b*) the Poisson approximation to the binomial distribution.

(*a*) Using the binomial distribution, $p = 0.03$, $q = 1 - p = 0.97$ and $n = 15$. Also

$$(q + p)^{15} = q^{15} + 15q^{14}p + \frac{15 \times 14}{2!}q^{13}p^2 + \cdots$$

these terms giving the probabilities of having 0, 1 or 2 defective items respectively. Taking the third term, the probability of having two defective items is

$\dfrac{15 \times 14}{2!}q^{13}p^2$, and substituting for q and p gives

$\dfrac{15 \times 14}{2!}(0.97)^{13}(0.03)^2$, i.e. $105 \times 0.673\,0 \times 0.000\,9$,

or $0.063\,6$

(*b*) For the Poisson approximation to the binomial distribution, $n = 15$, $p = 0.03$, so $\lambda = np = 0.45$. The probability of having $0, 1, 2, \ldots$ defective items is given by the terms $e^{-\lambda}$, $\lambda e^{-\lambda}$, $\dfrac{\lambda^2 e^{-\lambda}}{2!}, \ldots$ respectively. Taking the third term, the probability of two defective items is

$\dfrac{\lambda^2 e^{-\lambda}}{2!}$, i.e. $\dfrac{0.45^2\, e^{-0.45}}{2}$, or $0.064\,6$

These results differ by less than 2 per cent and when n becomes 50 or more and np is less than 5, the difference between the binomial and the Poisson approximation to the binomial distribution is barely detectable. It can be seen that the calculations in (*b*) are easier than those in (*a*) and the difference in the ease of calculations becomes more noticeable as n becomes larger.

The principal use of the Poisson distribution is to determine the probability of an event happening where the probability of the event not happening is not known. For example, the probability of a particular machine breaking down can be determined by noting the number of times it breaks down in a certain period (this is the expectation). It is not known how many times it did not break down in that period.

The Poisson distribution can be stated as follows:

If the chance of an event occurring at any instant is constant and the expectation of the event occuring in a period of time is λ, then the probability

*of the event occurring 0, 1, 2, 3, ... times is given by the successive terms of
the expansion of*

$$e^{-\lambda}\left(1 + \lambda + \frac{\lambda^2}{2!} + \frac{\lambda^3}{3!} + \cdots\right), \textit{ taken from left to right.}$$

For example, if between 10 and 11 o'clock in the morning, the average
number of telephone calls received by the switchboard of a company is 4 per
minute, the Poisson distribution can be used to determine the probability
that in any particular minute, say, 3 calls will arrive. The number of calls
which did not arrive is not known, hence the binomial distribution cannot
be used. The probability of receiving 0, 1, 2 or 3 calls is given by the terms
$e^{-\lambda}$, $\lambda e^{-\lambda}$, $\dfrac{\lambda^2 e^{-\lambda}}{2!}$ or $\dfrac{\lambda^3 e^{-\lambda}}{3!}$, respectively, where λ is the expectation of a call
arriving and is 4 per minute. The probability that there will be 3 calls is
given by the 4th term, i.e. $\dfrac{\lambda^3 e^{-\lambda}}{3!}$. Substituting $\lambda = 4$ gives:

$$\frac{\lambda^3 e^{-\lambda}}{3!} = \frac{4^3 e^{-4}}{3!} = 0.195$$

That is, the probability of 3 calls arriving in any particular minute is 0.195.
This type of information can be used to determine the number of lines
required by a switchboard in order to keep the probability of a person getting
an engaged tone, when ringing the switchboard, to within certain limits.

Worked problems on the Poisson distribution

Problem 1. If 2 per cent of the electric light bulbs produced by a
company are defective, determine the probability that in a sample
of 60 bulbs: (*a*) 3 bulbs, (*b*) not more than 3 bulbs, and (*c*) at least
2 bulbs, will be defective.

Since n is large and p is small, the Poisson approximation to
the binomial distribution can be used. The expectation λ is np, i.e.
60×0.02 or 1.2. The probability of having 0, 1, 2, 3, ... defective
bulbs is given by the terms $e^{-\lambda}$, $\lambda e^{-\lambda}$, $\dfrac{\lambda^2 e^{-\lambda}}{2!}$, $\dfrac{\lambda^3 e^{-\lambda}}{3!}$, ...
respectively.

(*a*) The probability of having 3 defective bulbs is given by $\dfrac{\lambda^3 e^{-\lambda}}{3!}$.
Substituting for λ gives

$$\frac{\lambda^3 e^{-\lambda}}{3!} = \frac{1.2^3 e^{-1.2}}{3 \times 2}$$

$$= 0.086\,7$$

That is, **the probability of having 3 defective bulbs is 0.086 7**

(b) The probability of not more than 3 bulbs being defective is the probability of there being no bulbs, 1 bulb, 2 bulbs and 3 bulbs defective, that is, the sum of the first 4 terms of the Poisson distribution. Now

$$e^{-\lambda} = e^{-1.2} = 0.3012$$

$$\lambda e^{-\lambda} = 1.2 \times 0.3012 = 0.3614$$

$$\frac{\lambda^2 e^{-\lambda}}{2!} = \frac{1.2^2 \times 0.3012}{2} = 0.2169$$

and $\dfrac{\lambda^3 e^{-\lambda}}{3!} = 0.0867$ from part (a)

Thus, the probability of having not more than 3 bulbs defective is given by the sum of these probabilities, i.e.

$$0.3012 + 0.3614 + 0.2169 + 0.0867 = 0.9662$$

That is, **the probability of having not more than 3 bulbs defective is 0.9662**

(c) The probability that at least 2 bulbs will be defective means two or more of the sample being defective. Since the total probability is unity, the probability of 2 or more being defective is {the total probability} less {the probability of having no defective bulbs and the probability of having one defective bulb}, i.e.

$$1 - e^{-\lambda} - \lambda e^{-\lambda}$$

From part (b) this is $1 - (0.3012 + 0.3614)$, i.e. 0.3374

Thus, **the probability of having two or more defective bulbs is 0.3374**

Problem 2. A team scores an average of 3 goals per match in 45 matches. Determine in how many matches they would expect to score: (a) 4 goals, and (b) less than 2 goals, assuming a Poisson distribution.

The probability of scoring 0, 1, 2, 3 or 4 goals is given by the successive terms of the expansion of $e^{-\lambda}\left(1 + \lambda + \dfrac{\lambda^2}{2!} + \dfrac{\lambda^3}{3!} + \cdots\right)$ taken from left to right. Hence:

Number of goals scored	0	1	2	3	4
Probability	$e^{-\lambda}$	$\lambda e^{-\lambda}$	$\dfrac{\lambda^2 e^{-\lambda}}{2!}$	$\dfrac{\lambda^3 e^{-\lambda}}{3!}$	$\dfrac{\lambda^4 e^{-\lambda}}{4!}$

The expectation of scoring, λ, is 3 per match, giving the value of $e^{-\lambda}$ as 0.0498

(*a*) The probability of scoring 4 goals in one match is given by
$\dfrac{\lambda^4 e^{-\lambda}}{4!}$, i.e. $\dfrac{3^4 \times 0.049\,8}{24}$ or $0.168\,1$

Thus the expectation of scoring 4 goals in any of the 45 matches is $0.168\,1 \times 45$, i.e. 7.565, which is 'rounded-up' to **8 matches**.

(*b*) The probability of scoring less than two goals in one match is the sum of the probabilities of scoring no goals and one goal, i.e. $e^{-\lambda} + \lambda e^{-\lambda}$. Substituting $\lambda = 3$ gives

$$e^{-\lambda} + \lambda e^{-\lambda} = 0.049\,8 + 3 \times 0.049\,8$$
$$= 0.199\,2$$

The expectation of scoring less than two goals in any of the 45 matches is $0.199\,2 \times 45$, i.e. 8.964, which is 'rounded-up' to **9 matches**.

Problem 3. Special drills of a certain diameter are kept in a machine shop store. A survey reveals that they are drawn from the store, on average, 1.5 times a day. Determine for a period of 200 working days:
(*a*) the number of days when none of the drills are used, and
(*b*) the number of days when four of the drills are in use, assuming the demand follows a Poisson distribution.

The probability of the demand being for 0, 1, 2, 3, 4, . . . drills is given by the successive terms of the expansion of

$$e^{-\lambda}\left(1 + \lambda + \frac{\lambda^2}{2!} + \frac{\lambda^3}{3!} + \frac{\lambda^4}{4!} + \cdots\right),\ \text{taken from left to right. Hence:}$$

Demand for drills	0	1	2	3	4
Probability	$e^{-\lambda}$	$\lambda e^{-\lambda}$	$\dfrac{\lambda^2 e^{-\lambda}}{2!}$	$\dfrac{\lambda^3 e^{-\lambda}}{3!}$	$\dfrac{\lambda^4 e^{-\lambda}}{4!}$

The daily expectation of there being a demand, λ, is 1.5, giving a value of $e^{-\lambda}$ of $0.223\,1$

(*a*) The probability of there being a demand for none of the drills in a day is given by $e^{-\lambda}$, i.e. $0.223\,1$. In 200 days, it is probable that no drills are used on $200 \times 0.223\,1 = 44.62$, i.e. **45 days** when 'rounded-up'.

(*b*) The probability of there being a demand for four drills on any day is given by $\dfrac{\lambda^4 e^{-\lambda}}{4!}$, i.e.
$$\frac{(1.5^4)(0.223\,1)}{(4)(3)(2)(1)} = 0.047\,06$$

In 200 days, it is probable that four drills are used on
$200 \times 0.047\,06 = 9.412$, i.e. **10 days** when 'rounded-up'.

Further problems on the Poisson distribution may found in the following Section 36.4 (Problems 16 to 28), page 630.

36.4 Further problems

Binomial distribution

Problems 1 to 5 refer to a box containing a large number of capacitors of which 70 per cent are within given tolerance values and the remainder are not. Determine the probabilities stated.

1. When three capacitors are drawn at random, determine the probability that: (*a*) two are not within, and (*b*) two are within the given tolerance values.   (*a*) [0.189] (*b*) [0.441]

2. When six capacitors are drawn at random, determine the probability that: (*a*) there are three not within, and (*b*) there are not more than two within the given tolerance values. (*a*) [0.185 2] (*b*) [0.070 5]

3. When nine capacitors are drawn at random, determine the probability that: (*a*) less than three are not within, and (*b*) six are within the given tolerance values. (*a*) [0.462 8] (*b*) [0.266 8]

4. When five capacitors are drawn at random, find the probability that: (*a*) more than four are not within, and (*b*) there are two not within the given tolerance values. (*a*) [0.002 4] (*b*) [0.308 7]

5. When seven capacitors are drawn at random, find the probability that: (*a*) at least two are within, and (*b*) at most three are not within the given tolerance values. (*a*) [0.996 2] (*b*) [0.874 0]

6. A machine produces 10 per cent defective components. In a sample of five drawn at random, find the probability of having: (*a*) three defective, (*b*) two defective, and (*c*) no defective components.   (*a*) [0.008] (*b*) [0.073] (*c*) [0.591]

7. A target is hit by a marksman once in four shots, on average. When firing seven shots, determine the probability of: (*a*) obtaining three hits, and (*b*) obtaining at least two hits. (*a*) [0.173 0] (*b*) [0.558 9]

8. A box contains 75 components and, when inspected, five are found to be defective. When six are chosen at random from the box, determine the probability of (*a*) having no defective components, and (*b*) having more than three defective

components in the sample.
(a) [0.661 0]   (b) [2.656 × 10^{-4}]

9. The probability of winning a prize at a fair is once in each
    eight tries, on average. Determine the probability of winning
    three prizes in nine tries.   [0.073 6]

10. The probability of passing an examination is 0.65. Determine
    the probability that out of eight students: (a) just three, (b)
    just five, and (c) just seven will pass the examination.
    (a) [0.080 8]   (b) [0.278 6]   (c) [0.137 3]

11. Six people are all the same age and in good health. Actuarial
    tables show that the probability that they will all be alive in
    25 years is $\frac{3}{7}$. Find the probability that in 25 years: (a) all
    six, (b) two, and (c) at least three will be alive.
    (a) [0.006 2]   (b) [0.293 8]   (c) [0.643 3]

12. The output of a machine has, on average, 94 per cent perfect
    components. Determine the probability that in a sample of
    four components, more than one will be imperfect.
    [0.019 9]

13. Resistors are packed in packets of ten and there are, on
    average, 2 per cent defective. Determine the probability of
    finding two defective resistors in any packet.   [0.015 3]

14. A large consignment of eggs is delivered to a shop and on
    average there is one broken egg in every four boxes of six
    eggs delivered. Determine the probability of a shopper
    having a box with two broken eggs in it.   [0.022 0]

15. A sampling inspection scheme samples ten components from
    each batch supplied. If any defective items are found, the
    batch is returned to the supplier. If the supplier's batches
    have, on average, 5 per cent defective components, determine
    the percentage of batches returned to the supplier.
    [40.13 per cent]

The Poisson distribution

16. The output of an automatic machine is inspected by taking
    samples of 60 items. If the probability of a defective item is
    0.001 5, find the probability of having: (a) two defective items,
    (b) more than two defective items in a sample.
    (a) [0.003 7]   (b) [1.35 × 10^{-4}]

17. If 4 per cent of the tyres produced by a company are
    defective, find the probability that in a sample of 40 tyres
    there will be no defective tyres.   [0.201 9]

18. In the previous problem, find the probability that the sample will contain one or two or three defective tyres.   [0.719 3]
19. Inspection shows that 90 per cent of the components produced by a process are perfect. Find the probability that in a sample of ten components chosen at random, two will be defective by using: (a) the binomial distribution, and (b) the Poisson distribution.   (a) [0.193 7]   (b) [0.183 9]
20. The probability of a person having an accident in a certain period of time is 0.001. Determine the probability that out of 2 000 people: (a) just three, and (b) more than two will have an accident.   (a) [0.180 4]   (b) [0.323 3]
21. In a two-hour period an average of $2\frac{1}{2}$ telephone calls per minute arrive at a switchboard. Determine the probabilities that: (a) one, (b) three, (c) less than five, and (d) more than six calls will arrive in any particular minute.
    (a) [0.205 2]   (b) [0.213 8]   (c) [0.891 2]   (d) [0.042 0]
22. Of 40 lathes in a machine shop, the outage due to breakdowns averages 0.8 per week. Determine the probability that in a given week, more than two lathes will break down.   [0.047 4]
23. In 30 games a team scores on average two goals per game. Determine the probability that they will score two goals in their next game.   [0.270 7]
24. A 600-page book contains an average of one error per page, distributed at random. Find the probability that there will be two or more errors on a page.   [0.264 2]
25. A company has, on average, 120 man-days lost due to sickness every 100 working days. Determine on how many days over a period of 50 working days they may expect: (a) no absence due to sickness, and (b) three or more men to be absent.   (a) [16]   (b) [7]
26. The deposition of grit particles from the atmosphere was measured by counting the numbers of particles deposited on 200 prepared cards in a specified time, and the following table was drawn up:

No. of particles	0	1	2	3	4	5	6
No. of cards	45	65	52	24	11	3	0

Calculate the mean and variance of this distribution, each correct to 1 place of decimals, and hence show that it is reasonable to assume that the deposition of grit particles is according to a Poisson distribution. Using the mean calculated above, calculate also the theoretical probabilities of obtaining 0, 1, 2, 3, 4, or 5 or more particles on any one card, and hence prepare a table similar to the experimental one above for the expected frequencies of grit particles on 200 cards placed at random.

$$
\left[
\begin{array}{l}
\text{mean} = \\
\text{variance} = 1.5 \\
\text{probabilities} \quad 0.223\,1 \;\; 0.334\,7 \;\; 0.251\,0 \;\; 0.126\,0 \;\; 0.047\,0 \;\; 0.014\,1 \\
\phantom{\text{probabilities} \quad} 45 \qquad\;\; 67 \qquad\;\; 50 \qquad\;\; 25 \qquad\;\; 10 \qquad\;\; 3
\end{array}
\right]
$$

27. If the probability of suffering side-effects from an infection is 10^{-3}, what is the expectation that out of 2 000 people: (a) three, (b) not more than two patients will suffer side-reactions? (a) [0.180 4]   (b) [0.323 3]

28. The mean number of breakdowns in a distillation plant is 2.5 per 5-day week. What is the probability of no breakdowns on a particular day? If the factory works 50 weeks in a year, on how many days may two or more breakdowns be expected?   [0.606 5, 23 days]

Chapter 37

The normal probability distribution

37.1   The normal probability curve

The binomial and Poisson distributions deal with the occurrence of distinct events, that is, they are discrete probability distributions. Many instances occur where distributions are continuous, particularly when data relates to measured quantities such as length, mass, time, electric current, temperature, luminous intensity and their derived units. There are many instances where data from these sources approximate to a distribution called the **normal** distribution.

When the sample number, n, in the binomial distribution is made very large and a histogram is drawn of the results, it is found that the shape of the diagram approaches a symmetrical mathematical curve. This curve has an equation of the form: $y = \dfrac{1}{\sigma} e^{-x^2/2\sigma^2}$, where σ is the standard deviation of the data and x and y are the coordinates of points on the curve. This curve is called the **normal probability curve**, the y-values the **probability density** and the x-values the **variable** or **variate**. Figure 37.1 shows that the normal probability curve is a symmetrical curve.

It is the most important of the various distribution curves. When a large number of values are taken of, say, peoples' heights, masses, intelligence quotients, or the sizes of parts produced by a machine or the marks gained in an examination, it is found that when the data is plotted as a frequency curve, a curve of similar shape to the normal probability curve is produced in each case.

If any normal probability curve is drawn and the standard deviation, σ, of the data calculated, then the area under the curve and the value of the standard deviation are related. It is found that if a vertical line is drawn to pass through the maximum value of the normal probability curve (the mean value since the curve is symmetrical) and two other vertical lines are drawn at a distance of one standard deviation on either side of this line, then in all cases:

the area enclosed by the curve and the vertical lines at distances $\pm 1\sigma$ from the mean value is about $66\frac{2}{3}\%$ of the total area under the curve (the true value being 68.26%).

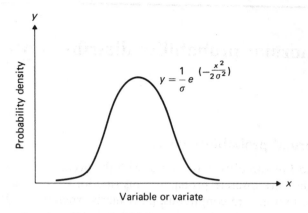

Fig. 37.1   The normal probability curve

Similarly, it is found that:

the area enclosed by the curve and the vertical lines at distances $\pm 2\sigma$ from the mean is about 95% of the total area under the curve (the true value being 95.44%).

Finally, it is found that:

the area enclosed by the curve and the vertical lines at distances $\pm 3\sigma$ from the mean is about $99\frac{3}{4}\%$ of the total area under the curve (the true value being 99.74%).

A normal probability curve showing these areas can be seen in Fig. 37.2.

The relationship between a normal probability curve and the area contained between the vertical lines drawn at various distances from the mean value forms the basis for such techniques as sampling and quality control. Because of this, it is worth becoming familiar with and committing to memory the approximate areas quoted in the last paragraph.

In histograms and distribution curves the area under the curve is proportional to frequency. In a normal probability curve, since the area between the curve and the vertical lines drawn at ± 1 standard deviation from the mean value is roughly $66\frac{2}{3}\%$ of the total area, it follows that $66\frac{2}{3}\%$ of the total frequency also lies between these values of standard deviation for normally distributed data.

For example, if the mean value of the heights of 50 people is 165 centimetres and the standard deviation is 5 centimetres, it is possible to predict how many people have heights between certain ranges. Taking plus or minus one standard deviation from the mean value gives $165 + 5 = 170$ centimetres and $165 - 5 = 160$ centimetres. Then $66\frac{2}{3}\%$ of the 50 people, i.e. 33 people, have heights between 160 and 170 centimetres. For plus or minus two standard deviations ($\sigma = 5$ cm, hence $2\sigma = 10$ cm), the range of heights

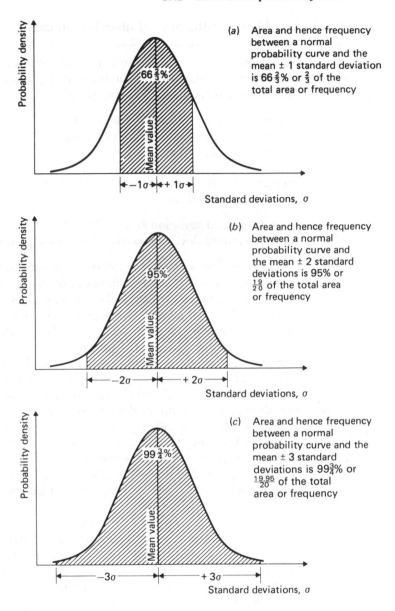

(a)   Area and hence frequency between a normal probability curve and the mean ± 1 standard deviation is $66\frac{2}{3}$% or $\frac{2}{3}$ of the total area or frequency

(b)   Area and hence frequency between a normal probability curve and the mean ± 2 standard deviations is 95% or $\frac{19}{20}$ of the total area or frequency

(c)   Area and hence frequency between a normal probability curve and the mean ± 3 standard deviations is $99\frac{3}{4}$% or $\frac{19.95}{20}$ of the total area or frequency

Fig. 37.2

is 165 + 10 or 175 centimetres and 165 − 10 or 155 centimetres. Thus, 95% of 50 people, i.e. 48 of the 50 people, have heights between 155 and 175 centimetres. Strictly, 95% of 50 people is 47.5 people but, for discrete data, results are invariably 'rounded-up' to the next largest possible value in the distribution.

Worked problems on the normal distribution curve

Problem 1. The mean mass of 200 people is 67 kilograms and the standard deviation is 7 kilograms. Assuming that the masses are normally distributed, determine how many people:

(a)   have a mass between 60 and 74 kilograms,
(b)   have a mass of more than 81 kilograms, and
(c)   have a mass between 53 and 88 kilograms.

(a) The mean value plus one standard deviation from the mean value is usually abbreviated to:

mean + 1 standard deviation = 67 + 7 = 74 kilograms.
Also, mean − 1 standard deviation = 67 − 7 = 60 kilograms.

The area enclosed by the normal probability curve and the mean ± 1 standard deviation is $66\frac{2}{3}\%$ or two-thirds of the total area. Since area is proportional to frequency, then two-thirds of the total frequency lies between 60 and 74 kilograms. Hence the number of people having a mass between 60 and 74 kilograms is 200 (the number of people) $\times \frac{2}{3}$, i.e. **134 people**.

(b) The mean $+2$ standard deviation value is $67 + (2 \times 7)$ or 81 kilograms. Ninety-five per cent or $\frac{19}{20}$ of the area under a normal probability curve is enclosed between the curve and the vertical lines drawn at ± 2 standard deviations from the mean. Hence 5% or $\frac{1}{20}$ of the area is either greater than the $+2$ standard deviation value or less than the -2 standard deviation value. Since a normal probability curve is symmetrical, then $2\frac{1}{2}\%$ or $\frac{1}{40}$ is greater than the $+2$ standard deviation value. For 200 people, $200 \times \frac{1}{40}$, i.e. **5 people**, have a mass of more than 81 kilograms, since frequency is proportional to area.

(c) Because approximately $66\frac{2}{3}\%$ of the total area under the normal probability curve is between the mean and ± 1 standard deviation, about 95% between the mean and ± 2 standard deviations, and about $99\frac{3}{4}\%$ between the mean and ± 3 standard deviations, the area under a normal curve (and hence the frequency) can be subdivided as shown in Fig. 37.2

The areas marked (A) are each half of $66\frac{2}{3}\%$, i.e. $33\frac{1}{3}\%$. The areas marked (B) are half of 95%, that is, $47\frac{1}{2}\%$, less the $33\frac{1}{3}\%$ already shown as area (A). Thus the area marked (B) is $(47\frac{1}{2} - 33\frac{1}{3})\%$ or $14\frac{1}{6}\%$. The areas marked (C) are half of $99\frac{3}{4}\%$, i.e. $49\frac{7}{8}\%$, less the areas marked (A) and (B), that is $(49\frac{7}{8} - 33\frac{1}{3} - 14\frac{1}{6})\%$, or $2\frac{3}{8}\%$. Using these values, the total area (and hence frequency) between 53 and 88 kilograms is obtained by adding the appropriate areas in

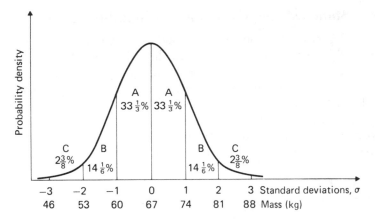

Fig. 37.3 Areas (or frequencies) between the normal probability curve and the mean ± 1, 2 or 3 standard deviations

Fig. 37.3, i.e.

$$(14\tfrac{1}{6} + 33\tfrac{1}{3} + 33\tfrac{1}{3} + 14\tfrac{1}{6} + 2\tfrac{3}{8})\% = 97\tfrac{3}{8}\%$$

(Alternatively, $99\tfrac{3}{4} - 2\tfrac{3}{8} = 97\tfrac{3}{8}\%$)

The number of people is $\dfrac{97\tfrac{3}{8}}{100} \times 200 = 194\tfrac{3}{4}$

As this is a discrete distribution, this must be rounded up to the next possible value, giving a result of **195 people** having a mass of between 53 and 88 kilograms.

Problem 2. A sample of 60 bolts produced by a machine was measured and the mean diameter was found to be 0.402 centimetres. The standard deviation of the sample was 0.000 5 centimetres. If only bolts having diameters between 0.401 and 0.403 centimetres were accepted, determine how many of the sample were rejected, assuming a normal distribution.

The range of bolts being accepted is 0.403 − 0.401, i.e. 0.002 centimetres. Since the standard deviation is 0.000 5 centimetres, this range corresponds to ± 2 standard deviations.

The area and also the frequency between a normal distribution curve and the mean ± 2 standard deviations is 95%. Hence the frequency outside of the range is 5% or $\tfrac{1}{20}$. Thus the number of bolts rejected is:

$$60 \times \tfrac{1}{20} = \textbf{3 bolts}$$

Further problems on the normal probability curve may be found in Section 37.4 (Problems 1 to 6), page 647.

37.2   Standardising the normal probability curve

Section 37.1 shows that approximate values of probability can be determined for normally distributed data, provided the values of the variable differ by exactly one, two or three standard deviations from the mean value. By standardising the normal probability curve, it is possible to find values of probability for any value of the variable and also to obtain actual, rather than approximate, results.

Normal probability curves differ from one another in four ways:

(i) The mean values can be different, as shown in Fig. 37.4

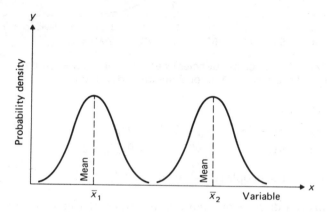

Fig. 37.4   Normal probability curves with the same standard deviation but having different mean values

(ii) The values of their standard deviations can be different, as shown in Fig. 37.5

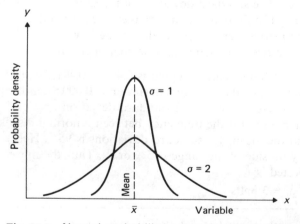

Fig. 37.5   Normal probability curves having the same mean value but with different values of standard deviation

(iii) The magnitude and units of the variables can be different.

(iv) The areas under their normal probability curves can be different.

The process of standardisation is to alter the normal probability curve to make it independent of these four items. This is achieved as follows:

(a) By writing $(x - \bar{x})$ for x in the equation of the normal probability curve, where $\bar{x}$ is the mean value, the origin of the graph is moved to $\bar{x}$. The equation now becomes

$$y = \frac{1}{\sigma} e^{-(x - \bar{x})^2/2\sigma^2}$$

All normal probability curves amended in this way have their mean values as the origin and hence the two curves shown in Fig. 37.4 would now sit on top of one another.

(b) The horizontal axis is scaled in standard deviations instead of the variable, x. This is achieved by using the relationship: $z = \dfrac{x - \bar{x}}{\sigma}$, where z is called the **normal standard variate**. For example, if the mean value of a distribution is, say, 147 centimetres and the value of the standard deviation is, say, 5 centimetres, then the variable value, say 152 centimetres, is $147 + 5$, that is, the mean value plus one standard deviation. Similarly, 137 centimetres is the mean value minus two standard deviations. Thus the x-axis is rescaled as the z-axis and some of the values are as shown:

x-axis (centimetres)	137	142	147	152	157
z-axis (standard deviations)	-2	-1	0	1	2

By introducing the normal standard variate, the origin of all normal probability curves is $\sigma = 0$ and hence they all have the same origin. In addition, the scaling of all normal probability curves is in standard deviations and all normal probability curves have the same scaling. Also, since standard deviation is a numerical value and can be independent of units, all normal probability curves can have the same units. Thus the introduction of the normal standard variates gives all normal probability curves a common value of unity for their standard deviation and can make them independent of units. Since $z = \dfrac{x - \bar{x}}{\sigma}$, then $\dfrac{(x - \bar{x})^2}{2\sigma^2}$, becomes $\dfrac{z^2}{2}$ and since, on the z-scale, σ has the value unity, the equation of the normal probability curve now becomes:

$$y = e^{-z^2/2}$$

The area between the normal probability curve and the z-axis is given by:

$$\int_{-\infty}^{\infty} e^{-z^2/2}\, dz$$

To evaluate this integral requires techniques not yet introduced (double integrals and the use of polar coordinates), giving its value as $\sqrt{(2\pi)}$. One of the requirements of a probability distribution curve is that the area beneath the curve is equal to unity, since the total area represents the total probability. To make the area under the normal probability curve unity, it is therefore necessary to divide the value of y by $\sqrt{(2\pi)}$. This gives the equation of the **standardised normal curve** as:

$$y = \frac{1}{\sqrt{(2\pi)}} e^{-z^2/2}$$

and by using it, any normally distributed data is represented by the same curve.

The area under the standardised normal curve represents probability and the area under the curve between limits, say z_1 and z_2, can be found by determining the value of:

$$\int_{z_2}^{z_1} \frac{1}{\sqrt{(2\pi)}} e^{-z^2/2}\, dz$$

Tables are available giving the values of this integral for z-values between 0.00 and 3.99 and one such table is Table 37.1, shown on page 641.

Worked problems on determining probabilities using a table of partial areas beneath the standardised normal curve

Problem 1. A certain machine produces components having a mean length of 15 centimetres. As a result of measuring samples of these components, it is found that the standard deviation is 0.2 centimetres. A test is carried out on a sample to check whether the data on the lengths of the components is normally distributed and it is found that this is so.
(a) Determine the number of components likely to have a length of less than 14.95 centimetres in a batch of 1 000 components.
(b) Determine the number of components likely to be between 14.95 and 15.15 centimetres long in a batch of 1 000 components.
(c) Determine the number of components likely to be larger than 15.43 centimetres long in a batch of 1 000 components.

(a) The z-value is given by $z = \dfrac{x - \bar{x}}{\sigma}$, where x is the variable value (14.95 cm), $\bar{x}$ the mean value (15 cm) and σ the value of the standard deviation (0.2 cm). Then

$$z = \frac{14.95 - 15}{0.2} = -0.25 \text{ standard deviations}$$

Table 37.1 Partial areas under the
standardised normal curve

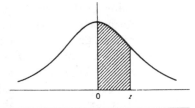

$z = \dfrac{x - \bar{x}}{\sigma}$	0	1	2	3	4 .	5	6	7	8	9
0.0	0.0000	0.0040	0.0080	0.0120	0.0159	0.0199	0.0239	0.0279	0.0319	0.0359
0.1	0.0398	0.0438	0.0478	0.0517	0.0557	0.0596	0.0636	0.0678	0.0714	0.0753
0.2	0.0793	0.0832	0.0871	0.0910	0.0948	0.0987	0.1026	0.1064	0.1103	0.1141
0.3	0.1179	0.1217	0.1255	0.1293	0.1331	0.1388	0.1406	0.1443	0.1480	0.1517
0.4	0.1554	0.1891	0.1628	0.1664	0.1700	0.1736	0.1772	0.1808	0.1844	0.1879
0.5	0.1915	0.1950	0.1985	0.2019	0.2054	0.2086	0.2123	0.2157	0.2190	0.2224
0.6	0.2257	0.2291	0.2324	0.2357	0.2389	0.2422	0.2454	0.2486	0.2517	0.2549
0.7	0.2580	0.2611	0.2642	0.2673	0.2704	0.2734	0.2760	0.2794	0.2823	0.2852
0.8	0.2881	0.2910	0.2939	0.2967	0.2995	0.3023	0.3051	0.3078	0.3106	0.3133
0.9	0.3159	0.3186	0.3212	0.3238	0.3264	0.3289	0.3315	0.3340	0.3365	0.3389
1.0	0.3413	0.3438	0.3451	0.3485	0.3508	0.3531	0.3554	0.3577	0.3599	0.3621
1.1	0.3643	0.3665	0.3686	0.3708	0.3729	0.3749	0.3770	0.3790	0.3810	0.3830
1.2	0.3849	0.3869	0.3888	0.3907	0.3925	0.3944	0.3962	0.3980	0.3997	0.4015
1.3	0.4032	0.4049	0.4066	0.4082	0.4099	0.4115	0.4131	0.4147	0.4162	0.4177
1.4	0.4192	0.4207	0.4222	0.4236	0.4251	0.4265	0.4279	0.4292	0.4306	0.4319
1.5	0.4332	0.4345	0.4357	0.4370	0.4382	0.4394	0.4406	0.4418	0.4430	0.4441
1.6	0.4452	0.4463	0.4474	0.4484	0.4495	0.4505	0.4515	0.4525	0.4535	0.4545
1.7	0.4554	0.4564	0.4573	0.4582	0.4591	0.4599	0.4608	0.4616	0.4625	0.4633
1.8	0.4641	0.4649	0.4656	0.4664	0.4671	0.4678	0.4686	0.4693	0.4699	0.4706
1.9	0.4713	0.4719	0.4726	0.4732	0.4738	0.4744	0.4750	0.4756	0.4762	0.4767
2.0	0.4772	0.4778	0.4783	0.4785	0.4793	0.4798	0.4803	0.4808	0.4812	0.4817
2.1	0.4821	0.4826	0.4830	0.4834	0.4838	0.4842	0.4846	0.4850	0.4854	0.4857
2.2	0.4861	0.4864	0.4868	0.4871	0.4875	0.4878	0.4881	0.4884	0.4882	0.4890
2.3	0.4893	0.4896	0.4898	0.4901	0.4904	0.4906	0.4909	0.4911	0.4913	0.4916
2.4	0.4918	0.4920	0.4922	0.4925	0.4927	0.4929	0.4931	0.4932	0.4934	0.4936
2.5	0.4938	0.4940	0.4941	0.4943	0.4945	0.4946	0.4948	0.4949	0.4951	0.4952
2.6	0.4953	0.4955	0.4956	0.4957	0.4959	0.4960	0.4961	0.4962	0.4963	0.4964
2.7	0.4965	0.4966	0.4967	0.4968	0.4969	0.4970	0.4971	0.4972	0.4973	0.4974
2.8	0.4974	0.4975	0.4976	0.4977	0.4977	0.4978	0.4979	0.4980	0.4980	0.4981
2.9	0.4981	0.4982	0.4982	0.4983	0.4984	0.4984	0.4985	0.4985	0.4986	0.4986
3.0	0.4987	0.4987	0.4987	0.4988	0.4988	0.4989	0.4989	0.4989	0.4990	0.4990
3.1	0.4990	0.4991	0.4991	0.4991	0.4992	0.4992	0.4992	0.4992	0.4993	0.4993
3.2	0.4993	0.4993	0.4994	0.4994	0.4994	0.4994	0.4994	0.4995	0.4995	0.4995
3.3	0.4995	0.4995	0.4995	0.4996	0.4996	0.4996	0.4996	0.4996	0.4996	0.4997
3.4	0.4997	0.4997	0.4997	0.4997	0.4997	0.4997	0.4997	0.4997	0.4997	0.4998
3.5	0.4998	0.4998	0.4998	0.4998	0.4998	0.4998	0.4998	0.4998	0.4998	0.4998
3.6	0.4998	0.4998	0.4999	0.4999	0.4999	0.4999	0.4999	0.4999	0.4999	0.4999
3.7	0.4999	0.4999	0.4999	0.4999	0.4999	0.4999	0.4999	0.4999	0.4999	0.4999
3.8	0.4999	0.4999	0.4999	0.4999	0.4999	0.4999	0.4999	0.4999	0.4999	0.4999
3.9	0.5000	0.5000	0.5000	0.5000	0.5000	0.5000	0.5000	0.5000	0.5000	0.5000

Using Table 37.1, the partial area beneath the standardised normal curve corresponding to a z-value of 0.25 is 0.098 7, the minus sign showing that it lies to the left of the zero z-value ordinate. This area is shown by the shaded area in Fig. 37.6(a), and is 0.098 7 of the total under the curve, the total area being unity.

Since the standardised normal curve is symmetrical about the ordinate drawn through $z = 0$, then the total area to the left of this line is 0.500 0, shown shaded in Fig. 37.6(b). The required area giving the probability of a component being less than 14.95 centimetres long is the area to the left of the $z = -0.25$ ordinate and is shown as the shaded area in Fig. 37.6(c). This area is the

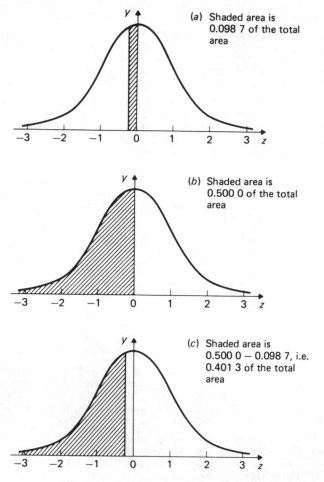

Fig. 37.6 Partial areas under the standardised normal curve (Problem 1)

shaded area in Fig. 37.6(b) minus the shaded area in Fig. 37.6(a), that is, $0.5000 - 0.0987$ or 0.4013. But the area under the standardised normal curve is proportional to probability, so the probability that a component will be less than 14.95 centimetres long is 0.4013. For 1 000 components, the expectation is 1000×0.4013, i.e. 401.3. Thus, there are likely to be **402 components** having a length of 14.95 centimetres or less, results usually being 'rounded-up' in statistics.

(b) The z-value for 14.95 centimetres is -0.25 standard deviations and the corresponding partial area is 0.0987 (see part (a)). The z-value for 15.15 centimetres is $\dfrac{15.15 - 15}{0.2}$, i.e. 0.75 standard deviations. Using Table 37.1, the partial area corresponding to this z-value is 0.2734. Hence the total partial area between $z = -0.25$ and $z = 0.75$ is $0.0987 + 0.2734$, i.e. 0.3721. For 1 000 components, the expectation of a component being between 14.95 and 15.15 centimetres long is 1000×0.3721, i.e. 372.1. Hence it is likely that **373 components** will be between 14.95 and 15.15 centimetres long.

(c) The z-value of 15.43 centimetres is given by

$$z = \frac{15.43 - 15}{0.2} = 2.15 \text{ standard deviations}$$

From Table 37.1, the partial area corresponding to a z-value of 2.15 is 0.4842. The total area to the right of the $z = 0$ ordinate is 0.5000. Hence the area to the right of the $z = 2.15$ ordinate is $0.5000 - 0.4842$, i.e. 0.0158, this being the probability of a component being larger than 15.43 centimetres. For 1 000 components, the expectation is 1000×0.0158, i.e. 15.8. Hence, there are likely to be **16 components** larger than 15.43 centimetres.

Problem 2. The heights of 500 people are normally distributed about a mean value of 172 centimetres. The standard deviation is 8 centimetres. Determine how many people: (a) are likely to be between 174 and 189 centimetres, and (b) smaller than 149 centimetres.

(a) The z-value for 174 centimetres is $\dfrac{174 - 172}{8}$, i.e. 0.25 standard deviations.
From Table 37.1, the corresponding area under the standardised normal curve is 0.0987

The z-value for 189 centimetres is $\dfrac{189 - 172}{8}$, i.e. 2.125 or 2.13 standard deviations correct to 2 decimal places. From Table 37.1, the corresponding area under the standardised normal curve for a z-value of 2.13 is 0.483 4. The area under the standardised normal curve between z-values of 2.13 and 0.25 is 0.483 4 − 0.098 7, i.e. 0.384 7. For 500 people, the expectation of being between 174 and 189 centimetres tall is 500 × 0.384 7, i.e. 192.35. Hence, there are likely to be **193 people** having heights between 174 and 189 centimetres.

(b) The z-value for 149 centimetres is $\dfrac{149 - 172}{8}$, i.e. $\dfrac{-23}{8}$ or −2.875, i.e. −2.88 correct to 2 decimal places. From Table 37.1, the partial area corresponding to a z-value of 2.88 is 0.498 0. The area under the standardised normal curve for the z-values of less than −2.88 is 0.500 0 − 0.498 0, i.e. 0.002. For 500 people, the expectation of being less than 149 centimetres tall is 500 × 0.002, i.e. 1 person. That is, **one person** is likely to be less than 149 centimetres tall.

Further problems on the standard normal curve may be found in Section 37.4 (Problems 7 to 22), page 649.

37.3   Testing the normality of a distribution

In the practical application of statistics, it should never be assumed that because a set of data is continuous, it is also normally distributed, or even approximates to a normal distribution. Various tests can be carried out to determine how accurately a set of data approximates to a normal distribution and the simplest test is to plot the data on specially ruled paper called **normal probability paper** or just **probability paper**.

This has a linear scale usually taken to be the horizontal axis, and probability is usually taken to be the vertical axis. The divisions of the probability scale are so spaced that when the percentage cumulative frequency is plotted against the upper class-boundaries of the variable, a straight line results when the data is normally distributed. The use of normal probability paper for testing the normality of a distribution is shown in the following worked problems.

Worked problems on testing the normality of a distribution

Problem 1. Use normal probability paper to test whether the data given below is normally distributed.

5. The heights of 1 000 people are normally distributed with a mean of 1.72 m and a standard deviation of 8 cm. Determine how many people will:
   (a) have heights between 1.48 and 1.96 m,
   (b) have heights of less than 1.72 m,
   (c) have heights of more than 1.80 m, and
   (d) have heights between 1.56 and 1.64 m
   (a) [998] (b) [500] (c) [167] (d) [142]
6. A set of 100 measurements is normally distributed with a mean of 157.8 m and a standard deviation of 1.4 m. Determine how many of the measurements will be:
   (a) between 155 and 162 m,
   (b) between 159.2 and 160.6 m,
   (c) more than 157.8 m, and
   (d) less than 156.4 m
   (a) [98] (b) [15] (c) [50] (d) [17]

The standardised normal probability curve

In Problems 7 to 13, use Table 37.1 on page 641 to determine the partial areas under the standardised normal curve for the z-values given.

7. Between $z = 0$ and $z = 1.35$ [0.411 5]
8. Between $z = -2.34$ and $z = 0$ [0.490 4]
9. Between $z = -0.46$ and $z = 2.21$ [0.663 6]
10. Between $z = 0.81$ and $z = 1.94$ [0.182 8]
11. Larger than $z = -1.28$ [0.899 7]
12. Less than $z = -0.6$ [0.274 3]
13. Less than $z = -1.44$ and larger than $z = 2.05$ [0.095 1]
14. The mean mass of a set of components is 151 kg and the standard deviation is 15 kg. Assuming the masses are normally distributed about the mean, determine for a sample of 500 components, how many are likely to have masses of:
    (a) between 120 and 155 kg,
    (b) more than 185 kg, and
    (c) less than 128 kg.
    (a) [294] (b) [6] (c) [32]
15. The mean diameter of a batch of bolts is 0.502 cm and the standard deviation of the batch is 0.05 mm. Bolts outside of the range of diameters 0.496 and 0.508 cm are rejected. Determine the percentage of bolts rejected assuming the diameters are normally distributed.   [23.02%]

16. Containers of a fluid for innoculation are filled with an average of 68 cm³ of a fluid and the standard volume deviation is 3 cm³. Assuming the contents of the containers to be normally distributed, determine in a sample of 300 containers, how many are likely to contain:
(a) more than 72 cm³,
(b) less than 64 cm³, and
(c) between 65 and 71 cm³
(a) [28]   (b) [28]   (c) [205]

17. Tablets contain on average 0.614 g of a drug, the standard deviation being 2.5 mg. Assuming the masses of the drug are normally distributed, determine the percentage of the tablets having:
(a) more than 0.617 g of the drug,
(b) less than 0.608 g of the drug, and
(c) between 0.610 and 0.618 g of the drug
(a) [11.51%]   (b) [0.82%]   (c) [89.04%]

18. A set of measurements is normally distributed. Determine the percentage which differ from the mean by:
(a) more than half of the standard deviation, and
(b) less than three-quarters of the standard deviation.
(a) [62.70%]   (b) [54.68%]

19. The contents of milk bottles are measured and the amount of milk the bottles contain is found to be normally distributed. Determine the percentage of bottles containing milk:
(a) within the range: mean ±2 standard deviations,
(b) outside of the range ±1.2 standard deviations, and
(c) greater than the mean −1.5 standard deviations
(a) [95.44%]   (b) [23.02%]   (c) [93.32%]

20. Tins are packed with an average of 1.0 kg of a compound and the masses are normally distributed about the average value. The standard deviation of a sample of the contents of the tins is 12 g. Determine the percentage of tins containing:
(a) less than 985 g
(b) more than 1 030 g,
(c) between 985 and 1 030 g
(a) [10.56%]   (b) [0.62%]   (c) [88.82%]

21. The weights of tablets in a bottle are normally distributed with a mean of 20 g and standard deviation of 2 g. Using the area under the standard normal curve, calculate the probability of a bottle containing tablets weighing between:
(a) 16 and 18 g, (b) 18 and 23 g
Note: extract from normal table:

Units of s. dev. from mean	0	0.5	1.0	1.5	2.0	2.5
Area under normal curve	0	0.1915	0.3413	0.4332	0.4772	0.4938

(a) [0.1359]   (b) [0.7745]

22. A chemical manufacturer produces aspirin tablets having a mean mass of 4 g and a standard deviation of 0.2 g. Assuming that the masses are normally distributed and that a tablet is chosen at random what is the probability that it: (a) has mass between 3.55 and 3.85 g, (b) differs from the mean by less than 0.35 g?
(c) If the tablets are packed in cartons of 400, how many in each carbon may be expected to have a mass of less than 3.7 g?
(a) [0.2144]   (b) [0.9198]   (c) [27]

Testing the normality of a distribution

In Problems 23 to 27 use normal probability paper to determine whether the distributions given are normally distributed.

23. The diameter of a sample of rivets produced by an automatic process.

Diameter mm	4.011	4.015	4.019	4.023	4.027	4.031	4.035
Frequency	13	40	79	114	116	82	40

[Yes]

24. The resistance of a sample of carbon resistors.

Resistance MΩ	1.28	1.29	1.30	1.31	1.32	1.33	1.34	1.35	1.36
Frequency	10	18	36	44	64	40	24	16	12

[No]

25. The mass of a sample of components produced by a casting process.

Mass kg	4.00	4.05	4.10	4.15	4.20	4.25
Frequency	22	78	150	260	280	440

Mass kg	4.30	4.35	4.40	4.45	4.50	4.55
Frequency	360	180	140	56	18	4

[Yes]

26. The volume output of a bottling machine.

Volume l	1.007	1.010	1.013	1.016	1.019	1.022	1.025	1.028
Frequency	2	5	12	17	14	6	3	1

[Yes]

27. The capacitance of a sample of rolled metal foil capacitors of nominal capacitance 3.75 μF.

Capacitance μF	3.72	3.73	3.74	3.75	3.76	3.77
Frequency	2	6	8	15	32	48

Capacitance μF	3.78	3.79	3.80	3.81	3.82	3.83
Frequency	39	25	18	12	4	1

[No]

28. The amount of lead in a given foodstuff was determined with the following results:

40–44 ppm	1 analysis
45–49 ppm	4
50–54 ppm	14
55–59 ppm	22
60–64 ppm	31
65–69 ppm	20
70–74 ppm	6
75–79 ppm	2

Prove graphically that the distribution is normal and find the mean and standard deviation.

[straight line with probability paper, mean = 60.6, standard deviation = 6.79]

Index